Essentials of Human Genomics

Essentials of Human Genomics

Editor: Alyssa Boyle

AMERICAN
MEDICAL PUBLISHERS
www.americanmedicalpublishers.com

AMERICAN
MEDICAL PUBLISHERS
www.americanmedicalpublishers.com

Cataloging-in-Publication Data

Essentials of human genomics / edited by Alyssa Boyle.
p. cm.
Includes bibliographical references and index.
ISBN 978-1-63927-249-5
1. Human genome. 2. Genomics. 3. Human genetics.
4. Medical genetics. I. Boyle, Alyssa.
RB155.5 .E87 2021
616.042--dc23

American Medical Publishers,
41 Flatbush Avenue,
1st Floor, New York,
NY 11217, USA

ISBN 978-1-63927-249-5 (Hardback)

Contents

Preface

The complete set of an organism's DNA is known as a genome. Human genomics is an interdisciplinary biological field that focuses on the evolution, structure, mapping, functioning and editing of human genomes. It also deals with the application of genomic approaches to enhance the understanding of human disease, variable drug reaction and drug discovery. Human genomes consist of DNA within cell nuclei and a small DNA molecule within individual mitochondria. These are classified as the nuclear genome and the mitochondrial genome. One of the primary goals of human genomics is the collective characterization of all of person's genes as well as their relations with each other. The topics included in this book on human genomics are of utmost significance and bound to provide incredible insights to readers. It traces the progress of this field and highlights some of its key concepts and applications. For all readers who are interested in this field, the case studies included in this book will serve as an excellent guide to develop a comprehensive understanding.

The information shared in this book is based on empirical researches made by veterans in this field of study. The elaborative information provided in this book will help the readers further their scope of knowledge leading to advancements in this field.

Finally, I would like to thank my fellow researchers who gave constructive feedback and my family members who supported me at every step of my research.

Editor

1

Estimating ancestral proportions in a multi-ethnic US sample: implications for studies of admixed populations

Orna Levran[1*], Olaoluwakitan Awolesi[1], Pei-Hong Shen[2], Miriam Adelson[3] and Mary Jeanne Kreek[1]

Abstract

This study was designed to determine the ancestral composition of a multi-ethnic sample collected for studies of drug addictions in New York City and Las Vegas, and to examine the reliability of self-identified ethnicity and three-generation family history data. Ancestry biographical scores for seven clusters corresponding to world major geographical regions were obtained using STRUCTURE, based on genotypes of 168 ancestry informative markers (AIMs), for a sample of 1,291 African Americans (AA), European Americans (EA), and Hispanic Americans (HA) along with data from 1,051 HGDP-CEPH 'diversity panel' as a reference. Self-identified ethnicity and family history data, obtained in an interview, were accurate in identifying the individual major ancestry in the AA and the EA samples (approximately 99% and 95%, respectively) but were not useful for the HA sample and could not predict the extent of admixture in any group. The mean proportions of the combined clusters corresponding to European and Middle Eastern populations in the AA sample, revealed by AIMs analysis, were 0.13. The HA subjects, predominantly Puerto Ricans, showed a highly variable hybrid contribution pattern of clusters corresponding to Europe (0.27), Middle East (0.27), Africa (0.20), and Central Asia (0.14). The effect of admixture on allele frequencies is demonstrated for two single-nucleotide polymorphisms (118A > G, 17 C > T) of the *mu* opioid receptor gene (*OPRM1*). This study reiterates the importance of AIMs in defining ancestry, especially in admixed populations.

Keywords: Ancestry informative markers, Hispanics, African Americans, Family history, Ancestry, Admixture

Introduction

The well-established genetic differences between ancestral populations may have an effect on disease prevalence and outcomes, as well as on drug response [1]. The presence of subgroups that differ in allele frequencies is relevant to public health and has numerous clinical implications. Analysis of population structure using a clustering algorithm can distinguish between populations based on DNA polymorphisms [1]. Admixture occurs when a new hybrid population is formed from formerly isolated populations [2]. Estimating the proportions of different ancestries in admixed populations is especially important in case–control association studies, since spurious associations may occur due to population substructure [3,4]. To avoid spurious associations, researchers can adjust the regression model in an association study for individual admixture by estimating individual ancestry with a set of ancestry informative markers (AIMs) that have high allele frequency differences between continental groups.

African and Hispanic Americans constitute a large part of the US population (approximately 12% each). The majority of the African American (AA) population is a recently admixed population (average of approximately six generations, >200 years ago) generated primarily by forced migration from Africa and a diverse range of admixture with European Americans (EA) and Native Americans (NA).

The term Hispanics/Latinos refers to a diverse population of Latin American descent. The Hispanic/Latino complex genetic structure reflects over five centuries of historical confluence of three major parental populations: Native American, European (primarily from the Iberian Peninsula and Southern Europe), and West African populations, with very large variations in ancestry proportions

* Correspondence: levrano@rockefeller.edu
[1]The Laboratory of the Biology of Addictive Diseases, The Rockefeller University, New York NY10065, USA

that is a result of country-specific migration history. Several studies have reported the admixture of this population [e.g., 5-14]. The Hispanic subgroups in the USA were shown to differ in the prevalence of several diseases compared with African and European Americans. Admixture analysis may show whether genetics explains these differences and may allow discriminating between socio-demographic and genetic contribution.

This study was designed to determine the ancestry proportions of a large multi-ethnic sample, recruited for studies of drug addiction in the USA, with a set of 168 AIMs and to examine the concordance of self-reported ethnicities and family history with AIMs data.

Materials and methods

Sample

The sample of 1,291 subjects is part of a larger cohort that was recruited for genetic study of drug addiction in the Laboratory of the Biology of Addictive Diseases at The Rockefeller University, New York. This set included only subjects who self-reported to be AA, EA, or Hispanic American (HA), which are the major groups in our cohort. Subjects were recruited at The Rockefeller University Hospital, The Weill Medical College of Cornell University, New York, The Manhattan Campus of the VA NY Harbor Health Care System, and the Dr. Miriam and Sheldon G. Adelson Clinic for Drug Abuse Treatment and Research, Las Vegas. They included former opiate addicts in methadone maintenance treatment ($n = 726$), cocaine and/or alcohol addicts ($n = 143$), and healthy volunteers ($n = 422$). A subset of this sample (e.g., EA and AA opiate addicts and healthy volunteers) was included in our previously reported case–control association studies [15,16].

Subjects completed a three-generation family history questionnaire developed in our laboratory as part of their ascertainment interview. The family history questionnaire includes questions on (1) place of birth (i.e., city, state, country, and region); (2) self-identified race, ethnicity, or cultural group; and (3) nationality. Each question was answered for self, parents, grandparents, and great-grandparents, when known. The questionnaire was filled in by an experienced clinician interviewer. No pre-set list of categories was provided. Responses were assigned into nine categories: African, AA, EA, HA, Asian, Caribbean African, NA, Other/ Mixed, and Unknown. Ambiguous responses were assigned by the authors based on geographical regions. All subjects signed an informed consent for genetic studies with a 99.7% consent rate [17]. The institutional review boards of The Rockefeller University, VA NY, and Cornell University approved the study. The Rockefeller University IRB also reviews the Adelson Clinic, Las Vegas.

DNA preparation and genotyping

DNA was extracted from blood using the standard salting-out method. Two hundred and fifty to five hundred ng DNA was used for genotyping, as described [16]. Genotyping of AIMs and *OPRM1* single-nucleotide polymorphisms (SNPs) was performed on a 1,536-plex GoldenGate™ Custom Panel (Illumina, San Diego, CA, USA) [18]. Genotyping was performed at The Rockefeller University Genomics Resource Center according to the manufacturer's protocol (Illumina). Analysis was performed using BeadStudio genotyping software (Illumina). Genotype data were filtered based on SNP call rates (>99.5%) and cluster separation, and 18 AIMs were excluded from the analysis because of poor cluster separation (Additional file 1: Table S1). Random samples (approximately 10%) were genotyped in duplicate with high reproducibility rate (99.9%).

Determination of individual ancestry by AIMs

The set of 186 unlinked AIMs was selected based on differences in allele frequency by at least 70% and 10-fold between at least two continental populations (from among European, African, and Asian populations of the HapMap project) (Additional file 1: Table S1) [19]. Genotypes of 168 AIMs with adequate quality were analyzed using STRUCTURE v.2.2 software [20] to obtain individual ancestry proportions (ancestry biographical score). Analysis was also performed for a known set of 1,051 subjects representing 51 worldwide populations (the Cell Line Panel of the Human Genome Diversity Project (HGDP)/Centre d'Etude du Polymorphisme Humain (CEPH)) [21].

The number of clusters (K) was defined by running the data with different K values and computing the probability of $K = n$ in the HGDP sample. The seven-factor solution was optimal and closely replicated the seven-factor solution found for the same 51 reference populations that matched major geographical regions [1,22].

In the current analysis, the entire sample was analyzed simultaneously with the reference set in an 'anchored' approach that was shown to yield a stable factor structure interpretable in the context of worldwide genetic diversity [22-24]. An ancestry biographical score for seven clusters corresponding to geographical regions: Africa, Europe, Middle East, Central Asia, Far East Asia, Oceania (islands of the tropical Pacific Ocean), and America, was estimated for each individual, with reference to the 1,051 individual panel, and sum to 1 [22].

Results

A set of 1,291 subjects was selected for this study based on self-report ethnicity. This set is part of a larger cohort that was recruited for genetic study of drug addiction from New York City and Las Vegas. The subjects belong to the following self-reported groups: AA ($n = 435$), EA ($n = 563$), and HA ($n = 293$). Ancestry biographical scores for seven clusters (K) were obtained for each subject using STRUCTURE analysis of 168 AIMs. The analysis was performed for each sample along with 1,051 HGDP samples as a reference (see the 'Materials and methods' section). Based on the HGDP sample, the seven factors correspond to the geographical regions of Africa, Europe, Middle East, Central Asia, Far East Asia, Oceania, and America, and they are named by the geographical regions hereafter, for simplicity. Figure 1a shows the relative ancestry contributions for each subject in each of the three groups. There is a very small contribution of clusters corresponding to Oceania and Far East Asia in this sample.

African Americans

Complete or almost complete (>10/14 possible answers) three-generation family history was available for 296 subjects (68%). Partial information was available for 87 subjects (20%), and no family data were available for 52 subjects including 3 adopted subjects (12%). Based on this data, the AA group can be divided into the following subgroups: AA for at least two generations with no report of another ancestry (41%), AA with some Caribbean ancestry, African Caribbean, AA with some Native American ancestry, AA with some European ancestry, mixed ancestry, new immigrants from Africa, and unknown ancestry (Table 1).

Based on AIMs analysis, the major ancestry contribution for the majority (96%) of the self-identified AA subjects is African, as is shown in Figure 1a. The second major contribution (0.26–0.40) is from Europe, Middle East, or Central Asia. The mean proportions of each of the major factors in the AA sample are shown in Table 1. The mean African contribution for the subgroups that were divided based on family history data was 0.80 for the subgroups of African Americans, African Caribbeans, some Caribbean ancestry, and some Native American ancestry. AIMs analysis detected a lower mean African contribution (0.70) in the sample that reported some European ancestry or mixed ancestry and a higher mean African contribution (0.94) in new immigrants from Africa, as expected, validating the sensitivity of this set of AIMs. Notably, there was a very small Native American contribution in all subgroups including those ($n = 59$) that reported some Native American ancestry.

Comparison between self-identification and AIMs data revealed 17 subjects for whom the contribution of the African factor was less than 50%. There were 13 subjects (3%) with an African contribution in the 0.25–0.5 range for which the other major contributing factor was European, Middle East, or Central Asia. There were four subjects (0.9%) with a <0.25 African contribution, but none of them had family history data.

Hispanic Americans

Complete or almost complete three-generation family history was available for 202 subjects (71%). Partial information was available for 53 subjects (18%), and no family data were available for 32 subjects (11%) including 1 adopted subject. Based on the three-generation family history data, the self-identified HA subjects can be divided to the following subgroups: Puerto Ricans (62%), mixed or other Caribbean Islands, Latin Americans, Caribbean/Latin Americans, Puerto Ricans with European ancestry, and no data (Table 1).

Based on AIMs analysis, the distribution of individual admixture estimates in the HA sample show a wide range of ancestry proportions (Figure 1a). Of the 293 self-reported HA, 150 subjects (51%) have a hybrid pattern of five ancestries (all with frequency <0.5), 52 subjects (18%) have major European ancestry, 53 (18%) subjects have a major Middle Eastern contribution, 22 subjects (8%) have a major African ancestry contribution, 13 subjects (4%) have a major Central Asian contribution, and 3 subjects (1%) have a major Native American contribution (Figure 1b). The mean proportions of each of the five major clusters are listed in Table 1. It is clear that the reported family history does not reflect the complex ancestry contribution indicated by analysis of AIMs. Although the mean contributions from Far East Asia and Oceania were low (<0.02), 33 subjects have a >0.05 contribution from Oceania and 24 subjects show a >0.05 contribution from Far East Asia. There is a higher contribution of Native American in the Latin American subgroup (0.19) and the mixed Caribbean/Latin American subgroup (0.13) compared to other subgroups supporting the sensitivity of this set of AIMs.

European Americans

AIMs analysis of the 563 self-reported European Americans showed that the major ancestry contribution was as follows (Figure 1a): 494 subjects (88%) have major European ancestry (>0.50), 33 (6%) subjects have a major Middle Eastern contribution, 12 (2%) subjects have a major Central Asian contribution, 4 (<1%) subjects have a major African contribution, 18 (3%) subjects have mixed contributions from Europe, Middle East, and Central Asia, and 2 subjects (<1%) show mixed contributions from Europe, Middle East, and Native America. The mean proportions of the major factors are shown in Table 1.

The comparison between family ancestry and AIMs in this group is beyond the scope of this study, as the specific AIMs were designed for continental populations and are of limited use in detecting substructure in

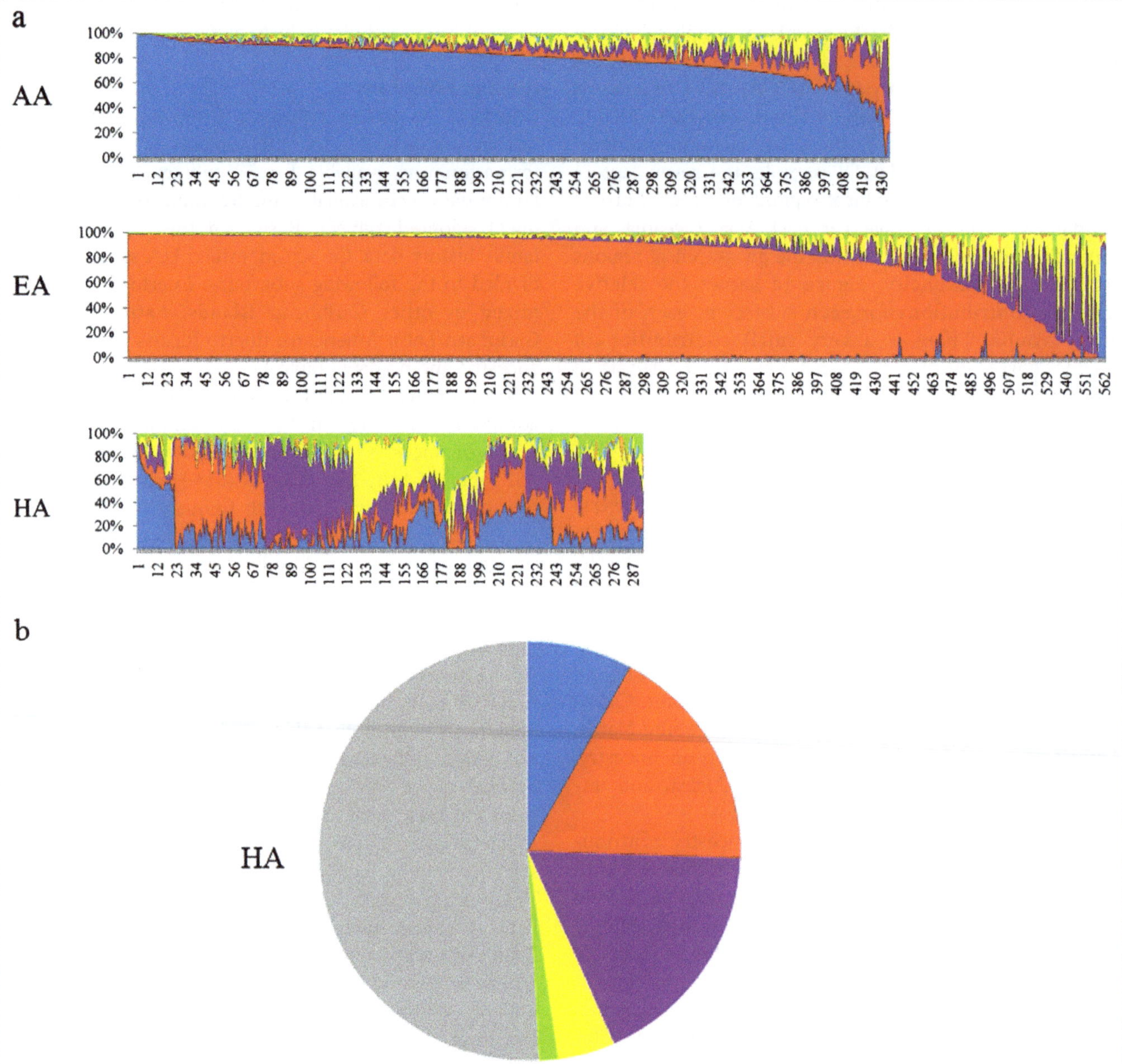

Figure 1 Individual admixture estimates and the distribution of the major ancestry contributions in the HA group. (**a**) Schematic representation of the individual admixture estimates using ($K = 7$). Each *vertical line* represents one individual, and subjects are displayed according to their predominant cluster contribution. The clusters correspond to the geographical regions based on the HGDP sample. Color code: Africa (blue), Europe (red), Middle East (purple), Central Asia (yellow), Far East Asia (cyan), Oceania (amber), and America (green). (**b**) The distribution of the major ancestry contributions (frequency > 0.5) in the HA group. Color code: no single major ancestry (gray); major ancestries: Europe (red), Middle East (purple), Africa (blue), Central Asia (yellow), and America (green).

closely related populations such as Europeans [19]. However, to identify potential conflicts between AIMs data and self-identified ethnicity, analysis of the family history data was performed for 34 subjects with a low European ancestry contribution indicated by AIMs analysis (<0.25). No family data were available for six subjects, including one adopted subject. As is shown in Table 2, out of the 18 subjects in which the major contribution indicated by AIMs analysis was from the Middle East, 6 subjects were Jewish (2 of them reported some non-European ancestry), 8 subjects were from Italy, Malta, Greece, or Iran, and 4 have no data available. We have recently showed a high contribution of the Middle East cluster in Israeli non-Ashkenazi Jewish subjects using the same set of AIMs [25]. The one subject, in whom a large Native American contribution was

Table 1 Self-identification, family history, and ancestry biographical score based on AIMs

Self-identified	Family history data	n	Clusters (K)[a]				
			1	2	3	4	5
			Africa	Europe	Middle East	Central Asia	America
African Americans							
	African American	179	0.83	0.05	0.05	0.05	0.01
	Some Caribbean ancestry	32	0.79	0.07	0.05	0.05	0.01
	African Caribbean	40	0.82	0.05	0.05	0.05	<0.01
	Some Native American ancestry	59	0.79	0.07	0.06	0.05	0.02
	Some European ancestry	17	0.70	0.10	0.12	0.05	<0.01
	Mixed ancestry	48	0.72	0.06	0.08	0.06	<0.01
	New immigrants from Africa	8	0.94	0.01	0.02	0.01	<0.01
	No family data	52	0.74	0.09	0.07	0.08	<0.01
	Total	*435*	*0.80*	*0.07*	*0.06*	*0.05*	*0.01*
	(SD)		(0.12)	(0.08)	(0.07)	(0.05)	(0.01)
Hispanic Americans							
	Puerto Rican	181	0.18	0.27	0.30	0.14	0.07
	Other Caribbean or mixed Caribbean	35	0.35	0.27	0.20	0.11	0.04
	Latin American	22	0.20	0.20	0.19	0.17	0.19
	Caribbean/Latin American	6	0.09	0.27	0.31	0.16	0.13
	Puerto Rican with some European ancestry	15	0.19	0.33	0.26	0.14	0.05
	No data	34	0.17	0.30	0.24	0.17	0.09
	Total	*293*	*0.20*	*0.27*	*0.27*	*0.14*	*0.08*
	(SD)		(0.18)	(0.22)	(0.24)	(0.16)	(0.11)
European Americans							
	Total	*563*	*0.01*	*0.81*	*0.10*	*0.06*	*0.01*
	(SD)		(0.01)	(0.17)	(0.13)	(0.08)	(0.01)

[a]Data for the clusters corresponding to Oceania and Far East Asia are not presented (<0.02). *SD* standard deviation.

indicated by AIMs analysis, reported some Latin American ancestry. We found no support in the family history for the subjects in whom a major Central Asian ($n = 9$) or African contribution ($n = 2$) was indicated by AIMs.

Table 2 Family ancestry data in European Americans with <0.25 European ancestry

Major ancestry (AIMs)	n	Family history	n
Middle East	18	Jewish	6
		Italy, Malta, Greece	7
		Iran	1
		No data	4
Mixed Native American/Middle East	1	Latin America	1
Central Asia	10	Adopted	1
		Europe	7
		No data	2
Africa	4	Europe	2
		No data	2
Middle East/Central Asia	1	Europe	1
Total	34		34

Allele frequencies of *mu* opioid receptor (*OPRM1*) SNPs in HA

To illustrate a situation of population-specific allele frequencies and their potential effect in admixed populations, allele and genotype frequencies were calculated for two missense polymorphisms (118A > G (rs1799971) and 17C > T (rs1799972)) in the *OPRM1* gene that plays an important role in opioid addiction [26]. Data from dbSNP, HapMap, and ALFRED show that the 118 G allele is common in Asians and Amerindians/Native Americans (0.35–0.5), is rare in African and Oceania populations, and occurs in moderate frequencies in European and Mexican populations (0.15–2.0). The 17C > T was not genotyped in the HapMap or ALFRED projects, but other studies and unpublished data from our laboratory indicate that it is African-specific (0.2–0.3) [27,28].

Analysis of genotype data from this study sample shows that the 17 T allele is rare in EA and HA (0.06) and its frequency in the AA group is 23%, in concordance with data from other studies (Table 3). The 118 G allele is rare in the AA group and has a frequency of

Table 3 Allele frequencies of *OPRM1* SNPs

Self-described	17 C > T (rs1799972)		118A > G (rs1799971)	
	C	T	A	G
EA	0.98	0.02	0.88	0.12
AA	0.77	0.23	0.97	0.03
HA	0.94	0.06	0.82	0.17

EA vs. HA $p < 0.004$.

12% in the EA group and 17% in the HA group. AIMs data demonstrate that all the carriers of the 118 G allele have some European, Asian, and/or Native American contribution that could explain its origin. The frequencies of the 17 T and 118 G alleles and their related genotypes in the HA sample were significantly higher than those of European Americans (17% vs. 12%, and 6% vs. 2%, respectively, $p < 0.004$ for chi-squared test) (Table 3). This difference can be explained by the contribution of African and Asian ancestries. Since the contributions of different ancestries vary between different HA samples, the frequencies of the 118 G and 17 T alleles found in this sample may not represent the frequency in other HA samples.

Discussion

In this study, we have used a panel of 168 AIMs to estimate the ancestry composition of a multi-ethnic US sample collected for studies of drug addictions in New York City and Las Vegas. We compared this information to self-identified ethnicity and family history data. This comparison revealed high concordance in the major ancestry between self-identified ethnicity and AIMs analysis in African Americans and European Americans in agreement with other studies [10,29-31]. However, self-identified ethnicity and family history data could not predict the degree of admixture that may have an effect on allele frequencies.

This study reiterates the complexity of the 'Hispanic/Latino' term. Our results are compatible with studies indicating a relatively high European contribution (>50%) in subgroups of this population [5,7-9,12-14]. The study emphasizes the importance of AIMs data in genetic studies of HA since the self-identified ethnicity and family history may not reveal the complex ancestry contributions of this group. A special scrutiny has to be used in case–control association studies in this population, and AIMs data should be used to correct for potential population stratification.

The major Hispanic subgroup in this study was Puerto Rican, a population that currently represents approximately 1.5% of the US population. The pattern of ancestral proportions may have clinical significance for specific diseases when a specific ancestry may have a protective effect based on alleles with higher frequency in this population. For example, a recent study of end-stage kidney disease in Hispanics from New York City reported an approximately 30% African contribution and a very small Native American contribution [9] emphasizing the difference between 'Mexican Hispanics' and 'Caribbean Hispanics.' The sample in the current study was collected in Las Vegas and New York City and has a small, unrepresentative proportion of Hispanics of Mexican origin; conversely, studies with a mix HA from the East, the South West, and the West coasts of the USA are expected to have even larger level of admixture.

This study confirmed the finding of other studies showing a highly diverse proportion of European ancestry in self-identified AAs (7–21%) [9,22,23,30,32,33]. This diversity can be explained in part by the historical 'one-drop rule' (which classified individuals with any level of African ancestry as 'African Americans'). It is clear that for the AA population, self-identified ethnicity is not sufficient to estimate the admixture level and a random AA sample may differ in admixture level from another sample, to an extent that will affect allele frequencies. The average of 7% of European admixture in this sample is compatible with some studies [22-24,33] but is lower than other studies [30,32,34,35]. This difference may be explained in part by the various numbers of defined clusters used in the different studies. Our STRUCTURE analysis was based on seven clusters, and the European cluster obtained in studies based on small number of clusters is most probably represented in our study by two clusters: Europe and Middle East. These clusters were found to be relatively close (population differentiation index $F_{st} = 0.005$) [36] and the Middle East cluster was shown to form a gradient across Europe [37]. Including the Middle East cluster in the total European contribution would result in a 13% contribution that is closer to the estimate by other studies. The difference between our estimates of ancestral proportions and other studies may also reflect recruitment from different US regions and the use of a different set of AIMs.

Our finding of a very low Native American contribution in this AA sample, based on AIMs analysis, is compatible with other studies [9,22,32] and may represent a conflict with some of the reported family history. It is also possible that this sample does not represent other AA groups in the USA. It most likely does not reflect a limitation in the AIMs set or the analysis since this cluster was clearly detected in HA.

In this study, we have shown the effect of admixture on allele frequencies of two SNPs in the *mu* opioid receptor gene, *OPRM1* (118A > G and 17C > T). The significantly higher frequency of the 118 G allele in this small random HA sample compared with the EA sample probably reflects the contribution of Asian and Native American ancestries. Similarly, the significantly higher frequency of the 17 T allele in the HA sample compared with the EA sample most probably reflects the contribution of African ancestry. This study reiterates the importance of AIMs in defining ancestry, especially in admixed populations and emphasizes the concept that Hispanic Americans is not a valid category in genetic research.

This study provides support for the robustness of this set of AIMs, as our results corroborate the results of other studies using this set [18,22-24]. The study demonstrates that computation of ethnic factor scores 'anchored, against worldwide genetic diversity (CEPH reference populations) yields a stable factor structure, allows comparisons between different datasets, and may permit combining data from different studies. This set of AIMs is especially useful in situations where large-scale genotyping is not available.

There are several limitations to this study: first, our sample does not represent the general US population, as it was derived from only two main locales (Las Vegas and New York City), with unrepresentative low proportion of Mexican Americans. Second, the specific AIMs used in this study were selected based on HapMap data (release #16c.1, 2005) [19] and as such are limited to allele frequency data from small samples of three main original HapMap populations (Northern and Western Europe (CEPH), Nigeria (Yoruba), and Han Chinese (Beijing)) and may not be suited for analysis of certain populations.

Albeit very promising, great care must be used in research of this kind to avoid misleading interpretations. Genetic ancestry estimates could help the dismissal of the concept of race, but may also support the notion of distinct human biological subgroups that may increase stigmatization and discrimination [38,39]. There is growing evidence that major health differences between populations involve gene-environment interaction, and, as such, their understanding will need not only genetic tools but also social/cultural information [22].

Competing interests

The authors declare that they have no competing interests.

Acknowledgments

We thank all the clinical staff including John Rotrosen, Paul Casadonte (VA New York Harbor Healthcare System), Shirley Linzy (The Adelson Clinic, LV), Lisa Borg, Brenda Ray, Elizabeth Ducat, and Dorothy Melia (The Rockefeller University) for patient recruitment and ascertainment; David Goldman and the Laboratory of Neurogenetics (NIAAA) for assistance with STRUCTURE analysis; Matthew Randesi, Connie Zhao, and Bin Zhang for technical assistance; and Susan Russo for proofreading the manuscript. This work was supported by the National Institutes of Health - National Institute of Drug Addiction [P60-05130 to MJK] and the Adelson Medical Research Foundation.

Author details

[1]The Laboratory of the Biology of Addictive Diseases, The Rockefeller University, New York NY10065, USA. [2]Laboratory of Neurogenetics, National Institute on Alcohol Abuse and Alcoholism, Bethesda, MD 20892, USA. [3]The Dr. Miriam and Sheldon G. Adelson Clinic for Drug Abuse Treatment and Research, Las Vegas, NV 89169, USA.

Authors' contributions

OL designed the study, carried out the genetic studies, performed data analysis and interpretation, and wrote and revised the manuscript. OA performed data analysis and interpretation. PHS performed STRUCTURE analysis. MA supervised subject recruitment, and ascertainment and acquisition of ancestry data of the LV sample. MJK supervised the overall subject recruitment, ascertainment and acquisition of ancestry data, and critically reviewed the manuscript. All authors read and approved the final manuscript.

References

1. Rosenberg NA, Pritchard JK, Weber JL, Cann HM, Kidd KK, Zhivotovsky LA, Feldman MW: **Genetic structure of human populations.** *Science* 2002, **298:**2381–2385.
2. Rosenberg NA, Huang L, Jewett EM, Szpiech ZA, Jankovic I, Boehnke M: **Genome-wide association studies in diverse populations.** *Nat Rev Genet* 2010, **11:**356–366.
3. Tian C, Gregersen PK, Seldin MF: **Accounting for ancestry: population substructure and genome-wide association studies.** *Hum Mol Genet* 2008, **17:**R143–R150.
4. Cardon LR, Palmer LJ: **Population stratification and spurious allelic association.** *Lancet* 2003, **361:**598–604.
5. Peralta CA, Li Y, Wassel C, Choudhry S, Palmas W, Seldin MF, Risch N, Siscovick D, Arnett D, Psaty B, Shlipak MG: **Differences in albuminuria between Hispanics and whites: an evaluation by genetic ancestry and country of origin: the multi-ethnic study of atherosclerosis.** *Circ Cardiovasc Genet* 2010, **3:**240–247.
6. Ruano G, Duconge J, Windemuth A, Cadilla CL, Kocherla M, Villagra D, Renta J, Holford T, Santiago-Borrero PJ: **Physiogenomic analysis of the Puerto Rican population.** *Pharmacogenomics* 2009, **10:**565–577.
7. Bonilla C, Shriver MD, Parra EJ, Jones A, Fernández JR: **Ancestral proportions and their association with skin pigmentation and bone mineral density in Puerto Rican women from New York city.** *Hum Genet* 2004, **115:**57–68.
8. Bryc K, Velez C, Karafet T, Moreno-Estrada A, Reynolds A, Auton A, Hammer M, Bustamante CD, Ostrer H: **Colloquium paper: genome-wide patterns of population structure and admixture among Hispanic/Latino populations.** *Proc Natl Acad Sci USA* 2010, **107**(Suppl 2):8954–8961.
9. Behar DM, Rosset S, Tzur S, Selig S, Yudkovsky G, Bercovici S, Kopp JV, Winkler CA, Nelson GW, Wasser WG, Skorecki K: **African ancestry allelic variation at the MYH9 gene contributes to increased susceptibility to non-diabetic end-stage kidney disease in Hispanic Americans.** *Hum Mol Genet* 2010, **19:**1816–1827.
10. DA, Young HJ, Arnett DK: **Comparing self-reported ethnicity to genetic background measures in the context of the Multi-Ethnic Study of Atherosclerosis (MESA).** *BMC Genet* 2011, **12:**28.
11. Wang S, Ray N, Rojas W, Parra MV, Bedoya G, Gallo C, Poletti G, Mazzotti G, Hill K, Hurtado AM, Camrena B, Nicolini H, Klitz W, Barrantes R, Molina JA, Freimer NB, Bortolini MC, Salzano FM, Petzl-Erler ML, Tsuneto LT, Dipierri JE, Alfaro EL, Bailliet G, Bianchi NO, Llop E, Rothhammer F, Excoffier L, Ruiz-Linares A: **Geographic patterns of genome admixture in Latin American Mestizos.** *PLoS Genet* 2008, **4:**e1000037.
12. Choudhry S, Burchard EG, Borrell LN, Tang H, Gomez I, Naqvi M, Nazario S, Torres A, Casal J, Martinez-Cruzado JC, Ziv E, Avila PC, Rodriguez-Cintron W, Risch NJ: **Ancestry-environment interactions and asthma risk among Puerto Ricans.** *Am J Respir Crit Care Med* 2006, **174:**1088–1093.
13. Choudhry S, Coyle NE, Tang H, Salari K, Lind D, Clark SL, Tsai HJ, Naqvi M, Phong A, Ung N, Matallana H, Avila PC, Casal J, Torres A, Nazario S, Castro R, Battle NC, Perez-Stable EJ, Kwok PY, Sheppard D, Shriver MD, Rodriguez-

Cintron W, Risch N, Ziv E, Burchard EG: **Population stratification confounds genetic association studies among Latinos.** *Hum Genet* 2006, **118**:652–664.

14. Via M, Gignoux CR, Roth LA, Fejerman L, Galanter J, Choudhry S, Toro-Labrador G, Viera-Vera J, Oleksyk TK, Beckman K, Ziv E, Risch N, Burchard EG, Martínez-Cruzado JC: **History shaped the geographic distribution of genomic admixture on the island of Puerto Rico.** *PLoS One* 2011, **6**:e16513.
15. Levran O, Londono D, O'Hara K, Randesi M, Rotrosen J, Casadonte P, Linzy S, Ott J, Adelson M, Kreek MJ: **Heroin addiction in African Americans: a hypothesis-driven association study.** *Genes Brain Behav* 2009, **8**:531–540.
16. Levran O, Londono D, O'Hara K, Nielsen DA, Peles E, Rotrosen J, Casadonte P, Linzy S, Randesi M, Ott J, Adelson M, Kreek MJ: **Genetic susceptibility to heroin addiction: a candidate gene association study.** *Genes Brain Behav* 2008, **7**:720–729.
17. Ray B, Jackson C, Ducat E, Ho A, Hamon S, Kreek MJ: **Effect of ethnicity, gender and drug use history on achieving high rates of affirmative informed consent for genetics research: impact of sharing with a national repository.** *J Med Ethics* 2011, **37**:374–379.
18. Hodgkinson CA, Yuan Q, Xu K, Shen PH, Heinz E, Lobos EA, Binder EB, Cubells J, Ehlers CL, Gelernter J, Mann J, Riley B, Roy A, Tabakoff B, Todd RD, Zhou Z, Goldman D: **Addictions biology: haplotype-based analysis for 130 candidate genes on a single array.** *Alcohol Alcohol* 2008, **43**:505–515.
19. Enoch MA, Shen PH, Xu K, Hodgkinson C, Goldman D: **Using ancestry-informative markers to define populations and detect population stratification.** *J Psychopharmacol* 2006, **20**:19–26.
20. Pritchard JK, Stephens M, Donnelly P: **Inference of population structure using multilocus genotype data.** *Genetics* 2000, **155**:945–959.
21. Cann HM, de Toma C, Cazes L, Legrand MF, Morel V, Piouffre L, Bodmer J, Bodmer WF, Bonne-Tamir B, Cambon-Thomsen A, Chen Z, Chu J, Carcassi C, Contu L, Du R, Excoffier L, Ferrara GB, Friedlaender JS, Groot H, Gurwitz D, Jenkins T, Herrera RJ, Huang X, Kidd J, Kidd KK, Langaney A, Lin AA, Mehdi SQ, Parham P, Piazza A, *et al*: **A human genome diversity cell line panel.** *Science* 2002, **296**:261–262.
22. Ducci F, Roy A, Shen PH, Yuan Q, Yuan NP, Hodgkinson CA, Goldman LR, Goldman D: **Association of substance use disorders with childhood trauma but not African genetic heritage in an African American cohort.** *Am J Psychiatry* 2009, **166**:1031–1040.
23. Liu J, Zhou Z, Hodgkinson CA, Yuan Q, Shen PH, Mulligan CJ, Wang A, Gray RR, Roy A, Virkkunen M, Goldman D, Enoch MA: **Haplotype-based study of the association of alcohol-metabolizing genes with alcohol dependence in four independent populations.** *Alcohol Clin Exp Res* 2011, **35**:304–316.
24. Enoch MA, Gorodetsky E, Hodgkinson C, Roy A, Goldman D: **Functional genetic variants that increase synaptic serotonin and 5-HT3 receptor sensitivity predict alcohol and drug dependence.** *Mol Psychiatry* 2011, **16**:1139–1146.
25. Levran O, Peles E, Hamon S, Randesi M, Adelson M, Kreek MJ: **CYP2B6 SNPs are associated with methadone dose required for effective treatment of opioid addiction.** *Addict Biol* 2011, doi:10.1111/j.1369-1600.2011.00349.x.
26. Kreek MJ, Bart G, Lilly C, LaForge KS, Nielsen DA: **Pharmacogenetics and human molecular genetics of opiate and cocaine addictions and their treatments.** *Pharmacol Rev* 2005, **57**:1–26.
27. Crystal HA, Hamon S, Randesi M, Cook J, Anastos K, Lazar J, Liu C, Pearce L, Golub E, Valcour V, Weber KM, Holman S, Ho A, Kreek MJ: **A C17T polymorphism in the mu opiate receptor is associated with quantitative measures of drug use in African American women.** *Addict Biol* 2012, **17**:181–191.
28. Gelernter J, Kranzler H, Cubells J: **Genetics of two mu opioid receptor gene (OPRM1) exon I polymorphisms: population studies, and allele frequencies in alcohol- and drug-dependent subjects.** *Mol Psychiatry* 1999, **4**:476–483.
29. Dumitrescu L, Ritchie MD, Brown-Gentry K, Pulley JM, Basford M, Denny JC, Oksenberg JR, Roden DM, Haines JL, Crawford DC: **Assessing the accuracy of observer-reported ancestry in a biorepository linked to electronic medical records.** *Genet Med* 2010, **12**:648–650.
30. Yaeger R, Avila-Bront A, Abdul K, Nolan PC, Grann VR, Birchette MG, Choudhry S, Burcahrd EG, Beckman KB, Gorroochurn P, Ziv E, Consedine NS, Joe AK: **Comparing genetic ancestry and self-described race in african americans born in the United States and in Africa.** *Cancer Epidemiol Biomarkers Prev* 2008, **17**:1329–1338.
31. Lee YL, Teitelbaum S, Wolff MS, Wetmur JG, Chen J: **Comparing genetic ancestry and self-reported race/ethnicity in a multiethnic population in New York City.** *J Genet* 2010, **89**:417–423.
32. Reiner AP, Ziv E, Lind DL, Nievergelt CM, Schork NJ, Cummings SR, Phong A, Burchard EG, Harris TB, Psaty BM, Kwok PY: **Population structure, admixture, and aging-related phenotypes in African American adults: the Cardiovascular Health Study.** *Am J Hum Genet* 2005, **76**:463–477.
33. Bensen JT, Xu Z, Smith GJ, Mohler JL, Fontham ET, Taylor JA: **Genetic polymorphism and prostate cancer aggressiveness: a case-only study of 1,536 GWAS and candidate SNPs in African-Americans and European-Americans.** *Prostate* 2012. doi:10.1002/pros.22532.
34. Fejerman L, Haiman CA, Reich D, Tandon A, Deo RC, John EM, Ingles SA, Ambrosone CB, Bovbjerg DH, Jandorf LH, Davis W, Ciupak G, Whittemore AS, Press MF, Ursin G, Bernstein L, Huntsman S, Henderson BE, Ziv E, Freedman ML: **An admixture scan in 1,484 African American women with breast cancer.** *Cancer Epidemiol Biomarkers Prev* 2009, **18**:3110–3117.
35. Murray T, Beaty TH, Mathias RA, Rafaels N, Grant AV, Faruque MU, Watson HR, Ruczinski I, Dunston GM, Barnes KC: **African and non-African admixture components in African Americans and an African Caribbean population.** *Genet Epidemiol* 2010, **34**:561–568.
36. Tian C, Kosoy R, Nassir R, Lee A, Villoslada P, Klareskog L, Hammarström L, Garchon HJ, Pulver AE, Ransom M, Gregersen PK, Seldin MF: **European population genetic substructure: further definition of ancestry informative markers for distinguishing among diverse European ethnic groups.** *Mol Med* 2009, **15**:371–383.
37. Kidd JR, Friedlaender FR, Speed WC, Pakstis AJ, De La Vega FM, Kidd KK: **Analyses of a set of 128 ancestry informative single-nucleotide polymorphisms in a global set of 119 population samples.** *Investig Genet* 2011, **2**:1.
38. Royal CD, Novembre J, Fullerton SM, Goldstein DB, Long JC, Bamshad MJ, Clark AG: **Inferring genetic ancestry: opportunities, challenges, and implications.** *Am J Hum Genet* 2010, **86**:661–673.
39. Caulfield T, Fullerton SM, Ali-Khan SE, Arbour L, Burchard EG, Cooper RS, Hardy BJ, Harry S, Hyde-Lay R, Kahn J, Kittles R, Koenig BA, Lee SSJ, Malinowski M, Ravitsky V, Sankar P, Scherer SW, Séguin B, Shickle D, Suarez-Kurtz G, Daar AS: **Race and ancestry in biomedical research: exploring the challenges.** *Genome Med* 2009, **1**:8.

2

Three-hour analysis of non-invasive foetal sex determination: application of Plexor chemistry

Adalgisa Pietropolli[1,2], Maria Vittoria Capogna[1,2], Raffaella Cascella[3,4*], Chiara Germani[5], Valentina Bruno[1,2], Claudia Strafella[3], Simona Sarta[1,2], Carlo Ticconi[1,2], Giusy Marmo[3], Sara Gallaro[3], Giuliana Longo[3], Luigi Tonino Marsella[3], Antonio Novelli[6], Giuseppe Novelli[3], Emilio Piccione[1,2] and Emiliano Giardina[3,5]

Abstract

Background: The knowledge of the individual genetic "status" in the prenatal era is particularly relevant in the case of positive family history for genetic diseases, in advanced maternal age and in the general screening for foetal abnormalities. In this context, here, we report an innovative molecular assay which utilizes the cell-free foetal DNA (cffDNA) as a source for the early and fast detection of the foetal sex. The study involved 132 pregnant women in their first 3 months of pregnancy, who agreed to give a blood sample. All the collected samples were immediately subjected to the separation of the plasma, which was utilized for the extraction of the cffDNA. Successively, the extracted cffDNA was analysed by a quantitative PCR (qPCR) method based on Plexor-HY chemistry, which is able to simultaneously identify, quantify and discriminate the autosomal DNA from the sex-linked DNA.

Results: Overall, the Plexor-HY assay demonstrated to be sensitive and specific for the determination of low-template DNA, such as the cffDNA. In fact, the Plexor-HY assay has been successfully performed in all the samples, identifying 70 males and 62 females. As the foetal sex can be provided in 120 min just by utilizing a maternal blood sample as cffDNA source, the assay represents a very fast, safe and non-invasive prenatal method.

Conclusions: The possibility of determining the foetal sex in the early prenatal life consents the application of our assay as a helpful screening test for subjects and families at risk of sex-linked disorders. Moreover, the early knowledge of the foetal sex may be of great help even for the specialist, who might promptly advise the patients concerning the foetal risk of inheriting sex-linked disorders and the clinical utility of performing an invasive prenatal diagnosis.

Background

Prenatal diagnosis represents a fundamental step for the identification of genetic pathologies, either caused by numeric and structural chromosomal aberrations, inherited DNA mutations (Mendelian disorders) or malformative parts [1]. In fact, prenatal knowledge of the genetic "status" of an individual is particularly relevant in the case of a positive family history for genetic diseases and advanced maternal age as well as in the general screening of foetal abnormalities [2]. A prenatal diagnosis is essentially composed of two steps: gynaecologic/genetic counselling and gynaecologic/laboratory practice. In particular, gynaecologic/genetic counselling is a service that provides information about the pros and cons of prenatal approaches in terms of the risk, waiting times and reliability of the results [3]. The gynaecologic/laboratory practice includes both prenatal techniques (invasive and non-invasive) and genetic approaches (karyotyping and molecular assays) [4]. To date, the invasive approaches (amniocentesis and chorionic villous sampling, CVS) are considered to be the gold standard techniques for prenatal testing, although these procedures can sometimes be fatal for the foetus (~0.2–0.5 %) [5, 6].

The non-invasive prenatal tests (NIPTs) allow for the gathering of information about the "status" of the foetus without affecting the structures that are primarily involved in foetal development/life [7]. The NIPT methods

* Correspondence: raffaellacascella@virgilio.it
[3]Department of Biomedicine and Prevention, School of Medicine, University of Rome "Tor Vergata", Montpelier Street, 1, 00133 Rome, Italy
[4]Emotest Laboratory, M. Licola Patria Street, 60, 80078 Pozzuoli, NA, Italy

are therefore safer in terms of abortion risk relative to the invasive methods. The NIPTs can be distinguished in diagnostic (ultrasonography) and non-diagnostic (biochemical screening and the analysis of the cell-free foetal DNA (cffDNA)) approaches [8, 9].

Analysis of cffDNA is the most recent non-diagnostic NIPT method. Although the source of cffDNA is not completely understood, it may originate from fragments of apoptotic placental cells (trophoblast cells) [10]. The cffDNA is highly detectable in the maternal blood, either in the plasma or in serum fractions at the fifth week of pregnancy [11, 12]. The concentration of cffDNA increases during the weeks of gestation and ranges from 3.3 to 69.4 to 76.9 to 769 copies/ml (accounting for 3 to 6 % of the free DNA present in the maternal circulation [13]. Given its characteristics, cffDNA represents as an excellent source of DNA that is applicable to the early prenatal tests [12]. On this subject, many innovative next-generation sequencing (NGS) platforms have been recently designed to evaluate chromosomal or DNA alterations with massive and high-throughput approaches. In addition, the NGS techniques are also able to discriminate foetal sex with high-accuracy and sensitive results. However, the effective application of NGS in clinical practice is limited by elevated costs in terms of the time and equipment required for the whole analysis [14]. Sex determination is, in fact, usually performed through the application of PCR-based techniques, either standard PCR or real-time PCR [15].

Knowledge of foetal sex can provide helpful information on the risk of X-linked diseases (Duchenne's dystrophy, fragile X disorder and haemophilia), metabolic conditions associated with development of ambiguous external genitalia and risk of congenital adrenal hyperplasia [15–17].

In this context, we describe a 3-h molecular assay that utilizes cffDNA as a source for the early detection of foetal sex.

Methods

This study was previously approved by the ethics committee of the University of Rome "Tor Vergata". The study was performed according to the tenets of the Declaration of Helsinki. The samples were recruited from the Department of Biomedicine and Prevention, Section of Obstetrics and Gynaecology at "Tor Vergata" University Hospital Rome, Italy. All of the participants were also subjected to invasive prenatal diagnosis (IPD) through traditional methods, in particular, amniocentesis or CVS (indications for IPD were advanced maternal age or abnormal results of NIPTs). The clinical data of the recruited samples are reported in Additional file 1. Signed informed consent was obtained from all of the participants before the blood sample collection. Peripheral blood samples (10 ml) were collected from 132 pregnant women during their first 3 months of pregnancy (from the 12th to 14th gestational weeks, in particular 44 women per class). The collected samples were utilized for the extraction of the cffDNA.

The cffDNA was analysed with a fluorescence-based quantitative PCR (qPCR) method to simultaneously identify, quantify and discriminate the autosomal DNA from the sex-linked DNA. The qPCR procedure was performed utilizing Plexor-HY chemistry (Promega), which exploits specific primers for the amplification of Y-chromosomal DNA, human autosomal DNA and an internal positive control (IPC). Additionally, the Plexor-HY assay provided a "standard DNA", which is a synthetic DNA (concentrated to 50 ng/µl) that is necessary to quantify the unknown samples (through the realization of a "standard curve" obtained by the serial dilution of the standard DNA from 10 to 0.032 ng/µl).

The specificity and reliability of the qPCR reaction is assessed by the generation of a dissociation curve and the calculation of the melting temperature (T_m) of the products after the amplification [18]. In our study, the dissociation curve and the T_m were crucial for the interpretation of the results and the correct determination of foetal sex.

The foetal sex of the samples was established according to the presence/absence of a fluorescence signal, which was emitted only in the presence of a Y-chromosomal DNA substrate. Therefore, the only sex-linked DNA that can be quantified is from the Y-chromosomal DNA and it is correlated with the amount of emitted-fluorescence. The quantity of X-chromosomal DNA cannot be specifically detected because the Plexor-HY system does not contain the primers for its amplification. The female sex is thereby deducible on the basis of the missing fluorescence (instead of the detection of a fluorescent signal) as well as of a "flat" dissociation curve (due to the absence of amplification). This aspect actually represents a limit of the method, which can be partially overcome by comparing the unknown samples with the control samples (with known sex).

Concerning the amplification/quantification of the autosomic DNA, it can be performed with all of the samples independently from gender.

All the technical details of the cffDNA extraction and of the Plexor-HY assay for the foetal sex determination have been reported in the Additional file 2.

Results

The molecular assay for the simultaneous determination of foetal sex and quantification of the autosomic/sex-linked DNA has been successfully performed in all 132 samples, identifying 70 males and 62 females. Figure 1a represents the melting curve from

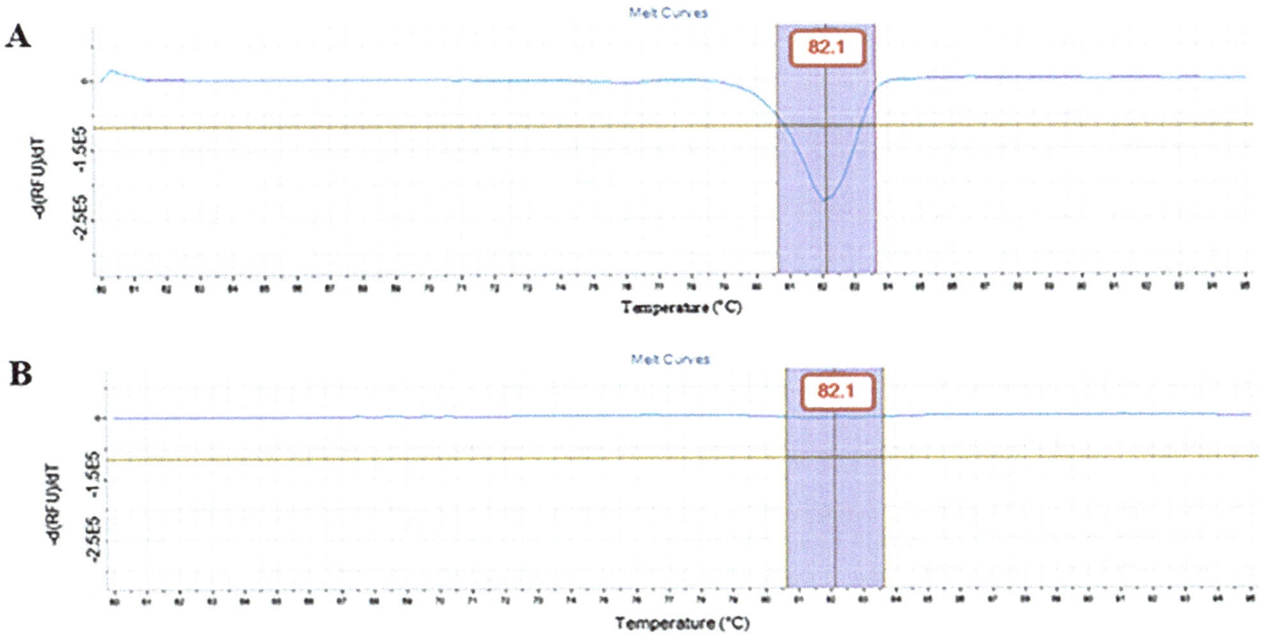

Fig. 1 a Result interpretation of the Plexor-HY assay on cffDNA. The detection of the fluorescence signal is indicative of the presence of Y-chromosomal DNA substrate (male foetal sex). **b** Result interpretation of the Plexor-HY assay on cffDNA. In this case, the absence of the fluorescence signal is interpreted as the presence of X-chromosomal DNA substrate (female foetal sex)

the amplification of the Y-chromosomal DNA (male foetal sex) sample with its specific T_m (82.1 °C) and fluorescence intensity. Figure 1b depicts the "flat" dissociation curve resulting from the absence of a fluorescent signal (no amplification) and indicates the presence of X-chromosomal DNA (female foetal sex).

Our method has been validated in terms of specificity and sensitivity for the determination of foetal sex by comparing our results with those obtained from traditional prenatal tests (i.e., amniocentesis, CVS and ultrasonography) on the same samples. The comparison between Plexor-HY technology and the common procedures showed a 100 % concordance, with no false positive/negative results (Table 1). In addition, the sensitivity of the Plexor-HY assay has been evaluated by the calculation of the minimum detectable concentration (limit of detection (LOD)) of cffDNA. The LOD has been measured testing cffDNA concentrations included in a range from 10 to 0.0001 ng/μl. These experiments showed that the LOD for Plexor-HY assay is 0.0001 ng/μl ± 0.5.

As expected, the amount of autosomic/sex-linked cffDNA was found to be increased from the 12th to the 14th gestational week. In fact, the autosomic DNA amount ranged from 1.42×10^{-2} to 1.86×10^{-1} pg/μl, while the sex-linked DNA (Y-chromosomal DNA) concentration varied from 0.009 to 0.09 pg/μl throughout the gestational weeks taken into consideration.

Discussion

Overall, the Plexor-HY assay was shown to be sensitive and specific for the determination of low-template DNA, such as cffDNA. Despite other qPCR methods, Plexor chemistry based on the incorporation of modified nucleotides, multicopy gene targets and generation of a dissociation curve made the assay highly performant and reliable for the analysis of cffDNA.

As the foetal sex can be determined within 120 min by utilizing a maternal blood sample as a DNA source, the assay is a very fast, safe and non-invasive prenatal method. In Fig. 2, the workflow depicts the different operational phases (counselling and laboratory practice) that characterize the test for early foetal sex determination by our method.

Given the possibility of applying the Plexor-HY assay for the identification of foetal sex in the first trimester (12th gestational week), it may represent a helpful screening test for inherited X-linked disorders, such as Duchenne's muscular dystrophy (DMD). The DMD is an X-linked recessive disorder with an incidence of 1/3300 male newborns. The clinical picture of the disease is characterized by the atrophy and progressive degeneration of the muscular fibres that lead to a deficit of muscular strength in the first 3–5 years of life. Concerning the genetic picture of the disease, several mutations in the DMD gene (Xp21.2, codifying for the dystrophin protein) have been identified as major causes of the pathologic phenotype.

Table 1 Concordance analysis of the foetal sex detected by cffDNA and the traditional prenatal approaches

Gestational week (w[a]) of blood sampling	cffDNA[b] (Foetal sex)	CVS[c] (Foetal sex)	Amniocentesis (Foetal sex)	Concordance (%)
12	44 (M[d] = 24; F[e] = 20)	32 (M = 17; F = 15)	12 (M = 7; F = 5)	100
13	44 (M = 22; F = 22)	20 (M = 8; F = 12)	24 (M = 14; F = 10)	100
14	44 (M = 24; F = 20)	9 (M = 5; F = 4)	35 (M = 19; F = 16)	100

[a]Gestational week
[b]Cell-free foetal DNA
[c]Chorionic villous sampling
[d]Males
[e]Females

On this subject, we evaluated the clinical utility of the Plexor-HY System in two families with a positive history for DMD. The blood samples of the two pregnant women were taken at their 12th gestational week to extract and analyse the cffDNA. The application of the Plexor-HY technology led to the identification of one female foetus and one male foetus, respectively. Given these results, the early prenatal identification of the foetal sex was useful to provide an initial indication about the risk of inheriting DMD. The results obtained by the Plexor-HY assay have been successively confirmed with the DNA analysis performed by an invasive procedure (CVS). These results suggested that our method may be employed to reassure patients about the risk of inheriting DMD in relation to foetal sex and eventually avoid further invasive procedures. However, it is important to remind that the test is able to determine the foetal sex, while the effective diagnosis of DMD can only be done by the traditional molecular diagnostic approaches.

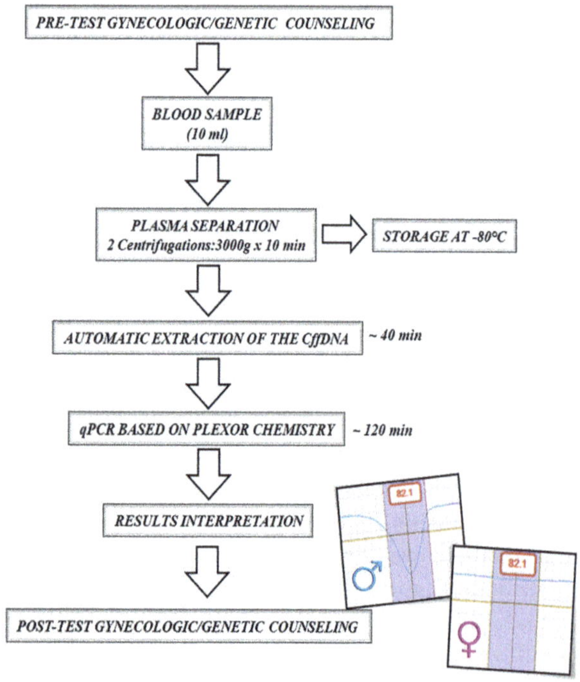

Fig. 2 Workflow for the foetal sex determination by the Plexor-HY assay on cffDNA

The determination of foetal sex by non-invasive cffDNA analysis may be thus validated as a rapid, safe and early prenatal screening test for subjects and families at risk of X-linked disorders, metabolic condition associated with ambiguous development of external genitalia and risk of congenital adrenal hyperplasia. This approach may be really advantageous for both the woman and the foetus, since the risk of abortion caused by the traditional invasive techniques is effectively abolished.

In this context, genetic counselling plays a fundamental role in providing the right support to the patient, both in the pre- and post-test phases. In fact, counselling before the test is necessary to elucidate the utility of the early prenatal screening of foetal sex, while post-test counselling has to explain the result of the test to enable the patient/couple to consciously take her/their own decision.

In conclusion, here, we report a 3-h molecular assay that utilizes cffDNA as the source for the early detection of the foetal sex. Our results showed that Plexor chemistry with cffDNA is a reproducible, sensitive, robust and non-invasive method for prenatal screening of the gender. Moreover, the early knowledge of foetal sex may be of great help even for the specialist, who might promptly advise the patients about the risks (sex-linked disorders) for the foetus and the clinical utility of performing an invasive prenatal diagnosis.

Ethics approval and consent to participate

The study has been approved by the Ethics Committee of the University of Rome "Tor Vergata". The study was performed according to the tenets of the Declaration of Helsinki.

Consent

The participants have signed the informed consent for publication of personal data.

Abbreviations

cffDNA: cell-free foetal DNA; CVS: chorionic villous sampling; DMD: Duchenne's muscular dystrophy; IPD: invasive prenatal diagnosis; NGS: next-generation sequencing; NIPTs: non-invasive prenatal tests; qPCR: quantitative PCR; T_m: melting temperature.

Competing interests

The authors declare that they have no competing interests.

Authors' contributions

All authors read and approved the final manuscript.

Funding

None to declare.

Author details

[1]Section of Gynecology and Obstetrics, Academic Department of Biomedicine and Prevention, "Tor Vergata" University Hospital, Oxford Street, 81, 00133 Rome, Italy. [2]Clinical Department of Surgery, "Tor Vergata" University Hospital, Oxford Street, 81, 00133 Rome, Italy. [3]Department of Biomedicine and Prevention, School of Medicine, University of Rome "Tor Vergata", Montpelier Street, 1, 00133 Rome, Italy. [4]Emotest Laboratory, M. Licola Patria Street, 60, 80078 Pozzuoli, NA, Italy. [5]Molecular Genetics Laboratory UILDM, Santa Lucia Foundation, Ardeatina Street, 306, 00142 Rome, Italy. [6]Bambino Gesù Children's Hospital, IRCCS, Sant'Onofrio Square, 4, Rome, Italy.

References

1. Verma IC, Puri RD. Global burden of genetic disease and the role of genetic screening. Semin Fetal Neonatal Med 2015;doi:10.1016/j.siny.2015.07.002.
2. Löwy I. How genetics came to the unborn: 1960-2000. Stud Hist Philos Biol Biomed Sci. 2014;47:154–62.
3. Colmant C, Morin-Surroca M, Fuchs F, Fernandez H, Senat MV. Non-invasive prenatal testing for fetal sex determination: is ultrasound still relevant? Eur J Obstet Gynecol Reprod Biol. 2013;171:197–04.
4. de Jong A, Maya I, van Lith JM. Prenatal screening: current practice, new developments, ethical challenges. Bioethics. 2015;29:1–8.
5. Young BK, Chan MK, Liu L, Basch RS. Amniotic fluid as a source of multipotent cells for clinical use. J Perinat Med 2015;doi:10.1515/jpm-2015-0152.
6. Ghidini A. Risk of miscarriage following amniocentesis and chorionic villus sampling. Ultrasound Obstet Gynecol. 2015;45:755.
7. Rebouché R. Non-invasive testing, non-invasive counseling. J Law Med Ethics. 2015;43:228–40.
8. Reddy UM, Filly RA, Copel JA. Prenatal imaging: ultrasonography and magnetic resonance imaging. Obstet Gynecol. 2008;112:145–57.
9. Liao GJ, Gronowski AM, Zhao Z. Non-invasive prenatal testing using cell-free fetal DNA in maternal circulation. Clin Chim Acta. 2014;428:44–50.
10. Hahn S, Huppertz B, Holzgreve W. Fetal cells and cell free fetal nucleic acids in maternal blood: new tools to study abnormal placentation? Placenta. 2005;26:515–26.
11. Lo YM, Corbetta N, Chamberlain PF, Rai V, Sargent IL, Redman CW, et al. Presence of fetal DNA in maternal plasma and serum. Lancet. 1997;350:485–7.
12. Everett TR, Chitty LS. Cell-free fetal DNA: the new tool in fetal medicine. Ultrasound Obstet Gynecol. 2015;45:499–07.
13. Lo YM, Tein MS, Lau TK, Haines CJ, Leung TN, Poon PM, et al. Quantitative analysis of fetal DNA in maternal plasma and serum: implications for noninvasive prenatal diagnosis. Am J Hum Genet. 1998;62:768–75.
14. Nepomnyashchaya YN, Artemov AV, Roumiantsev SA, Roumyantsev AG, Zhavoronkov A. Non-invasive prenatal diagnostics of aneuploidy using next-generation DNA sequencing technologies, and clinical considerations. Clin Chem Lab Med. 2013;51:1141–54.
15. Kolialexi A, Tounta G, Apostolou P, Vrettou C, Papantoniou N, Kanavakis E, et al. Early non-invasive detection of fetal Y chromosome sequences in maternal plasma using multiplex PCR. Eur J Obstet Gynecol Reprod Biol. 2012;161:34–7.
16. Wu D, Hou Q, Li T, Chu Y, Guo Q, Kang B, et al. The use of cffDNA in fetal sex determination during the first trimester of pregnancy of female DMD carriers. Intractable Rare Dis Res. 2012;1:157–60.
17. Rijnders RJ, van der Schoot CE, Bossers B, de Vroede MA, Christiaens GC. Fetal sex determination from maternal plasma in pregnancies at risk for congenital adrenal hyperplasia. Obstet Gynecol. 2001;98:374–8.
18. Frackman S, Ekenberg S, Hoffmann K, Krenke B, Sprecher C, Storts D. Plexor™ technology: a new chemistry for real-time PCR. Promega Notes. 2005;90:2–4.

3

Organization, evolution and functions of the human and mouse Ly6/uPAR family genes

Chelsea L. Loughner[1], Elspeth A. Bruford[2], Monica S. McAndrews[3], Emili E. Delp[1], Sudha Swamynathan[1] and Shivalingappa K. Swamynathan[1,4,5,6,7*]

Abstract

Members of the lymphocyte antigen-6 (Ly6)/urokinase-type plasminogen activator receptor (uPAR) superfamily of proteins are cysteine-rich proteins characterized by a distinct disulfide bridge pattern that creates the three-finger Ly6/uPAR (LU) domain. Although the Ly6/uPAR family proteins share a common structure, their expression patterns and functions vary. To date, 35 human and 61 mouse Ly6/uPAR family members have been identified. Based on their subcellular localization, these proteins are further classified as GPI-anchored on the cell membrane, or secreted. The genes encoding Ly6/uPAR family proteins are conserved across different species and are clustered in syntenic regions on human chromosomes 8, 19, 6 and 11, and mouse Chromosomes 15, 7, 17, and 9, respectively. Here, we review the human and mouse Ly6/uPAR family gene and protein structure and genomic organization, expression, functions, and evolution, and introduce new names for novel family members.

Keywords: Ly6/uPAR family, LU domain, Three-finger domain, uPAR, Lymphocytes, Neutrophils

Introduction

The lymphocyte antigen-6 (Ly6)/urokinase-type plasminogen activator receptor (uPAR) superfamily of structurally related proteins is characterized by the LU domain, an 80 amino acid domain containing ten cysteines arranged in a specific spacing pattern that allows distinct disulfide bridges which create the three-fingered (3F) structural motif [1, 2]. Ly6/uPAR proteins were first identified in the mouse over 35 years ago using antisera against lymphocytes [3]. Human homologs were subsequently isolated, leading to the recognition that they represent a well-conserved family with wide-ranging expression patterns and important functions. The fully annotated human and mouse genomes contain 35 and 61 Ly6/uPAR family members, respectively. Research over the last decade has begun to unravel the important functions of the encoded proteins. In this review, we provide an overview of the Ly6/uPAR gene family and their genomic organization, evolution, as well as functions, and provide a nomenclature system for the newly identified members of this family.

* Correspondence: Swamynathansk@upmc.edu
[1]Department of Ophthalmology, University of Pittsburgh School of Medicine, Pittsburgh, USA
[4]Department of Cell Biology, University of Pittsburgh School of Medicine, Pittsburgh, USA
Full list of author information is available at the end of the article

Inclusion and approved nomenclature for novel Ly6/uPAR family members

Although Ly6/uPAR family members are related by their structure, absence of a uniform naming convention resulted in arbitrary nomenclature for these genes as they were discovered. As many of the currently approved gene symbols for Ly6/uPAR family members (e.g., *CD59* and *PLAUR*) have been widely used in the scientific literature for many years, we have refrained from a family-wide attempt to standardize their well-established names, avoiding the potential for additional confusion. In compiling this update, we came across many novel members of the Ly6/uPAR gene family, especially in the mouse genome, that did not yet have a systematic name. We named these novel family members in line with the Ly6/uPAR genes that they are most related to, based on a phylogenetic analysis (see below) using either the established LY6/Ly6# root for those that fell within the

LY6 clades, or the LYPD/Lypd# (LY6/PLAUR domain-containing) root for those outside the LY6 clades. The new symbols for these genes (1 human and 18 mouse), approved by the HGNC (HUGO Gene Nomenclature Committee) [4, 5] and MGNC (Mouse Genomic Nomenclature Committee) [6], are listed in Tables 1 and 2, respectively. We use the newly approved names for these genes in the rest of this update. HGNC have also created a gene family web-page for the human Ly6/uPAR family members (http://www.genenames.org/cgi-bin/genefamilies/set/1226).

Genomic organization of the Ly6/uPAR gene family

The Ly6/uPAR gene family currently includes 35 well-characterized human members, while the mouse gene family is considerably larger with 61 genes. Information including the name, chromosomal location, numbers of exons and LU domains for human and mouse family members is summarized in Tables 1 and 2, respectively. Twelve human Ly6 genes are clustered together within a short span of about 500 kb on chromosome 8 (8q24) (moving outward from the center of the chromosome: *PSCA*, *LY6K*, *SLURP1*, *LYPD2*, *LYNX1/SLURP2*, *LY6D*, *GML*, *LY6E*, *LY6L*, *LY6H*, and *GPIHBP1*) (http://genome-euro.ucsc.edu). The syntenic region on mouse Chromosome 15 (15D3-15E1) contains *Psca*, *Slurp1*, *Lypd2*, *Slurp2*, *Lynx1*, *Ly6d*, *Ly6g6g*, *Ly6k*, *Gml*, *Gml2*, *Ly6m*, *Ly6e*, *Ly6i*, *Ly6a*, *Ly6c1*, *Ly6c2*, *Ly6a2*, *Ly6g*, *Ly6g2*, *Ly6f*, *Ly6l*, *Ly6h*, and *Gpihbp1*. Other smaller clusters are seen on human chromosome 19 (19q13) (*LYPD4*, *CD177*, *TEX101*, *LYPD3*, *PINLYP*, *PLAUR*, *LYPD5*, and *SPACA4* with syntenic region on mouse Chromosome 7 containing *Lypd5*, *Plaur*, *Pinlyp*, *Lypd3*, *Tex101*, *Lypd10*, *Lypd11*, *Cd177*, *Lypd4*, and *Spaca4*), human chromosome 11 (11q24.2) (*ACRV1*, *PATE1*, *PATE2*, *PATE3*, and *PATE4* with syntenic region on mouse Chromosome 9 containing *Pate4*, *Pate2*, *Pate13*, *Pate3*, *Pate1*, *Pate10*, *Pate7*, *Pate6*, *Pate5*, *Pate12*, *Pate11*, *Pate9*, *Pate8*, *Pate14*, and *Acrv1*), and human chromosome 6 (6p21) (*LY6G6C*, *LY6G6D*, *LY6G6F*, *LY6G5C*, and *LY6G5B* with syntenic region in the MHC class III region of the mouse Chromosome 17 containing *Ly6g6c*, *Ly6g5c*, *Ly6g5b*, *Ly6g6d*, *Ly6g6f*, and *Ly6g6e*), while the remaining family members are found on other chromosomes (Tables 1 and 2).

Typical Ly6/uPAR gene structure

Ly6/uPAR family members typically contain one LU domain, with the exception of *LYPD3* [7] and *CD177* [8] which contain two, and *PLAUR* [9], which contains three direct repeats of the LU domain (Tables 1 and 2). The mouse *Cd177* differs from its human ortholog in that it contains four direct repeats of the LU domain. A typical Ly6/uPAR family gene consists of three exons and two introns (Fig. 1a), with the signal peptide being encoded in the first exon. The mature polypeptide is encoded by the last two exons, with the GPI-anchor domain encoded by the third exon.

Based on their subcellular localization, Ly6/uPAR family members are further subdivided into two groups: membrane-tethered (through a GPI-anchor domain) or secreted (lacking the GPI-anchor domain). GPI-anchored Ly6/uPAR family members tend to congregate on lipid rafts on the cell surface, which promotes their interactions with other proteins. A fraction of the GPI-anchored Ly6/uPAR family proteins such as PLAUR are secreted after their GPI-anchor domain is proteolytically cleaved [10–12]. Experimental evidence supports the presence of a GPI-anchoring signal peptide in a majority of Ly6/uPAR family members, while it is absent in a few (Table 3). For those with no experimental evidence, the GPI-anchor signal predictor 'PredGPI' program (http://gpcr.biocomp.unibo.it/predgpi/) [13] predicted the presence of a GPI-anchor signal within mouse and human LYPD8 and LY6L, and in mouse LYPD10, LYPD11, LYPD9, LY6F, and LY6M, while predicting its absence in mouse and human LYPD4, LY6G6F, and PINLYP, and in mouse GML2 and LY6G6 (Table 3).

Structure of the LU domain

The Ly6/uPAR family members have a well-conserved LU domain with a characteristic three-finger structure formed by disulfide bridges connecting the conserved cysteine residues in a specific pattern. LU domains are topologically similar to the three-finger structure of snake venom neurotoxins, which have three β-sheet loops fixed in space by virtue of their unique disulfide bridges. The structure of the extracellular region of CD59 was first solved by 2D NMR methods [14, 15] and further refined by crystallography [16] revealing it to be a flat, disk-shaped molecule consisting of a two-stranded beta-sheet finger loosely packed against a protein core formed by a three-stranded beta-sheet and a short helix.

Alignment of LU domain amino acid sequences of selected human LY6/UPAR proteins performed using ProbCons (http://toolkit.tuebingen.mpg.de) revealed the location of conserved cysteines (Fig. 1b). Five well-conserved disulfide bridges between cysteine pairs 3 and 26, 6 and 13, 19 and 39, 45 and 63, and 64 and 69 stabilize the hydrophobic core, from which three β-sheet-based fingers protrude (Fig. 1b). The sequence of the amino acids exposed at the tips of each finger as well as the length of each of the fingers is variable, providing the three-finger motif with the flexibility for a wide

Table 1 Name, chromosomal location, number of exons, and LU domains for human Ly6/uPAR family genes

Approved gene symbol (NCBI accession #)	Gene name	Alias(es)	Genomic location	Number of	
				Exons	LU Domains
ACRV1 (NP_001603.1)	Acrosomal vesicle protein 1	*SP-10; SPACA2; D11S4365*	11q24.2	5	1
CD177 (BAE93254.1)	CD177 molecule (Human Neutrophil Alloantigen 2A)	*NB1; PRV1; HNA2A; PRV-1; HNA-2a; NB1 GP*	19q13.2	9	2
CD59 (CAG46523.1)	CD59 molecule, complement regulatory protein	*1 F5; EJ16; EJ30; EL32; G344; MIN1; MIN2; MIN3; MIRL; HRF20; MACIF; MEM43; MIC11; MSK21; 16.3A5; HRF-20; MAC-IP; p18-20*	11p13	5	1
GML (EAW82296.1)	Glycosylphosphatidylinositol anchored molecule like	*LY6DL*	8q24.3	5	1
GPIHBP1 (AAH35810.2)	Glycosylphosphatidylinositol anchored high density lipoprotein-binding protein 1	*HYPL1D; GPI-HBP1*	8q24.3	4	1
LY6D (AAH31330.1)	Lymphocyte antigen 6 complex, locus D	*E48; Ly-6D*	8q24	3	1
LY6E (AAH03392.1)	Lymphocyte antigen 6 complex, locus E	*RIGE; SCA2; RIG-E; SCA-2; TSA-1*	8q24.3	4	1
LY6G5B (CAC85543.1)	Lymphocyte antigen 6 complex, locus G5B	*G5b; C6orf19*	6p21.3	3	1
LY6G5C (CAC85542.1)	Lymphocyte antigen 6 complex, locus G5C	*G5C; NG33; C6orf20; LY6G5CA; LY6G5CB*	6p21	3	1
LY6G6C (EAX03491.1)	Lymphocyte antigen 6 complex, locus G6C	*G6c; NG24; C6orf24*	6p21	3	1
LY6G6D (CAC85540.1)	Lymphocyte antigen 6 complex, locus G6D	*G6D; NG25; LY6-D; MEGT1; C6orf23*	6p21.3	3	1
LY6G6F (AAI37213.1)	Lymphocyte antigen 6 complex, locus G6F	*G6F; NG32; LY6G6D; C6orf21*	6p21.33	6	1
LY6H (BAA34115.1)	Lymphocyte antigen 6 complex, locus H	*NMLY6*	8q24.3	4	1
LY6K (AAI17145.1)	Lymphocyte antigen 6 complex, locus K	*CT97; ly-6 K; URLC10; HSJ001348*	8q24.3	3	1
LY6L[a] (XP_011544859.1)	Lymphocyte antigen 6 complex, locus L	*LOC101928108*	8q24.3	5	1
LYNX1 (NP_803429.1)	Ly6/neurotoxin 1		8q24.3	4	1
LYPD1 (EAX11675.1)	Ly6/PLAUR domain containing 1	*PHTS; LYPDC1; LYNX2*	2q21.2	3	1
LYPD2 (EAW82307.1)	Ly6/PLAUR domain containing 2	*LYPDC2; UNQ430*	8q24.3	3	1
LYPD3 (EAW57190.1)	Ly6/PLAUR domain containing 3	*C4.4A*	19q13.31	5	2
LYPD4 (AAH34629.1)	Ly6/PLAUR domain containing 4	*Sperm Membrane Receptor*	19q13.2	5	1
LYPD5 (EAW57232.1)	Ly6/PLAUR domain containing 5	*PRO4356*	19q13.31	5	1
LYPD6 (AAH47013.1)	Ly6/PLAUR domain containing 6		2q23.2	5	1
LYPD6B (AAH18203.1)	Ly6/PLAUR domain containing 6B	*CT116; LYPD7*	2q23.2	7	1
LYPD8 (NP_001278212.1)	Ly6/PLAUR domain containing 8		1q44	7	1
PATE1 (AAI07045.1)	Prostate and testis expressed 1	*PATE*	11q24.2	5	1
	Prostate and testis expressed 2	*PATE-M; UNQ3112; C11orf38*	11q24.2	6	1

Table 1 Name, chromosomal location, number of exons, and LU domains for human Ly6/uPAR family genes *(Continued)*

PATE2 (AAI44527.1)					
PATE3 (NP_001123355.3)	Prostate and testis expressed 3	*HEL-127; PATE-DJ*	11q24.2	3	1
PATE4 (NP_001138346.1)	Prostate and testis expressed 4	*PATE-B*	11q24.2	3	1
PINLYP (NP_001180550.1)	Phospholipase A2 inhibitor and Ly6/PLAUR domain containing		19q13.31	6	1
PLAUR (CAG33233.1)	Plasminogen activator, urokinase receptor	*CD87; UPAR; URKR; U-PAR*	19q13	7	3
PSCA (AAH65183.1)	Prostate stem cell antigen	*PRO232*	8q24.2	3	1
SLURP1 (AAT01436.1)	Secreted Ly6/PLAUR domain containing 1	*ARS; MDM; ANUP; ArsB; LY6LS*	8q24.3	3	1
SLURP2 (NP_803253.1)	Secreted Ly6/PLAUR domain containing 2	*LYNX1, isoform B precursor*	8q24.3	4	1
SPACA4 (AAQ88753.1)	Sperm acrosome associated 4	*SAMP14*	19q13.33	1	1
TEX101 (EAW57189.1)	Testis expressed 101	*SGRG; CT131; GTPR867; NYD-SP8; PRO1884; SPATA44; TES101RP*	19q13.31	9	1

[a]Novel gene named in this study

range of intermolecular interactions. In addition to the LU domain, Ly6/uPAR family proteins possess a well-conserved LeuXxxCysXxxXxxCys motif at the amino-terminus and CysCysXxxXxxXxxXxxCysAsn motif at the carboxyl-terminus (Fig. 1b). Functional relevance of these motifs is not yet known.

Most Ly6/uPAR family proteins maintain the ten cysteines characteristic of the LU domain, with some notable exceptions. In PLAUR, which consists of three LU domains (designated D1, D2 and D3), only domain D2 is fully intact with ten cysteines, while domains D1 and D3 have seven and eight cysteines, respectively. Isoforms of proteins such as human LY6G5C maintain conservation throughout the LU domain in almost every isoform. In contrast, different isoforms of human LYNX1 maintain the necessary cysteines, but little else is conserved (Fig. 1b).

Expression of Ly6/uPAR family genes

The expression pattern, interacting factors, and cellular functions of the mouse and human Ly6/uPAR family members are summarized in Table 3. Expression of Ly6/uPAR proteins is (i) widespread and variable across diverse cell types and tissues, (ii) tightly regulated in a spatiotemporal manner, and (iii) often correlated with cellular differentiation. Although the Ly6/uPAR family protein structures are well-conserved across species, their expression patterns tend to vary, indicating divergence among their regulatory networks. Many Ly6/uPAR family members are expressed in hematopoietic precursors in a lineage-specific fashion making them useful cell surface markers for leukocytes, facilitating identification of individual leukocyte subgroups [17–19]. For example, mouse myeloid differentiation marker LY6G (also called Gr-1) is expressed by the myeloid lineage cells in a developmentally regulated manner in the bone marrow. Anti-LY6G antibodies are routinely used to identify neutrophils in the mouse but not humans as there is no human ortholog for *Ly6g*. Ly6/uPAR family members are generally up-regulated during inflammatory conditions or infections and in cancerous cells, with a notable exception of SLURP1, which is invariably downregulated in pro-inflammatory conditions [9, 20–24].

Functions of Ly6/uPAR family proteins

Commensurate with their varied expression patterns, Ly6/uPAR proteins have a wide range of functions in cell proliferation, migration, cell-cell interaction, immune cell maturation, macrophage activation, and cytokine production. They typically exert their influence by targeting nicotinic acetylcholine receptors (nAChRs) (reviewed in [1]). GPI-anchored Ly6/uPAR proteins lacking a cytoplasmic tail are unable to directly participate in intracellular signaling but can initiate signaling by interacting with other transmembrane proteins. Such interactions of GPI-anchored proteins are further facilitated by their tendency to congregate in lipid rafts on the cell surface, where other signaling molecules also are enriched. While GPI-anchored Ly6/uPAR proteins control signaling through interaction with their ligand(s), secreted Ly6/uPAR proteins may serve as agonists for

Table 2 Name, chromosomal location, number of exons and LU domains for mouse Ly6/uPAR family genes

Approved gene symbol (NCBI Accession #)	Gene name	Alias(es)	Genomic location	Number of	
				Exons	LU domains
Acrv1 (EDL25409.1)	Acrosomal vesicle protein 1	*Msa63; SP-10*	9 A4	4	1
Cd177 (AAH27283.1)	CD177 antigen	*Pdp3; 1190003K14Rik*	7 A3	17	4
Cd59a (AAL04433.1)	CD59a antigen	*Cd59; AA987121; protectin*	2 54.53 cM	6	1
Cd59b (AAL04434.1)	CD59b antigen		2 E2	9	1
Gml (CAB57316.1)	Glycosylphosphatidylinositol anchored molecule like	*HemT3; EG625599*	15 D3	5	1
Gml2 (AAI19338.1)	Glycosylphosphatidylinositol anchored molecule like 2	*HemT; Hemt1; 1700057K19Rik*	15 D3	7	1
Gpihbp1 (NP_081006.1)	Glycosylphosphatidylinositol anchored high density lipoprotein-binding protein 1	*GPI-HBP1; 1110002J19Rik*	15 E1	4	1
Ly6a (AAH02070.1)	Lymphocyte antigen 6 complex, locus A	*TAP; Sca1; Sca-1; Ly-6A.2; Ly-6A/E; Ly-6E.1*	15 34.29 cM	4	1
Ly6a2[a] (XP_006543305.1)	Lymphocyte antigen 6 complex, locus A2	*Ly6A-2/E-1, I830127L07Rik*	15 D3	4	1
Ly6c1 (AAH10764.1)	Lymphocyte antigen 6 complex, locus C1	*Ly6c; Ly-6C; Ly-6C1; AA682074; AA959465*	15 34.29 cM	6	1
Ly6c2 (NP_001092687.1)	Lymphocyte antigen 6 complex, locus C2	*Ly-6C2; Ly-6C.2*	15 D3	4	1
Ly6d (EDL29445.1)	Lymphocyte antigen 6 complex, locus D	*Thb; Ly61; Ly-61*	15 D3	3	1
Ly6e (CAJ18452.1)	Lymphocyte antigen 6 complex, locus E	*Ly67; Tsa1; RIG-E; Sca-2; TSA-1*	15 D3	5	1
Ly6f (EDL29474.1)	Lymphocyte antigen 6 complex, locus F		15 D3	4	1
Ly6g (NP_001297367.1)	Lymphocyte antigen 6 complex, locus G	*Gr1; Gr-1; Ly-6G*	15 D3	4	1
Ly6g2[a] (AAH25446.1)	Lymphocyte antigen 6 complex, locus G2	*BC025446*	15 D3	6	1
Ly6g5b (CAC85549.1)	Lymphocyte antigen 6 complex, locus G5B		17 B1	3	1
Ly6g5c (CAC85548.1)	Lymphocyte antigen 6 complex, locus G5C	*G5c; NG33*	17 B1	3	1
Ly6g6c (AAI16367.1)	Lymphocyte antigen 6 complex, locus G6C	*G6c; NG24; AU016360; 1110003M04Rik*	17 B1	3	1
Ly6g6d (CAC85545.1)	Lymphocyte antigen 6 complex, locus G6D	*G6d; G6f; NG25; NG32; MEGT1; A930024F17Rik*	17 B1	4	1
Ly6g6e (AAI38778.1)	Lymphocyte antigen 6 complex, locus G6E	*G6e; 2310011I02Rik*	17 B2	3	1
Ly6g6f (AAI72069.1)	Lymphocyte antigen 6 complex, locus G6F	*CJ068215*	17 B1	6	1
Ly6g6g[a] (EDL29446.1)	Lymphocyte antigen 6 complex, locus G6G	*D730001G18Rik*	15 D3	3	1
Ly6h (AAH28758.1)	Lymphocyte antigen 6 complex, locus H		15 E1	6	1
Ly6i (AAI45088.1)	Lymphocyte antigen 6 complex, locus I	*Ly-6I; Ly-6 M; AI789751*	15 D3	6	1
Ly6k (AAH49723.1)	Lymphocyte antigen 6 complex, locus K	*mLy-6 K; 2410015A16Rik; 3110035B01Rik*	15	4	1

Table 2 Name, chromosomal location, number of exons and LU domains for mouse Ly6/uPAR family genes *(Continued)*

Ly6l[a] (XP_006521679.1)	Lymphocyte antigen 6 complex, locus L	*Gm20654*	15 34.37 cM	3	1
Ly6m[a] (EDL29458.1)	Lymphocyte antigen 6 complex, locus M	*2010109I03Rik*	15 D3	3	1
Lynx1 (AAF16899.1)	Ly6/neurotoxin 1	*AI838844*	15 D3	4	1
Lypd1 (AAH58599.1)	Ly6/PLAUR domain containing 1	*Lynx2; Lypdc1; AI853408; 2700050C12Rik; C530008O16Rik*	1 E3	6	1
Lypd2 (AAI32410.1)	Ly6/PLAUR domain containing 2	*VLL; Lypdc2; 0610005K03Rik*	15 E1	4	1
Lypd3 (AAH16549.1)	Ly6/PLAUR domain containing 3	*C4.4a; 2310061G07Rik*	7 A3	5	2
Lypd4 (AAH49744.1)	Ly6/PLAUR domain containing 4	*4933400F01Rik*	7 A3	5	1
Lypd5 (AAI07188.1)	Ly6/PLAUR domain containing 5	*2210003I03Rik*	7 A3	5	1
Lypd6 (AAH70462.1)	Ly6/PLAUR domain containing 6	*E130115E03Rik*	2 C1.1	7	1
Lypd6b (AAI26944.1)	Ly6/PLAUR domain containing 6B	*AW049525; 2310010M24Rik*	2 C1.1	14	1
Lypd8 (NP_001077353.1)	Ly6/PLAUR domain containing 8	*2210415F13Rik*	11 B1.3	8	1
Lypd9[a] (AAH48595.1)	Ly6/PLAUR domain containing 9	*4930504O13RikGm524*	11 B1.3	4	1
Lypd10[a] (BC049730.1)	Ly6/PLAUR domain containing 10	*BC049730*	7 A3	8	1
Lypd11[a] (NP_808261.1)	Ly6/PLAUR domain containing 11	*Gm4763,EG210155*	7 A3	9	1
Pate1 (NP_001186882.1)	Prostate and testis expressed 1	*Pate*	9 A4	5	1
Pate2 (NP_001028593.1)	Prostate and testis expressed 2	*Gm846; Pate-M; mANLP1*	9 A4	10	1
Pate3 (NP_001161064.1)	Prostate and testis expressed 3	*Pate-dj*	9 A4	3	1
Pate4 (AAI20767.1)	Prostate and testis expressed 4	*Svs7; Pate-B; 9530004K16Rik*	9 A4	4	1
Pate5[a] (NP_084139.1)	Prostate and testis expressed 5	*9230110F15Rik, Pate-A, mANLP3*	9 A4	3	1
Pate6[a] (NP_080869.1)	Prostate and testis expressed 6	*D730048I06Rik, Pate-C, mANLP2*	9 A4	3	1
Pate7[a] (NP_001161145.1)	Prostate and testis expressed 7	*Pate-E, Gm17727*	9 A4	3	1
Pate8[a] (NP_001161056.1)	Prostate and testis expressed 8	*Pate-G,Gm17689*	9 A4	3	1
Pate9[a] (NP_001028955.1)	Prostate and testis expressed 9	*Pate-H, EG434396, Gm5615*	9 A4	3	1
Pate10[a] (ACD81927.1)	Prostate and testis expressed 10	*Pate-N, Gm17677*	9 A4	3	1
Pate11[a] (ACD81928.1)	Prostate and testis expressed 11	*Pate-P, Gm9513*	9 A4	4	1
Pate12[a] (NP_001161058.1)	Prostate and testis expressed 12	*Pate-Q, EG639025, Gm7257*	9 A4	4	1
Pate13[a] (XP_006510783.1)	Prostate and testis expressed 13	*9230113P08Rik*	9 A4	3	1

Table 2 Name, chromosomal location, number of exons and LU domains for mouse Ly6/uPAR family genes *(Continued)*

Pate14[a] (NP_001028497.1)	Prostate and testis expressed 14	*A630095E13Rik, Gm191; Sslp1*	9 A4	4	1
Pinlyp (NP_001032220.1)	Phospholipase A2 inhibitor and Ly6/PLAUR domain containing	*2310033E01Rik*	7 A2-A3	6	1
Plaur (NP_035243.1)	Plasminogen activator, urokinase receptor	*Cd87; uPAR; u-PAR*	7 A3	11	3
Psca (EDL29439.1)	Prostate stem cell antigen	*2210408B04Rik*	15 D3	3	1
Slurp1 (EDL29441.1)	Secreted Ly6/PLAUR domain containing 1	*ARS; ArsB; Slurp-1; AI415082; 1110021N19Rik*	15 D3	3	1
Slurp2[a] (AAI15612.1)	Secreted Ly6/PLAUR domain containing 2	*2300005B03Rik*	15 D3	3	1
Spaca4 (AAH48608.1)	Sperm acrosome associated 4	*Samp14; AV043694; 1700008E09Rik*	7 B4	1	1
Tex101 (AAH48475.1)	Testis expressed 101	*AI429076; TES101RP; 1700008H15Rik*	7 A3	6	1

[a]Novel genes named in this study

other receptors including nAChR and/or competing scavengers of their ligands [1, 20, 21, 25–27]. Many Ly6/uPAR family members have a prominent role in neutrophils (Table 3) [28]. Below, we summarize the functions of a few well-studied members.

Prostate and testis expressed genes

Human chromosome 11 contains 5 prostate and testis expressed (PATE) genes while the syntenic region on murine Chromosome 9 contains 15 genes [29]. Recent evidence demonstrates that PATE proteins are much more predominantly expressed in the epididymis with a significantly lower expression in the prostate and testis, suggesting that their names are misnomers [30]. PATE proteins secreted by epithelial cells to the epididymal lumen facilitate spermatozoan maturation as they leave the testis and travel through the epididymis. PATE proteins localized in the sperm head assist in sperm-oolemma fusion and penetration [31]. Defects in PATE1 result in decreased sperm motility in aged men and young asthenozoospermia patients, revealing the molecular basis for the decline in sperm quality with age [32]. PATE4 is abundantly expressed in the mouse prostate, spermatozoa, and seminal vesicles. *Pate4−/−* mice remain fertile and do not display any histological abnormalities [33]. PATE proteins are also expressed in neuron-rich tissues, where they function by modulating nAChR activities [29].

Plasminogen activator, urokinase receptor

Also known as the urokinase-type plasminogen activator receptor (uPAR), plasminogen activator, urokinase receptor (PLAUR) is the most well-studied family member [9]. It is widely expressed in different cell types and plays a key regulatory role in cell surface plasminogen activation, influencing many normal and pathologic processes [9, 23]. PLAUR consists of three direct repeats of the LU domain, which together bind urokinase-type plasminogen activator (PLAU/uPA) in both the pro-protein and mature forms. PLAUR (i) expression is regulated by KLF4 [34] and is upregulated in cancer cells [35, 36] and in response to pro-inflammatory conditions [37], (ii) facilitates neutrophil recruitment in response to bacterial infection [38], (iii) facilitates clearance of Borrelia infection [39], and (iv) interacts with multiple partners including vitronectin and different integrins. Although the bulk of PLAUR exists as GPI-anchored, some of it is known to be secreted as "soluble uPAR" (suPAR), the expression level of which is correlated with disease conditions [10–12, 40].

PLAUR is a multi-functional protein with important roles in regulating cell-matrix interaction, motility, and immune response. PLAUR expression levels directly correlate with the invasive potential of endometrial carcinomas, suggesting that it is a valuable prognostic marker for aggressive endometrial tumors [35]. PLAUR expression is normally low in healthy glomeruli and is elevated in glomeruli from individuals with focal segmental glomerulosclerosis, consistent with its role in regulating renal permeability [41]. PLAUR is required for neutrophil recruitment into alveoli and lungs in response to *S. pneumoniae* infection [42]. *Plaur−/−* macrophages display an enhanced ability to engulf wild-type neutrophils, but *Plaur−/−* neutrophils do not, suggesting that PLAUR plays an essential role in recognition and clearance of neutrophils [43]. *Plaur−/−* mice exhibit abnormal interneuron migration from the ganglionic eminence, and reduced interneurons in the frontal and parietal cortex [44, 45].

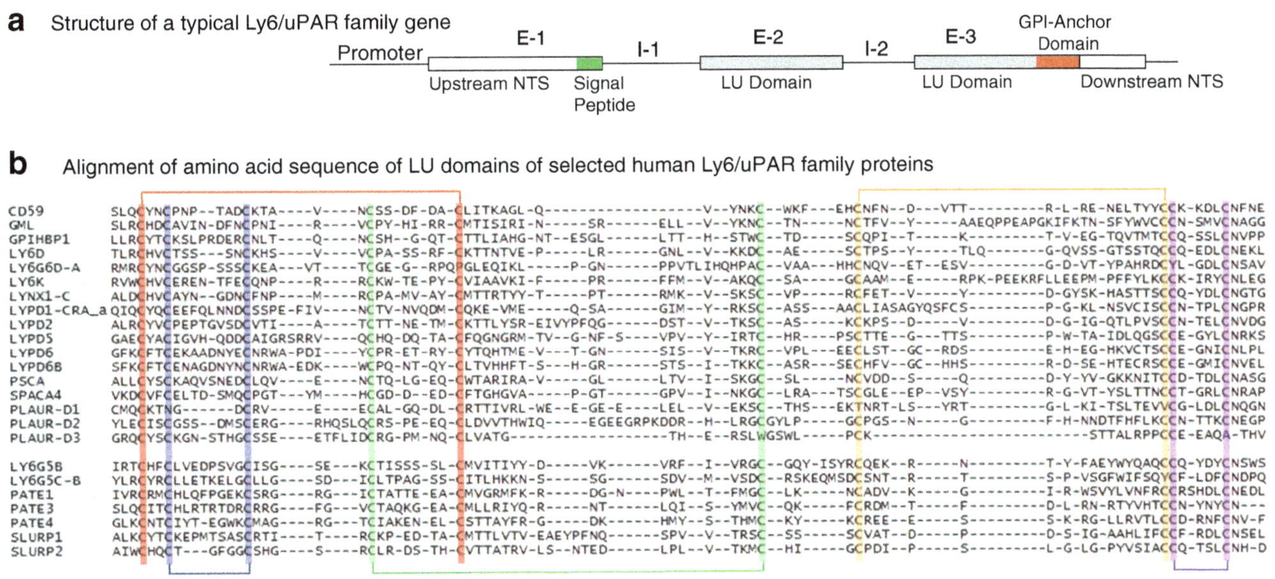

Fig. 1 Structure of a typical Ly6/UPAR family gene and alignment of amino acid sequences of selected LU domains. **a** Structure of a typical LY6/UPAR family gene, showing three exons (E-1, E-2 and E-3), two introns (I-1 and I-2), and the location of signal peptide as well as GPI-anchor domain. **b** Alignment of LU domain amino acid sequences of selected human LY6/UPAR proteins. GPI-anchored (*top*) and secreted (*bottom*) proteins are clustered together, with an empty line in between. Alignments were performed using ProbCons in Bioinformatics toolkit provided by Max-Planck Institute for Developmental Biology (http://toolkit.tuebingen.mpg.de). Conserved cysteines linked by non-variant disulfide bridges are highlighted in similar colors. Isoforms are denoted with a *dash* and the isoform name (e.g. LY6G6D-A). LYNX1-C and SLURP2 are precursor forms of the final protein. NTS, Non-translated sequence

CD177

Expressed by neutrophils, neutrophilic metamyelocytes, and myelocytes, CD177 mediates neutrophil migration across the endothelium by binding PECAM1 (CD31). Anti-CD177 antibodies inhibit neutrophil transmigration across the endothelial monolayer, potentially by interfering with an interaction between CD177 and PECAM1 [46]. Mutations in *CD177* or its dysregulated expression are associated with myeloproliferative diseases, secondary to a gain-of-function mutation in *JAK2* [8]. Exposure of human neutrophils to pulmonary endotoxin results in strong upregulation of *CD177* [47]. Expression of *CD177* mRNA is highly upregulated following endotoxin exposure. Overexpression of CD177 is a biomarker for thrombocythemia patients with elevated risk of thromboembolic complications [8]. While human *CD177* contains nine exons that encode a protein with two LU domain repeats, mouse *Cd177* is substantially larger with 17 exons that encode a larger protein with four LU domain repeats. Surprisingly, *Cd177−/−* mice displayed no discernible phenotype or any change in immune cells, other than decreased neutrophil counts in peripheral blood [47]. Absence of CD177 had no significant impact on CXCL1/KC- or fMLP-induced mouse neutrophil migration, but led to significant cell death [47].

Complement regulatory protein CD59

CD59 is an essential regulatory protein that protects hematopoietic and neuronal cells against complement-mediated osmolytic pore formation by binding C8 and/or C9 and inhibiting the incorporation of C9 into the membrane attack complex [17, 48–51]. CD2-mediated CD59 stimulation results in secretion of IL1A (IL-1α), IL6, and CSF2 (GM-CSF) in keratinocytes [52]. Inadequate complement regulation is associated with age-related macular degeneration [53]. Mutations in *CD59* cause uncontrolled complement activation in hemolytic anemia, thrombosis, and cerebral infarction in paroxysmal nocturnal hemoglobinuria [54]. The mouse genome contains two homologs of *CD59*, termed *Cd59a*, and *Cd59b*. Mouse CD59B has approximately a sixfold higher specific activity than CD59A and is considered a true ortholog of human CD59. *Cd59a* deficiency exacerbated the skin disease and lymphoproliferative characteristic of the MRL/lpr murine lupus model suggesting that CD59A inhibits systemic autoimmunity in the MRL/lpr lupus model through a complement-independent mechanism [55]. Consistent with its higher specific activity, *Cd59b−/−* mice display a stronger phenotype including hemolytic anemia, anisopoikilocytosis, echinocytosis, schistocytosis, hemoglobinuria with hemosiderinuria, and platelet activation [56]. *Cd59b−/−* males suffer from progressive loss of fertility after 5 months of age [56].

Prostate stem cell antigen

Prostate stem cell antigen (PSCA) is a 123 amino acid protein with an N-terminal signal sequence, and a C-

Table 3 Expression patterns, interacting factors and cellular functions of mouse Ly6/uPAR family proteins

Approved protein symbol	Human ortholog?	Expression pattern	Interacting factors	Cellular function	Gene knockout?	GPI-anchor?	References
ACRV1	Yes	Spermatazoa (acrosomes)	TARDBP (TDP-43)	Sperm-oolemma binding/penetration	No	No	[29, 31, 109–111]
CD177	Yes	Neutrophils	PR3, PECAM1	Neutrophil activation	Yes	Yes	[46, 47, 112–114]
CD59A	No	Neurons, T cells, splenic macrophages, spermatids	MAC Complex	Complement inhibitor	Yes	Yes	[115–118]
CD59B	Yes (as CD59)	Lymphocytes, granulocytes, platelets, erythrocytes, activated T cells, spermatazoa, splenic macrophages, mature spermatozoa	C8, C9	Complement inhibition, cell-cell adhesion	Yes	Yes	[17, 48, 56, 67, 118–120]
GML	Yes	Cancer cells	TRP53	Apoptosis	No	Yes	[121, 122]
GML2	No	Hematopoetic stem cells	Unknown	Unknown	No	Predicted No	[123]
GPIHBP1	Yes	Capillary endothelial cells	LPL	Chylomicron processing, LPL transport	Yes	Yes	[62, 63, 124]
LY6A	No	Hematopoetic stem cells, B cells, T cells, DCs	IFN-γ, MMPs	T cell activation and proliferation controller	Yes	Yes	[28, 125–129]
LY6A2	No	Unknown	Unknown	Unknown	No	Predicted Yes	
LY6C1	No	T cells, NK cells, monocytes, neutrophils, DCs, bone marrow myeloid cells	Type I and II interferons, LFA-1	$CD8^+$ T cell migration	No	Yes	[28]
LY6C2	No	T cells, NK cells, monocytes, neutrophils, DCs, bone marrow myeloid cells	Type I and II interferons, LFA-1	$CD8^+$ T cell migration	No	Yes	[28]
LY6D	Yes	Keratinocytes, B cells	Camptothecin, mitomycin C, carboplatin, hydroxyurea, aphidicolin	Cell adhesion, B cell specification	No	Yes	[130–134]
LY6E	Yes	Myeloid cells, thymocytes	Retinoic acid, INF1	Monocyte inhibitor, T cell development	No	Yes	[135–139]
LY6F	No	Non-lymphoid tissues	Unknown	Unknown	No	Predicted Yes	[140]
LY6G	No	Neutrophils, granulocytes	LTB_4, β2-integrins	Neutrophil recruitment	No	Yes	[28, 109, 141]
LY6G2	No	Unknown	Unknown	Unknown	No	Predicted No	
LY6G5B	Yes	Stomach, placenta	Unknown	Unknown	No	No	[142, 143]
LY6G5C	Yes	Testis, fetal liver/spleen/retina/heart/thymus, MS lesions, germinal center B cells, carcinoid lung	Unknown	Hematopoietic cell differentiation	No	No	[142, 143]
LY6G6C	Yes	Fibroblasts	Unknown	Filipodia functions	No	Yes	[142, 143]
LY6G6D	Yes	Fibroblasts	Unknown	Filipodia functions	No	Yes	[142, 143]
LY6G6E	No	Undifferentiated megakaryocyte-like cells	Unknown	$\alpha_4\beta_2$ potentiator, cell adhesion and migration	No	Yes	[105, 142]
LY6G6F	Yes	K562 acute myeloblastic leukemia cells	GRB2, GRB7	Unknown	No	Predicted No	[144]
LY6G6G	No	Unknown	Unknown	Unknown	No	Predicted No	
LY6H	Yes	Exocrine and endocrine epithelial cells, CNS	Unknown	Unknown	No	Yes	[145, 146]

Table 3 Expression patterns, interacting factors and cellular functions of mouse Ly6/uPAR family proteins *(Continued)*

LY6I	No	Monocytes, granulocytes, $CD19^+$ B cells, thymocytes and T cells	Unknown	Unknown	No	Yes	[28, 147]
LY6K	Yes	Keratinocytes, squamous cell carcinomas	JUND, FOSL1, RAS	Invasive cell metastasis	Yes	Yes	[148–152]
LY6L	Yes	Unknown	Unknown	Unknown	No	Predicted Yes	
LY6M	No	Unknown	Unknown	Unknown	No	Predicted Yes	
LYNX1	Yes	CNS neurons	$\alpha_4\beta_2$ and α_7 nAChRs	nAChR modulator	Yes	Yes	[25, 66, 73, 153]
LYPD1	Yes	Airway epithelial cells, embryonic tissue, CNS and PNS neurons	$\alpha_4\beta_2$, $\alpha_4\beta_4$, and α_7 nAChRs	α7 nAChR modulator, tumor suppression	Yes	Yes	[103, 104, 154, 155]
LYPD2	Yes	Unknown	$\alpha_4\beta_2$ nAChRs	nAChR Modulator	No	Yes	[105]
LYPD3	Yes	Stratum spinosum keratinocytes, cancer cells	$\alpha_6\beta_4$, MMP14	Metastasis, EMT, wound healing	No	Yes	[156–160]
LYPD4	Yes	Unknown	Unknown	Unknown	No	Predicted No	
LYPD5	Yes	Stratum granulosum keratinocytes	AP-1	Unknown	No	Yes	[156, 161]
LYPD6	Yes	Ubiquitously	WNT3A	Tumor suppression	No	Yes	[162–164]
LYPD6B	Yes	Testis, lung, stomach, prostate	AP-1	PKC signal transduction pathway	No	Yes	[165]
LYPD8	Yes	Unknown	Unknown	Unknown	No	Predicted Yes	
LYPD9	No	Unknown	Unknown	Unknown	No	Predicted Yes	
LYPD10	No	Unknown	Unknown	Unknown	No	Predicted Yes	
LYPD11	No	Unknown	Unknown	Unknown	No	Predicted Yes	
PATE1	Yes	Leydig cells, testicular germ cells, prostatic epithelial cells principle cells, spermatazoa,	INCA1	Spermatazoa/egg fusion and penetration, spermatazoa motility	No	No	[29, 32, 166–168]
PATE2	Yes	Spermatazoa, testicular germ cells, neuronal tissue	nAChR	nAChR modulator	No	No	[29, 31]
PATE3	Yes	Testis	nAChR	nAChR modulator	No	No	[29]
PATE4	Yes	Spermatazoa, prostatic apical epithelial cells, spinal cord tissue	α_7 nAChR	nAChR modulator	Yes	No	[29, 33]
PATE5	No	Male reproductive organs	Androgens, lumicrine testicular factors	Unknown	No	No	[29, 30]
PATE6	No	Male reproductive organs	Androgens	nAChR regulator	No	No	[29, 30]
PATE7	No	Male reproductive organs	Androgens	Unknown	No	No	[29, 30]
PATE8	No	Male reproductive organs, skeletal muscle	Androgens	Unknown	No	No	[29, 30]
PATE9	No	Male reproductive organs	Unknown	Unknown	No	No	[29, 30]
PATE10	No	Male reproductive organs	Androgens, lumicrine testicular factors	Unknown	No	No	[29, 30]

Table 3 Expression patterns, interacting factors and cellular functions of mouse Ly6/uPAR family proteins *(Continued)*

PATE11	No	Placenta	α4β2 nAChR	$\alpha_4\beta_2$ nAChR regulator	No	No	[29, 30]
PATE12	No	Placenta	Unknown	Unknown	No	No	[29, 30]
PATE13	No	Male reproductive organs	Androgens	Unknown	No	No	[29]
PATE14	No	Luminal epithelium of seminal vesicles	Androgens	Unknown	No	No	[169]
PINLYP	Yes	Unknown	Unknown	Unknown	No	Predicted No	
PLAUR	Yes	Airway epithelial cells, trophoblast cells, cancerous cells, HUVEC, pan-epithelial	uPA, Mac-1, caveolin, SERPINE1, vitronectin, kininogen, thrombospondin, α2-macroglobulin receptor, MRC2, cation-independent mannose 6-phosphate/insulin-like growth factor II receptor	Fibrinolysis, matrix remodeling, cell migration, growth factor activation, integrin regulation, tumor cell invasion, cell adhesion	Yes	Yes	[23, 170–185]
PSCA	Yes	Prostate basal cells, epithelial cells of prostate, urinary bladder, kidney, skin, esophagus, stomach, and placenta, cortical cells	SNAI2, α4 nAChR	Tumor suppression, oncogene, nAChR modulator	Yes	Yes	[26, 57–59, 61, 150, 186–188]
SLURP1	Yes	Keratinocytes, mucocutaneous and aerodigestive epithelia, some C-fiber neurons	α_7 nAChR, KLF4	Tumor suppression, nAChR modulator	Yes	No	[20, 76, 79, 94, 99, 189–192]
SLURP2	Yes	Oral and epidermal keratinocytes	α_3 nAChR	nAChR modulator, blocks apoptosis	Yes	No	[99–102, 193]
SPACA4	Yes	Spermatazoa	Unknown	Spermatazoa-egg binding/fusion	No	Yes	[194]
TEX101	Yes	Spermatazoa, cancerous tissues	Progesterone, PLAU	Spermatazoaatogenesis, spermatazoa-egg interaction, protease suppressor	Yes	Yes	[151, 195–198]

terminal GPI-anchoring sequence [57]. It was initially identified as a prostate-specific cell surface antigen in normal male tissues and found to be highly expressed in human prostate cancer [58, 59]. Later studies have revealed it to be more widely expressed. A genome-wide association study of Japanese patients with gastric cancer revealed that genetic variation in *PSCA* is associated with susceptibility to diffuse-type gastric cancer [60]. *Psca−/−* mice are viable, and fertile, with similar rates of spontaneous or radiation-induced primary epithelial tumor formation as the wild-type mice. However, *Psca −/−* mice display an increased frequency of metastasis suggesting that PSCA may play a role in limiting tumor progression, and deletion of *Psca* promotes tumor migration and metastasis [61].

GPI-anchored high density lipoprotein-binding protein 1

GPI-anchored high density lipoprotein-binding protein 1 (GPIHBP1) is an endothelial cell protein expressed on the luminal face of capillaries in brown adipose tissue, heart, lung, and liver. GPIHBP1 binds high density lipoprotein and provides a platform for lipoprotein lipase (LPL)-mediated processing of chylomicron lipoprotein particles which transport dietary lipids from the intestines to other locations in the body. GPIHBP1 mutations that affect its ability to bind LPL or chylomicrons are associated with chylomicronemia [62–64]. *Gpihbp1−/−* mice cannot transport lipoprotein lipase to the capillary lumen, resulting in mislocalization of lipoprotein lipase within tissues, defective lipolysis of triglyceride-rich lipoproteins, and chylomicronemia [62–64]. Defective lipolysis causes reciprocal metabolic perturbations in *Gpihbp1−/−* mouse adipose tissue and liver. The essential fatty acid content of triglycerides is decreased and lipid biosynthetic gene expression is increased in adipose tissue, while the opposite changes occur in the liver [65].

Ly6/neurotoxin-1

As an allosteric modulator of nAChR function, Ly6/neurotoxin-1 (LYNX1) serves as a cholinergic brake that limits neuronal plasticity, balancing neuronal activity, and survival in the adult visual cortex [25, 66–68]. LYNX1 also inhibits SRC activation, suppressing mucin expression in the airway epithelium [69]. The *LYNX1* gene is positioned in close proximity to *SLURP2*, leading to the mistaken idea that they are alternatively spliced isoforms of the same gene, a theory which was disproved recently [70]. *LYNX1* is one of the genes that has shown accelerated evolution in humans relative to other primates, correlating with the increased brain size and complexity [71]. The juvenile brain exhibits high plasticity which is severely restricted in adulthood. Adult *Lynx1−/−* mice exhibited visual cortex plasticity similar to that of juveniles, suggesting that LYNX1 serves as a break for cortical plasticity [68]. Using the mouse model, it was demonstrated that LYNX1 plays a modulatory role in the aging brain, and that soluble LYNX1 may be useful for adjusting cholinergic-dependent plasticity and learning mechanisms [72–74].

Secreted Ly6/urokinase-type plasminogen activator receptor-related protein 1

Secreted Ly6/urokinase-type plasminogen activator receptor-related protein 1 (SLURP1) is expressed in a variety of cells including immune cells [75], sensory neurons [76], and epithelial cells [77–80], and secreted into plasma, saliva, sweat, urine, and tears [22, 81]. *SLURP1* is downregulated in corneal neovascularization [82], asthmatic lungs [83], Barrett's esophagus [84, 85], malignant melanomas [86], and squamous cell carcinomas [87, 88]. Mutations or deletions in *SLURP1* cause autosomal recessive palmoplantar hyperkeratotic disorder 'mal de Meleda' [78, 81, 89–93]. SLURP1 is structurally similar to the snake and frog cytotoxin α-bungarotoxin, and acts as a CHRNA7 (α7nAChR)-ligand, regulating keratinocytes through cholinergic pathways [78, 94]. It modulates signal transduction, activation of the immune response, and cell adhesion, and blocks malignant transformation [75, 79, 95–97]. SLURP1 is proposed to modulate acetylcholine signaling through CHRNA7 [98]. We have documented that SLURP1 serves as an important immunomodulatory molecule at the ocular surface by acting as a soluble scavenger of urokinase (PLAU) [20–22]. *Slurp1−/−* mice develop signs of palmoplantar keratoderma including elevated keratinocyte proliferation, accumulation of lipid droplets in the stratum corneum, and defective epidermal barrier function reminiscent of mal de Meleda. *Slurp1−/−* mice also display decreased adiposity, low plasma lipid levels, and a neuromuscular abnormality (hind-limb clasping), suggesting additional functions for SLURP1 [99].

Secreted Ly6/urokinase-type plasminogen activator receptor-related protein 2 (SLURP2)

SLURP2 is expressed by human epidermal and oral keratinocytes, from where it is secreted into sweat and saliva, respectively [100]. *SLURP2* expression is strongly induced in psoriatic skin lesions possibly by IL22, and is blocked by IFNG [70, 101]. SLURP2 blocks the effect of acetylcholine by binding CHRNA3 (α3nAChR), and delays keratinocyte differentiation and prevents apoptosis [100]. Although the *SLURP2* and *LYNX1* genes are closely linked leading to a mistaken idea that they are isoforms, it is now clear that they constitute separate transcription units that are differently regulated [70]. *Slurp2−/−* mice also develop

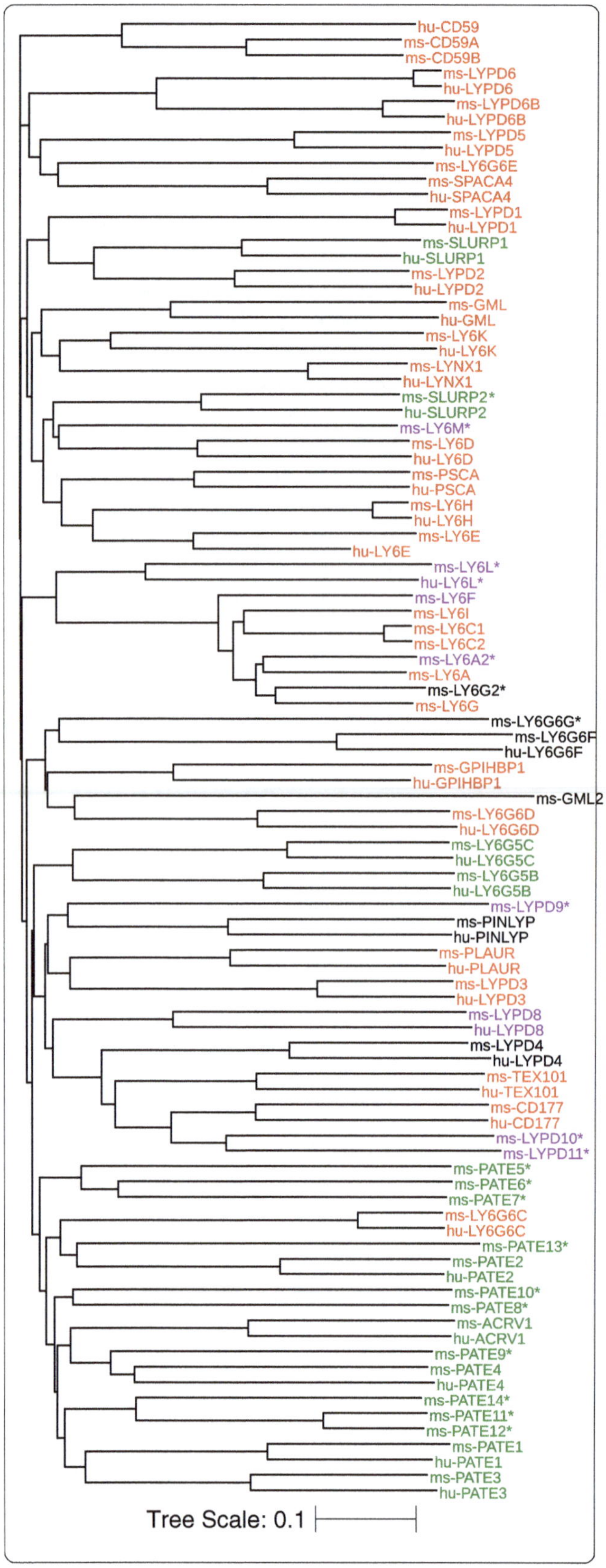

Fig. 2 Phylogram revealing the evolutionary relationship among mouse (ms-) and human (hu-) Ly6/uPAR family proteins. The phylogram was generated using the amino acid sequences in Clustal-Omega web-based program [106, 107] (http://www.ebi.ac.uk/Tools/msa/clustalo/). The display was generated using the methods described [108]. The length of each branch from the most recent branch point indicates the evolutionary distance, or the relative period of time the protein has been in its current state. Known GPI-anchored Ly6/uPAR family proteins are shown in *red*, and those secreted (without GPI-anchor) are shown in *green*. Those predicted to contain a GPI-anchor (but not yet experimentally proven) are in *purple*, and those predicted to not contain a GPI-anchor sequence (but not yet experimentally proven) are in *black*. Novel genes named in this study are indicated with an asterisk (*)

signs of palmoplantar keratoderma and neuromuscular abnormality (hind-limb clasping) reminiscent of those seen in *Slurp1−/−* mice [99, 102].

Ly6/Plaur domain containing 1

Ly6/Plaur domain containing 1 (LYPD1), also known as LYNX2, is a prototoxin gene that is expressed in postmitotic central and peripheral neurons including subpopulations of motor neurons, sensory neurons, interneurons, and neurons of the autonomous nervous system [103]. LYPD1 is expressed at high levels in anxiety associated brain areas and plays an important role in regulating anxiety by binding and modulating neuronal nicotinic receptors [104, 105]. Ablation of *Lypd1* alters the actions of nicotine on glutamatergic signaling in the prefrontal cortex, resulting in elevated anxiety-like behaviors [104].

Evolution of Ly6/uPAR family proteins

Ly6/uPAR family genes are conserved across species suggesting that they are evolutionarily ancient. Organization of the genes in this family in clusters on multiple chromosomes suggests that both gene duplications and translocations have played a role in their evolution. Comparison of the mouse and human Ly6/uPAR family genes reveals that while there are many orthologs, some Ly6 genes are only present in the mouse. The Ly6 gene complexes on human chromosomes 8, 19, 11, and 6 are syntenic with their counterparts on mouse Chromosomes 15, 7, 9, and 17, respectively, suggesting that these gene complexes were already present in their common ancestor. There are no human orthologs for the subcluster of murine Ly6 genes *Ly6i, Ly6a, Ly6c1, Ly6c2, Ly6a2, Ly6g, Ly6g2,* and *Ly6f* on Chromosome 15, and *Pate10, Pate7, Pate6, Pate5, Pate12, Pate11, Pate9, Pate8,* and *Pate14* on Chromosome 9, suggesting that these regions may have arisen in the mouse through gene duplication after evolutionary divergence of these two species. What their functions are in the murine neutrophils and epididymis, respectively, where they are abundantly expressed, and how they are compensated in the corresponding human tissues, remains to be determined.

In order to evaluate the evolutionary relatedness of LU family proteins, we generated a phylogram by multiple sequence alignment of their amino acid sequences using web-based Clustal-Omega software, and visualized it with web-based software from Interactive Tree of Life (Fig. 2) [106–108]. Where multiple isoforms exist, we only used the sequence of the longest isoform. Analysis of the phylogenetic relationship among human and mouse Ly6/uPAR family proteins revealed that (i) human LY6K and mouse GML2 are the most ancestral Ly6 proteins with the longest unbranched streak in these two species, (ii) human and mouse LYPD6 are the most recent addition to the family closely followed by mouse LY6C1 and LY6C2, (iii) most of the secreted family proteins (with the notable exception of SLURP1 and SLURP2) form a separate cluster distinct from the GPI-anchored proteins, and (iv) several mouse PATE proteins (PATE4, 5, 6, 7, 8, 9, 10, 13, and 14) have long unbranched streaks suggesting that they have ancient origin and that the important function(s) that they serve have not changed much (Fig. 2).

Concluding remarks

In this gene family update, we have summarized the current literature on the organization, expression patterns, functions, and evolution of human and mouse Ly6/uPLAR family genes. In addition, we identified and named many novel Ly6/uPAR family members. Considering that Ly6/uPLAR family members play critical roles in regulating immunological and physiological responses to infections and varying environmental conditions, it is imperative that we understand them in greater detail. Their involvement in regulating a wide range of functions such as progression of inflammation, complement activity, neuronal activity, angiogenesis, wound healing, and cancer growth indicates that Ly6/uPAR family members will be useful therapeutic targets. Additional insight into (i) the biological functions of individual family proteins, (ii) signaling cascades that regulate their expression and functions, and (iii) the identity of their interacting partners is expected to herald new modalities for diagnosis and treatment of diverse diseases.

Competing interests

C.L. Loughner, None; E. A. Bruford, None; M.S. McAndrews, None; E.E. Delp, None; S. Swamynathan (Patent); S.K. Swamynathan, (Patent).

Authors' contributions

CLL, EAB, MSM, EED, SS, and SKS were involved in drafting the manuscript and revising it critically for important intellectual content. Each author has participated sufficiently in the work to take public responsibility for appropriate portions of the content. All authors read and approved the final manuscript. Each author has agreed to be accountable for all aspects of the work in ensuring that questions related to the accuracy or integrity of any part of the work are appropriately investigated and resolved.

Acknowledgements

Work in the Swamynathan lab was supported by R01EY022898 grant from the National Eye Institute (NEI), National Institutes of Health (NIH) (SKS), NEI Core Grant P30 EY08098, unrestricted grants from Research to Prevent Blindness, the Eye and Ear Foundation of Pittsburgh and the PA Lions Eye Research Sight Conservation Foundation (SKS), and "Fight for Sight/The Eye Bank for Sight" (EED). EAB is supported by NIH Grant U41HG003345 and Wellcome Trust grant 099129/Z/12/Z. MSM is supported by program project grant HG000330 from the National Human Genome Research Institute (NHGRI), NIH.

Author details

[1]Department of Ophthalmology, University of Pittsburgh School of Medicine, Pittsburgh, USA. [2]HUGO Gene Nomenclature Committee (HGNC), European Bioinformatics Institute (EMBL-EBI), Wellcome Genome Campus, Hinxton, Cambridgeshire CB10 1SD, UK. [3]Mouse Genome Informatics, The Jackson Laboratory, Bar Harbor, ME 04609, USA. [4]Department of Cell Biology, University of Pittsburgh School of Medicine, Pittsburgh, USA. [5]McGowan Institute of Regenerative Medicine, University of Pittsburgh, Pittsburgh, USA. [6]Fox Center for Vision Restoration, University of Pittsburgh School of Medicine, Pittsburgh, USA. [7]Eye and Ear Institute, University of Pittsburgh School of Medicine, 203 Lothrop Street, Room 1025, Pittsburgh, PA 15213, USA.

References

1. Tsetlin VI. Three-finger snake neurotoxins and Ly6 proteins targeting nicotinic acetylcholine receptors: pharmacological tools and endogenous modulators. Trends in Pharmacological Sciences. 2015;36(2):109–23.
2. Gumley TP, McKenzie IF, Sandrin MS. Tissue expression, structure and function of the murine Ly-6 family of molecules. Immunology and cell biology. 1995;73(4):277–96.
3. McKenzie IF, Gardiner J, Cherry M, Snell GD. Lymphocyte antigens: Ly-4, Ly-6, and Ly-7. Transplantation proceedings. 1977;9(1):667–9.
4. Wain HM, Bruford EA, Lovering RC, Lush MJ, Wright MW, Povey S. Guidelines for human gene nomenclature. Genomics. 2002;79(4):464–70.
5. Gray KA, Seal RL, Tweedie S, Wright MW, Bruford EA. A review of the new HGNC gene family resource. Human genomics. 2016;10(1):6.
6. Eppig JT, Blake JA, Bult CJ, Kadin JA, Richardson JE, Mouse Genome Database G. The Mouse Genome Database (MGD): facilitating mouse as a model for human biology and disease. Nucleic Acids Res. 2015;43(Database issue):D726–736.
7. Wurfel J, Seiter S, Stassar M, Claas A, Klas R, Rosel M, Marhaba R, Savelyeva L, Schwab M, Matzku S, et al. Cloning of the human homologue of the metastasis-associated rat C4.4A. Gene. 2001;262(1–2):35–41.
8. Stroncek DF. Neutrophil-specific antigen HNA-2a, NB1 glycoprotein, and CD177. Current opinion in hematology. 2007;14(6):688–93.
9. Smith HW, Marshall CJ. Regulation of cell signalling by uPAR. Nat Rev Mol Cell Biol. 2010;11(1):23–36.
10. Sloand EM. Soluble urokinase activator receptor (suPAR) in stem cell mobilization. Blood. 2005;105(5):1847–8.
11. Eugen-Olsen J, Giamarellos-Bourboulis EJ. suPAR: the unspecific marker for disease presence, severity and prognosis. International journal of antimicrobial agents. 2015;46 Suppl 1:S33–34.
12. Backes Y, van der Sluijs KF, Mackie DP, Tacke F, Koch A, Tenhunen JJ, Schultz MJ. Usefulness of suPAR as a biological marker in patients with systemic inflammation or infection: a systematic review. Intensive care medicine. 2012;38(9):1418–28.
13. Pierleoni A, Martelli PL, Casadio R. PredGPI: a GPI-anchor predictor. BMC bioinformatics. 2008;9:392.
14. Fletcher CM, Harrison RA, Lachmann PJ, Neuhaus D. Sequence-specific 1H-NMR assignments and folding topology of human CD59. Protein science : a publication of the Protein Society. 1993;2(12):2015–27.
15. Kieffer B, Driscoll PC, Campbell ID, Willis AC, van der Merwe PA, Davis SJ. Three-dimensional solution structure of the extracellular region of the complement regulatory protein CD59, a new cell-surface protein domain related to snake venom neurotoxins. Biochemistry. 1994;33(15):4471–82.
16. Huang Y, Fedarovich A, Tomlinson S, Davies C. Crystal structure of CD59: implications for molecular recognition of the complement proteins C8 and

C9 in the membrane-attack complex. Acta crystallographica Section D, Biological crystallography. 2007;63(Pt 6):714–21.

17. Davies A, Simmons DL, Hale G, Harrison RA, Tighe H, Lachmann PJ, Waldmann H. CD59, an LY-6-like protein expressed in human lymphoid cells, regulates the action of the complement membrane attack complex on homologous cells. J Exp Med. 1989;170(3):637–54.
18. Simmons PJ, Zannettino AC, Harrison-Findik D, Swart B, Tomlinson S, Hill B, Javni JA. A novel epitope of CD59 expressed by primitive human hematopoietic progenitors. Experimental hematology. 2001;29(12):1474–83.
19. Hill B, Rozler E, Travis M, Chen S, Zannetino A, Simmons P, Galy A, Chen B, Hoffman R. High-level expression of a novel epitope of CD59 identifies a subset of CD34+ bone marrow cells highly enriched for pluripotent stem cells. Experimental hematology. 1996;24(8):936–43.
20. Swamynathan S, Buela KA, Kinchington P, Lathrop KL, Misawa H, Hendricks RL, Swamynathan SK. Klf4 regulates the expression of Slurp1, which functions as an immunomodulatory peptide in the mouse cornea. Invest Ophthalmol Vis Sci. 2012;53(13):8433–46.
21. Swamynathan S, Swamynathan SK. SLURP-1 modulates corneal homeostasis by serving as a soluble scavenger of urokinase-type plasminogen activator. Invest Ophthalmol Vis Sci. 2014;55(10):6251–61.
22. Swamynathan S, Delp EE, Harvey SA, Loughner CL, Raju L, Swamynathan SK. Corneal expression of SLURP-1 by age, sex, genetic strain, and ocular surface health. Invest Ophthalmol Vis Sci. 2015;56(13):7888–96.
23. Blasi F, Carmeliet P. uPAR: a versatile signalling orchestrator. Nat Rev Mol Cell Biol. 2002;3(12):932–43.
24. Flanagan K, Modrusan Z, Cornelius J, Chavali A, Kasman I, Komuves L, Mo L, Diehl L. Intestinal epithelial cell up-regulation of LY6 molecules during colitis results in enhanced chemokine secretion. J Immunol. 2008;180(6):3874–81.
25. Ibanez-Tallon I, Miwa JM, Wang HL, Adams NC, Crabtree GW, Sine SM, Heintz N. Novel modulation of neuronal nicotinic acetylcholine receptors by association with the endogenous prototoxin lynx1. Neuron. 2002;33(6):893–903.
26. Jensen MM, Arvaniti M, Mikkelsen JD, Michalski D, Pinborg LH, Hartig W, Thomsen MS. Prostate stem cell antigen interacts with nicotinic acetylcholine receptors and is affected in Alzheimer's disease. Neurobiology of aging. 2015;36(4):1629–38.
27. Wang H, Yu M, Ochani M, Amella CA, Tanovic M, Susarla S, Li JH, Yang H, Ulloa L, Al-Abed Y, et al. Nicotinic acetylcholine receptor alpha7 subunit is an essential regulator of inflammation. Nature. 2003;421(6921):384–8.
28. Lee PY, Wang JX, Parisini E, Dascher CC, Nigrovic PA. Ly6 family proteins in neutrophil biology. J Leukoc Biol. 2013;94(4):585–94.
29. Levitin F, Weiss M, Hahn Y, Stern O, Papke RL, Matusik R, Nandana SR, Ziv R, Pichinuk E, Salame S, et al. PATE gene clusters code for multiple, secreted TFP/Ly-6/uPAR proteins that are expressed in reproductive and neuron-rich tissues and possess neuromodulatory activity. J Biol Chem. 2008;283(24): 16928–39.
30. Turunen HT, Sipila P, Pujianto DA, Damdimopoulos AE, Bjorkgren I, Huhtaniemi I, Poutanen M. Members of the murine Pate family are predominantly expressed in the epididymis in a segment-specific fashion and regulated by androgens and other testicular factors. Reproductive Biology and Endocrinology. 2011;9(1):128.
31. Margalit M, Yogev L, Yavetz H, Lehavi O, Hauser R, Botchan A, Barda S, Levitin F, Weiss M, Pastan I, et al. Involvement of the prostate and testis expression (PATE)-like proteins in sperm-oocyte interaction. Human reproduction. 2012;27(5):1238–48.
32. Liu FJ, Liu X, Han JL, Wang YW, Jin SH, Liu XX, Liu J, Wang WT, Wang WJ. Aged men share the sperm protein PATE1 defect with young asthenozoospermia patients. Human reproduction. 2015;30(4):861–9.
33. Heckt T, Keller J, Reusch R, Hartmann K, Krasemann S, Hermans-Borgmeyer I, Amling M, Schinke T. No obvious phenotypic abnormalities in mice lacking the Pate4 gene. Biochem Biophys Res Commun. 2016;469(4):1069–74.
34. Wang H, Yang L, Jamaluddin MS, Boyd DD. The Kruppel-like KLF4 transcription factor, a novel regulator of urokinase receptor expression, drives synthesis of this binding site in colonic crypt luminal surface epithelial cells. J Biol Chem. 2004;279(21):22674–83.
35. Foca C, Moses EK, Quinn MA, Rice GE. Differential mRNA expression of urokinase-type plasminogen activator, plasminogen activator receptor and plasminogen activator inhibitor type-2 in normal human endometria and endometrial carcinomas. Gynecologic oncology. 2000; 79(2):244–50.
36. Memarzadeh S, Kozak KR, Chang L, Natarajan S, Shintaku P, Reddy ST, Farias-Eisner R. Urokinase plasminogen activator receptor: prognostic biomarker for endometrial cancer. Proc Natl Acad Sci U S A. 2002;99(16):10647–52.
37. Coleman JL, Gebbia JA, Benach JL. Borrelia burgdorferi and other bacterial products induce expression and release of the urokinase receptor (CD87). J Immunol. 2001;166(1):473–80.
38. Gyetko MR, Sud S, Kendall T, Fuller JA, Newstead MW, Standiford TJ. Urokinase receptor-deficient mice have impaired neutrophil recruitment in response to pulmonary Pseudomonas aeruginosa infection. J Immunol. 2000;165(3):1513–9.
39. Hovius JW, Bijlsma MF, van der Windt GJ, Wiersinga WJ, Boukens BJ, Coumou J, Oei A, de Beer R, de Vos AF, van 't Veer C, et al. The urokinase receptor (uPAR) facilitates clearance of Borrelia burgdorferi. PLoS Pathog. 2009;5(5), e1000447.
40. Hodges GW, Bang CN, Wachtell K, Eugen-Olsen J, Jeppesen JL. suPAR: a New biomarker for cardiovascular disease? The Canadian journal of cardiology. 2015;31(10):1293–302.
41. Wei C, Moller CC, Altintas MM, Li J, Schwarz K, Zacchigna S, Xie L, Henger A, Schmid H, Rastaldi MP, et al. Modification of kidney barrier function by the urokinase receptor. Nat Med. 2008;14(1):55–63.
42. Rijneveld AW, Levi M, Florquin S, Speelman P, Carmeliet P, van Der Poll T. Urokinase receptor is necessary for adequate host defense against pneumococcal pneumonia. J Immunol. 2002;168(7):3507–11.
43. Park YJ, Liu G, Tsuruta Y, Lorne E, Abraham E. Participation of the urokinase receptor in neutrophil efferocytosis. Blood. 2009;114(4):860–70.
44. Dewerchin M, Nuffelen AV, Wallays G, Bouche A, Moons L, Carmeliet P, Mulligan RC, Collen D. Generation and characterization of urokinase receptor-deficient mice. J Clin Invest. 1996;97(3):870–8.
45. Powell EM, Mars WM, Levitt P. Hepatocyte growth factor/scatter factor is a motogen for interneurons migrating from the ventral to dorsal telencephalon. Neuron. 2001;30(1):79–89.
46. Sachs UJ, Andrei-Selmer CL, Maniar A, Weiss T, Paddock C, Orlova VV, Choi EY, Newman PJ, Preissner KT, Chavakis T, et al. The neutrophil-specific antigen CD177 is a counter-receptor for platelet endothelial cell adhesion molecule-1 (CD31). J Biol Chem. 2007;282(32):23603–12.
47. Xie Q, Klesney-Tait J, Keck K, Parlet C, Borcherding N, Kolb R, Li W, Tygrett L, Waldschmidt T, Olivier A, et al. Characterization of a novel mouse model with genetic deletion of CD177. Protein & cell. 2015;6(2):117–26.
48. Meri S, Morgan BP, Davies A, Daniels RH, Olavesen MG, Waldmann H, Lachmann PJ. Human protectin (CD59), an 18,000-20,000 MW complement lysis restricting factor, inhibits C5b-8 catalysed insertion of C9 into lipid bilayers. Immunology. 1990;71(1):1–9.
49. Meri S, Morgan BP, Wing M, Jones J, Davies A, Podack E, Lachmann PJ. Human protectin (CD59), an 18-20-kD homologous complement restriction factor, does not restrict perforin-mediated lysis. J Exp Med. 1990;172(1):367–70.
50. Chang CP, Husler T, Zhao J, Wiedmer T, Sims PJ. Identity of a peptide domain of human C9 that is bound by the cell-surface complement inhibitor, CD59. J Biol Chem. 1994;269(42):26424–30.
51. Hochsmann B, Schrezenmeier H. Congenital CD59 deficiency. Hematology/ oncology clinics of North America. 2015;29(3):495–507.
52. Naderi S, Hofmann P, Seiter S, Tilgen W, Abken H, Reinhold U. CD2-mediated CD59 stimulation in keratinocytes results in secretion of IL-1alpha, IL-6, and GM-CSF: implications for the interaction of keratinocytes with intraepidermal T lymphocytes. International Journal of Molecular Medicine. 1999;3(6):609–14.
53. Ebrahimi KB, Fijalkowski N, Cano M, Handa JT. Decreased membrane complement regulators in the retinal pigmented epithelium contributes to age-related macular degeneration. J Pathol. 2013;229(5):729–42.
54. Brodsky RA. Paroxysmal nocturnal hemoglobinuria. Blood. 2014;124(18): 2804–11.
55. Miwa T, Zhou L, Maldonado MA, Madaio MP, Eisenberg RA, Song WC. Absence of CD59 exacerbates systemic autoimmunity in MRL/lpr mice. J Immunol. 2012;189(11):5434–41.
56. Qin X, Krumrei N, Grubissich L, Dobarro M, Aktas H, Perez G, Halperin JA. Deficiency of the mouse complement regulatory protein mCd59b results in spontaneous hemolytic anemia with platelet activation and progressive male infertility. Immunity. 2003;18(2):217–27.
57. Saeki N, Gu J, Yoshida T, Wu X. Prostate stem cell antigen: a Jekyll and Hyde molecule? Clinical Cancer Research. 2010;16(14):3533–8.
58. Reiter RE, Gu Z, Watabe T, Thomas G, Szigeti K, Davis E, Wahl M, Nisitani S, Yamashiro J, Le Beau MM, et al. Prostate stem cell antigen: a cell surface

marker overexpressed in prostate cancer. Proc Natl Acad Sci U S A. 1998; 95(4):1735–40.

59. Gu Z, Thomas G, Yamashiro J, Shintaku IP, Dorey F, Raitano A, Witte ON, Said JW, Loda M, Reiter RE. Prostate stem cell antigen (PSCA) expression increases with high gleason score, advanced stage and bone metastasis in prostate cancer. Oncogene. 2000;19(10):1288–96.
60. Study Group of Millennium Genome Project for C, Sakamoto H, Yoshimura K, Saeki N, Katai H, Shimoda T, Matsuno Y, Saito D, Sugimura H, Tanioka F, et al. Genetic variation in PSCA is associated with susceptibility to diffuse-type gastric cancer. Nat Genet. 2008;40(6):730–40.
61. Moore ML, Teitell MA, Kim Y, Watabe T, Reiter RE, Witte ON, Dubey P. Deletion of PSCA increases metastasis of TRAMP-induced prostate tumors without altering primary tumor formation. The Prostate. 2008;68(2):139–51.
62. Beigneux AP, Davies BS, Gin P, Weinstein MM, Farber E, Qiao X, Peale F, Bunting S, Walzem RL, Wong JS, et al. Glycosylphosphatidylinositol-anchored high-density lipoprotein-binding protein 1 plays a critical role in the lipolytic processing of chylomicrons. Cell Metabolism. 2007;5(4):279–91.
63. Beigneux AP, Franssen R, Bensadoun A, Gin P, Melford K, Peter J, Walzem RL, Weinstein MM, Davies BS, Kuivenhoven JA, et al. Chylomicronemia with a mutant GPIHBP1 (Q115P) that cannot bind lipoprotein lipase. Arteriosclerosis, Thrombosis, and Vascular Biology. 2009;29(6):956–62.
64. Franssen R, Young SG, Peelman F, Hertecant J, Sierts JA, Schimmel AW, Bensadoun A, Kastelein JJ, Fong LG, Dallinga-Thie GM, et al. Chylomicronemia with low postheparin lipoprotein lipase levels in the setting of GPIHBP1 defects. Circulation Cardiovascular Genetics. 2010;3(2):169–78.
65. Weinstein MM, Goulbourne CN, Davies BS, Tu Y, Barnes 2nd RH, Watkins SM, Davis R, Reue K, Tontonoz P, Beigneux AP, et al. Reciprocal metabolic perturbations in the adipose tissue and liver of GPIHBP1-deficient mice. Arteriosclerosis, Thrombosis, and Vascular Biology. 2012;32(2):230–5.
66. Miwa JM, Stevens TR, King SL, Caldarone BJ, Ibanez-Tallon I, Xiao C, Fitzsimonds RM, Pavlides C, Lester HA, Picciotto MR, et al. The prototoxin lynx1 acts on nicotinic acetylcholine receptors to balance neuronal activity and survival in vivo. Neuron. 2006;51(5):587–600.
67. Miwa T, Song WC. Membrane complement regulatory proteins: insight from animal studies and relevance to human diseases. Int Immunopharmacol. 2001;1(3):445–59.
68. Morishita H, Miwa JM, Heintz N, Hensch TK. Lynx1, a cholinergic brake, limits plasticity in adult visual cortex. Science. 2010;330(6008):1238–40.
69. Fu XW, Rekow SS, Spindel ER. The ly-6 protein, lynx1, is an endogenous inhibitor of nicotinic signaling in airway epithelium. American Journal of Physiology Lung Cellular and Molecular Physiology. 2012;303(8):L661–668.
70. Moriwaki Y, Takada K, Tsuji S, Kawashima K, Misawa H. Transcriptional regulation of SLURP2, a psoriasis-associated gene, is under control of IL-22 in the skin: a special reference to the nested gene LYNX1. Int Immunopharmacol. 2015;29(1):71–5.
71. Dorus S, Vallender EJ, Evans PD, Anderson JR, Gilbert SL, Mahowald M, Wyckoff GJ, Malcom CM, Lahn BT. Accelerated evolution of nervous system genes in the origin of Homo sapiens. Cell. 2004;119(7):1027–40.
72. Miwa JM, Walz A. Enhancement in motor learning through genetic manipulation of the Lynx1 gene. PLoS One. 2012;7(11):e43302.
73. Lyukmanova EN, Shulepko MA, Buldakova SL, Kasheverov IE, Shenkarev ZO, Reshetnikov RV, Filkin SY, Kudryavtsev DS, Ojomoko LO, Kryukova EV, et al. Water-soluble LYNX1 residues important for interaction with muscle-type and/or neuronal nicotinic receptors. J Biol Chem. 2013;288(22):15888–99.
74. Bukhari N, Burman PN, Hussein A, Demars MP, Sadahiro M, Brady DM, Tsirka SE, Russo SJ, Morishita H. Unmasking proteolytic activity for adult visual cortex plasticity by the removal of Lynx1. The Journal of Neuroscience. 2015;35(37):12693–702.
75. Moriwaki Y, Yoshikawa K, Fukuda H, Fujii YX, Misawa H, Kawashima K. Immune system expression of SLURP-1 and SLURP-2, two endogenous nicotinic acetylcholine receptor ligands. Life Sciences. 2007;80(24–25):2365–8.
76. Moriwaki Y, Watanabe Y, Shinagawa T, Kai M, Miyazawa M, Okuda T, Kawashima K, Yabashi A, Waguri S, Misawa H. Primary sensory neuronal expression of SLURP-1, an endogenous nicotinic acetylcholine receptor ligand. Neuroscience Research. 2009;64(4):403–12.
77. Horiguchi K, Horiguchi S, Yamashita N, Irie K, Masuda J, Takano-Ohmuro H, Himi T, Miyazawa M, Moriwaki Y, Okuda T, et al. Expression of SLURP-1, an endogenous alpha7 nicotinic acetylcholine receptor allosteric ligand, in murine bronchial epithelial cells. J Neurosci Res. 2009;87(12):2740–7.
78. Mastrangeli R, Donini S, Kelton CA, He C, Bressan A, Milazzo F, Ciolli V, Borrelli F, Martelli F, Biffoni M, et al. ARS Component B: structural characterization, tissue expression and regulation of the gene and protein (SLURP-1) associated with Mal de Meleda. Eur J Dermatol. 2003;13(6):560–70.
79. Arredondo J, Chernyavsky AI, Grando SA. SLURP-1 and -2 in normal, immortalized and malignant oral keratinocytes. Life Sciences. 2007;80(24–25):2243–7.
80. Norman B, Davis J, Piatigorsky J. Postnatal gene expression in the normal mouse cornea by SAGE. Invest Ophthalmol Vis Sci. 2004;45(2): 429–40.
81. Favre B, Plantard L, Aeschbach L, Brakch N, Christen-Zaech S, de Viragh PA, Sergeant A, Huber M, Hohl D. SLURP1 is a late marker of epidermal differentiation and is absent in Mal de Meleda. J Invest Dermatol. 2007; 127(2):301–8.
82. Jia C, Zhu W, Ren S, Xi H, Li S, Wang Y. Comparison of genome-wide gene expression in suture- and alkali burn-induced murine corneal neovascularization. Mol Vis. 2011;17:2386–99.
83. Narumoto O, Horiguchi K, Horiguchi S, Moriwaki Y, Takano-Ohmuro H, Shoji S, Misawa H, Yamashita N, Nagase T, Kawashima K. Down-regulation of secreted lymphocyte antigen-6/urokinase-type plasminogen activator receptor-related peptide-1 (SLURP-1), an endogenous allosteric alpha7 nicotinic acetylcholine receptor modulator, in murine and human asthmatic conditions. Biochem Biophys Res Commun. 2010;398(4):713–8.
84. Kimchi ET, Posner MC, Park JO, Darga TE, Kocherginsky M, Karrison T, Hart J, Smith KD, Mezhir JJ, Weichselbaum RR, et al. Progression of Barrett's metaplasia to adenocarcinoma is associated with the suppression of the transcriptional programs of epidermal differentiation. Cancer Res. 2005;65(8):3146–54.
85. Stairs DB, Nakagawa H, Klein-Szanto A, Mitchell SD, Silberg DG, Tobias JW, Lynch JP, Rustgi AK. Cdx1 and c-Myc foster the initiation of transdifferentiation of the normal esophageal squamous epithelium toward Barrett's esophagus. PLoS One. 2008;3(10):e3534.
86. Talantov D, Mazumder A, Yu JX, Briggs T, Jiang Y, Backus J, Atkins D, Wang Y. Novel genes associated with malignant melanoma but not benign melanocytic lesions. Clinical Cancer Research. 2005;11(20):7234–42.
87. Hu N, Clifford RJ, Yang HH, Wang C, Goldstein AM, Ding T, Taylor PR, Lee MP. Genome wide analysis of DNA copy number neutral loss of heterozygosity (CNNLOH) and its relation to gene expression in esophageal squamous cell carcinoma. BMC Genomics. 2010;11:576.
88. Reis PP, Waldron L, Perez-Ordonez B, Pintilie M, Galloni NN, Xuan Y, Cervigne NK, Warner GC, Makitie AA, Simpson C, et al. A gene signature in histologically normal surgical margins is predictive of oral carcinoma recurrence. BMC Cancer. 2011;11:437.
89. Eckl KM, Stevens HP, Lestringant GG, Westenberger-Treumann M, Traupe H, Hinz B, Frossard PM, Stadler R, Leigh IM, Nurnberg P, et al. Mal de Meleda (MDM) caused by mutations in the gene for SLURP-1 in patients from Germany, Turkey, Palestine, and the United Arab Emirates. Hum Genet. 2003;112(1):50–6.
90. Fischer J, Bouadjar B, Heilig R, Huber M, Lefevre C, Jobard F, Macari F, Bakija-Konsuo A, Ait-Belkacem F, Weissenbach J, et al. Mutations in the gene encoding SLURP-1 in Mal de Meleda. Hum Mol Genet. 2001;10(8):875–80.
91. Hu G, Yildirim M, Baysal V, Yerebakan O, Yilmaz E, Inaloz HS, Martinez-Mir A, Christiano AM, Celebi JT. A recurrent mutation in the ARS (component B) gene encoding SLURP-1 in Turkish families with mal de Meleda: evidence of a founder effect. J Invest Dermatol. 2003;120(6):967–9.
92. Ward KM, Yerebakan O, Yilmaz E, Celebi JT. Identification of recurrent mutations in the ARS (component B) gene encoding SLURP-1 in two families with mal de Meleda. J Invest Dermatol. 2003;120(1):96–8.
93. Marrakchi S, Audebert S, Bouadjar B, Has C, Lefevre C, Munro C, Cure S, Jobard F, Morlot S, Hohl D, et al. Novel mutations in the gene encoding secreted lymphocyte antigen-6/urokinase-type plasminogen activator receptor-related protein-1 (SLURP-1) and description of five ancestral haplotypes in patients with Mal de Meleda. J Invest Dermatol. 2003;120(3):351–5.
94. Arredondo J, Chernyavsky AI, Webber RJ, Grando SA. Biological effects of SLURP-1 on human keratinocytes. J Invest Dermatol. 2005;125(6):1236–41.
95. Arredondo J, Chernyavsky AI, Grando SA. Overexpression of SLURP-1 and −2 alleviates the tumorigenic action of tobacco-derived nitrosamine on immortalized oral epithelial cells. Biochem Pharmacol. 2007;74(8):1315–9.
96. Chimienti F, Hogg RC, Plantard L, Lehmann C, Brakch N, Fischer J, Huber M, Bertrand D, Hohl D. Identification of SLURP-1 as an epidermal neuromodulator explains the clinical phenotype of Mal de Meleda. Hum Mol Genet. 2003;12(22):3017–24.
97. Grando SA. Basic and clinical aspects of non-neuronal acetylcholine: biological and clinical significance of non-canonical ligands of epithelial nicotinic acetylcholine receptors. J Pharmacol Sci. 2008;106(2):174–9.

98. Adeyo O, Oberer M, Ploug M, Fong LG, Young SG, Beigneux AP. Heterogeneity in the properties of mutant secreted lymphocyte antigen 6/urokinase receptor-related protein 1 (SLURP1) in Mal de Meleda. The British Journal of Dermatology. 2015;173(4):1066–9.
99. Adeyo O, Allan BB, Barnes 2nd RH, Goulbourne CN, Tatar A, Tu Y, Young LC, Weinstein MM, Tontonoz P, Fong LG, et al. Palmoplantar keratoderma along with neuromuscular and metabolic phenotypes in Slurp1-deficient mice. J Invest Dermatol. 2014;134(6):1589–98.
100. Arredondo J, Chernyavsky AI, Jolkovsky DL, Webber RJ, Grando SA. SLURP-2: a novel cholinergic signaling peptide in human mucocutaneous epithelium. J Cell Physiol. 2006;208(1):238–45.
101. Tsuji H, Okamoto K, Matsuzaka Y, Iizuka H, Tamiya G, Inoko H. SLURP-2, a novel member of the human Ly-6 superfamily that is up-regulated in psoriasis vulgaris. Genomics. 2003;81(1):26–33.
102. Allan CM, Procaccia S, Tran D, Tu Y, Barnes 2nd RH, Larsson M, Allan BB, Young LC, Hong C, Tontonoz P, et al. Palmoplantar keratoderma in Slurp2-deficient mice. J Invest Dermatol. 2016;136(2):436–43.
103. Dessaud E, Salaun D, Gayet O, Chabbert M, de Lapeyriere O. Identification of lynx2, a novel member of the ly-6/neurotoxin superfamily, expressed in neuronal subpopulations during mouse development. Molecular and Cellular Neurosciences. 2006;31(2):232–42.
104. Tekinay AB, Nong Y, Miwa JM, Lieberam I, Ibanez-Tallon I, Greengard P, Heintz N. A role for LYNX2 in anxiety-related behavior. Proc Natl Acad Sci U S A. 2009;106(11):4477–82.
105. Wu M, Puddifoot CA, Taylor P, Joiner WJ. Mechanisms of inhibition and potentiation of alpha4beta2 nicotinic acetylcholine receptors by members of the Ly6 protein family. J Biol Chem. 2015;290(40):24509–18.
106. McWilliam H, Li W, Uludag M, Squizzato S, Park YM, Buso N, Cowley AP, Lopez R. Analysis Tool Web Services from the EMBL-EBI. Nucleic Acids Res. 2013;41(Web Server issue):W597–600.
107. Sievers F, Wilm A, Dineen D, Gibson TJ, Karplus K, Li W, Lopez R, McWilliam H, Remmert M, Soding J, et al. Fast, scalable generation of high-quality protein multiple sequence alignments using Clustal Omega. Molecular Systems Biology. 2011;7:539.
108. Letunic I, Bork P. Interactive Tree Of Life v2: online annotation and display of phylogenetic trees made easy. Nucleic Acids Res. 2011;39(Web Server issue):W475–478.
109. Palfree RG. Ly-6-domain proteins—new insights and new members: a C-terminal Ly-6 domain in sperm acrosomal protein SP-10. Tissue antigens. 1996;48(2):71–9.
110. Hamatani T, Tanabe K, Kamei K, Sakai N, Yamamoto Y, Yoshimura Y. A monoclonal antibody to human SP-10 inhibits in vitro the binding of human sperm to hamster oolemma but not to human Zona pellucida. Biology of Reproduction. 2000;62(5):1201–8.
111. Foster JA, Klotz KL, Flickinger CJ, Thomas TS, Wright RM, Castillo JR, Herr JC. Human SP-10: acrosomal distribution, processing, and fate after the acrosome reaction. Biology of Reproduction. 1994;51(6):1222–31.
112. von Vietinghoff S, Tunnemann G, Eulenberg C, Wellner M, Cristina Cardoso M, Luft FC, Kettritz R. NB1 mediates surface expression of the ANCA antigen proteinase 3 on human neutrophils. Blood. 2007;109(10):4487–93.
113. Bauer S, Abdgawad M, Gunnarsson L, Segelmark M, Tapper H, Hellmark T. Proteinase 3 and CD177 are expressed on the plasma membrane of the same subset of neutrophils. J Leukoc Biol. 2007;81(2):458–64.
114. Stroncek DF, Skubitz KM, McCullough JJ. Biochemical characterization of the neutrophil-specific antigen NB1. Blood. 1990;75(3):744–55.
115. Mead RJ, Neal JW, Griffiths MR, Linington C, Botto M, Lassmann H, Morgan BP. Deficiency of the complement regulator CD59a enhances disease severity, demyelination and axonal injury in murine acute experimental allergic encephalomyelitis. Laboratory Investigation. 2004;84(1):21–8.
116. Holt DS, Botto M, Bygrave AE, Hanna SM, Walport MJ, Morgan BP. Targeted deletion of the CD59 gene causes spontaneous intravascular hemolysis and hemoglobinuria. Blood. 2001;98(2):442–9.
117. Longhi MP, Sivasankar B, Omidvar N, Morgan BP, Gallimore A. Cutting edge: murine CD59a modulates antiviral CD4+ T cell activity in a complement-independent manner. J Immunol. 2005;175(11):7098–102.
118. Qin X, Miwa T, Aktas H, Gao M, Lee C, Qian YM, Morton CC, Shahsafaei A, Song WC, Halperin JA. Genomic structure, functional comparison, and tissue distribution of mouse Cd59a and Cd59b. Mammalian Genome. 2001;12(8):582–9.
119. Jiang H, Pillai S. Complement regulatory proteins on the sperm surface: relevance to sperm motility. American Journal of Reproductive Immunology. 1998;39(4):243–8.
120. Fenichel P, Cervoni F, Hofmann P, Deckert M, Emiliozzi C, Hsi BL, Rossi B. Expression of the complement regulatory protein CD59 on human spermatozoa: characterization and role in gametic interaction. Molecular Reproduction and Development. 1994;38(3):338–46.
121. Furuhata T, Tokino T, Urano T, Nakamura Y. Isolation of a novel GPI-anchored gene specifically regulated by p53; correlation between its expression and anti-cancer drug sensitivity. Oncogene. 1996;13(9):1965–70.
122. Ueda K, Miyoshi Y, Tokino T, Watatani M, Nakamura Y. Induction of apoptosis in T98G glioblastoma cells by transfection of GML, a p53 target gene. Oncology Research. 1999;11(3):125–32.
123. Xue H, O'Neill D, Morrow J, Bank A. A novel mouse gene, HemT, encoding an hematopoietic cell-specific transcript. Gene. 1999;231(1–2):49–58.
124. Holmes RS, Cox LA. Comparative studies of glycosylphosphatidylinositol-anchored high-density lipoprotein-binding protein 1: evidence for a eutherian mammalian origin for the GPIHBP1 gene from an LY6-like gene. 3 Biotech. 2012;2(1):37–52.
125. van de Rijn M, Heimfeld S, Spangrude GJ, Weissman IL. Mouse hematopoietic stem-cell antigen Sca-1 is a member of the Ly-6 antigen family. Proc Natl Acad Sci U S A. 1989;86(12):4634–8.
126. Bailey B, Fransioli J, Gude NA, Alvarez Jr R, Zhang X, Gustafsson AB, Sussman MA. Sca-1 knockout impairs myocardial and cardiac progenitor cell function. Circulation Research. 2012;111(6):750–60.
127. Hsu YC, Mildenstein K, Hunter K, Tkachenko O, Mullen CA. Acute lymphoid leukemia cells with greater stem cell antigen-1 (Ly6a/Sca-1) expression exhibit higher levels of metalloproteinase activity and are more aggressive in vivo. PLoS One. 2014;9(2):e88966.
128. Epting CL, King FW, Pedersen A, Zaman J, Ritner C, Bernstein HS. Stem cell antigen-1 localizes to lipid microdomains and associates with insulin degrading enzyme in skeletal myoblasts. J Cell Physiol. 2008;217(1):250–60.
129. Stanford WL, Haque S, Alexander R, Liu X, Latour AM, Snodgrass HR, Koller BH, Flood PM. Altered proliferative response by T lymphocytes of Ly-6A (Sca-1) null mice. J Exp Med. 1997;186(5):705–17.
130. Reese JT, Mehta H, Chappell CH, Bamezai A. Downregulated expression of Ly-6-ThB on developing T cells marks CD4+CD8+ subset undergoing selection in the thymus. Developmental Immunology. 2001;8(2):107–21.
131. Brakenhoff RH, Gerretsen M, Knippels EM, van Dijk M, van Essen H, Weghuis DO, Sinke RJ, Snow GB, van Dongen GA. The human E48 antigen, highly homologous to the murine Ly-6 antigen ThB, is a GPI-anchored molecule apparently involved in keratinocyte cell-cell adhesion. J Cell Biol. 1995; 129(6):1677–89.
132. Inlay MA, Bhattacharya D, Sahoo D, Serwold T, Seita J, Karsunky H, Plevritis SK, Dill DL, Weissman IL. Ly6d marks the earliest stage of B-cell specification and identifies the branchpoint between B-cell and T-cell development. Genes Dev. 2009;23(20):2376–81.
133. Kurosawa M, Jeyasekharan AD, Surmann EM, Hashimoto N, Venkatraman V, Kurosawa G, Furukawa K, Venkitaraman AR, Kurosawa Y. Expression of LY6D is induced at the surface of MCF10A cells by X-ray irradiation. The FEBS Journal. 2012;279(24):4479–91.
134. Dullens HF, Hilgers J, Van Basten CD, De Weger RA, De Heer E, Den Otter W. Cell surface antigen phenotypes and enzyme expression patterns of two murine T-cell lymphomas derived from early and/or mature thymus cells. Leukemia Research. 1982;6(3):425–8.
135. Mao M, Yu M, Tong JH, Ye J, Zhu J, Huang QH, Fu G, Yu L, Zhao SY, Waxman S, et al. RIG-E, a human homolog of the murine Ly-6 family, is induced by retinoic acid during the differentiation of acute promyelocytic leukemia cell. Proc Natl Acad Sci U S A. 1996;93(12):5910–4.
136. Xu X, Qiu C, Zhu L, Huang J, Li L, Fu W, Zhang L, Wei J, Wang Y, Geng Y, et al. IFN-stimulated gene LY6E in monocytes regulates the CD14/TLR4 pathway but inadequately restrains the hyperactivation of monocytes during chronic HIV-1 infection. J Immunol. 2014;193(8):4125–36.
137. MacNeil I, Kennedy J, Godfrey DI, Jenkins NA, Masciantonio M, Mineo C, Gilbert DJ, Copeland NG, Boyd RL, Zlotnik A. Isolation of a cDNA encoding thymic shared antigen-1. A new member of the Ly6 family with a possible role in T cell development. J Immunol. 1993;151(12):6913–23.
138. Shan X, Bourdeau A, Rhoton A, Wells DE, Cohen EH, Landgraf BE, Palfree RG. Characterization and mapping to human chromosome 8q24.3 of Ly-6-related gene 9804 encoding an apparent homologue of mouse TSA-1. J Immunol. 1998;160(1):197–208.
139. Capone MC, Gorman DM, Ching EP, Zlotnik A. Identification through bioinformatics of cDNAs encoding human thymic shared Ag-1/stem cell

Ag-2. A new member of the human Ly-6 family. J Immunol. 1996;157(3): 969–73.
140. Fleming TJ, O'HUigin C, Malek TR. Characterization of two novel Ly-6 genes. Protein sequence and potential structural similarity to alpha-bungarotoxin and other neurotoxins. J Immunol. 1993;150(12):5379–90.
141. Wang JX, Bair AM, King SL, Shnayder R, Huang YF, Shieh CC, Soberman RJ, Fuhlbrigge RC, Nigrovic PA. Ly6G ligation blocks recruitment of neutrophils via a beta2-integrin-dependent mechanism. Blood. 2012;120(7):1489–98.
142. Mallya M, Campbell RD, Aguado B. Characterization of the five novel Ly-6 superfamily members encoded in the MHC, and detection of cells expressing their potential ligands. Protein Science. 2006;15(10):2244–56.
143. Mallya M, Campbell RD, Aguado B. Transcriptional analysis of a novel cluster of LY-6 family members in the human and mouse major histocompatibility complex: five genes with many splice forms. Genomics. 2002;80(1):113–23.
144. De Vet EC, Aguado B, Campbell RD. Adaptor signalling proteins Grb2 and Grb7 are recruited by human G6f, a novel member of the immunoglobulin superfamily encoded in the MHC. Biochem J. 2003;375(Pt 1):207–13.
145. Horie M, Okutomi K, Taniguchi Y, Ohbuchi Y, Suzuki M, Takahashi E. Isolation and characterization of a new member of the human Ly6 gene family (LY6H). Genomics. 1998;53(3):365–8.
146. Apostolopoulos J, Chisholm LJ, Sandrin MS. Identification of mouse Ly6H and its expression in normal tissue. Immunogenetics. 1999;49(11–12):987–90.
147. Pflugh DL, Maher SE, Bothwell AL. Ly-6 superfamily members Ly-6A/E, Ly-6C, and Ly-6I recognize two potential ligands expressed by B lymphocytes. J Immunol. 2002;169(9):5130–6.
148. Kong HK, Yoon S, Park JH. The regulatory mechanism of the LY6K gene expression in human breast cancer cells. J Biol Chem. 2012;287(46): 38889–900.
149. Choi SH, Kong HK, Park SY, Park JH. Metastatic effect of LY-6 K gene in breast cancer cells. International Journal of Oncology. 2009;35(3):601–7.
150. de Nooij-van Dalen AG, van Dongen GA, Smeets SJ, Nieuwenhuis EJ, Stigter-van Walsum M, Snow GB, Brakenhoff RH. Characterization of the human Ly-6 antigens, the newly annotated member Ly-6 K included, as molecular markers for head-and-neck squamous cell carcinoma. International Journal of Cancer Journal International Du Cancer. 2003; 103(6):768–74.
151. Fujihara Y, Okabe M, Ikawa M. GPI-anchored protein complex, LY6K/TEX101, is required for sperm migration into the oviduct and male fertility in mice. Biology of Reproduction. 2014;90(3):60.
152. von Allmen CE, Bauer M, Dietmeier K, Buser RB, Gwerder M, Muntwiler S, Utzinger S, Saudan P, Bachmann MF, Beerli RR. Identification of Ly-6K as a novel marker for mouse plasma cells. Molecular Immunology. 2008;45(10): 2727–33.
153. Miwa JM, Ibanez-Tallon I, Crabtree GW, Sanchez R, Sali A, Role LW, Heintz N. lynx1, an endogenous toxin-like modulator of nicotinic acetylcholine receptors in the mammalian CNS. Neuron. 1999;23(1):105–14.
154. Fu XW, Song PF, Spindel ER. Role of Lynx1 and related Ly6 proteins as modulators of cholinergic signaling in normal and neoplastic bronchial epithelium. Int Immunopharmacol. 2015;29(1):93–8.
155. Yu DH, Fan W, Liu G, Nguy V, Chatterton JE, Long S, Ke N, Meyhack B, Bruengger A, Brachat A, et al. PHTS, a novel putative tumor suppressor, is involved in the transformation reversion of HeLaHF cells independently of the p53 pathway. Exp Cell Res. 2006;312(6):865–76.
156. Kriegbaum MC, Clausen OP, Laerum OD, Ploug M. Expression of the Ly6/uPAR-domain proteins C4.4A and Haldisin in non-invasive and invasive skin lesions. The Journal of Histochemistry and Cytochemistry. 2015;63(2):142–54.
157. Oshiro R, Yamamoto H, Takahashi H, Ohtsuka M, Wu X, Nishimura J, Takemasa I, Mizushima T, Ikeda M, Sekimoto M, et al. C4.4A is associated with tumor budding and epithelial-mesenchymal transition of colorectal cancer. Cancer Science. 2012;103(6):1155–64.
158. Smith BA, Kennedy WJ, Harnden P, Selby PJ, Trejdosiewicz LK, Southgate J. Identification of genes involved in human urothelial cell-matrix interactions: implications for the progression pathways of malignant urothelium. Cancer Res. 2001;61(4):1678–85.
159. Ngora H, Galli UM, Miyazaki K, Zoller M. Membrane-bound and exosomal metastasis-associated C4.4A promotes migration by associating with the alpha(6)beta(4) integrin and MT1-MMP. Neoplasia. 2012;14(2):95–107.
160. Kriegbaum MC, Jacobsen B, Hald A, Ploug M. Expression of C4.4A, a structural uPAR homolog, reflects squamous epithelial differentiation in the adult mouse and during embryogenesis. The Journal of Histochemistry and Cytochemistry. 2011;59(2):188–201.
161. Gardsvoll H, Kriegbaum MC, Hertz EP, Alpizar-Alpizar W, Ploug M. The urokinase receptor homolog Haldisin is a novel differentiation marker of stratum granulosum in squamous epithelia. The Journal of Histochemistry and Cytochemistry. 2013;61(11):802–13.
162. Zhang Y, Lang Q, Li J, Xie F, Wan B, Yu L. Identification and characterization of human LYPD6, a new member of the Ly-6 superfamily. Molecular Biology Reports. 2010;37(4):2055–62.
163. Ozhan G, Sezgin E, Wehner D, Pfister AS, Kuhl SJ, Kagermeier-Schenk B, Kuhl M, Schwille P, Weidinger G. Lypd6 enhances Wnt/beta-catenin signaling by promoting Lrp6 phosphorylation in raft plasma membrane domains. Developmental Cell. 2013;26(4):331–45.
164. Darvas M, Morsch M, Racz I, Ahmadi S, Swandulla D, Zimmer A. Modulation of the Ca2+ conductance of nicotinic acetylcholine receptors by Lypd6. European Neuropsychopharmacology. 2009;19(9):670–81.
165. Ni J, Lang Q, Bai M, Zhong C, Chen X, Wan B, Yu L. Cloning and characterization of a human LYPD7, a new member of the Ly-6 superfamily. Molecular Biology Reports. 2009;36(4):697–703.
166. Bera TK, Maitra R, Iavarone C, Salvatore G, Kumar V, Vincent JJ, Sathyanarayana BK, Duray P, Lee BK, Pastan I. PATE, a gene expressed in prostate cancer, normal prostate, and testis, identified by a functional genomic approach. Proc Natl Acad Sci U S A. 2002;99(5):3058–63.
167. Soler-Garcia AA, Maitra R, Kumar V, Ise T, Nagata S, Beers R, Bera TK, Pastan I. The PATE gene is expressed in the accessory tissues of the human male genital tract and encodes a secreted sperm-associated protein. Reproduction. 2005;129(4):515–24.
168. Rolland T, Tasan M, Charloteaux B, Pevzner SJ, Zhong Q, Sahni N, Yi S, Lemmens I, Fontanillo C, Mosca R, et al. A proteome-scale map of the human interactome network. Cell. 2014;159(5):1212–26.
169. Li SH, Lee RK, Lin MH, Hwu YM, Lu CH, Chen YJ, Chen HC, Chang WH, Chang WC. SSLP-1, a secreted Ly-6 protein purified from mouse seminal vesicle fluid. Reproduction. 2006;132(3):493–500.
170. Collen D. The plasminogen (fibrinolytic) system. Thrombosis and Haemostasis. 1999;82(2):259–70.
171. Stewart CE, Sayers I. Urokinase receptor orchestrates the plasminogen system in airway epithelial cell function. Lung. 2013;191(2):215–25.
172. Carmeliet P, Moons L, Lijnen R, Baes M, Lemaitre V, Tipping P, Drew A, Eeckhout Y, Shapiro S, Lupu F, et al. Urokinase-generated plasmin activates matrix metalloproteinases during aneurysm formation. Nat Genet. 1997; 17(4):439–44.
173. Estreicher A, Muhlhauser J, Carpentier JL, Orci L, Vassalli JD. The receptor for urokinase type plasminogen activator polarizes expression of the protease to the leading edge of migrating monocytes and promotes degradation of enzyme inhibitor complexes. J Cell Biol. 1990;111(2):783–92.
174. Lyons RM, Gentry LE, Purchio AF, Moses HL. Mechanism of activation of latent recombinant transforming growth factor beta 1 by plasmin. J Cell Biol. 1990;110(4):1361–7.
175. Houck KA, Leung DW, Rowland AM, Winer J, Ferrara N. Dual regulation of vascular endothelial growth factor bioavailability by genetic and proteolytic mechanisms. J Biol Chem. 1992;267(36):26031–7.
176. Renaud SJ, Macdonald-Goodfellow SK, Graham CH. Coordinated regulation of human trophoblast invasiveness by macrophages and interleukin 10. Biology of Reproduction. 2007;76(3):448–54.
177. Sidenius N, Blasi F. The urokinase plasminogen activator system in cancer: recent advances and implication for prognosis and therapy. Cancer Metastasis Reviews. 2003;22(2–3):205–22.
178. Colman RW, Pixley RA, Najamunnisa S, Yan W, Wang J, Mazar A, McCrae KR. Binding of high molecular weight kininogen to human endothelial cells is mediated via a site within domains 2 and 3 of the urokinase receptor. J Clin Invest. 1997;100(6):1481–7.
179. Higazi AA, Upson RH, Cohen RL, Manuppello J, Bognacki J, Henkin J, McCrae KR, Kounnas MZ, Strickland DK, Preissner KT, et al. Interaction of single-chain urokinase with its receptor induces the appearance and disappearance of binding epitopes within the resultant complex for other cell surface proteins. Blood. 1996;88(2):542–51.
180. Czekay RP, Kuemmel TA, Orlando RA, Farquhar MG. Direct binding of occupied urokinase receptor (uPAR) to LDL receptor-related protein is required for endocytosis of uPAR and regulation of cell surface urokinase activity. Mol Biol Cell. 2001;12(5):1467–79.
181. Behrendt N, Jensen ON, Engelholm LH, Mortz E, Mann M, Dano K. A urokinase receptor-associated protein with specific collagen binding properties. J Biol Chem. 2000;275(3):1993–2002.

182. Nykjaer A, Christensen EI, Vorum H, Hager H, Petersen CM, Roigaard H, Min HY, Vilhardt F, Moller LB, Kornfeld S, et al. Mannose 6-phosphate/insulin-like growth factor-II receptor targets the urokinase receptor to lysosomes via a novel binding interaction. J Cell Biol. 1998;141(3):815–28.
183. Bugge TH, Flick MJ, Danton MJ, Daugherty CC, Romer J, Dano K, Carmeliet P, Collen D, Degen JL. Urokinase-type plasminogen activator is effective in fibrin clearance in the absence of its receptor or tissue-type plasminogen activator. Proc Natl Acad Sci U S A. 1996;93(12):5899–904.
184. Moller LB, Pollanen J, Ronne E, Pedersen N, Blasi F. N-linked glycosylation of the ligand-binding domain of the human urokinase receptor contributes to the affinity for its ligand. J Biol Chem. 1993;268(15):11152–9.
185. Pass J, Gardsvoll H, Lund LR, Dano K, Hoyer-Hansen G. Generation of antibodies to the urokinase receptor (uPAR) by DNA immunization of uPAR knockout mice: membrane-bound uPAR is not required for an antibody response. Scandinavian Journal of Immunology. 2003;58(3):298–305.
186. Taeb J, Asgari M, Abolhasani M, Farajollahi MM, Madjd Z. Expression of prostate stem cell antigen (PSCA) in prostate cancer: a tissue microarray study of Iranian patients. Pathology, Research and Practice. 2014;210(1):18–23.
187. Bahrenberg G, Brauers A, Joost HG, Jakse G. Reduced expression of PSCA, a member of the LY-6 family of cell surface antigens, in bladder, esophagus, and stomach tumors. Biochem Biophys Res Commun. 2000;275(3):783–8.
188. Wang L, Sang Y, Tang J, Zhang RH, Luo D, Chen M, Deng WG, Kang T. Down-regulation of prostate stem cell antigen (PSCA) by Slug promotes metastasis in nasopharyngeal carcinoma. J Pathol. 2015;237(4):411–22.
189. Adermann K, Wattler F, Wattler S, Heine G, Meyer M, Forssmann WG, Nehls M. Structural and phylogenetic characterization of human SLURP-1, the first secreted mammalian member of the Ly-6/uPAR protein superfamily. Protein Science. 1999;8(4):810–9.
190. Chernyavsky AI, Galitovskiy V, Shchepotin IB, Grando SA. Anti-inflammatory effects of the nicotinergic peptides SLURP-1 and SLURP-2 on human intestinal epithelial cells and immunocytes. BioMed Research International. 2014;2014:609086.
191. Chernyavsky AI, Arredondo J, Galitovskiy V, Qian J, Grando SA. Upregulation of nuclear factor-kappaB expression by SLURP-1 is mediated by alpha7-nicotinic acetylcholine receptor and involves both ionic events and activation of protein kinases. Am J Physiol Cell Physiol. 2010;299(5):C903–911.
192. Lyukmanova EN, Shulepko MA, Kudryavtsev D, Bychkov ML, Kulbatskii DS, Kasheverov IE, Astapova MV, Feofanov AV, Thomsen MS, Mikkelsen JD, Shenkarev ZO, Tsetlin VI, Dolgikh DA, Kirpichnikov MP. Human Secreted Ly-6/uPAR Related Protein-1 (SLURP-1) Is a Selective Allosteric Antagonist of α7 Nicotinic Acetylcholine Receptor. PLoS One. 2016;11(2):e0149733.
193. Tang T, Li L, Tang J, Li Y, Lin WY, Martin F, Grant D, Solloway M, Parker L, Ye W, et al. A mouse knockout library for secreted and transmembrane proteins. Nature Biotechnology. 2010;28(7):749–55.
194. Shetty J, Wolkowicz MJ, Digilio LC, Klotz KL, Jayes FL, Diekman AB, Westbrook VA, Farris EM, Hao Z, Coonrod SA, et al. SAMP14, a novel, acrosomal membrane-associated, glycosylphosphatidylinositol-anchored member of the Ly-6/urokinase-type plasminogen activator receptor superfamily with a role in sperm-egg interaction. J Biol Chem. 2003;278(33):30506–15.
195. Teng X, Yang J, Xie Y, Ni Z, Hu R, Shi L, Lin Z, Hu L, Zhao G, Ding X, et al. A novel spermatogenesis-specific uPAR gene expressed in human and mouse testis. Biochem Biophys Res Commun. 2006;342(4):1223–7.
196. Yin L, Chung CM, Huo R, Liu H, Zhou C, Xu W, Zhu H, Zhang J, Shi Q, Wong HY, et al. A sperm GPI-anchored protein elicits sperm-cumulus cross-talk leading to the acrosome reaction. Cellular and Molecular Life Sciences. 2009;66(5):900–8.
197. Yin LL, Chung CM, Chen J, Fok KL, Ng CP, Jia RR, Ren X, Zhou J, Zhang T, Zhao XH, et al. A suppressor of multiple extracellular matrix-degrading proteases and cancer metastasis. Journal of Cellular and Molecular Medicine. 2009;13(9B):4034–41.
198. Kurita A, Takizawa T, Takayama T, Totsukawa K, Matsubara S, Shibahara H, Orgebin-Crist MC, Sendo F, Shinkai Y, Araki Y. Identification, cloning, and initial characterization of a novel mouse testicular germ cell-specific antigen. Biology of Reproduction. 2001;64(3):935–45.

4

Genetic risk factors for restenosis after percutaneous coronary intervention in Kazakh population

Elena V. Zholdybayeva[1*], Yerkebulan A. Talzhanov[1], Akbota M. Aitkulova[1], Pavel V. Tarlykov[1], Gulmira N. Kulmambetova[1], Aisha N. Iskakova[1,5], Aliya U. Dzholdasbekova[2], Olga A. Visternichan[3], Dana Zh. Taizhanova[3] and Yerlan M. Ramanculov[4,1]

Abstract

Background: After coronary stenting, the risk of developing restenosis is from 20 to 35 %. The aim of the present study is to investigate the association of genetic variation in candidate genes in patients diagnosed with restenosis in the Kazakh population.

Methods: Four hundred fifty-nine patients were recruited to the study; 91 patients were also diagnosed with diabetes and were excluded from the sampling. DNA was extracted with the salting-out method. The patients were genotyped for 53 single-nucleotide polymorphisms. Genotyping was performed on the QuantStudio 12K Flex (Life Technologies). Differences in distribution of BMI score among different genotype groups were compared by analysis of variance (ANOVA). Also, statistical analysis was performed using R and PLINK v.1.07. Haplotype frequencies and LD measures were estimated by using the software Haploview 4.2.

Results: A logistic regression analysis found a significant difference in restenosis rates for different genotypes. *FGB* (rs1800790) is significantly associated with restenosis after stenting (OR = 2.924, $P = 2.3\text{E}{-}06$, additive model) in the Kazakh population. *CD14* (rs2569190) showed a significant association in the additive (OR = 0.08033, $P = 2.11\text{E}{-}09$) and dominant models (OR = 0.05359, $P = 4.15\text{E}{-}11$). *NOS3* (rs1799983) was also highly associated with development of restenosis after stenting in additive (OR = 20.05, $P = 2.74\ \text{E}{-}12$) and recessive models (OR = 22.24, $P = 6.811\text{E}{-}10$).

Conclusions: Our results indicate that *FGB* (rs1800790), *CD14* (rs2569190), and *NOS3* (rs1799983) SNPs could be genetic markers for development of restenosis in Kazakh population. Adjustment for potential confounder factor BMI gave almost the same results.

Keywords: Coronary heart disease, Restenosis, SNP, Genotyping

Background

Coronary heart disease (CHD) is a disease characterized by reduced blood supply to the heart muscle. Narrowing of the coronary arterial lumen due to atherosclerosis is the primary cause in 97–98 % of CHD cases. Coronary heart disease has the highest rate of death and serious complications among all forms of cardiovascular disease. An estimated 17.5 million people died from cardiovascular diseases in 2012, representing 31 % of all global deaths. Of these deaths, an estimated 7.4 million were due to CHD and 6.7 million were due to stroke [1]. It should be noted that CHD usually affects the population aged between 35 and 65 years. In addition, CHD represents the most important cause of sudden cardiac deaths. Together with cerebrovascular diseases, CHD accounts for 64 % of all cardiovascular deaths [2].

According to the Ministry of Healthcare and Social Development of the Republic of Kazakhstan prior to year 2012, CHD morbidity has been constantly increasing in 2009–2012 with exception of 2013 when a minor decrease in the morbidity was observed [3]. In 2013,

* Correspondence: lenazhol@gmail.com
[1]National Center for Biotechnology, 13/5, KorgalzhinskoeHighway, Astana, Kazakhstan

there were 59,799 new cases of CHD registered in Kazakhstan and the morbidity rate reached 500.6 cases per 100,000 population, compared to 445.6 cases in 2011 and 507.4 cases in 2012 [3].

Advances in medicine have led to the emergence of novel methods of CHD treatment, such as angioplasty or coronary stenting. The first use of coronary stenting in clinical practice was in 1986 [4]. Primary percutaneous coronary intervention (PCI) has become a well-established strategy for patients with coronary heart disease [5]. Nowadays, endovascular methods for the reestablishment of coronary blood flow preserve the lives and health of hundreds of thousands of people around the world. Nevertheless, there is a possibility that during the first 6 months to 1 year after successful coronary stenting, a symptomatic relapse of angina may occur due to the development of restenosis. Reoccurrence of stenosis is a major limitation to the effectiveness of the stenting, and even the use of drug-eluting stents does not solve the problem completely [6]. After coronary stenting, the restenosis rate is 20–30 %. The use of second generation drug-eluting stents has reduced this rate, but the development of restenosis after implantation remains a serious clinical problem [7].

Restenosis can occur for many different reasons. The pathophysiological mechanisms of restenosis have not yet been fully explained, but it is believed that those mechanisms include inflammation, proliferation, and matrix remodeling. Over the years, many predictive clinical, biological, genetic, epigenetic, lesion-related, and procedural risk factors for restenosis have been identified. Those factors are useful in the risk stratification of patients and also contribute to our understanding of this condition [8]. In this sense, the search for new predictor factors in the development of restenosis is topical.

Currently, the genetic factors of restenosis have been studied mostly in European populations. The ethnical variability of genetic markers is well known, as shown in the results of the GENetic DEterminants of Restenosis (GENDER) study discussed in the article by Verschuren et al. The GENDER databank contains the genotypic data of 2,571,586 single-nucleotide polymorphism (SNPs) from 295 cases with restenosis and 571 matched controls. The set that included all 36 reported genes in the literature was indeed significantly associated with restenosis in the GENDER study ($P = 0.024$). Subsequent analyses of the individual genes demonstrated that this association was determined by 6 of the 36 genes [9]. As a result of the GENDER study, the selected SNPs have been associated with the risk of developing restenosis in European populations.

Based on literature review, candidate genes for restenosis were not studied or validated in Kazakh population. That is why the purpose of the current study is to look for associations of genetic variation in candidate genes in patients diagnosed with restenosis after percutaneous coronary intervention in the Kazakh population.

Methods

Study population

There were initially 459 patients with diagnosed CHD recruited to the study. Of these patients, 91 were also diagnosed with diabetes and were excluded from the sampling, since several studies have shown that diabetes is an independent risk factor for restenosis and may introduce bias to the interpretation of results [10]. Anthropometrical and biochemical characteristics were gathered for the population sample comprising 368 patients (299 males and 69 females). There were 99 case subjects, those who manifested in-stent restenosis within 6 months after stenting and 269 control subjects and those who did not developed restenosis after stenting. The study protocol was approved by the Ethics Committee of the National Center for Biotechnology. All subjects were ethnic Kazakhs.

Genotyping

Whole blood samples of 368 patients (9 ml) were collected into tubes containing 50 mmol/l ethylenediaminetetraacetic acid (disodium salt). DNA was extracted with the salting-out method. [11]. Genotyping of the extended panel of polymorphisms of candidate genes was performed on the QuantStudio 12K Flex (Life Technologies). The total reaction volume was 5 µl, containing 2.5 µl of 2× OppenArray Real-time master mix, and 2.5 µl of DNA concentration of 50 ng/µl. Temperature conditions were 10 min at 93 °C; cycling for 45 s at 93 °C, 13 s at 94 °C, and 2.14 min at 53.5 °C for 50 cycles, followed by incubation at 25 °C for 2 min. Data analysis was performed using the software package TaqMan Genotyper Software v.1.3.

Table 1 shows a panel of 53 SNPs used in the current study. All SNPs were selected based on the results of the GENDER study.

Statistical analysis

A pairwise correlation matrix was made to check for multicollinearity between measured variables. A stepwise regression was performed to evaluate significance of potential confounders. A logistic regression analysis with adjustment for potential confounder was used to test for differences between statuses of restenosis according to genotyping. Associations between each SNP and development of the restenosis were tested according to three genetic models: additive (cumulative effect of each additional variant allele), dominant (homozygous wild-type vs. variant allele-carrying genotype) and recessive (wild-type allele carrying genotype vs. homozygous variant genotype). Every SNP that reached statistically significant level of $P < 0.0001$ were reported. Significant SNPs were further annotated using RegulomeDB that harbors information for known

Table 1 Description of SNPs included in the study

Gene	Polymorphism	Locus
Adrenergic beta-2-receptor (ADRB2)	rs1042713	5q31–q32
Advanced glycosylation end product-specific receptor (AGER)	rs2070600	6p21.3
Advanced glycosylation end product-specific receptor (AGER)	rs1800624	6p21.3
Angiotensin II receptor, type 1 (AGTR1)	rs5186	3q24
Angiotensin II receptor, type 1 (AGTR1)	rs5182	3q24
Butyrylcholinesterase (BCHE)	rs1803274	3q26.1–q26.2
Chemokine (C–C motif) ligand 11 (CCL11)	rs4795895	17q21.1–q21.2
Cluster of differentiation 14 (CD14)	rs2569190	5q31.1
Cyclin-dependent kinase inhibitor 1B (p27, Kip1, CDKN1B)	rs34330	12p13.1–p12
Collagen, type III, alpha 1 (Col3A1)	rs1800255	2q31
Colony stimulating factor 2 (CSF2)	rs25882	5q31.1
Chemokine (C-X3-C motif) receptor 1 (CX3CR1)	rs3732379	3p21.3
Cytochrome b-245, alpha polypeptide (CYBA)	rs4673	16q24
Cytochrome P450, family2, subfamily C, polypeptide 19 (CYP2C19)	rs12248560	10q24
Fibrinogen beta chain (FGB)	rs1800790	4q28
Fibrinogen beta chain (FGB)	rs1044291	4q28
Coagulation factor V (F5)	rs6025	1q23
Glutathione peroxidase 1 (GPX1)	rs8179164	3p21.3
Integrin, beta 2 (ITGB2)	rs235326	21q22.3
Lipoprotein lipase (LPL)	rs328	8p22
Matrix metallopeptidase 12 (MMP12)	rs12808148	11q22.3
Matrix metallopeptidase 12 (MMP12)	rs17099726	20q11.2–q13.1
Matrix metallopeptidase 12 (MMP12)	rs2276109	20q11.2–q13.1
Methylenetetrahydrofolate reductase (NAD(P)H) MTHFR)	rs1801133	1p36.3
Nitric oxide synthase 3 (NOS3)	rs2070744	7q36
Nitric oxide synthase 3 (NOS3)	rs1799983	7q36
K(lysine) acetyltransferase2B (KAT2B, PCAF)	rs2948080	3p24
K(lysine) acetyltransferase2B (KAT2B, PCAF)	rs6776870	3p24
K(lysine) acetyltransferase2B (KAT2B, PCAF)	rs2929404	3p24
K(lysine) acetyltransferase2B (KAT2B, PCAF)	rs17796904	3p24
Peroxisome proliferator-activated receptor gamma (PPARG)	rs3856806	3p25
C-ros oncogene1, receptor tyrosine kinase (ROS1)	rs529038	6q22
Thrombomodulin (THBD)	rs1042579	20p11.2
Thrombospondin 4 (THBS4)	rs1866389	5q13
Thrombopoietin (THPO)	rs6141	3q27
Tumor protein p53 (TP53)	rs1042522	17p13.1
Transferrin (TF)	rs1799899	3p
Uncoupling protein 3 (UCP3)	rs1800849	11q13.4
Vitamin D receptor (VDR)	rs11568820	12q13.11
Vitamin D receptor (VDR)	rs11574027	12q13.11
Vitamin D receptor (VDR)	rs11574077	12q13.11
Tumor necrosis factor α (TNFα)	rs1800629	6p21.3
Tumor necrosis factor α (TNFα)	rs361525	6p21.3
Interleukin 1 receptor antagonist (IL1RN)	rs419598	2q14.2
Interleukin 1α (IL1A)	rs1800587	2q12–q21
Interleukin 1 β (IL1B)	rs1143627	2q13–q21
Interleukin 4 (IL4)	rs2243250	5q23–31
Interleukin 6 (IL6)	rs1800796	7p21
Interleukin 8 (IL8)	rs 4073	4q12–q13
Interleukin 10 (IL10)	rs1800871	1q31–q32
Interleukin 10 (IL10)	rs1800872	1q31–q32
Interleukin 10 (IL10)	rs1800896	1q31–q32
Interleukin 10 (IL10)	rs3024498	1q31–q32

and predicted regulatory elements [12]. Annotation of functional variation in personal genomes using RegulomeDB. Additionally, a differences in distribution of BMI score among different genotype groups were compared by one-way analysis of variance (one-way ANOVA). For ANOVA test a default threshold of 0.05 was used to report significance. Statistical analysis was performed using R and PLINK [13, 14].

LD statistical analysis was performed using Haploview 4.2. For block generations, Hardy-Weinberg *P* value cutoff 0.001 was used [15]. We ignored SNPs that minor allele frequencies (MAF) of less than 0.001. For block generations, the confidence intervals default algorithm was used.

Results

A total of 368 patients participated in the study. Body mass index (BMI), the blood levels of cholesterol, low-density lipoproteins (LDL), and high-density lipoproteins (HDL) were recorded for every patient at the time of participation. Table 2 shows summary statistics of the measured variables.

Anthropometrical and measured biochemical characteristics for potential confounders were evaluated before performing a test for association between genotype distribution and restenosis status. The distributions of the measured traits fitted the normality assumption and were included in the analysis as is without any transformation. Construction of the correlation matrix revealed that in our dataset, cholesterol and LDL are

Table 2 Anthropometrical and biochemical characteristics of the population sample ($n = 368$)

Restenosis (case/control)	99/269
Gender (male/female)	299/69
Age (years)	58.61 ± 11.67
BMI (kg/m^2)	28.35 ± 4.53
Cholesterol (mmol/l)	4.93 ± 1.24
LDL (mmol/l)	3.05 ± 1.18
HDL (mmol/l)	1.24 ± 0.63

BMI body mass index, *LDL* low-density lipoproteins, *HDL* high-density lipoproteins

highly correlated with each other (Table 3). In order to avoid multicollinearity, cholesterol was excluded from further analysis.

The remaining traits were used to build a statistical model that predicts the development of restenosis as a main outcome. Age, LDL, and HDL contribute insignificantly to restenosis variability in our dataset based on the results from stepwise regression analysis. The final statistical model included BMI as a single confounding factor that may affect the association between genotype and restenosis status.

There were 368 patients available to collect whole blood for further genetic analysis. After applying quality control filters, the final dataset contained 268 patients with genotype information for 48 SNPs. Four SNPs were excluded based on the Hardy Weinberg equilibrium test ($P \leq 0.001$). One SNP was excluded because of its low genotyping rate.

SNPs that reached a significant level $P < 0.0001$ were reported. The results from logistic regression revealed that *FGB* (rs1800790) SNP was associated with restenosis after stenting. Notably that homozygous A allele carriers of the rs1800790 SNP are 24 times less likely to develop restenosis after stenting compared to other allele carries. *CD14* (rs2569190) SNP showed highly significant correlation development of restenosis. Based on results, carrying homozygous G allele is a risk factor for development of restenosis. Finally, a missense mutation rs1799983 SNP mapped to *NOS3* was also highly associated with the development of restenosis after stenting. Both additive and recessive models showed similar results, suggesting that carrying additional G allele for rs1799983 SNP is a protective factor against developing restenosis. Summary for the results of logistic regression analysis is shown in Table 4.

Table 3 Pairwise correlation matrix of measured variables

	BMI	HDL	LDL	Cho	Age
Age	−0.072	−0.067	−0.061	−0.081	1
Cho	0.402	0.364	0.739	1	
LDL	0.322	0.492	1		
HDL	0.163	1			
BMI	1				

Association of genotype of the candidate genes 53 SNP with the restenosis was examined with one-way ANOVA by comparing the mean scores of BMI to the genotype. According to ANOVA, BMI distribution had significant differences with genotypes of rs6025 SNP in coagulation factor V also known as *F5* ($P = 0.00643$) in control group. rs419598 SNP mapped to *IL1RN* was positively associated with BMI in the control group ($P = 0.0299$). Finally, rs4795895 SNPs mapped to *CCL11* gene showed significant differences with BMI in group of patients ($P = 0.012$).Genotypes of other SNPs were not significantly different.

Using Haploview 4.2 software, LD statistics results for the Kazakh population were obtained (Fig. 1) (HW *P* cutoff, 0.001; MAF, 0.001). As a result, two haplotype blocks were defined: one block consisting of two SNPs, i.e., rs5182 and rs5186 (block 1, chromosome 3); one block consisting of two SNPs, i.e., rs1800871 and rs1800896 (block 2, chromosome 1). The haplotype frequencies in the studied population are presented in Table 5. Haplotype CC (rs5182-rs5186, block 1) was associated with the risk of developing restenosis (OR 2.17; 95 % CI; 1.33–3.53, $P = 0.002$) and the haplotype GC (rs1800871-rs1800896, block 2) was associated with restenosis (OR 1.51; CI, 1.01–2.26, $P = 0.04$).

Discussion

The present study included Kazakhs that are Turkic people of the northern parts of Central Asia (largely Kazakhstan). From the historic point of view and because of scarce genetic data, it was concluded that Kazakh population was formed as a result of admixture of the European and Asian populations [16]. Case and controls were diagnosed with CHD. BMI was increased in both groups. Our results showed that BMI is a single confounding factor that may affect the association between genotype and restenosis status. In turn, genotype polymorphisms of *F5*, *IL1RN*, *CCL11* genes significantly influence on increased body mass index.

The present study focuses on 48 SNPs in 36 candidate genes that were previously reported as genetic risk factors for restenosis. Association of hemostatic gene polymorphisms with restenosis after coronary stent placement was the first genetic risk to be described [17, 18]. The association of selected SNPs in inflammation-related genes with restenosis is also well documented [19–21]. In addition, a number of candidate genes in the renin-angiotensin hormone system and the endothelial nitric oxide synthase (*eNOS*, Glu298Aps and −786T\C) are also involved in this process [22–24]. New molecular markers of restenosis are constantly emerging. For example, in a recent study, SNPs in *VDR* (vitamin D-dependent receptor) gene were

Table 4 Results from logistic regression analysis

CHR	SNP	BP	Type	Genes	Genotype	Mean	*N*	Additive		Dominante		Recessive	
								OR	*P*	OR	*P*	OR	*P*
4	rs1800790	155483708	Intron	FGB	GG	0.18	201						
					GA	0.17	42	2.967	1.62E-06			24.95	7.74E-07
					AA	0.86	21	2.924	2.31E-06	2.725	0.001843	24.36	8.95E-07
5	**rs2569190**	140012916	Intron	CD14	GG	0.47	115						
					GA	0.08	88	0.08694	2.20E-09	0.05848	5.31E-11		
					AA	0	57	0.08033	2.11E-09	0.05359	4.15E-11		
7	rs1799983	150696111	Missense mutation	NOS3	TT	0.81	31						
					GT	0.35	93	20.56	1.86E-12			23.6	2.87E-10
					GG	0	137	20.05	2.74E-12			22.24	6.81E-10

bold – snp that also had functional effect

red – p values unadjusted

green – p values adjusted for BMI

*Bold*snp that also had functional effect, *red P* values unadjusted, *green P* values adjusted for BMI

considered as risk markers of restenosis [25]. Fragoso et al. reported that transforming growth factor-1β (rs1800469) was associated with the risk of developing restenosis after coronary stenting in Mexican patients [26]. This SNP was not investigated in our study.

Our study has revealed that blood coagulation fibrinogen factor I (*FGB*), monocyte differentiation antigen CD14 (*CD14*), and nitric oxide synthase 3 (*NOS3*) genes are among the factors associated with the risk of restenosis in the studied population.

Monocytes play a central role in restenosis after balloon angioplasty and stent implantation. Monocytes migrate into the damaged area either as a direct response or through the release of platelet-derived factors. Activated monocytes release large amounts of proinflammatory cytokines, which cause vasoconstriction and non-specific recruitment, proliferation, and activation of other cells including vascular smooth muscle cells in the vascular wall. The activation of monocytes/macrophages, endothelial cells, and smooth muscle cells mediated by CD14 and/or CD14 may play an important role in the restenosis processes [27].

Functional C(-260)→T polymorphism in the promoter of the *CD14* gene has been reported to be associated with CHD but data have yielded conflicting results. In the meta-analysis of Zhang et al., TT genotype is associated with ischemic heart disease in the East Asian population but not in the European or Indian populations [28]. Two previous studies have investigated the role of CD14 in the development of restenosis, one being a prospective study by Zee et al. [29] in 779 patients and the other being a prospective study by Shimada et al. [27] in 129 Japanese patients. They found the −260T/T genotype to be a risk factor for restenosis. But the GENDER study found that T allele was not associated with restenosis in the European population [9]. Also, it was showed that the CD14+CD16+CX3CR1+monocytes might have a role in-stent restenosis following coronary implantation of bare-metal stents in patients with acute myocardial infarction [30].

The results of our study have revealed that *CD14* promoter polymorphism remained statistically significant in the additive (OR = 0.08033, *P* = 2.11E−09) and dominant models (OR = 0.05359, *P* = 4.15E−11). Therefore, the T allele at position −260 of *CD14* gene is a risk allele for restenosis in Kazakh population. Population stratification based on ethnicities may lead to inconsistency, especially when both allele frequencies and incidence rates of the diseases vary across ethnic groups.

Fibrinogen (factor I) is a glycoprotein synthesized by the liver. It consists of three polypeptides Aα, Bβ, and γ coded by the alpha (*FGA*), beta (*FGB*), and gamma (*FGG*) genes, respectively. Fibrinogen is an important component of the coagulation cascade and a major determinant of

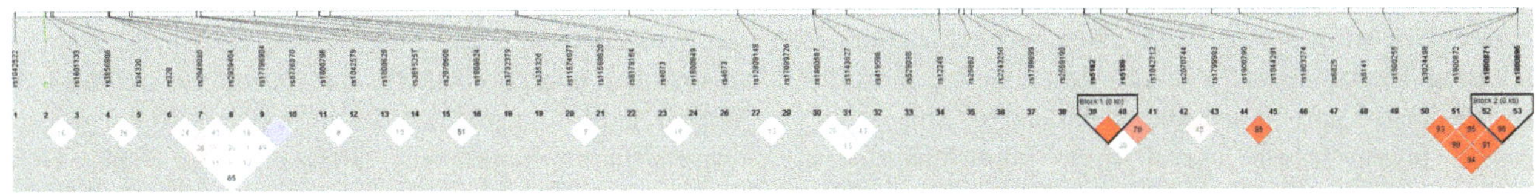

Fig. 1 LD SNP plot. The LD is displayed according to standard color schemes, with *bright red* for very strong LD (LOD > 2, D′ = 1), *light red* (LOD > 2, D′ < 1) and *blue* (LOD < 2, D′ = 1) for intermediate LD, and *white* (LOD < 2, D′ < 1) for no LD

Table 5 Frequencies (%) of *AGTR* haplotypes [rs5182(573C>T), rs5186 (1166A>C)] and *IL10* haplotypes [rs1800871(−819 C>T), and rs1800896 (−1082 A>G)] in patient with and without restenosis

		With restenosis ($n = 99$)	Without restenosis ($n = 269$)	OR	95 % CI	*P* value
Locus	Haplotype	Hf	Hf			
Block1	TA	0.634	0.669	–	–	NS
rs5182|rs5186	CA	0.162	0.221	–	–	NS
573C>T|1166A>C	CC	0.205	0.110	2.17	1.33–3.53	0.002
Block 2	AT	0.455	0.493	–	–	NS
rs1800871|rs1800896	GC	0.340	0.238	1.51	1.01–2.26	0.04
819 C>T|−1082 A>G	GT	0.198	0.266	–	–	NS

blood viscosity and platelet aggregation [31]. Polymorphisms of *FGB* was shown to be associated with coronary heart disease [32, 33]. In a study by Völzke in 2004, there was no association between the β-fibrinogen −455G/A and the risk of restenosis after PTCA or recurrent restenosis after re-PTCA [34]. In the GENDER study, a multicenter prospective study, the association of gene polymorphism *FGB* −455G>A(rs1800790) with the risk of restenosis was not found in the European population [9]. In the study by Oguri in 2007, the association of *FGB* −455G>A (rs1800790) with restenosis was shown in the Japanese population [35]. Our results suggest that genotype *FGB* −455G>A (rs1800790) is significantly associated with restenosis after stenting (OR = 2.924, $P = 2.3E-06$, additive model) in the Kazakh population. Interestingly that the same SNP in dominant model gave close *P* value, but much higher OR = 24.36, suggesting that allele A is a recessive protective factor against development of restenosis. Thus, it can be concluded that the relationship between genetic polymorphisms of *FGB −455G>A* (rs1800790) and the development of restenosis is not universal between different ethnic groups. It is also interesting to observe that Japanese and Kazakhs are alike in turns of effect from polymorphisms of *FGB −455G>A* (rs1800790).

Nitric oxide synthase 3 (NOS3) locates at chromosome 7q36, and it encodes endothelial nitric oxide synthase (eNOS), which can generate nitric oxide (NO) in endothelial cells. Endothelial NO is a key determinant of vascular homeostasis, and it can also participate in vascular repair. Dysfunction of any of these processes may result in atherosclerotic and thrombotic diseases [36–38]. The association between several polymorphisms of the *NOS3* gene and CAD and restenosis risks has been previously studied [39–41]. In the meta-analysis by Zhang et al., 18 case-control studies with 2994 cases and 3130 controls, including 13 studies of East-Asia descendents, 5 studies of non-East-Asian descendents indicated that *eNOS* 894G/T polymorphism may play an important role in CHD development among Asian population [42]. In 2010, Li et al. performed a meta-analysis involving 20 studies relating non-Asian population and 3 studies relating Asian population and found significant association of T allele in *eNOS* 894G/T with CHD in non-Asian population [43]. Other studies showed that homozygosity of Glu298Asp and 786T/C polymorphisms of the *NOS3* gene represented an independent risk factor for in-stent restenosis, and the 894G/T polymorphism of *NOS3* gene was associated with an increased risk of death and/or myocardial infarction within 1 year after stent placement [24, 44, 45]. Shuvalova et al. showed that minor allele of polymorphism 298G/T of the *eNOS* gene (rs1799983) is associated with an increased risk of in-stent restenosis [46].

A number of polymorphisms have been identified in the *NOS3* gene among which two polymorphisms in the promoter region (−786C/T), and one in the exon (894G/T or Glu298Asp) was studied. Only one of these polymorphisms *NOS3* (rs1799983) has been found to be significantly associated with restenosis in our study OR = 20.05, $P = 2.74E-12$ in additive and OR = 22.24, $P = 6.81E-10$ in recessive models. Thus, the present results demonstrated that there might be a significant association between the *NOS3* polymorphism (rs1799983) and restenosis after PCI in the Kazakh population.

In the linkage disequilibrium analysis, *AGTR1* risk (CC) haplotype and *IL10* (GC) for developing restenosis were detected in our study. Su et al. studied 16 *AGTR1* polymorphisms. Based on the linkage disequilibrium pattern among these SNPs, six polymorphisms were selected as haplotype tagging SNPs and further were genotyped. SNP analyses indicated that GTC haplotype (rs275650, rs2276736, rs5182) associated with the risk of developing myocardial infarction [47]. Koch et al. investigated the possibility that single-nucleotide polymorphisms of the genes encoding *TNF* (_863C/A, _308G/A), *LT-a* (252G/A), and *IL10* (_1082G/A, _819C/T, and _592C/A) are associated with the incidence of restenosis, death, or myocardial infarction (MI) after coronary stenting [48]. With regard to the *IL10* polymorphisms, they observed three different haplotypes, _1082G/_819C/_592C (GCC), ACC, and ATA, with relative frequencies of 0.45, 0.29, and 0.26, respectively. Koch et al. have not detected a significant correlation between restenosis and the frequency of the haplotypes [48]. But in our study, it was identified that GC haplotype of *IL10* (819 C>T, −1082 A>G) was associated with restenosis. Thus, individuals with the risk haplotype have compromised function of inflammatory reactions.

Conclusion

In conclusion, the present study examined the association between 48 SNPs and restenosis in the Kazakh population. Our results indicate that BMI is a single confounding factor and *FGB* (rs1800790), *CD14* (rs2569190), and *NOS3* (rs1799983). SNPs could be genetic markers for the development of restenosis in the Kazakh population. Genotyping of these polymorphisms may be used in predicting the risk of restenosis in the Kazakh population.

Abbreviations

ANOVA, analysis of variance; BMI, body mass index; CHD, coronary heart disease; GENDER, GENetic DEterminants of Restenosis; HDL, high-density lipoproteins; LDL, low-density lipoproteins, PCI, percutaneous coronary intervention; SNP, single-nucleotide polymorphism

Funding

This work was supported by a grant #0112PK00361 from the Ministry of Education and Science of the Republic of Kazakhstan.

Authors' contributions

EVZh and YMR conceived and designed the study, analyzed the data, interpreted the results, and wrote the paper. AMA and PVT carried out experimental work. AUD, OAV, and DZhT were involved in the clinical data collection and sampling YAT, GNK, and ANI performed the statistical analysis. All authors have read and approved the final manuscript.

Competing interests

All authors declare no support from any organization or the submitted work, no financial relationships with any organizations that might have an interest in the submitted work in the previous 3 years, and no other relationships or activities that could appear to have influenced the submitted work.

Consent for publication

Not applicable.

Author details

[1]National Center for Biotechnology, 13/5, KorgalzhinskoeHighway, Astana, Kazakhstan. [2]National Scientific Medical Research Center, 42 Abylai Khan Avenue, Astana, Kazakhstan. [3]Karaganda State Medical University, 40, Gogol Street, Karaganda, Kazakhstan. [4]School of Science and Technology, Nazarbayev University, 53 Kabanbay Batyr Ave, Astana, Kazakhstan. [5]Al-Farabi Kazakh National University, Almaty, Kazakhstan.

References

1. World Health Organization. Cardiovascular diseases. http://www.who.int/mediacentre/factsheets/fs317/en/. Accessed 9 February 2016.
2. Tardif J-C. Coronary artery disease in 2010. Eur Heart J. 2010;12(suppl C):C2–C10.
3. Medical statistics [In Russian]. http://www.medinfo.kz/medstat.jsp. Accessed 9 February 2016.
4. Sigwart U, Puel J, Mirkovitch V, Joffre F, Kappenberger L. Intravascular stents to prevent occlusion and restenosis after transluminal angioplasty. N Engl J Med. 1987;316(12):701–6.
5. Fischman D, Leon M, Baim D, Schatz R, Savage M, Penn I, Detre K, Veltri L, Ricci D, Nobuyoshi M, et al. A randomized comparison of coronary-stent placement and balloon angioplasty in the treatment of coronary artery disease. Stent Restenosis Study Investigators. N Engl J Med. 1994;331(8):496–501.
6. Spertus J, Nerella R, Kettlekamp R, House J, Marso S, Borkon A, Rumsfeld J. Risk of restenosis and health status outcomes for patients undergoing percutaneous coronary intervention versus coronary artery bypass graft surgery. Circulation. 2005;111(6):768–73.
7. Otsuki S, Sabate M. Drug-eluting stents and acute myocardial infarction: a lethal combination or friends? World J Cardiol. 2014;6(9):929–38.
8. Jukema J, Verschuren J, Ahmed T, Quax P. Restenosis after PCI. Part 1: pathophysiology and risk factors. Nat Rev Cardiol. 2011;9(1):53–62.
9. Verschuren J, Trompet S, Postmus I, Sampietro M, Heijmans B, Houwing-Duistermaat J, Slagboom P, Jukema J. Systematic testing of literature reported genetic variation associated with coronary restenosis: results of the GENDER Study. PLoS One. 2012;7(8):3.
10. Mercado N, Boersma E, Wijns W, Gersh B, Morillo C, de Valk V, van Es G, Grobbee D, Serruys P. Clinical and quantitative coronary angiographic predictors of coronary restenosis: a comparative analysis from the balloon-to-stent era. J Am Coll Cardiol. 2001;38(3):645–52.
11. Miller S, Dykes D, Polesky H. A simple salting out procedure for extracting DNA from human nucleated cells. Nucleic Acids Res. 1988;16(3):1215.
12. Boyle AP, Hong EL, Hariharan M, Cheng Y, Schaub MA, Kasowski M, Karczewski KJ, Park J, Hitz BC, Weng S, Cherry JM, Snyder M. Annotation of functional variation in personal genomes using RegulomeDB. Genome Res. 2012;22(9):1790–7.
13. The R project for statistical computing. https://www.r-project.org/. Accessed 9 February 2016.
14. Purcell S, Neale B, Todd-Brown K, Thomas L, Ferreira MA, Bender D, Maller J, Sklar P, de Bakker PI, Daly MJ, Sham PC. PLINK: a tool set for whole-genome association and population-based linkage analyses. Am J Hum Genet. 2007; 81(3):559–75.
15. Barrett JC, Fry B, Maller J, Daly MJ. Haploview: analysis and visualization of LD and haplotype maps. Bioinformatics (Oxford, England). 2005;21(2):263–5.
16. Comas D, Calafell F, Mateu E, Perez-Lezaun A, Bosch E, Martinez-Arias R, Clarimon J, Facchini F, Fiori G, Luiselli D, Pettener D, Bertranpetit J. Trading genes along the silk road: mtDNA sequences and the origin of central Asian populations. Am J Hum Genet. 1998;63(6):1824–38.
17. Kastrati A, Dirschinger J, Schomig A. Genetic risk factors and restenosis after percutaneous coronary interventions. Herz. 2000;25(1):34–46.
18. Ortlepp J, Hoffmann R, Killian A, Lauscher J, Merkelbach-Brese S, Hanrath P. The 4G/5G promotor polymorphism of the plasminogen activator inhibitor-1 gene and late lumen loss after coronary stent placement in smoking and nonsmoking patients. Clin Cardiol. 2001;24(9):585–91.
19. Toutouzas K, Colombo A, Stefanadis C. Inflammation and restenosis after percutaneous coronary interventions. Eur Heart J. 2004;25(19):1679–87.
20. Monraats P, Kurreeman F, Pons D, Sewgobind V, de Vries F, Zwinderman A, de Maat M, Doevendans P, de Winter R, Tio R, Waltenberger J, Huizinga T, Eefting D, Quax P, Frants R, van der Laarse A, van der Wall E, Jukema J. Interleukin 10: a new risk marker for the development of restenosis after percutaneous coronary intervention. Genes Immun. 2007;8(1):44–50.
21. Tashiro H, Shimokawa H, Sadamatsu K, Aoki T, Yamamoto K. Role of cytokines in the pathogenesis of restenosis after percutaneous transluminal coronary angioplasty. Coron Artery Dis. 2001;12(2):107–13.
22. Ryu S, Cho E, Park H, Im E, Jang Y, Shin G, Shim W, Cho S. Renin-angiotensin-aldosterone system (RAAS) gene polymorphism as a risk factor of coronary in-stent restenosis. Yonsei Med J. 2002;43(4):461–72.
23. Wijpkema JS, van Haelst PL, Monraats PS, Bruinenberg M, Zwinderman AH, Zijlstra F, van der Steege G, de Winter RJ, Doevendans PA, Waltenberger J, Jukema JW, Tio RA. Restenosis after percutaneous coronary intervention is associated with the angiotensin-II type-1 receptor 1166A/C polymorphism but not with polymorphisms of angiotensin-converting enzyme, angiotensin-II receptor, angiotensinogen or heme oxygenase-1. Pharmacogenet Genomics. 2006;16(5):331–7.
24. Gomma A, Elrayess M, Knight C, Hawe E, Fox K, Humphries S. The endothelial nitric oxide synthase (Glu298Asp and -786 T > C) gene polymorphisms are associated with coronary in-stent restenosis. Eur Heart J. 2002;23(24):1955–62.
25. Monraats P, Fang Y, Pons D, Pires N, Pols H, Zwinderman A, de Maat M, Doevendans P, DeWinter R, Tio R, Waltenberger J, Frants R, Quax P, van der Laarse A, van der Wall E, Uitterlinden A, Jukema J. Vitamin D receptor: a new risk marker for clinical restenosis after percutaneous coronary intervention. Expert Opin Ther Targets. 2010;14(3):243–51.

26. Fragoso JM, Zuniga-Ramos J, Arellano-Gonzalez M, Alvarez-Leon E, Villegas-Torres BE, Cruz-Lagunas A, Delgadillo-Rodriguez H, Pena-Duque MA, Martinez-Rios MA, Vargas-Alarcon G. The T29C (rs1800470) polymorphism of the transforming growth factor-beta1 (TGF-beta1) gene is associated with restenosis after coronary stenting in Mexican patients. Exp Mol Pathol. 2015;98(1):13–7.
27. Shimada K, Miyauchi K, Mokuno H, Watanabe Y, Iwama Y, Shigekiyo M, Matsumoto M, Okazaki S, Tanimoto K, Kurata T, Sato H, Daida H. Promoter polymorphism in the CD14 gene and concentration of soluble CD14 in patients with in-stent restenosis after elective coronary stenting. Int J Cardiol. 2004;94(1):87–92.
28. Zhang HF, Zhong BL, Zhu WL, Xie SL, Qiu LX, Zhu LG, Wang Y, Lei L. CD14 C-260 T gene polymorphism and ischemic heart disease susceptibility: a HuGE review and meta-analysis. Genet Med. 2009;11(6):403–8.
29. Zee RY, Hoh J, Cheng S, Reynolds R, Grow MA, Silbergleit A, Walker K, Steiner L, Zangenberg G, Fernandez-Ortiz A, Macaya C, Pintor E, Fernandez-Cruz A, Ott J, Lindpainter K. Multi-locus interactions predict risk for post-PTCA restenosis: an approach to the genetic analysis of common complex disease. Pharmacogenomics J. 2002;2(3):197–201.
30. Liu Y, Imanishi T, Ikejima H, Tsujioka H, Ozaki Y, Kuroi A, Okochi K, Ishibashi K, Tanimoto T, Ino Y, Kitabata H, Akasaka T. Association between circulating monocyte subsets and in-stent restenosis after coronary stent implantation in patients with ST-elevation myocardial infarction. Circ J. 2010;74(12):2585–91.
31. Scott EM, Ariens RA, Grant PJ. Genetic and environmental determinants of fibrin structure and function: relevance to clinical disease. Arterioscler Thromb Vasc Biol. 2004;24(9):1558–66.
32. Reiner AP, Carty CL, Carlson CS, Wan JY, Rieder MJ, Smith JD, Rice K, Fornage M, Jaquish CE, Williams OD, Tracy RP, Lewis CE, Siscovick DS, Boerwinkle E, Nickerson DA. Association between patterns of nucleotide variation across the three fibrinogen genes and plasma fibrinogen levels: the Coronary Artery Risk Development in Young Adults (CARDIA) study. J Thromb Haemost. 2006;4(6):1279–87.
33. Martiskainen M, Pohjasvaara T, Mikkelsson J, Mantyla R, Kunnas T, Laippala P, Ilveskoski E, Kaste M, Karhunen PJ, Erkinjuntti T. Fibrinogen gene promoter -455 A allele as a risk factor for lacunar stroke. Stroke. 2003;34(4):886–91.
34. Volzke H, Grimm R, Robinson DM, Wolff B, Schwahn C, Hertwig S, Motz W, Rettig R. Candidate genetic markers and the risk of restenosis after coronary angioplasty. Clin Sci. 2004;106(1):35–42.
35. Oguri M, Kato K, Hibino T, Yokoi K, Segawa T, Matsuo H, Watanabe S, Nozawa Y, Murohara T, Yamada Y. Identification of a polymorphism of UCP3 associated with recurrent in-stent restenosis of coronary arteries. Int J Mol Med. 2007;20(4):533–8.
36. Voetsch B, Jin RC, Loscalzo J. Nitric oxide insufficiency and atherothrombosis. Histochem Cell Biol. 2004;122(4):353–67.
37. Palmer RM, Ashton DS, Moncada S. Vascular endothelial cells synthesize nitric oxide from L-arginine. Nature. 1988;333(6174):664–6.
38. Radomski MW, Palmer RM, Moncada S. The anti-aggregating properties of vascular endothelium: interactions between prostacyclin and nitric oxide. Br J Pharmacol. 1987;92(3):639–46.
39. Galluccio E, Piatti P, Citterio L, Lucotti PC, Setola E, Cassina L, Oldani M, Zavaroni I, Bosi E, Colombo A, Alfieri O, Casari G, Reaven GM, Monti LD. Hyperinsulinemia and impaired leptin-adiponectin ratio associate with endothelial nitric oxide synthase polymorphisms in subjects with in-stent restenosis. Am J Physiol Endocrinol Metab. 2008;294(5):E978–986.
40. Antoniades C, Tousoulis D, Vasiliadou C, Pitsavos C, Chrysochoou C, Panagiotakos D, Tentolouris C, Marinou K, Koumallos N, Stefanadis C. Genetic polymorphism on endothelial nitric oxide synthase affects endothelial activation and inflammatory response during the acute phase of myocardial infarction. J Am Coll Cardiol. 2005;46(6):1101–9.
41. Cam SF, Sekuri C, Tengiz I, Ercan E, Sagcan A, Akin M, Berdeli A. The G894T polymorphism on endothelial nitric oxide synthase gene is associated with premature coronary artery disease in a Turkish population. Thromb Res. 2005;116(4):287–92.
42. Zhang K, Bai P, Shi S, Zhou B, Wang Y, Song Y, Rao L, Zhang L. The G894T polymorphism on endothelial nitric oxide synthase gene is associated with increased coronary heart disease among Asia population: evidence from a Meta analysis. Thromb Res. 2012;130(2):192–7.
43. Li J, Wu X, Li X, Feng G, He L, Shi Y. The endothelial nitric oxide synthase gene is associated with coronary artery disease: a meta-analysis. Cardiology. 2010;116(4):271–8.
44. Suzuki T, Okumura K, Sone T, Kosokabe T, Tsuboi H, Kondo J, Mukawa H, Kamiya H, Tomida T, Imai H, Matsui H, Hayakawa T. The Glu298Asp polymorphism in endothelial nitric oxide synthase gene is associated with coronary in-stent restenosis. Int J Cardiol. 2002;86(1):71–6.
45. Gorchakova O, Koch W, von Beckerath N, Mehilli J, Schomig A, Kastrati A. Association of a genetic variant of endothelial nitric oxide synthase with the 1 year clinical outcome after coronary stent placement. Eur Heart J. 2003;24(9):820–7.
46. Shuvalova YA, Kaminnyi AI, Meshkov AN, Shirokov RO, Samko AN. Association between polymorphisms of eNOS and GPx-1 genes, activity of free-radical processes and in-stent restenosis. Mol Cell Biochem. 2012;370(1-2):241–9.
47. Su S, Chen J, Zhao J, Huang J, Wang X, Chen R, Gu D. Angiotensin II type I receptor gene and myocardial infarction: tagging SNPs and haplotype based association study. The Beijing atherosclerosis study. Pharmacogenetics. 2004;14(10):673–81.
48. Koch W, Tiroch K, von Beckerath N, Schomig A, Kastrati A. Tumor necrosis factor-alpha, lymphotoxin-alpha, and interleukin-10 gene polymorphisms and restenosis after coronary artery stenting. Cytokine. 2003;24(4):161–71.

5

Targeted next generation sequencing identifies novel NOTCH3 gene mutations in CADASIL diagnostics patients

Neven Maksemous, Robert A. Smith, Larisa M. Haupt and Lyn R. Griffiths*

Abstract

Background: Cerebral autosomal dominant arteriopathy with subcortical infarcts and leukoencephalopathy (CADASIL) is a monogenic, hereditary, small vessel disease of the brain causing stroke and vascular dementia in adults. CADASIL has previously been shown to be caused by varying mutations in the *NOTCH3* gene. The disorder is often misdiagnosed due to its significant clinical heterogeneic manifestation with familial hemiplegic migraine and several ataxia disorders as well as the location of the currently identified causative mutations. The aim of this study was to develop a new, comprehensive and efficient single assay strategy for complete molecular diagnosis of *NOTCH3* mutations through the use of a custom next-generation sequencing (NGS) panel for improved routine clinical molecular diagnostic testing.

Results: Our custom NGS panel identified nine genetic variants in *NOTCH3* (p.D139V, p.C183R, p.R332C, p.Y465C, p. C597W, p.R607H, p.E813E, p.C977G and p.Y1106C). Six mutations were stereotypical CADASIL mutations leading to an odd number of cysteine residues in one of the 34 *NOTCH3* gene epidermal growth factor (EGF)-like repeats, including three new typical cysteine mutations identified in exon 11 (p.C597W; c.1791C>G); exon 18 (p.C977G; c. 2929T>G) and exon 20 (p.Y1106C; c.3317A>G). Interestingly, a novel missense mutation in the *CACNA1A* gene was also identified in one CADASIL patient. All variants identified (novel and known) were further investigated using in silico bioinformatic analyses and confirmed through Sanger sequencing.

Conclusions: NGS provides an improved and effective methodology for the diagnosis of CADASIL. The NGS approach reduced time and cost for comprehensive genetic diagnosis, placing genetic diagnostic testing within reach of more patients.

Keywords: AmpliSeq Custom Panel, CADASIL, Next-generation sequencing, *NOTCH3*

Background

The stroke syndrome CADASIL [MIM 125310] (cerebral autosomal dominant arteriopathy with subcortical infarcts and leukoencephalopathy) disorder results in neuronal white matter abnormalities and is characterised by a variety of symptoms including, vascular degeneration, recurrent subcortical ischaemic strokes, progressive cognitive decline, dementia, migraine with aura (22 % of patients) and premature death [1]. The unique deposition of granular osmiophilic material (GOM) in systemic and brain vasculature differentiates CADASIL patients from those suffering similar hereditary vascular disorders [2]. CADASIL is often misdiagnosed due to its significant clinically heterogeneic manifestation with familial hemiplegic migraine and several ataxia disorders, as these disorders have an autosomal dominant mode of inheritance and share clinical characteristics such as hemiplegic migraine, migraine with typical aura and progressive ataxia [3–6]. Mutations implicated in CADASIL have been identified on chromosome 19, specifically within *NOTCH3* (MIM 600276), which encodes a transmembrane receptor primarily expressed in vascular smooth muscle cells. *NOTCH3* located at 19p13 is 33 exons long and spans approximately 7 kb [4]. Currently, at least 200 mutations resulting in an odd number of cysteine residues are known to be associated with CADASIL. These

* Correspondence: lyn.griffiths@qut.edu.au
Genomics Research Centre, Institute of Health and Biomedical Innovation (IHBI), School of Biomedical Sciences, Queensland University of Technology (QUT), Q Block, 60 Musk Ave, Kelvin Grove Campus, Brisbane 4059, Queensland, Australia

mutations all occur in exons 2–24 of *NOTCH3* that encode 34 epidermal growth factor (EGF)-like repeats in the extracellular domain of the NOTCH3 protein. The large number of exons combined with their high GC content makes comprehensive sequencing of this gene with traditional Sanger sequencing (SS) expensive and time consuming. With the advent of next-generation sequencing (NGS), the sequencing of target genes, regions, exomes or whole genomes, provides cost-effective, high-throughput screening suitable for molecular diagnostics enabling detection of a wide array of mutations with sensitivity and specificity. Here, we have performed targeted gene sequencing using a custom five-gene NGS panel [7], encompassing the coding sequences, 20–100 bp exon/intron boundaries and the 5′ and 3′UTR regions of *NOTCH3* in 44 patients.

Results

NGS-panel sequencing output

The sequencing output data from the Ion Torrent PGM was analysed using the Ion Torrent platform-specific software Torrent Suite V3.6 (Thermo Fisher Scientific, Scoresby, Victoria, Australia). The 44 samples were sequenced using seven different Ion 316 chips, to generate an average sequencing of 3,303,300 total reads, 477.4 Mb total bases sequenced, and 472.2 Mb with 99 % of bases aligned to the human complete genome (hg19) per Ion 316 chip. For all samples sequenced, the average read depth across the target region was 560.65×, while the average percentage of target bases covered at 20× or greater was 96 % and the average uniformity of coverage was 90.64 %.

Sequencing data analysis

Comprehensive screening for *NOTCH3* using the AmpliSeq Custom NGS panel [7] (Thermo Fisher Scientific, Scoresby, Victoria, Australia) for targeted gene sequencing was conducted on 44 patients, previously screened for standard sequencing exons (3 and 4) and/or (2,11, 18 and 19) by SS and classified as being negative for known mutations.

Initial analysis using the IonReporter software (Thermo Fisher Scientific, Scoresby, Victoria, Australia) identified 42 variants scattered over *NOTCH3* among the 44 patients. An overview of all variants detected in our study is shown in Additional file 1: Table S1 online. Of these, nine particularly notable genetic variants were identified in 10 patients (22.7 %) out of the 44 subjects: five novel potential mutations (NOTCH3:NM_000435:exon4:c.416A>T:p.D139V, exonl 1:c.1791C>G:p.C597W and c.1820G>A: p.R607H, exon18: c.2929T>G,: p.C977G; and exon20: c.3317A>G: p.Y1106C); three previously reported disease-causing missense mutations (NOTCH3:NM_000435:exon4:c.547T>Cp.C183R [8], exon 6:c.994C>T, p.R332C [9, 10]; and exon9: c.1394A>G: p.Y465R [11]) and one novel synonymous genetic variant in NOTCH3:NM_000435:exon16:c.2439G>A: p.E813E [Tables 1 and 2]. Clinical information for all samples was not available [see Additional file 2: Table S2]; however, the following clinical parameters were attributed to the relevant samples in our cohort: (i) white matter abnormalities were seen in patients with E813E, C183R and R332C mutations; (ii) positive skin biopsy signs were reported in patients with C977G, Y1106C, Y465C and C183R and (iii) a family history of dementia and/or stroke was reported in patients with E813E, C977G, Y1106C, C183R, R332C and R465C mutations.

Molecular genetic testing using the custom NGS panel encompassing five genes (*NOTCH3, CACNA1A, ATP1A2, SCN1A* and *TRESK* genes) identified two remarkable variants in case C-36 (CACNA1A:NM_023035:c.832G>T:-p.A278S and SCN1A:NM_006920:c.3924A>T: p.E1308D [Table 1]). These mutations correspond to highly conserved amino acid residues according to four in silico prediction tools (PhyloP of score of >2.0, PolyPhen2 HVar of score >0.7, and MutationTaster with a damaging effect and GERP++ score above 5).

In addition to these variants, nine rare single nucleotide polymorphisms (SNPs) in the *NOTCH3* gene with minor allele frequency (MAF) ≤0.1 % were observed in nine patients with no other causative mutation found in *NOTCH3* [Additional file 1: Table S1 online and Table 3]. One patient (case C-3) was shown to carry two rare amino acid changing variants p.S497L and p.A1020P in exons 9 and 19 of *NOTCH3*, respectively. All nine SNPs were further assessed by seven in silico prediction programmes with three of these variants (p.S497L, p.P496L and p.Y220Y) shown to have a damaging effect by MutationTaster.

All variants detected by NGS and reported in this study were visually confirmed using Integrative Genomics Viewer (IGV v2.3) software [12] and compared with NCBI reference sequences [13]. In order to verify the accuracy of potential novel mutations identified by NGS, SS was performed for all samples with the five non-synonymous variants along with the synonymous new variant showing complete consistency (100 %) between the two methods [Fig. 1].

Analyses of the potential functional significance of the six novel *NOTCH3* genetic variants identified the C597W, C977G and Y1106C missense mutations to be pathogenic by six of the seven genetic prediction software programmes (PhyloP, SIFT, PolyPhen2, MutationTaster, AGVGD and PhD-SNP) [Table 1].

Finally, we compared the potential functional significance of the three known pathogenic missense mutations (C183R, R332C and Y465C) with the six novel genomic variants identified using the same seven in silico software programmes. The C183R and R332C mutations showed a high potential damaging effect when analysed by all seven programmes used; in contrast, the

Table 1 Variants of unknown significance of *NOTCH3*, *CACNA1A* and *SCN1A* genes identified in eight suspicious CADASIL patients. RefSeq NM_000435, NM_001127221 and NM_001165963

Sample ID	Gene	Gender	Age	Exon	EGF-repeat	Codon change (FWD)	Protein change	PhyloP	SIFT	PolyPhen2 HVar	MutationTaster	GERP++	AGVGD	PhD-SNP
C-11	NOTCH3	F	42	4	3	c.416A>T	p.Asp139Val	C (1.89)	T (0.06)	P (0.499)	D	5.02	C65	Non-neutral
C-4	NOTCH3	M	67	11	15	c.1791C>G	p.Cys597Trp	N (0.434)	D (0)	D (1.0)	D	2.25	C65	Non-neutral
C-15	NOTCH3	F	52	11	15	c.1820G>A	p.Arg607His	C (2.21)	T (0.54)	B (0.0.026)	D	3.22	C25	Neutral
C-24	NOTCH3	M	54	16	21	c.2439G>A	p.Glu813Glu	–	–	–	D (splice site changes)	–	–	–
C-10 and C-44	NOTCH3	F	74, 52	18	25	c.2929T>G	p.Cys977Gly	C (2.04)	D (0)	D (1.0)	D	5.36	C65	Non-neutral
C-6	NOTCH3	F	51	20	28	c.3317A>G	p.Tyr1106Cys	C (1.92)	D (0)	D (0.998)	D	5.08	C65	Non-neutral
C-36	CACNA1A	M	60	6	–	c.832G>T	p.Ala278Ser	C (2.46)	T (0.13)	D (0.963)	D	5.27	NA	Neutral
C-36	SCN1A	M	60	20	–	c.3924A>T	p.Glu1308Asp	C (2.08)	T (0.37)	P (0.727)	D	5.46	NA	Neutral

PhyloP, SIFT, Polyphen-2, MutationTaster, GERP++, AGVGD and PhD-SNP are functional prediction scores in which increasing values indicate a more damaging effect except SIFT score <0.05 has damaging effect
Abbreviations: *C* conserved, *N* not-conserved or neutral, *D* damaging or deleterious, *P* possible damaging, *T* tolerated, *B* benign, *NA* not applicable

Table 2 Variants of known *NOTCH3* mutations identified in three patients by NGS. RefSeq NM_000435.2

Sample ID	Gender	Age	Exon	EGF-repeat	Codon change	Protein change	PhyloP	SIFT	PolyPhen2 HVar	MutationTaster	GERP++	AGVGD	PhD-SNP	Snp138
C-20	F	54	4	4	c.547T>C	p.Cys183Arg	C (1.82)	D (0)	D (1)	D	4.32	C65	Non-neutral	
C-42	F	47	6	8	c.994C>T	p.Arg332Cys	C (2.46)	D (0.03)	D (1)	D	4.6	C65	Non-neutral	rs137852641
C-14	M	33	9	11	c.1394A>G	p.Tyr465Cys	N (−0.833)	T (0.08)	P (0.886)	D	−4.41	C65	Non-neutral	

PhyloP, SIFT, Polyphen-2, MutationTaster, GERP++, AGVGD and PhD-SNP are functional prediction scores in which increasing values indicate a more damaging effect except SIFT score <0.05 has damaging effect
Abbreviations: *C* conserved, *N* not-conserved or neutral, *D* damaging or deleterious, *P* possible damaging, *T* tolerated

Table 3 Rare variants identified by NGS in the *NOTCH3* gene. RefSeq NM_000435.2

Patient ID	Locus.	Ref	Location	Codon change	Protein change	PhyloP	SIFT	PolyPhen	LRT	MutationTaster	GERP++	AGVGD	PhD-SNP	dbSNP	MAF
C31	chr19:15281342	T	Ex27	c.4914A>G	WT (p.Glu1638Glu)	0.25				Poly				rs149222385	0.001
C13, C28, C33	chr19:15290236	G	Ex21	c.3399C>A	p.His1133Gln	−2.55	D	P (0.68)	NA	Poly	−8.6	C15	Neutral	rs112197217	0.005
C3, C12	chr19:15291576	C	Ex19	c.3058G>C	p.Ala1020Pro	0.57	T (0.19)	B (0.054)	N (0.006)	Poly	1.44	C25	Neutral	rs35769976	0.083
C3	chr19:15296513	C	In12			0.5				Poly				rs147014533	0.006
C3	chr19:15299048	G	Ex9	c.1490C>T	p.Ser497Leu	2.35	T (0.33)	B (0.036)	N (0.018)	D	5.04	C65	Neutral	rs114207045	0.006
C34	chr19:15299051	G	Ex9	c.1487C>T	p.Pro496Leu	2.35	T (0.1)	p (0.883)	N (0.007)	D	5.04	C65	Disease	rs11670799	0.005
C9	chr19:15302790	G	Ex4	c.660C>T	WT (p.Tyr220Tyr)	−1.85				D				rs114457076	0.001
C3	chr19:15308287	G	In2			−0.16				Poly				rs188132716	0.006
C16	chr19:15308288	G	In2			−0.98				Poly				rs202151374	0.003

PhyloP, SIFT, Polyphen-2, Mutation Taster, GERP++, AGVGD, and PhD-SNP are functional prediction scores in which increasing values indicate a more damaging effect except SIFT score <0.05 has damaging effect
Abbreviations: *B* benign, *C* conserved, *D* damaging or deleterious, *Ex* exon, *In* intron, *NA* not applicable, *N* not-conserved or neutral, *P* possible damaging, *Poly* polymorphic, *T* tolerated, *WT* wild type

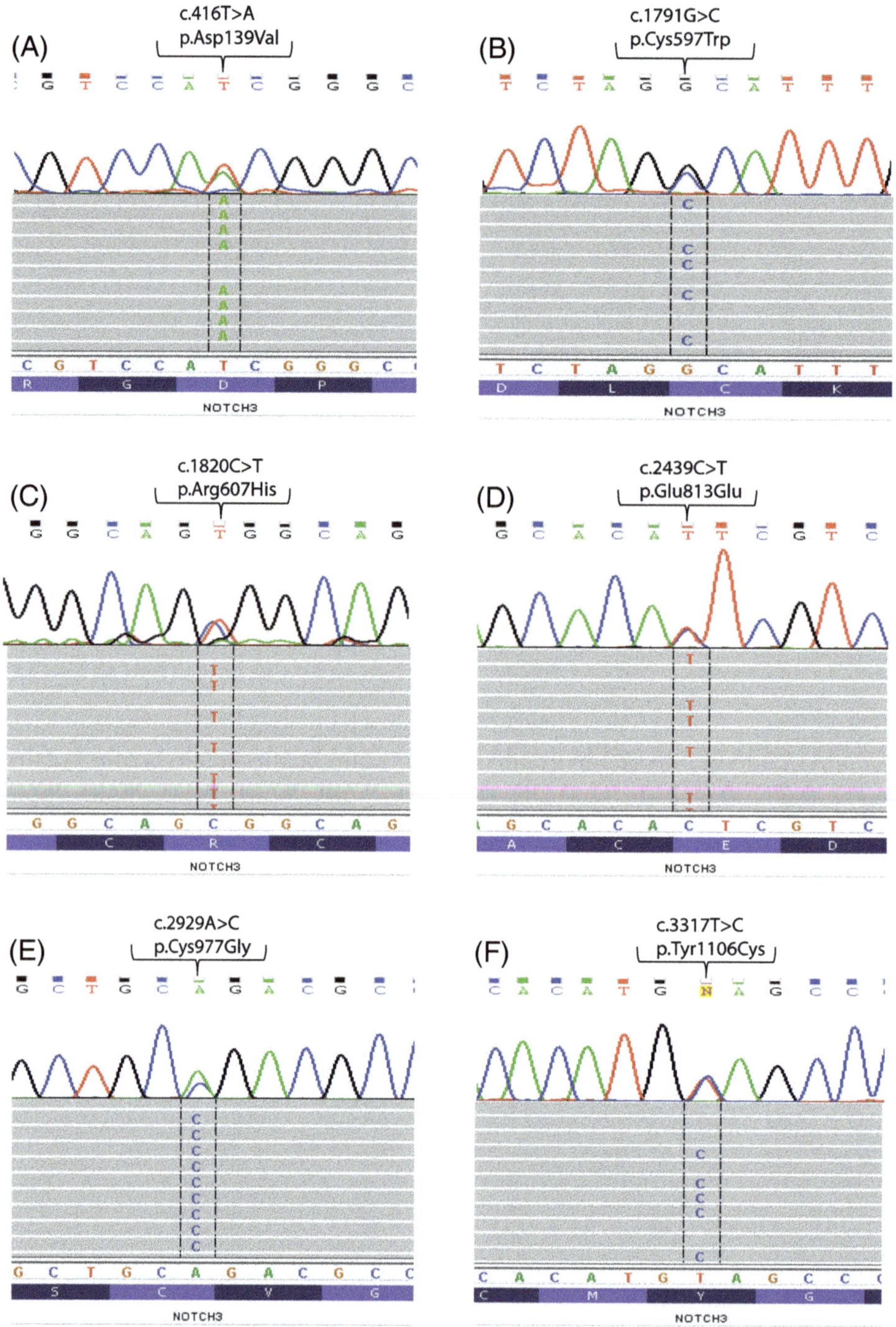

Fig. 1 Sequences (reverse complement) of the six novel genetic variants in *NOTCH3* identified by NGS. The figure shows the six heterozygous exonic point variants: **a** p.D139V in exon 4, **b** p.C597W in exon 11, **c** p.R607H in exon 11, **d** synonymous variant p.E813E in exon 16, **e** p.C977G in exon 18 and **f** p.Y1106C in exon 20 identified in this study. Only bases non-concordant with consensus sequence are displayed in the target reads with the integrative genomics viewer IGV [12]. The normal nucleotide and protein sequences are depicted at the *bottom* and *top* of the figure

Y465C mutation showed a tolerated or benign effect in four of the seven in silico programmes [Table 2].

Discussion

Molecular genetic testing is an essential tool for accurate CADASIL diagnosis. Several diagnostic approaches have been used for CADASIL, in particular the use of skin biopsy to detect unusual NOTCH3 expression. However, despite the widespread use of biopsy testing, the low sensitivity of this method in CADASIL diagnosis has been reported [14]. In addition, previous work by Markus et al. tested the sensitivity of single strand conformation polymorphism (SSCP) analysis for detecting *NOTCH3* mutations, with an effective success rate of 80 to 85 % [14]. More recently, He et al. reported that varying and population-dependant results in the effectiveness of using the pre-genetic "CADASIL scale" screening tool which evaluates clinical presentations and neuroimaging data in an effort to minimise *NOTCH3* gene testing [15, 16]. As such, current diagnosis relies on the screening of all exons by sequencing to identify mutations in *NOTCH3*.

We have previously demonstrated the efficiency of our NGS panel for detecting known and novel mutations in a cohort of episodic ataxia patients and increasing the rate of mutation detection by 48 % [7]. We have now utilised this custom targeted massively parallel NGS panel to examine the coding sequences, intron/exon boundaries including 20–100 bases of flanking intronic nucleotides and the 5′ and 3′UTR regions of *NOTCH3* in a cohort of 44 patients with clinically suspicious CADASIL.

Targeted gene sequencing analysis efficiently identified nine novel genetic variants in *NOTCH3*, of which five non-synonymous mutations (p.D139V, p.C597W, p.R607H, p.C977G and p.Y1106C) and one synonymous variant (p.E813E) have not been previously described. In addition, three missense mutations previously reported as pathogenic (C183R [8], R332C [9, 10] and Y465C [11]) but not previously identified in our diagnostics cohort were also detected [17]. In total, six typical CADASIL mutations involving cysteine alterations were identified in seven patients (15.9 %) out of 44 subjects, a detection rate higher than previously reported by Fernandez et al. and Bianchi et al. [18, 19].

Interestingly, previous studies have revealed differences in the spectrum of *NOTCH3* mutations between Asian and Italian populations and populations of Caucasian ethnicity [18, 20, 21]. Our results also showed no evidence of strong clustering of *NOTCH3* mutations in specific exons. The variants identified in this study occur in seven different exons (4, 6, 9, 11, 16, 18 and 20) within the EGF-like repeat regions of the gene. The patient cohort encompasses different ethnic backgrounds, reflecting the diversity of the Australian population. This highlights the potential confounding factor in nations of multiple ethnic backgrounds, where mutations may occur at multiple sites making molecular diagnosis difficult and time consuming if using traditional SS methodologies. While exonic clustering in ethnic groups is likely due to founder effects, de novo mutations resulting in mutations in ethnically homogenous populations are possible. In this instance, the use of SS may still miss mutations in a proportion of patients suggesting that screening of all coding regions in *NOTCH3* is of benefit for the comprehensive molecular diagnosis of CADASIL.

Six of the missense mutations identified were stereotypical CADASIL mutations, resulting in a loss or gain of one of the six cysteine residues (4, 8, 11, 15, 25 and 28) of the EGF-like repeats located in the extracellular domain of *NOTCH3* [4]. Any mutation within the cysteine residues (a gain or loss) leads to an odd number of cysteine residues and result in impaired dimerisation of NOTCH3 or formation of inappropriate disulphide bonds causing aberrant NOTCH3 signaling [22, 23]. As such, these three mutations (p.C597W, p.C977G and p.Y1106C) were considered to be disease-causing and associated with the pure and typical pathogenetic mechanisms of CADASIL [4, 24]. The substitution of the p.C597S has been previously identified in an Arabic family [25], while the substitution of p.C977S has been reported in a Chinese patient [26] with both mutations found to be associated with CADASIL pathogenesis.

We also observed two novel amino acid substitutions (p.D139V and p.R607H) not directly involving cysteine residues, predicted to be possibly damaging and benign, respectively. As discussed by Roy et al. [17], there is some controversy over the classification non-cysteine residue altering variants and their significance to CADASIL. Several *NOTCH3* alterations that do not affect cysteine residues have been reported in families with CADASIL, which may involve other disruptions to protein function, though these may result in changes that effectively change cysteine residue availability [27–32]. It is worth noting that the predicted score for p.D139V by the SIFT programme (0.06) was more deleterious than the known pathogenic mutation p.Y465C, with a score of 0.08, technically considered to be benign. A SIFT score from 0 to 0.05 indicates that the amino acid change has a damaging effect. Further investigation of this mutation is warranted to determine the effect of these non-cysteine affecting changes on NOTCH3 function as well as on mediating signal transduction for vascular development and inducing the pathology of CADASIL. This provides new insights into the diagnosis of and pathomechanisms causing CADASIL.

The last novel synonymous variant we identified (p.E813E) was predicted by the MutationTaster programme to cause the gain of an RNA splicing donor site. This gain may result in altered protein function and therefore, despite being silent, this variant could be a real mutation causing CADASIL. Direct functional evaluation of NOTCH3 in this patient is needed to confirm this hypothesis, but such studies were not able to be performed at this time.

The in silico analysis tools to analyse the detected variants also revealed some interesting potential ramifications of the previously identified p.Y465C mutation. In 2003, Razvi et al. described this amino acid substitution as a mutation causing CADASIL [11]. In contrast, during our analysis, the computational tools predicted this amino acid change as tolerated or benign. The PhyloP score of 0.0272 and SIFT score of 0.08 (damaging score <0.05) suggest this amino acid is not conserved. The mutation is a classical CADASIL mutation; however, as stated by Joutel et al. "mutations can be unambiguously classified as pathogenic when they lead to an uneven number of cysteine residues in one of the 34 EGFR domains constituting the extracellular domain of the receptor" [24]. This discrepancy between evolutionary conservation and functional correlation models suggests caution when using functional prediction software in assigning a role to missense mutations involving cysteine residues in *NOTCH3*. The in vivo effect of amino acid substitutions should be the final arbiter for precisely describing their role in causing CADASIL, but as such tests are laborious to undertake, they are rarely performed for diagnoses. Careful consideration of the symptomatic profile may be useful in such cases and in the future when sufficient mutation data has accumulated offering clinicians more precision in ascribing the functional role of mutations in CADASIL.

Interestingly, in this study, patient C-36 demonstrated compound heterozygosity for two missense mutations in the *CACNA1A* and *SCN1A* genes (not normal target genes for CADASIL screening) [Table 1]. Mutations within these two ion channel genes are associated with various autosomal dominant disorders: hemiplegic migraine, episodic ataxia type 2, spinocerebellar ataxia type 6 and epilepsy with previously reported overlapping symptoms among these disorders [33–35]. It is worth noting that the p.E1297D mutation in *SCN1A* gene was previously reported in an Italian family with idiopathic childhood epilepsy [36]. The linkage between *CACNA1A* and *SCN1A* gene mutations and CADASIL has not previously been reported; therefore, an ongoing study in our lab will investigate the effect of these two variants/genes on CADASIL disease pathophysiology.

In terms of the clinical classification of the detected genetic variants, the full available evidence needs to be considered. Typical CADASIL mutations involve the addition or elimination of a cysteine residue in one of the 34 NOTCH3 gene epidermal growth factor (EGF)-like repeats, resulting in mismatched disulphide bridging and altered protein function, a hypothesis which has been borne out by observational and functional studies [37, 38]. Under the current American College of Medical Genetics and Genomics (ACMG) guidelines for variant classification, functional studies supporting a damaging effect for a variant on gene function constitute strong evidence for pathogenicity. Each of the cysteine altering variants also has multiple moderate and supporting lines of evidence. These include presence in a disease-associated functional domain; presence at a loci where another pathogenic mutation is known (as determined by searching HGMD, LOVD and VEP databases); absence from controls in population databases (1000 Genomes, dbSNP, ExAC); being the kind of variant (missense SNV) associated with the disease; presence in an individual with a clear phenotype; cosegregation with disease in family members (only for patients C-10 and C-44) and multiple in silico analyses predicting pathogenicity. This combination of evidence is sufficient to characterise them as pathogenic or disease-causing mutations according to the ACMG guidelines [39].

For the non-cysteine altering *NOTCH3* variants, there is less information available. Family segregation analysis and clinical information were not available for patients C-11 (p.D39V) or C-15 (p. R607H). Despite being novel amino acid changing variants in loci where disease-causing mutations are known to exist and/or functional domains, there is insufficient strength of evidence to classify either variant as pathogenic or likely pathogenic. Additionally, both these patients had complex phenotypes that do not precisely map to CADASIL, and share features of episodic ataxia or familial hemiplegic migraine, indicating a potential overlapping pathophysiology or comorbidities with these disorders. Thus, these variants should be classified as variants of uncertain significance (VOUS) according to the ACMG guidelines. Patient C-24 with the synonymous variant (p.E318E) had family history indicative of CADASIL, but no other supporting evidence, though neither does the variant have any criteria for being classified as benign. This variant has also been classified as a VOUS.

Patient C-36, who bears variants in both *CACNA1A* and *SCN1A* also, had no family members available for further investigation. Neither variant has sufficient evidence to indicate direct pathogenicity, despite being in regions of these genes known to harbour disease-causing mutations. Additionally, their presence in a gene which causes symptoms overlapping with CADASIL indicates a possible complex pathophysiology that requires more research. Hence, these variants have been classified as VOUS [39].

Finally, nine rare variants were identified in nine patients with no other causative mutation in *NOTCH3* [Table 3]. Of these, three amino acid changing variants (p.S497L, p.A1020P and p.H1133Q) were recently reported by Abramycheva et al. [15] as normal polymorphisms in Russian CADASIL patients. However, in this study, patient C-3 was found to carry non-cysteine *NOTCH3* gene variants (p.S496L and p.A1020P). As yet, a comparison of the effect of these two non-cysteine variants on the pathogenic mechanisms of CADASIL or CADASIL-like phenotype [16] to a single non-cysteine variant on disease pathogenesis has not been functionally tested.

We have identified classical CADASIL-causing mutations as well as a number of amino acid changing variants that have uncertain causative effects on this disease. The study of a larger population cohort of cases including symptomatic detail will likely provide more clinical and molecular information about their impact as well as the potential effect of any rare SNPs. Most interestingly, our results indicate that there may be other CADASIL gene/genes yet to be identified for inclusion in future diagnostic arrays.

Conclusions

NGS technologies provide an effective method for CADASIL and related disease diagnosis. Sequencing large but targeted regions of interest of pooled DNA from multiple samples is a promising tool for the discovery of both known and novel variants associated with disease. Compared with traditional SS, the NGS platform provides increased accuracy along with reduced time and assay costs necessary to perform routine genetic diagnosis of CADASIL in ethnically heterogeneous populations, putting such testing within reach of more patients.

Materials and methods

Patients

Forty-four patients with a suspected clinical diagnosis of CADASIL were re-screened using the NGS approach. Patients referred to the Genomics Research Centre (GRC) diagnostic laboratory for CADASIL molecular testing through neurologists from Australia and New Zealand and showed no mutations when using SS in our standard exon sequencing (3 and 4) at the first stage and (2, 11, 18 and 19) second stage [17]. Re-sequencing of the 44 patients was based on the clinical information had provided (i.e. positive skin biopsy results for CADASIL or white matter changes in their MRI) indicating that CADASIL-causing mutations may be present.

Molecular analysis

Ion AmpliSeq custom panel design

The AmpliSeq design target used in this report comprised the coding exons, exon/intron junctions and UTR regions of the *NOTCH3* gene. The Ampliseq automated primer design tool (http://www.ampliseq.com) was used to design primers covering 92.79 % of the desired target area (8071 bp) aligned to the reference human genome (hg 19). The missing regions include a 175 bp region in exon 1 (position 15311617-15311792 on chromosome 19) and a 407 bp region in exon 24 (position 15288427-15288834 on chromosome 19). The remainder of the 33 exons in the *NOTCH3* gene were included at 100 % coverage.

Library preparation

Genomic DNA was previously purified from peripheral blood samples using standard extraction conditions using Qiagen QIAamp DNA Blood Midi Kits as recommended by the manufacturer. The Qubit dsDNA High Sensitivity (HS) Assay Kit (Thermo Fisher Scientific, Scoresby, Victoria, Australia) was used to ensure accuracy of DNA concentration input (10 ng/μL) to NGS library construction.

Library preparation was performed using the Ion AmpliSeq library kit 2.0 (Thermo Fisher Scientific, Scoresby, Victoria, Australia) according to the standard protocol (Cat. no. 4480441, Rev. 4.0). Briefly, for the multiplex PCR amplification, 10 ng of each genomic DNA sample was amplified using the optimised modification method generated in our laboratory allowing each primer pool to be amplified as a 5-μL reaction, rather than a 20-μL reaction (protocol is available upon request). This was performed using 1 μL of 5× Ion AmpliSeq HiFi Master Mix, 2.5 μL of 2× AmpliSeq Custom primer pool, 0.5 μL nuclease-free water and 1 μL (10 ng/μL) of DNA. The reaction mix was heated for 2 min at 99 °C for enzyme activation, followed by 18 two-step cycles of 99 °C for 15 s and 60 °C for 4 min, ending with a holding period at 10 °C.

After cycling, the two 5 μL/reaction pools for each sample were combined into a single well with a total volume 10 μL. The pooled amplified samples were partially digested using 1 μL FuPa enzyme per sample at 50 °C for 10 min and 55 °C for 10 min followed by enzyme inactivation at 60 °C for 20 min. To enable multiple sample libraries to be loaded per chip, 2 μL of a unique diluted Barcode Adapter mix including Ion Xpress Barcode (numbered 1-16) and Ion P1 Adaptor at standard volumes was ligated to the end of the digested amplicons using 1 μL DNA ligase for 30 min at 22 °C followed by ligase inactivation for 10 min at 72 °C. The resulting unamplified adaptor-ligated libraries were purified using the 22.5 μL Agencourt AMPure XP system (Beckman Coulter, Brea, CA, USA) followed by addition of 75 μL freshly prepared 70 % ethanol to each library.

After purification, the amplicon libraries were further amplified to enrich material for accurate quantification using 25 μL Platinum PCR SuperMix High Fidelity and 1 μL of library Amplification Primer Mix (Ion AmpliSeq library kit 2.0, Thermo Fisher Scientific, Scoresby, Victoria, Australia), at 98 °C for 2 min followed by five two-step cycles of 98 °C for 15 s and 60 °C for 1 min. The amplified amplicon libraries were then purified using 12.5 μL Agencourt AMPure XP Reagent followed by a second purification step with 30 μL AMPure XP and 75 μL of freshly prepared 70 % ethanol added to each library. The concentration and size of amplicons was then determined using an Agilent BioAnalyzer DNA High-Sensitivity chip (Agilent Technologies, Santa Clara,

CA, USA), according to manufacturers' instructions. After quantification, each library was diluted to a concentration of ~10 pM prior to template preparation. Subsequently, libraries ($n = 16$) were pooled in equimolar amounts prior to further processing.

Template preparation (emulsion PCR) and sequencing

Emulsion PCR, emulsion breaking and enrichment (template preparation) were performed using the Ion PGM OT2 200 Template Kit (Thermo Fisher Scientific, Scoresby, Victoria, Australia), according to the manufacturers' instructions (part no. 4480974 Rev. 4.0).

After preparation of the ISPs, sequencing was performed with an Ion Torrent Personal Genome Machine (PGM) system using Ion Sequencing 200 Kit V2 and an Ion 316 Chip (Thermo Fisher Scientific, Scoresby, Victoria, Australia) according to the manufacturers' procedures (Cat. no.4482006 Rev.1.0).

Bioinformatic analyses

The Ion Torrent PGM sequence data was mapped to the complete human genome (hg19) by the Ion Torrent Suite software and Torrent Server along with Torrent Mapping Alignment Program optimised to Ion Torrent data. The bam format file generated by Torrent Suite was uploaded and visualised for human examination using Integrative Genomics Viewer (IGV) 2.3 software [12]. The Ion Reporter software 4.0 (Thermo Fisher Scientific, Scoresby, Victoria, Australia) was used to analyse data from Torrent PGM. The software identifies variants and performs automated annotation on Ion PGM data. Variants were classified into simple categories, summarised into a report which included links to appropriate databases for known variants.

DNA and protein sequences from NGS and SS were compared with the NCBI reference sequences [13] and the UCSC genome browser [40]. All rs ID numbers, locations, allele frequencies and genotypes for all variants were determined based on SNPs reported in the dbSNP database [41] and further analysed in the 1000 Genomes data set. To predict the effect of non-synonymous single nucleotide substitutions on protein structure, function or phenotype, we used the wANNOVAR programme [42, 43] which included the use of five functional prediction software programmes for non-synonymous variants (PhyloP [44], SIFT [45], PolyPhen2 [46], MutationTaster [47] and GERP++ [48]). In silico prediction programmes including AGVGD [49] and PhD-SNP [50] were also used to predict causative variants. For synonymous variants and variants in non-coding regions, the MutationTaster [47] software alone was used. All variants detected were examined for associated information in the public databases (at a minimum, dbSNP, OMIM, LOVD, 1000 Genomes and HGMD) and in the published literature.

Sanger sequencing (SS)

All detected novel mutations by NGS were further investigated by SS. Molecular analysis of the *NOTCH3* gene was performed according to a previously described protocol [17]. Briefly, genomic DNA was extracted using Qiagen QIAamp DNA Blood Midi kits. DNA was amplified by PCR to screen the exons containing novel mutations and was performed with the primers shown in Additional file 3: Table S3 online. PCR amplification for all exons were conducted as previously described [17], and cycling protocols is available for all exons upon request. PCR products were purified using Affymetrix ExoSap-IT reagent (ExoSap-IT, USB Corporation, Staufen, Germany) and directly sequenced for both sense and antisense strands using Big Dye Terminator V3.1 (Applied Biosystems, Foster City, CA, USA) on an ABI 3500 Genetic Analyser (Applied Biosystems) according to established procedures. Sequences were analysed with Chromas 2.33 software (Technelysium, Brisbane, Queensland, Australia).

Abbreviations

CADASIL: Cerebral autosomal dominant arteriopathy with subcortical infarcts and leukoencephalopathy; DNA: Deoxyribonucleic acid; EGF: Epidermal growth factor; GERP: Genomic evolutionary rate profiling; GOM: Granular osmiophilic material; GRC: The Genomics Research Centre; IGV: Integrative Genomics Viewer; NATA: National Association of Testing Authorities, Australia; NCBI: National Centre for Biotechnology Information; NGS: Next-generation sequencing; *NOTCH3*: Notch, Drosophila, Homolog of, 3; PCR: Polymerase chain reaction; PGM: Personal Genome Machine; SNPs: Single nucleotide polymorphisms; UTR: Untranslated region

Acknowledgements

The authors express their gratitude to all neurologists for referring patients and supplying clinical data to our NATA accredited diagnostic lab at the Genomics Research Centre, IHBI, QUT. We thank A/Prof Rod Lea and Dr Miles Benton for help with bioinformatic analysis, interpretation and advice.

Funding

Neven Maksemous was supported by a QUT Postgraduate Scholarship. This work was supported by an Australian International Science Linkages grant and by infrastructure purchased with Australian Government EIF Super Science Funds as part of the Therapeutic Innovation Australia - Queensland Node project and by the Migraine Research Foundation, NY, USA.

Authors' contributions

NM performed the experimental, sequence analysis research, data analysis and drafted the manuscript with input from all authors. LMH contributed to the interpretation of the data, read, edited and approved the manuscript. LRG and RAS conceived the project, supervised research and analysis, read, edited and approved the manuscript.

Competing interests

The authors declare that they have no competing interests.

Consent for publication

Not applicable.

References

1. Chabriat H, et al. Clinical spectrum of CADASIL: a study of 7 families. Cerebral autosomal dominant arteriopathy with subcortical infarcts and leukoencephalopathy. Lancet. 1995;346(8980):934–9.
2. Ruchoux MM, et al. Presence of ultrastructural arterial lesions in muscle and skin vessels of patients with CADASIL. Stroke. 1994;25(11):2291–2.
3. Chabriat H, et al. Autosomal dominant migraine with MRI white-matter abnormalities mapping to the CADASIL locus. Neurology. 1995;45(6):1086–91.
4. Joutel A, et al. Notch3 mutations in CADASIL, a hereditary adult-onset condition causing stroke and dementia. Nature. 1996;383(6602):707–10.
5. Thomsen LL, Olesen J, Russell MB. Increased risk of migraine with typical aura in probands with familial hemiplegic migraine and their relatives. Eur J Neurol. 2003;10(4):421–7.
6. Vedeler C, Bindoff L. A family with atypical CADASIL. J Neurol. 2011;258(10):1888–9.
7. Maksemous N, et al. Next-generation sequencing identifies novel CACNA1A gene mutations in episodic ataxia type 2. Mol Genet Genomic Med. 2016;4(2):211–22.
8. Dichgans M, et al. Small in-frame deletions and missense mutations in CADASIL: 3D models predict misfolding of Notch3 EGF-like repeat domains. Eur J Hum Genet. 2000;8(4):280–5.
9. Oliveri RL, et al. A novel mutation in the Notch3 gene in an Italian family with cerebral autosomal dominant arteriopathy with subcortical infarcts and leukoencephalopathy: genetic and magnetic resonance spectroscopic findings. Arch Neurol. 2001;58(9):1418–22.
10. Tang SC, et al. Arg332Cys mutation of NOTCH3 gene in the first known Taiwanese family with cerebral autosomal dominant arteriopathy with subcortical infarcts and leukoencephalopathy. J Neurol Sci. 2005;228(2):125–8.
11. Razvi SS, et al. Diagnostic strategies in CADASIL. Neurology. 2003;60(12): 2019–20. author reply 2020.
12. Thorvaldsdottir H, Robinson JT, Mesirov JP. Integrative Genomics Viewer (IGV): high-performance genomics data visualization and exploration. Brief Bioinform. 2013;14(2):178–92.
13. Pruitt KD, et al. NCBI Reference Sequences (RefSeq): current status, new features and genome annotation policy. Nucleic Acids Res. 2012; 40(Database issue):D130–5.
14. Markus HS, et al. Diagnostic strategies in CADASIL. Neurology. 2002;59(8):1134–8.
15. Abramycheva N, et al. New mutations in the Notch3 gene in patients with cerebral autosomal dominant arteriopathy with subcortical infarcts and leucoencephalopathy (CADASIL). J Neurol Sci. 2015;349(1-2):196–201.
16. He D, et al. The comparisons of phenotype and genotype between CADASIL and CADASIL-like patients and population-specific evaluation of CADASIL scale in China. J Headache Pain. 2016;17:55.
17. Roy B, et al. Two novel mutations and a previously unreported intronic polymorphism in the NOTCH3 gene. Mutat Res. 2012;732(1-2):3–8.
18. Bianchi S, et al. CADASIL in central Italy: a retrospective clinical and genetic study in 229 patients. J Neurol. 2015;262(1):134–41.
19. Fernandez A, et al. A next-generation sequencing of the NOTCH3 and HTRA1 Genes in CADASIL Patients. J Mol Neurosci. 2015;56(3):613–6.
20. Adib-Samii P, et al. Clinical spectrum of CADASIL and the effect of cardiovascular risk factors on phenotype: study in 200 consecutively recruited individuals. Stroke. 2010;41(4):630–4.
21. Kim YE, et al. Spectrum of NOTCH3 mutations in Korean patients with clinically suspicious cerebral autosomal dominant arteriopathy with subcortical infarcts and leukoencephalopathy. Neurobiol Aging. 2014;35(3):726 e1–6.
22. Joutel A, et al. Pathogenic mutations associated with cerebral autosomal dominant arteriopathy with subcortical infarcts and leukoencephalopathy differently affect Jagged1 binding and Notch3 activity via the RBP/JK signaling Pathway. Am J Hum Genet. 2004;74(2):338–47.
23. Wang W, et al. Notch3 signaling in vascular smooth muscle cells induces c-FLIP expression via ERK/MAPK activation. Resistance to Fas ligand-induced apoptosis. J Biol Chem. 2002;277(24):21723–9.
24. Joutel A. Loss-of-function mutation in the NOTCH3 gene: simply a polymorphism? Hum Mutat. 2013;34(11):v.
25. Bohlega S. Novel mutation of the notch3 gene in arabic family with CADASIL. Neurol Int. 2011;3(2):e6.
26. Lee YC, et al. Cerebral autosomal dominant arteriopathy with subcortical infarcts and leukoencephalopathy: two novel mutations in the NOTCH3 gene in Chinese. J Neurol Sci. 2006;246(1-2):111–5.
27. Brass SD, et al. Case records of the Massachusetts General Hospital. Case 12-2009. A 46-year-old man with migraine, aphasia, and hemiparesis and similarly affected family members. N Engl J Med. 2009;360(16):1656–65.
28. Kim Y, et al. Characteristics of CADASIL in Korea: a novel cysteine-sparing Notch3 mutation. Neurology. 2006;66(10):1511–6.
29. Mazzei R, et al. A novel Notch3 gene mutation not involving a cysteine residue in an Italian family with CADASIL. Neurology. 2004;63(3):561–4.
30. Santa Y, et al. Genetic, clinical and pathological studies of CADASIL in Japan: a partial contribution of Notch3 mutations and implications of smooth muscle cell degeneration for the pathogenesis. J Neurol Sci. 2003;212(1-2):79–84.
31. Scheid R, et al. Cysteine-sparing notch3 mutations: cadasil or cadasil variants? Neurology. 2008;71(10):774–6.
32. Uchino M, et al. Cerebral autosomal dominant arteriopathy with subcortical infarcts and leukoencephalopathy (CADASIL) and CADASIL-like disorders in Japan. Ann N Y Acad Sci. 2002;977:273–8.
33. Herman-Bert A, et al. Mapping of spinocerebellar ataxia 13 to chromosome 19q13.3-q13.4 in a family with autosomal dominant cerebellar ataxia and mental retardation. Am J Hum Genet. 2000;67(1):229–35.
34. Ophoff RA, et al. Familial hemiplegic migraine and episodic ataxia type-2 are caused by mutations in the Ca2+ channel gene CACNL1A4. Cell. 1996;87(3):543–52.
35. Wallace RH, et al. Neuronal sodium-channel alpha1-subunit mutations in generalized epilepsy with febrile seizures plus. Am J Hum Genet. 2001;68(4): 859–65.
36. Orrico A, et al. Mutational analysis of the SCN1A, SCN1B and GABRG2 genes in 150 Italian patients with idiopathic childhood epilepsies. Clin Genet. 2009; 75(6):579–81.
37. Arboleda-Velasquez JF, et al. CADASIL mutations impair Notch3 glycosylation by Fringe. Hum Mol Genet. 2005;14(12):1631–9.
38. Opherk C, et al. CADASIL mutations enhance spontaneous multimerization of NOTCH3. Hum Mol Genet. 2009;18(15):2761–7.
39. Richards S, et al. Standards and guidelines for the interpretation of sequence variants: a joint consensus recommendation of the American College of Medical Genetics and Genomics and the Association for Molecular Pathology. Genet Med. 2015;17(5):405–24.
40. Dreszer TR, et al. The UCSC Genome Browser database: extensions and updates 2011. Nucleic Acids Res. 2012;40(Database issue):D918–23.
41. Sherry ST, et al. dbSNP: the NCBI database of genetic variation. Nucleic Acids Res. 2001;29(1):308–11.
42. Chang X, Wang K. wANNOVAR: annotating genetic variants for personal genomes via the web. J Med Genet. 2012;49(7):433–6.
43. Wang K, Li M, Hakonarson H. ANNOVAR: functional annotation of genetic variants from high-throughput sequencing data. Nucleic Acids Res. 2010;38(16):e164.
44. Pollard KS, et al. Detection of nonneutral substitution rates on mammalian phylogenies. Genome Res. 2010;20(1):110–21.
45. Ng PC, Henikoff S. Predicting deleterious amino acid substitutions. Genome Res. 2001;11(5):863–74.
46. Adzhubei IA, et al. A method and server for predicting damaging missense mutations. Nat Methods. 2010;7(4):248–9.
47. Schwarz JM, et al. MutationTaster evaluates disease-causing potential of sequence alterations. Nat Methods. 2010;7(8):575–6.
48. Davydov EV, et al. Identifying a high fraction of the human genome to be under selective constraint using GERP++. PLoS Comput Biol. 2010;6(12):e1001025.

6

The Israeli National Genetic database

Joël Zlotogora[1] and George P. Patrinos[2,3*]

Abstract

Background: The Israeli National and Ethnic Mutation database (http://server.goldenhelix.org/israeli) was launched in September 2006 on the *ETHNOS* software to include clinically relevant genomic variants reported among Jewish and Arab Israeli patients. In 2016, the database was reviewed and corrected according to ClinVar (https://www.ncbi.nlm.nih.gov/clinvar) and ExAC (http://exac.broadinstitute.org) database entries. The present article summarizes some key aspects from the development and continuous update of the database over a 10-year period, which could serve as a paradigm of successful database curation for other similar resources.

Results: In September 2016, there were 2444 entries in the database, 890 among Jews, 1376 among Israeli Arabs, and 178 entries among Palestinian Arabs, corresponding to an ~4× data content increase compared to when originally launched. While the Israeli Arab population is much smaller than the Jewish population, the number of pathogenic variants causing recessive disorders reported in the database is higher among Arabs (934) than among Jews (648). Nevertheless, the number of pathogenic variants classified as founder mutations in the database is smaller among Arabs (175) than among Jews (192). In 2016, the entire database content was compared to that of other databases such as ClinVar and ExAC. We show that a significant difference in the percentage of pathogenic variants from the Israeli genetic database that were present in ExAC was observed between the Jewish population (31.0%) and the Israeli Arab population (20.6%).

Conclusions: The Israeli genetic database was launched in 2006 on the *ETHNOS* software and is available online ever since. It allows querying the database according to the disorder and the ethnicity; however, many other features are not available, in particular the possibility to search according to the name of the gene. In addition, due to the technical limitations of the previous ETHNOS software, new features and data are not included in the present online version of the database and upgrade is currently ongoing.

Keywords: Israel, Arabs, Database, Genetic disorders, Founder effect

Background

The Israeli population includes Jewish and Arab communities, in each of which genetic disorders have been reported in a relatively high frequency [1]. Among Jews, the late Professor Richard Goodman compiled relatively frequent genetic disorders and published a seminal book "Genetic disorders among the Jewish people" [2]. On the basis of this book, a catalog of genetic disorders in the Jewish population was created in 1998 and later, a catalog of genetic disorders in the non-Jewish population in Israel was added. The data included in these two catalogs were updated at least once a year. In 2006, this compilation gave rise to the Israeli National Genetic database [3], a freely available online resource for genetic services in Israel. This database has resulted from a customized version of the *ETHNOS* software, adapted to accommodate large datasets and to support both menu- and keyword-based queries. Ever since, the Israeli national and ethnic mutation database (NEMDB) has become a useful online resource for genetic services in Israel.

In 2016, the Israeli population included 8,556,000 citizens of whom 74.8% were Jews and 20.8% Arabs [4]. Among Israeli Arabs, 83.1% were Muslim, 9.7%

* Correspondence: gpatrinos@upatras.gr
[2]Department of Pharmacy, University of Patras School of Health Sciences, University Campus, Rion, 26504 Patras, Greece
[3]Department of Bioinformatics, School of Medicine and Health Sciences, Erasmus University Medical Center Rotterdam, Rotterdam, The Netherlands

Christian Arabs, and 7.8% Druze. The Bedouins that have been a nomad population represent one fifth of the Muslim Arabs and in a majority live in the Negev desert. Jewish communities emigrated from many countries in the world joining the Jewish population living in Israel. While differences existed between the various regions in Europe where the Ashkenazi Jews were living, it is difficult to distinguish subgroups among them. On the other hand, most of the other Jewish communities remained distinct.

Implementation

The Israeli National Genetic database (http://server.goldenhelix.org/israeli) includes causative genomic variants that were characterized among affected patients either Israeli Arabs or Jews living in Israel and in the diaspora. Palestinian Arabs living under the Palestinian Authority, or for whom details on the exact origin were not available, were recently added as a distinct population. All recessive mutations are included while dominant or X-linked mutations are recorded in the database either if they are founder mutations, or if they were reported in several families, or if the disorder is relatively frequent. Since the launch of the database, each entry consists of the name of the disorder (usually as used in the Online Mendelian Inheritance in Man (OMIM); [https://www.ncbi.nlm.nih.gov/omim]), the name of the gene and its OMIM number, and the name of the mutation, whether it was reported as a single allele, a family, several families or is a founder mutation. In addition, the entry includes details on the origin of the patient if Jewish either Ashkenazi or the country of origin and if Arab the religion either Druze Christian or Muslim and the locality of origin. The Muslim Arabs known to be Bedouins are entered as a separate subgroup. The frequency of the mutation in the specific population is recorded when available.

The information included in the database is updated according to publications in the scientific literature and personal communications. When details needed for the database are not included in the original report, the authors are contacted in order to complete the data if available. Since the end of 2010, the source of the data was added to each new entry and when possible, this was done also for older data.

In 2016, the entire database was reviewed and corrected according to ClinVar (https://www.ncbi.nlm.nih.gov/clinvar) and ExAC (http://exac.broadinstitute.org) database entries. Data from ClinVar were added to the database for each mutation including, when available, genomic location, rs number, pathogenicity, and OMIM reference for the mutation. The frequency in ExAC as calculated on 60,706 unrelated individuals was added for the variants documented in ExAC [5].

Results

Database entries

In September 2016, there were 2444 entries in the database, 890 among Jews, 1376 among Israeli Arabs, and 178 entries among Palestinian Arabs. The data on Palestinian Arabs are scarce since added only recently and therefore were not included in the present analysis. Among the 890 entries in Jews, there were 783 that were different mutations; the other entries were cases of a same pathogenic variant present in more than one community. There were 357 entries among the Ashkenazi Jews, 86 entries among Moroccan Jews, 71 entries among Iraqi Jews, 57 entries among Iranian Jews, 47 entries among Yemenite Jews, and 260 entries among specific smaller communities. In addition, there were 98 entries in three Jewish subgroups for which details were not available: non-Ashkenazi (38 entries), Sephardi (11 entries), or North African (21 entries) and a group of Jews of unknown origin (28 entries). These data are summarized in Table 1.

Among the Israeli Arabs, there were 1376 entries, 99 being duplicated since a same pathogenic variant was reported in more than one locality in patients from a same religious group. There were 102 unique entries among Christian Arabs and 108 among Druze. Among the Muslim Arabs, there were 814 unique entries with an additional 253 entries in Muslims known to be of Bedouin origin. Among the 1277 non-duplicated entries, the village of origin was known in 1075 cases.

Frequency of the pathogenic variants as reported in ExAC

Among the 783 unique entries in Jews, the pathogenic variant was present in ExAC in 247 entries (31.8%) and the frequency of the pathogenic variant was 0.5% or more for 44 of them (5.8%). Among the 977 unique entries in Israeli Arabs, the pathogenic variant was present in ExAC in 201 cases (20.6%) and the frequency of the pathogenic variant was 0.5% or more for 22 of them (2.2%).

Details on the presence of the pathogenic variant and their frequency in ExAC in the different Jewish and Israeli Arab communities are given in Table 1.

Autosomal recessive disorders in the Israeli NEMDB

Subsequently, we sought to analyze the number of pathogenic variants that were reported among patients affected with autosomal recessive disorders among Jews and Israeli Arabs. Among the Jews, there were 648 different pathogenic variants in a total of 306 different genes, from which 192 being founder mutations (29.6%). In 135 out of the 306 genes, more than one pathogenic variant was reported (44.1%), and in 77 out of these genes, at least one was a founder mutation. Among the Israeli Arabs, there were 934 different pathogenic variants in a total of 473 different genes, 175 being founder

Table 1 Data content statistics from the entire Israeli NEMDB

	Entries	Unique entries	ExAC (% unique entries)	ExAC >0.5 per 1000
Jews	890	783	247 (31.8%)	44 (5.8%)
Ashkenazi	357	357	136 (38.9%)	28 (7.8%)
Morocco	86	86	34 (39.5%)	8 (9.3%)
Iraq	71	71	20 (28.2%)	8 (11.7%)
Iran	57	57	19 (33.3%)	2 (2.8%)
Yemen	46	46	8 (17.0%)	1 (2.1%)
Israeli Arabs	1376	977	201 (20.6%)	22 (2.2%)
Christian	102	102	19 (18.6%)	4 (3.9%)
Muslim	814	814	151 (18.6%)	19 (2.3%)
Bedouins	253	253	36 (14.2%)	4 (1.6%)
Druze	108	108	19 (17.6%)	1 (1.9%)

The total numbers of entries among Jews include similar entries reported in more than one community, and entries in small communities are not included in the Table. Among Arabs, the total number of entries includes duplicated entries in which a same pathogenic variant was reported in more than one locality in patients from a same religious group and similar entries reported in more than one community

mutations (18.7%). In 151 out of the 473 genes, more than one pathogenic variant was reported (31.9%), and in 61 out of these genes, at least one was a founder mutation. The same parameters were looked for in the different Jewish and Israeli Arab communities, and the data are summarized in Table 2.

Pathogenic variants present in more than one community
We then looked for pathogenic variants that were present in more than one community. In 52 instances, the same pathogenic variant was found among Jews and Arabs out of which 37 variants were present in the ExAC database (71.2%). Twenty-five out of these 37 variants (67.6%) were found in Ashkenazi Jews, from which 23 variants were reported in the ExAC database (92%).

In 17 instances, the same pathogenic variant was found in geographically close Jewish communities, including 6 among North African communities and 11 among Eastern Jewish communities. In 32 cases, the same pathogenic variant was found in geographically distant Jewish communities, 17 of which were documented in ExAC (53.1%). In 19 of these 32 cases, the pathogenic variant was found in Ashkenazi Jews, 11 of which were also documented in ExAC (57.9%).

In 16 cases, a pathogenic variant was reported in more than one Arab religious community but not among Jews. Among these 16 cases, 6 pathogenic variants were reported in ExAC (37.5%).

Table 2 Autosomal recessive disorders reported in affected patients

	Unique entries	Number of genes	Genes with >1 pathogenic variant[a]	Founder mutations[b]	Founder in >1 mutation[c]
JEWS	648	306	135 (44.1%)	192 (29.6%)	77 (57%)
Ashkenazi	283	161	61 (37.9%)	66 (23.3%)	44 (72.1%)
Morocco	80	48	17 (35.4%)	32 (40%)	13 (76.5%)
Iraq	50	44	11 (22.9%)	23 (46%)	8 (72.7%)
Iran	62	43	3 (7%)	23 (37.1%)	3 (100%)
Yemen	37	31	4 (13.3%)	7 (18.9%)	2 (50%)
Israeli Arabs	934	473	151 (31.9%)	175 (18.7%)	61 (40.4%)
Christian	58	38	7 (17.5%)	6 (10.3%)	3 (48.3%)
Muslim	553	335	89 (26.6%)	89 (16.1%)	33 (37.1%)
Bedouins	170	140	25 (17.9%)	86 (50.6%)	20 (80%)
Druze	75	58	12(20.7%)	19 (25.3%)	7 (58.3%)

The total numbers of entries among Jews include similar entries reported in more than one community, and entries in small communities are not included in the Table. Among Arabs, the total number of entries includes duplicated entries in which a same pathogenic variant was reported in more than one locality in patients from a same religious group and similar entries reported in more than one community

[a]The percentage in parenthesis is the number of genes with more than one pathogenic variant out of the number of genes

[b]The percentage in parenthesis is the number of founder mutations out of the number of unique entries

[c]The percentage in parenthesis is the number of genes with more than one founder mutation out of the number of genes with more than one pathogenic variant

Discussion

The technology revolution in genomic analysis changed the ability for diagnosis and characterization of genetic diseases in the last decades. While the late Richard M Goodman in 1979 delineated only 11 monogenic disorders relatively frequent in the Ashkenazi Jews [2], in September 2016, founder mutations were reported in the Ashkenazim responsible for 62 autosomal recessive disorders, 7 dominant disorders, and one X-linked disease. Since the creation of the Israeli NEMDB in September 2006, the number of database records has been almost quadrupled. For instance, new founder mutations were added in 36 genes among the Ashkenazi Jews in 10 years since the launch of the Israeli NEMDB.

The Ashkenazi represents the largest Jewish community in Israel and in the diaspora, and therefore, the observation that most of the entries among Jews (63%) are in this community was not unexpected. However, while the Israeli Arab population is much smaller than the Jewish population, the number of pathogenic variants causing recessive disorders reported in the database is higher among Arabs (934) than among Jews (648). Nevertheless, the number of pathogenic variants classified as founder mutations in the database is smaller among Arabs (175) than among Jews (192), which can be explained by pathogenic variant expansion in isolated populations.

Whole genome sequencing of random individuals and their parents have demonstrated that every individual is born with 44–82 de novo single-nucleotide mutations [6], and therefore, in a defined population, many of the newborns are carriers of new variants responsible for recessive diseases. In a previous study, the fate of recessive mutations was followed in an Israeli Muslim village in which the families are large and close consanguinity is frequent [7]. In this village, a new variant, occurring de novo or being introduced by marriage of a carrier from another village, may appear in homozygosity in a patient already after three generations. In such isolated populations, some of the new variants spread within the kindred of the first carrier, either randomly or as the result of a selective advantage, and later may become founder mutations within the community. A change in the marriage patterns such as marriages outside of the isolate and smaller size of the families will reduce the number of patients affected due to founder mutations. In parallel, some of the founder mutations that were present only in isolated communities will appear at a lower frequency in the whole population. Before the creation of the state of Israel, the Jewish communities were isolated one from the other because of geographical distances and from the surrounding populations by the preference of marriages within the religious community and often within the family. As a result, among Jews, founder mutations are found in the database in each of the different communities. Indeed, few of these founder Jewish mutations are present in more than one community, mainly since they occurred before the dispersion of the Jews. For instance, the mutation p.F301L in F11 which is nowadays frequent among Ashkenazi Jews and Jews from Iraq was probably present among Jews already 2.5 millennia ago [8]. Among the Israeli Arabs, the preference of marriages within the close family and within the religious community that was responsible for their isolation is still predominant nowadays. Therefore, among the Israeli Arabs, founder mutations are still mostly limited to single villages or tribes. There are three exceptions in the database: one is the p.T322X mutation in the *ERCC8* gene causing Cockayne syndrome that was shown to be frequent in the entire Christian Arab community and has been reported in Christian Arabs from Lebanon [9]. The second exception is the p.S52_G55del mutation in the *TBCE* gene causing hypoparathyroidism and mental retardation that originated among Bedouins in Saudi Arabia who are at the origin of the Israeli Bedouin Arabs [10]. The third exception is a mutation present in all the Israeli Arab Bedouins p.P615Sfs*12 in the *NTRK1* gene causing congenital insensitivity to pain with anhidrosis but has not been reported in other populations and therefore probably occurred more recently [11]. In Israel, many changes are occurring both in the Jewish and Arab populations in particular of the marriage patterns that while they remain within the religious community are more often outside the isolate. This is particularly frequent among Jews since consanguinity became rare since the creation of the State of Israel and intercommunity marriages became frequent [12]. Among Arabs, the changes are slower and consanguinity is still preferred, but marriages outside the village became more and more frequent [13]. In both cases, these changes are expected to scatter the founder mutations. The ultimate result that may be expected is that some founder mutations will remain and will become Israeli Jewish or Israeli Arab mutations existing in each population at lower frequency.

A significant difference in the percentage of pathogenic variants from the Israeli NEMDB that were present in ExAC was observed between the Jewish population (31.8%) and the Israeli Arab population (20.6%). The difference between the Ashkenazi Jews and the Israeli Arabs was expected since the European population including Ashkenazi Jews is better represented in ExAC than the Middle East population [5]. However, the Jews from Morocco, Iran, and Iraq have a percentage of mutations present in ExAC in the same range as the Ashkenazi Jews even though these communities originated from populations that are not well represented in ExAC. This may in part be explained by the observation that several variants found among Ashkenazi Jews and present in ExAC were also characterized in other Jewish communities. For instance, among the 34 variants found in Moroccan Jews that were present in ExAC, 8 were variants that are also found among Ashkenazi Jews. Similarly, among the mutations common to Jews and Israeli Arabs, those reported among Ashkenazi Jews were almost always present in ExAC (92%).

Conclusion

The Israeli NEMDB was launched in 2006 on the ETHNOS software and is available online ever since. It allows querying the database according to the disorder and the ethnicity; however, many other features are not available, in particular the possibility to search according to the name of the gene. In addition, due to the technical limitations of the previous ETHNOS software on which the Israeli NEMDB runs, new features and data are not included on the present online version of the database such as the source of the data, genomic location of the variant, it rs number, and OMIM reference or the effect of the mutation as described in the ClinVar database. Upgrade of the Israeli NEMDB is currently ongoing in order to include all the data described in the article and allow new querying possibilities.

Abbreviations

NEMDB: National and Ethnic Mutation Database; OMIM: Online Mendelian Inheritance in Man

Acknowledgements

The authors cordially thank the Israeli NEMDB user community for their feedback which allowed us to keep the database as updated and complete as possible.

Funding

Part of this work was partly funded by the Golden Helix Foundation and by European Commission grants (FP7-200754; GEN2PHEN and FP7-305444; RD-Connect) to GPP.

Authors' contributions

JZ summarized and analyzed the data and drafted the manuscript. JZ and GPP discussed the data and performed critical revisions to the manuscript. JZ and GPP read and approved the final manuscript.

Competing interests

The authors declare that they have no competing interests.

Consent for publication

Not applicable.

Author details

[1]Faculty of Medicine, Hebrew University, Jerusalem, Israel. [2]Department of Pharmacy, University of Patras School of Health Sciences, University Campus, Rion, 26504 Patras, Greece. [3]Department of Bioinformatics, School of Medicine and Health Sciences, Erasmus University Medical Center Rotterdam, Rotterdam, The Netherlands.

References

1. Zlotogora J. Genetics and genomic medicine in Israel. Mol Genet Genomic Med. 2014;2:85–94.
2. Goodman RM. Genetic disorders among the Jewish people. Baltimore: Johns Hopkins University Press; 1979.
3. Zlotogora J, van Baal S, Patrinos GP. Documentation of inherited disorders and mutation frequencies in the different religious communities in Israel in the Israeli National Genetic Database. Hum Mutat. 2007;28:944–9.
4. Statistical Abstracts of Israel. Central bureau of statistics Jerusalem. 2016.
5. Lek M, Karczewski KJ, Minikel EV, Samocha KE, Banks E, Fennell T, et al. Analysis of protein-coding genetic variation in 60,706 humans. Nature. 2016;536:285–91.
6. Acuna-Hidalgo R, Veltman JA, Hoischen A. New insights into the generation and role of de novo mutations in health and disease. Genome Biol. 2016;17:241.
7. Zlotogora J, Hujerat Y, Barges S, Shalev SA, Chakravarti A. The fate of 12 recessive mutations in a single village. Ann Hum Genet. 2007;71:202–8.
8. Shpilberg O, Peretz H, Zivelin A, Yatuv R, Chetrit A, Kulka T, Stern C, Weiss E, Seligsohn U. One of the two common mutations causing factor XI deficiency in Ashkenazi Jews (type II) is also prevalent in Iraqi Jews, who represent the ancient gene pool of Jews. Blood. 1995;15(85):429–32.
9. Khayat M, Hardouf H, Zlotogora J, Shalev SA. High carriers frequency of an apparently ancient founder mutation p.Tyr322X in the ERCC8 gene responsible for Cockayne syndrome among Christian Arabs in Northern Israel. Am J Med Genet A. 2010;152A:3091–4.
10. Parvari R, Hershkovitz E, Grossman N, Gorodischer R, Loeys B, Zecic A, Mortier G, Gregory S, Sharony R, Kambouris M, Sakati N, Meyer BF, Al Aqeel AI, Al Humaidan AK, Al Zanhrani F, Al Swaid A, Al Othman J, Diaz GA, Weiner R, Khan KT, Gordon R, Gelb BD. HRD/Autosomal Recessive Kenny-Caffey Syndrome Consortium. Mutation of TBCE causes hypoparathyroidism-retardation-dysmorphism and autosomal recessive Kenny-Caffey syndrome. Nat Genet. 2002;32:448–52
11. Shatzky S, Moses S, Levy J, Pinsk V, Hershkovitz E, Herzog L, Shorer Z, Luder A, Parvari R. Congenital insensitivity to pain with anhidrosis (CIPA) in Israeli-Bedouins: genetic heterogeneity, novel mutations in the TRKA/NGF receptor gene, clinical findings, and results of nerve conduction studies. Am J Med Genet. 2000;92:353–60.
12. Cohen T, Vardi-Saliternik R, Friedlander Y. Consanguinity, intracommunity and intercommunity marriages in a population sample of Israeli Jews. Ann Hum Biol. 2004;31:38–48.
13. Na'amnih W, Romano-Zelekha O, Kabaha A, Rubin LP, Bilenko N, Jaber L, Honovich M, Shohat T. Continuous decrease of consanguineous marriages among Arabs in Israel. Am J Hum Biol. 2015;27:94–8.

7

Genetic variation in neurodegenerative diseases and its accessibility in the model organism *Caenorhabditis elegans*

Yiru Anning Wang[1,2], Jan Edward Kammenga[2] and Simon Crawford Harvey[1*]

Abstract

Background: Neurodegenerative diseases (NGDs) such as Alzheimer's and Parkinson's are debilitating and largely untreatable conditions strongly linked to age. The clinical, neuropathological, and genetic components of NGDs indicate that neurodegeneration is a complex trait determined by multiple genes and by the environment.

Main body: The symptoms of NGDs differ among individuals due to their genetic background, and this variation affects the onset and progression of NGD and NGD-like states. Such genetic variation affects the molecular and cellular processes underlying NGDs, leading to differential clinical phenotypes. So far, we have a limited understanding of the mechanisms of individual background variation. Here, we consider how variation between genetic backgrounds affects the mechanisms of aging and proteostasis in NGD phenotypes. We discuss how the nematode *Caenorhabditis elegans* can be used to identify the role of variation between genetic backgrounds. Additionally, we review advances in *C. elegans* methods that can facilitate the identification of NGD regulators and/or networks.

Conclusion: Genetic variation both in disease genes and in regulatory factors that modulate onset and progression of NGDs are incompletely understood. The nematode *C. elegans* represents a valuable system in which to address such questions.

Keywords: Neurodegenerative diseases, Natural variation, Quantitative genetics, *C. elegans*

Background

Neurodegenerative diseases (NGDs) cause disability and premature death, primarily among older people. These chronic and fatal illnesses include Alzheimer's disease (AD), Parkinson's disease (PD), Huntington's diseases (HD), spinocerebellar ataxia, prion diseases (PrD) (i.e., transmissible spongiform encephalopathies), frontotemporal dementia (FTD), and amyotrophic lateral sclerosis (ALS) [88]. Most NGDs are age dependent with their incidence increasing with advancing age. This makes understanding NGDs increasingly important given the recent increase in human lifespan seen in many countries. Many disease-causing mutations underlying different NGDs have phenotypic effects that result from the misfolding of proteins and/or mitochondrial dysfunctions leading to widespread damage in different parts of the nervous system (Fig. 1). This damage leads to a range of symptoms, and the overlap between symptoms in different NGDs can make it difficult to precisely diagnose patients. Diagnosis is further complicated by the individual variability apparent from the onset of disease [99]. For example, PD is characterized by motor issues, including tremor, slowing of movement, and an unstable gait, and by cognitive symptoms, but only some patients suffer from cognitive impairment and develop dementia [20].

The rate of disease progression (i.e., the duration of a given neuropathological stage) and clinical presentation also vary from one patient to another. Young onset PD patients, for example, often have a more frequent family history of PD and a more variable survival rate relative to those without the familial history [79]. Several studies of the amyloid-β protein aggregates, which cause AD, also indicate that the existence of distinct shapes in beta amyloid peptide (Aβ) aggregates, 40 residue Aβ (Aβ40)

* Correspondence: simon.harvey@canterbury.ac.uk
[1]Biomolecular Research Group, School of Human and Life Science, Canterbury Christ Church University, Canterbury CT1 1QU, UK
Full list of author information is available at the end of the article

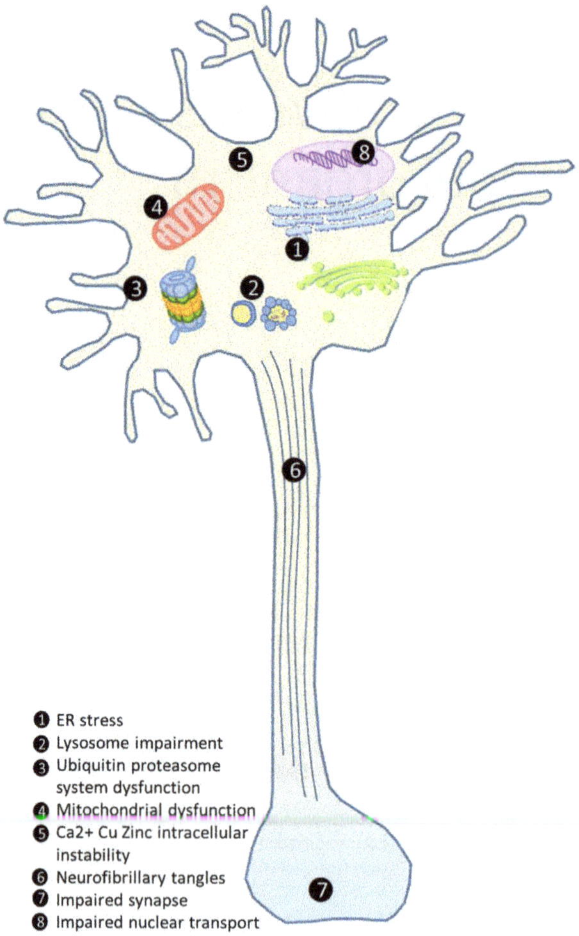

Fig. 1 Schematic diagram of the pathological mechanisms associated with neurodegeneration in neurons. Eight main events exist in the cellular biological processes related to neurodegeneration: *1* The protein-misfolding process inhibits ER to Golgi trafficking and alters ER-associated degradation, inducing the ER stress [55]; *2* Damaged lysosomes disrupt processes that maintains lysosomal homeostasis [78]; *3* Accumulation of misfolded proteins generates positive feedback exacerbating other effects [15]; *4* Dysfunctional mitochondrial maintenance directly affects mitochondrial biogenesis and induction of autophagy [75], e.g., the production of ROS and ATP in the cell; *5* Altered homeostasis generates excessive influx of calcium, copper, and zinc, due to ER and mitochondrial stresses [78]; *6* Plaques and neurofibrillary tangles, due to tau phosphorylation and aggregates, reduce neurotransmitter release, which also *7* weaken synaptic strength [15]; *8* Signaling pathways in the stressed organelle or the cytoplasm induce the transductions of the signals to the nucleus, which provoke DNA damage [10, 80]

and 42 residue Aβ fibril structures [51], and identify the distinct strain-specific traits (defined as "strainness") of the forms of AD by the different conformation of the aggregates [91, 104].

There is however limited knowledge about the mechanisms that determine individual variation. Individuals that carry the same mutation in the same disease-causing gene may display a range of different clinical symptoms. For example, assessing 6-year change in verbal memory, processing speed, and executive function in AD identified effects of MS4A4E, CLU, and NME8 in whites and of ZCWPW1 and CDS33 variants in African Americans. For MS4A4E and CLU, this association was only significant in individuals bearing at least one *APOE ϵ4* allele (AD risk gene) [9]. Apart from lifestyle and environment, these individual differences are caused by the unique genetic background of each person. The genetic background, thus, could be defined as the genetic makeup of all alleles that interact with the disease related to the "disease-causing" mutation.

Detailed mechanistic studies into background modifiers of NGDs are however difficult to conduct at the individual level in humans for ethical and technical reasons. Model organisms such as yeast, insects, worms, fish, and rodents are therefore critical to furthering our understanding of differences between individual in NGDs. This use is facilitated by the range of methods and approaches available to construct transgenic models of human diseases in model species with different genetic backgrounds [1, 19, 32, 53]. They are likewise an important resource for investigating the genetic underpinnings of quantitative traits, including complex disease phenotypes. Here, we review the onset and progression of NGDs in the context of genetic background and illustrate how work on the model species *Caenorhabditis elegans* can illuminate the underlying mechanisms of individual variation.

NGD phenotypes depend on the genetic background

Individual genetic backgrounds differ from thousands to millions of genetic variants that will range from single-nucleotide polymorphisms (SNPs) to, potentially very large, copy number variants (CNVs) [92, 109]. This genetic variation is a major determinate of differences in predisposition to disease [1, 41], where risk-increasing variants are numerous, display intricate patterns of interaction with each other as well as with non-genetic variables, and unlike classical Mendelian ("monogenic") disorders will often exhibit no simple mode of inheritance. Differential phenotypes for most NGDs arise from multiple genetic variants and their interaction with each other, as well as environmental factors. Hence, the genetics of these diseases is considered "complex" based on the heterogeneity in pathology and the disease polygenicity [8].

Genetic background can affect an important feature shared by different NGDs, specifically the formation of cerebral deposits of misfolded protein aggregates (also called prion-like proteins). For example, AD is characterized by the presence of neurofibrillary tangles and beta amyloid peptide (Aβ) in neural plaques, which are abnormal accumulations of microtubule-associated protein tau in a hyperphosphorylated state [82]. Similarly, the accumulation of proteins with polyglutamine-rich extensions is characteristic of HD and associated polyglutamine diseases [13], while PD involves the loss of dopaminergic neurons and the presence of Lewy bodies and Lewy neurites that are the aggregates of the synaptic protein alpha-synuclein. However, exactly how genetic variation modifies and affects specific parts of the NGD pathways is mostly unclear. We suggest that the genetic variation associated with NGDs can be grouped into two classes. Firstly, variation present directly in disease genes. Secondly, variation in regulatory factors that modulate onset and progression of the NGDs. Critically, this distinction separates those variations that are causative of disease from those that modify the disease but that on their own, cannot cause NGD.

In an example of the first such class of variants, multiple rare mutations in amyloid precursor protein (APP), or in the presenilin-1 and 2 genes (*PSEN1* and *PSEN2*), can cause early-onset AD (Fig. 3). Similarly, multiple mutations in leucine-rich repeat kinase 2 (*LRRK2*), a gene associated with PD, are known to be related to the sporadic late-onset form of the disease. Here, the G2385R and R1628P *LRRK2* variants are validated risk factors for PD in Asian populations, while the G2019S variant has been identified in different populations worldwide [57]. Examples of the second class of variants are those known to act in the IIS/mTOR pathway (Figs. 2 and 3). This pathway modulates response to a range of stresses and in the NGDs has been linked to a range of pathological processes (Fig. 1 and Table 1). For instance, Baleriola et al. [4] reported moderate eIF2α activation by Aβ and a greater frequency of ATF4 (the eIF2α effector) transcripts were identified in axons in the brain of AD patients. ATF4 is also known to activate the Parkin gene in PD and also to be related to stress pathways [55]. ATF4 therefore induces transcriptional expression of genes mediated by the UPR, including genes involved in amino acid metabolism, resistance to oxidative stress, and the proapoptotic transcription factor *CHOP*, which are related to disease processing [4, 24]. However, rare coding variation in ATF4 has been also found with pathway impairment in patients with sporadic cervical dystonia [69]. Additionally, Chaudhry et al. [11], using gene-based analyses for late-onset AD, revealed associations with the WT1, ZC3H12C, DLGAP2, and GPR1 genes, suggesting a possible role in AD pathogenesis.

Genetic variants therefore produce a broad spectrum of biological effects. This can be seen in analysis of natural variants affecting late-onset AD (LOAD), where more than 20 genes involved in a range of processes including metabolism, inflammation, synaptic activity, and intracellular trafficking have been identified [45]. CNVs are also associated with variation in NGD phenotypes, with, for example, a rare duplication of the amyloid-β protein precursor linked to early-onset AD [35]. Thus, the question of how genetic variants affect mechanisms in diseases such as AD and PD remains challenging. Here, model species like the nematode *C. elegans* might offer a solution and provide more insight into the mechanism(s) underlying individual variation in NGD disease phenotypes.

Natural variation associated with complex traits in *C. elegans* NGD models

C. elegans is a globally distributed nematode, and its level of natural genetic diversity is similar to human genetic variation [18]. Studying NGDs in a genetically tractable model species, such as *C. elegans*, allows detailed insight into the molecular pathogenesis. *C. elegans* models of AD, PD, HD, and other several NGDs have been established [22, 30, 31, 50, 84]. These models involve the transgenic expression of human genes under the control of a *C. elegans* promotor, with the resulting protein often linked to GFP or YFP. Applying fluorescent proteins (e.g., YFP) allows tracking of target proteins over time and the visualization of aggregation in vivo via the observation of the fluorescent foci (e.g., [101]). Such studies of AD using transgenic *C. elegans* have been carried out for many years. More recently, McColl et al. [53] have generated an improved model with human DA-Aβ 1–42 under the control of the muscle-specific promoter *unc-54* promotor. In this line, the full-length Aβ1-42 (the predominant Aβ species in human brain) is expressed in worm body wall muscle cells, and its oligomerizations and aggregations develop and result in severe. This fully penetrant, age progressive paralysis also shows more rapid than that caused from Aβ3-42 expression [53].

These transgenic models have all been generated in the N2 strain of *C. elegans* and have been widely used in screens for gene function analysis. However, such analyses in a single background have constrained the analysis and detection of natural allelic variants associated with complex traits. The most striking demonstration of this is a comparison of the RNAi phenotypes of ~1400 genes between N2 and CB4856 that showed that ~20% of genes differ in the severity of phenotypes between just these two genetic backgrounds [103]. Crucially, the natural variation found in *C. elegans* is also sufficient to cause significant changes in signaling pathways both at the gene expression (transcript and protein abundance) and phenotypic levels [87]. Accordingly, these could contribute to understanding how allelic variation affects gene expression, at the level of translation, in the multiple pathways and/or networks of complex traits involved in development and disease progression. For comparative investigations on mammalian genetics (including metabolism, aging, cancer, and neurodegeneration diseases), we therefore suggest that taking the different genetic backgrounds of the worm into consideration could be

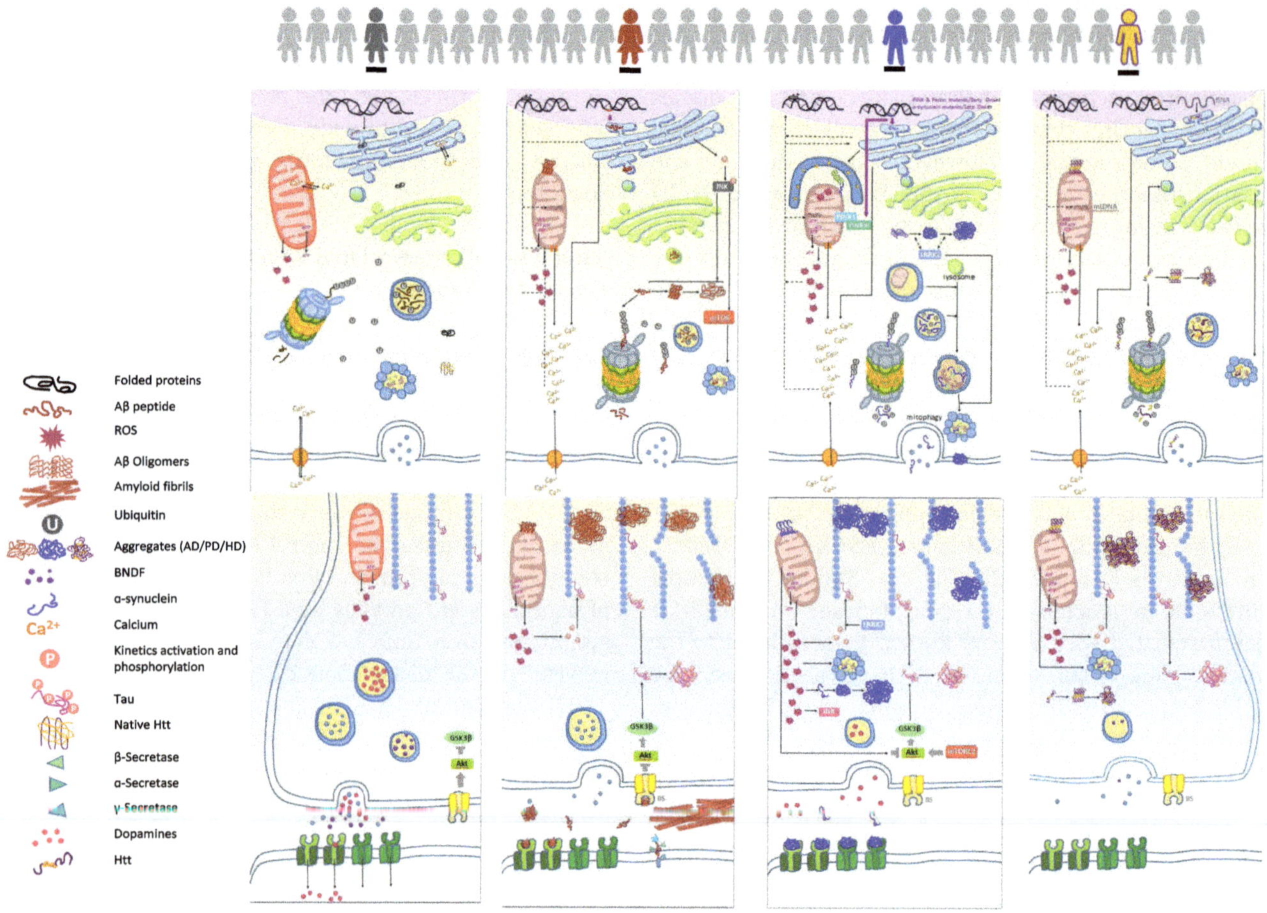

Fig. 2 The molecular processes implicated in neurodegeneration in the neurons of a normal healthy individual (*gray*) and in AD (*red*), PD (*blue*), and HD (*yellow*) patients. The schematic neuron is divided into the soma and axonal terminal bouton. For simplicity, postsynaptic/dendritic events are not included. Misfolded proteins aggregate first into oligomers and then into higher-molecular-weight insoluble protofibrils and further aggregates [17, 37, 104]. In AD patients with mutations in APP, PSEN1, PSEN2, or APOE, the JNK pathway is activated, increasing levels of phospho-JNK in neurons. This mediates the phosphorylation of APP and FoxO-dependent autophagy [52, 63]. Moreover, the soluble Aβ oligomers activate the mTOR pathways again promoting autophagy [52]. Further phosphorylation of tau and impaired Aβ activate the IIS/Akt pathways and affect cognitive function and synaptic plasticity [38]. In PD, mutations in PINK and Parkin (related to EOPD) or α-synuclein mutations (LOPD) lead to the inhibition of α-synuclein degradation as well as accumulation of autophagic vacuoles, which result in neuronal death [16, 96]. Misfolded α-synuclein also interacts with membranes and mitochondria, causing calcium dysregulation and a reduction of mitochondrial activity. This results in mtDNA damage as well as impairments to the ubiquitin-proteasome system (UPS) and mitophagy. The significant pathological etiology of HD is the enlargement of the polyglutamine (polyQ) domain within the HTT protein's N-terminus [110]. MLH1 (MutL homolog 1) and an SNP within a nuclear factor-κB binding site (Nf-KB) in the HTT promoter play a role in the altered onset of HD. In comparison with AD and PD, proteasome efficiency is strongly reduced in HD patients. Meanwhile, the polyQ domain of mutant HTT contributes increased toxicity by attracting and binding to other cytoplasmic and nuclear structures that contain polyglutamine (reviewed by [6]). Additionally, a major loss of brain-derived neurotrophic factor (BDNF) protein has been shown in HD and may due to the deficits in BDNF delivery and/or loss of BDNF gene transcription by mutant Htt [23, 39]

valuable. This approach would be likely to clarify proteostasis mechanisms and more fully reveal the relationships between genotype and phenotype. For example, in *Drosophila*, Chow et al. [14] showed 114 lines from the sequenced *Drosophila* Genetic Reference Panel of wild-derived inbred strains that exhibit the high heterogeneity in survival under endoplasmic reticulum (ER) stress conditions, and showed that 17 of 25 tested candidate genes were active in the putative response to ER stress [14].

Genetic and phenotypic differences between *C. elegans* populations have been identified with many traits shown to be variable between isolates (e.g., [2, 33, 40, 42, 90, 102]; for review, see [67]). Next to the canonical wild type Bristol N2, the most extensively characterized *C. elegans* isolate is the Hawaiian strain CB4856. This isolate is extensively diverged from N2, with a recent de novo assembly of the CB4856 genome suggesting that many variations have been maintained by balancing

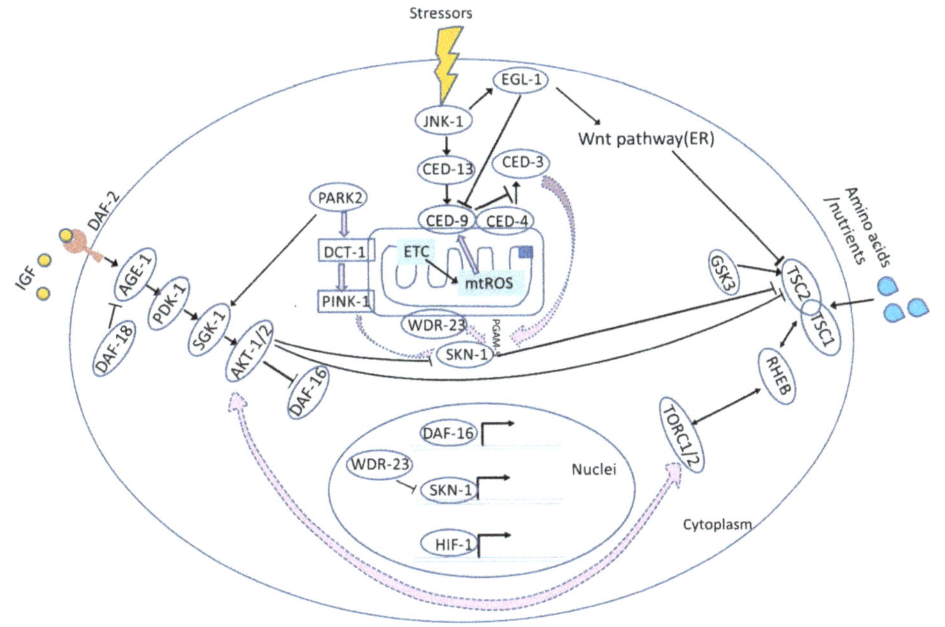

Fig. 3 *C. elegans* cellular pathways and their crosstalk, relating to aging and the stress response. Shown (from *left to right*) are the insulin/insulin-like growth factor (IIS) signaling pathway, the mitochondrial signaling pathways, and the mechanistic target of rapamycin (TOR) pathway. *Arrows* indicate positive regulatory events and *bars* indicate inhibitory interactions. *Purple block arrows* represent interactions between the different pathways, whereas *dashed purple block arrows* indicate possible indirect interactions. The oval molecules and their corresponding mammalian homologs involved in IIS [28, 56, 62, 72] are as below: AGE-1/PI3K, phospatidylinositol-3-kinase; PDK-1, 3-phosphoinositide-dependent kinase 1 ortholog; SGK-1, a serine/threonine protein kinase that is orthologous to the mammalian serum- and glucocorticoid-inducible kinases (SGKs); Akt/PKB, the serine/threonine kinase; DAF-16/FOXO, forkhead box O (FOXO) transcription factor. Mitochondrial dysfunctions [90–96] are associated with apoptotic/programmed cell death (PCD), aberrant autophagic regulation, endoplasmic reticulum dysfunction, and intracellular calcium, including c-Jun N-terminal kinase (JNK) subgroup of mitogen-activated protein (MAP) kinases; CED-4, CED-9, and EGL-1 belong to a conserved genetic pathway to regulate apoptosis during *C. elegans* development [63]; PINK-1, a predicted serine/threonine kinase which is similar with human *PINK1*, PTEN-induced kinase-1; *SKN-1/Nrf*, skin in excess transcription factor 1/NF-E2-related factor; *mtROS*, mitochondrial reactive oxygen species; *ATP*, adenosine-5′-triphosphate; *HIF-1*, hypoxia-inducible transcription factor 1. Major molecules in TOR pathways include *TSC1/2*, tuberous sclerosis complexes 1 and 2; *RHEB*, Ras homolog enriched in brain; *TOR*, target of rapamycin kinase; *GSK3*, glycogen synthase kinase ortholog. See text in Table 1 for further details

selection over long evolutionary timescales [94]. Wild *C. elegans* populations have also been shown to display significant local adaptation to their environment at multiple levels from the genotype and transcriptome [102]. Given this genetic and phenotypic variation, it is likely that variable natural genetic backgrounds of *C. elegans* will harbor abundant genetic variation that will modify the severity of NGD phenotypes. Wider analysis of more recently isolated wild isolates is also important as recent research has revealed specific adaptations to laboratory conditions that exist in the canonical wild-type Bristol N2 [5, 54, 66].

Studies into genetic background effects on NGDs in *C. elegans* have focused on a polyglutamine model. Here, Gidalevitz et al. [27] looked at the effects of natural genetic variation on susceptibility to aggregation and to toxicity [27]. This work introgressed a polyQ40 transgene from an N2 background into three wild genetic backgrounds, including the California-derived isolate DR1350, the Madeira isolate JU258, and the genetically distant Hawaiian isolate CB4856. A series of markers of pathology progression, such as onset of toxicity aggregation and the differential cell-specific susceptibility to aggregation, showed wide variation among the new introgression lines and between polyQ40 carrying recombinant inbred lines (RILs) derived from DR1350 and N2. This indicates that natural variation in genetic background can control resistance to misfolded protein aggregations and act to bind its associated cellular dysfunctions, which encourages further dissection on natural genetic variation susceptibility to age-related protein homeostasis in disease mechanisms.

Li et al. [49] mapped a large fraction of the *C. elegans* protein-protein interaction network. This was extended when the initial version of a human interactome map came out, adding more than 300 new connections to over 100 disease-associated proteins, including proteins related to NGD [73]. Lejeune et al. [48] accomplished a large-scale RNA interference screen in *C. elegans* strains that express N-terminal huntingtin in touch receptor neurons. Then, a subset of high-confidence modifier genes in pathways of interest in HD was identified by

Table 1 Pathways regulating longevity, stress, and disease responses

Insulin/Insulin-like growth factor (IIS) signaling pathway. The insulin-like receptor *DAF-2* is one of the main molecular components of IIS pathway in *C. elegans*, which is activated by binding insulin-like peptides (IPs) [44]. *DAF-2* activation by IPs results in recruitment and activation of AKT and SGK-1 (which are activated by phosphorylation of PDK-1 at the plasma membrane) to phosphorylate and sequester DAF-16 in the cytosol [95]. When activated by stresses, dephosphorylated DAF-16 enter the nucleus and regulates the transcription of a large number of genes involved in resistance to abiotic and biotic stresses, dauer larvae formation, metabolism, and longevity [63, 98]. DAF-16 can become dephosphorylated by the absence of the IP ligand(s), by inhibition from upstream pathway members—e.g., DAF-18 opposes the activity of PDK-1 and AKT-1 via inhibition of AGE-1—or by mutations in upstream genes, leading to lifespan extension [28, 56, 62, 72]. Strikingly, downregulation of IIS via *daf-2* or *age-1* mutations in *C. elegans* delays the formation of small aggregates in nematode models of NGDs and also slows the onset and decreases the severity of associated pathologies in a *daf-16*-dependent manner [17].

Mechanistic target of rapamycin (mTOR) pathway. The kinase mTOR integrates nutrient and anabolic signals to promote growth [52, 106, 112]. The TSC1/TSC2 complex acts as a negative regulator of mTORCs upstream, RHEB, but can be activated by AKT kinase (which belongs to IIS pathway) [36]. The mTORC signaling regulates a large number of developmental processes. TORC1 signaling transactivates/represses PH4-4 that induces pro-survival factors expression for life extension under nutrient restriction [85]. Moreover, TORC1 can increase protein synthesis and extends lifespan by both activating the ribosomal subunit S6 kinase (S6K) and inhibiting 4E-BP1 (a negative regulator of translation) [81, 97]. Meanwhile, investigations in *C. elegans* as well as other organisms have shown that autophagy is induced by inhibition of TORC1 [46]. An inadequate level of autophagy has been implicated in physiological responses to exercise and aging as well as in pathophysiological processes, such as cancer and metabolic and neurodegenerative disorders [47]. Although mammalian TORC2 (mTORC2) signaling shows insensitivity to nutrients in comparison to mTORC1, it does respond to growth factors like insulin through a poorly defined mechanism that requires PI3K (the *C. elegans* homolog is AGE-1) [111]. Moreover, mTORC2 directly activates AKT and SCK-1 which regulate cellular processes such as metabolism, survival, apoptosis, growth, and proliferation through the phosphorylation of several effectors [47]. In addition, TORC2 controls the cell cycle-dependent polarization of the actin cytoskeleton, which is involved in complex liposome mechanism, inducing several processes required for cancer cell growth, survival, and proliferation. Functioning as a nutrient sensor—detecting nutrients and amino acids—mTOR has a complex influence on several crucial cellular functions and shows clear effects on aging, protein synthesis and autophagy, and the homeostasis pathways that play a key role in the mechanisms that affect NGDs.

Mitochondrial signaling pathway. Mitochondria play important roles in aging and disease through apoptotic/programmed cell death (PCD), aberrant autophagic regulation, endoplasmic reticulum dysfunction, and intracellular calcium (see review [108]). The intrinsic apoptosis machinery of *C. elegans* comprises CED-9, CED-4, and CED-3, which is conserved from nematode to vertebrates and stimulates a protective response to mitochondrial dysfunction. JNK-1 promotes the expression of EGL-1 and CED-13. EGL-1 is required for all apoptosis, but CED-13 is required for pro-longevity signaling through the intrinsic pathway, i.e., mitochondrial reactive oxygen species (mROS) signaling pathway [80, 107]. Several stress factors, such as low IIS, heat stress, mitochondrial stress, and oxidative stress, among others, have been found to simulate the autophagic degradation of mitochondria (mitophagy). Besides sharing crucial regulatory factors with the general autophagy pathways, stressful conditions, specific components are also recruited for mitochondrial degradation. Outer mitochondrial membrane kinase PINK1 and its recruitment the cytosolic E3 ubiquitin ligase PARK2/parkin (also on the organelle) as well as their downstream signaling mediator, DCT-1, participate in mitophagy induction, mitochondrial homoeostasis protection, and survival promotion under stress conditions [65]. Mutations in PINK1 and PARK2 result in recessive familial forms of human Parkinson's disease and

Table 1 Pathways regulating longevity, stress, and disease responses *(Continued)*

correlate with mitochondrial dysfunction in mouse models [13, 58]. Similar to DAF-16, the transcription of SKN-1, which plays a key regulator in mitochondrial signaling pathway, has an influence on longevity and stress responses as well as NGDs [12, 107]. SKN-1 activation can be triggered by increased numbers of damaged mitochondria as well as increased cytoplasmic calcium levels upon mitophagy inhibition [65]. However, phosphorylation of SKN-1 by GSK-3 and kinases downstream from the *DAF-2* insulin-like pathway (AKT-1, AKT-2, and SGK-1) negatively regulate SKN-1 nuclear accumulation and activation [12, 98]. Prohibitin PHB complex, PHB1/PHB2, affect mitochondrial metabolism by sensing free radicals and/or mediating ROS production to influence longevity [3]. Interestingly, another negative regulator of SNK-1, WDR-23, has been found to interact with and regulate nuclear accumulation of SKN-1 but function independently of DAF-16 [12, 89].

Furthermore, Ca2+ released from the endoplasmic reticulum (ER) can also lead to ROS accumulation. The perturbation of ER Ca2+ homeostasis causes mitochondrial dysfunction, activating the mitochondrial-mediated apoptotic pathway, which has also been implicated in neuronal death in AD mouse model [59, 65]. Additionally, protein-folding stress at the ER stimulates the unfolded protein response (UPR), which is involved in the pathogenesis of many human diseases. For example, the basal activity of the UPR is beneficial under normal conditions but accelerates the pathology caused by toxic Ab protein in a *C. elegans* model of AD [74].

network-based analysis, including metabolic, neurodevelopmental, and pro-survival pathways. These results and those of similar large-scale analyses support the investigation of human disease pathways using *C. elegans* as a model (Table 1).

C. elegans offers the opportunity to investigate the genetic background modifiers affecting complex disease pathways (Table 2). For example, comparisons of RAS/MAPK signaling—a pathway critical in many complex human diseases, including AD [43]—across two different genetic backgrounds, N2 and CB4856, as well as their derived recombinant inbred lines identified the polymorphic monoamine oxidase *amx-2* (MAOA) as a negative regulator of RAS/MARK [77]. It was found that MAOA's effects on RAS/MAPK signaling are produced by its effects on the level of the serotoninmetabolite 5 hydroxyindoleacetic acid (5-HIAA), the first endogenous small molecule identified to act as a systemic inhibitor of RAS/MAPK signaling [77].

Genetic networks in stress and aging leverage the detection of NGD modifiers

NGDs are associated with other complex diseases (e.g., cancers). In part, this is a consequence of the involvement of the cellular systems that deal with various types of stress. For instance, variation in the genes that regulate the brain's molecular response to oxidative stress are associated with differential neural vulnerability to the damaging effects of amyloid-β [34]. Here, oxidative phosphorylation takes place within mitochondria, to meet the elevated energy demands of neurons (Figs. 1, 2, and 3). However, this can also accelerate the accumulated oxidative damage, which could trigger impaired

Table 2 Quantitative genetic studies of genomic and phenotypic variation

The majority of quantitative trait locus (QTL) mapping approaches require inbred strains that have different alleles at loci affecting variation in the trait of interest and a polymorphic molecular marker linkage map. A cross between two genetically distinct parental strains or a series of crosses among parental lines produces recombinant inbred lines (RILs) or other segregating populations, such as introgression lines (IL). In most cases, the final panel is repeatedly inbred to obtain isogenic, and thus genetically stable, lines. As a consequence of random recombination events, each RIL combines different parts of the two (or more) parental genomes. In most previous research on *C. elegans*, the Hawaiian CB4856 strain—a strain that is highly divergent from N2 at many loci—and N2 have been used to detect QTLs [7, 70]. Various N2xCB4856 RILs and ILs present variable transgressive traits, and large genetic and phenotypic differences for a wide range of traits including reproduction, growth, and gene expression.

QTL mapping has however a number of shortcomings. Since the location of identified QTLs is indicated only by looking at which markers give the greatest differences between genotype class averages, QTL effects are often poor estimates of the true allele(s) effect. Moreover, when QTLs are far from markers—as has often been the case in earlier work when the markers are widely spaced—the mapping resolution is low. Furthermore, the network of genes as well as by environmental factors can regulate the complex dynamic process, such as disease progressing. Therefore, the QTLs or nucleotides (QTNs) that underlie a complex dynamic trait are expected to be characterized. Despite functional mapping provides a useful quantitative and testable framework for assessing the interplay between gene actions or interactions and developmental changes [105], further investigations are still needed through its opportunity to study development in a comprehensive manner and to study the dynamic network of genes that determine the physiology of an individual organism over time.

For understanding genetic pathways and gene networks, mapping of gene expression levels as quantitative trait loci (called eQTLs) has also been increasingly used for determining how a given variant affects gene expression. Sequence variants in eQTLs affect the expression of a gene, while other QTLs impact on any given trait of interest except from the trait causing by gene expression [71]. Owing to available measurement of either traits or phenotypes, eQTLs have been identified by evaluating gene expression in panels of genetically distinct genotyped individuals [29, 71].

mitochondrial energy production and upregulate oxidative phosphorylation causing further DNA damage and significant levels of neuronal apoptosis [21].

A large number of genome-wide association studies (GWAS) in humans have identified pathways modulating the rate of aging and simultaneously influencing multiple age-related diseases such as NGDs [68, 76]. These GWAS studies have been invaluable in identifying and characterizing genetic variants associated with variation in disease phenotypes. For example, Seshadri et al. [83] identified 38 SNPs within 10 loci by integrating data from multiple studies and also identified CLU, PICALM, and CR1 as novel genes for late-onset AD (LOAD). A similar meta-analysis of GWAS studies of PD identified six more risk loci associated with the disease [60], while a combination of GWAS and whole-exome sequencing (WES) identified variants in the CEL gene/locus associated with PD [86]. Similarly, N'Songo et al. [61] identified ABCA7 missense variants that play a role in conferring AD risk in African Americans by performing WES with previously identified AD GWAS loci. According to their highlighted allelic heterogeneity at this locus, they also suggested the presence of additional AD risk variants in MS4A6A, PTK2B, and ZCWPW1.

NGDs are also strongly associated with aging, indicating that there is a link between protein misfolding and aging. This is very clear when the ILS (FOXO) pathway, a representative aging-related pathway, is considered. In *C. elegans*, this pathway ultimately controls the localization of DAF-16—a FOXO transcription factor that regulates a large number of genes involved in abiotic and biotic stress resistance, metabolism, and longevity—with dephosphorylation of DAF-16 allowing it to enter the nucleus [32]. The phosphorylation of DAF-16 is controlled by the insulin/IGF-1 transmembrane receptor ortholog *DAF-2*, with reductions in *daf-2* activity resulting in *daf-16*-dependent increases in lifespan. Loss of function *daf-2* mutations also increase lifespan in PD mutants of *C. elegans*, delaying the accumulation of small aggregates of alpha-synuclein in the body wall muscle and rescuing deficiencies in resistance to different stresses [17]. In humans, the age of onset of HD has a strong association with the length of polyQ expansion in the huntingtin protein but also varies between individuals with the same repeat length. In laboratory mouse strains, different genetic backgrounds can induce differential somatic expansion of CAG repeat.

Perspectives: what is the genetic architecture of variation in NGD responses in *C. elegans*?

C. elegans quantitative genetics (Table 2) has yielded considerable insights into understanding complex human disease pathways, but there has been limited work in extending this to the analysis of NGDs. To this end, powerful and serviceable datasets relating to causal genetic variant exploration based on genomic analysis have been collected and can be obtained from WormQTL^{HD} (human disease) www.WormQTL-HD.org [100]. WormQTL^{HD} allows for systematically investigating phenotypic expression of *C. elegans* at levels equivalent to those of human diseases, by catalyzing integration of reported disease candidate gene associations, gene orthologue data, molecular profiles, phenotypic variations, and QTL results [93, 100]. Thus, this will support relevant available meta-datasets for human-worm studies and database exploration.

Taken together, *C. elegans,* a powerful model organism, can be used to study how variation affects the onset and progression of protein-misfolding disease and how the susceptibility to proteotoxicity performs in genetically diverse but phenotypically general individuals. This may explain the higher propensity to aggregation of the mutant disease-related proteins, in order to uncovering onset and phenotypes of disease patterns [27]. Regardless of the toxicity and aggregation of misfolded proteins, other factors leading to the variable complex traits could possibly include multiple additive or allele interactions, with the consequence of underlying intervention strategies at onset and/or in progression of disease. In addition, Paaby et al. [64] revealed cryptic genetic variations (CGVs) in the gene networks of *C. elegans* embryogenesis. CGVs are silent alleles in general and can be only activated to influence phenotype, when other functional genes are perturbed. The seemingly omnipresent cryptic-effect loci segregate at intermediate frequencies in the wild. However, they reveal low developmental pleiotropy. In the specific perturbations, e.g., changes of the molecular, cellular, or developmental processes that govern its phenotypic expression in complex metabolic human diseases, CGVs are required to be revealed [42, 56]; for reviews, see [25, 26]. Above all, the natural variation play an important role in neurodegeneration, involving inherently plastic genetic and molecular pathways, and might allow for description of complex etiology and implementation in eluding the harmful influences.

Conclusions

Given the molecular conservation in neuronal signaling between *C. elegans* and vertebrates, including humans, this nematode is a valuable model species for studying NGD pathways and the alleles that affect them. Despite progress in identification of several AD- and PD-related genes, the effect of natural alleles underlying protein misfolding in these diseases remains mostly unclear. Here, we reviewed how natural variation could influence the severity of disease phenotypes. Experiments that rely solely on induced mutants in Bristol N2 limit the ability to explore how naturally varying alleles alter signaling pathways. Thus, further research should go beyond classic mutant screens on the genetic pathway analysis of complex traits, i.e., phenotypic differences among individuals, to also consider the allelic interactions in different genetic backgrounds. Consequently, the predictive nematode models of human genetic diseases could provide a more complete genetic and molecular understanding of how genetic variation shapes gene expression and cell biology for personalized genomic medicine.

Abbreviations

5-HIAA: Serotoninmetabolite 5 hydroxyindoleacetic acid; AD: Alzheimer's disease; ALS: Amyotrophic lateral sclerosis; APP: Amyloid precursor protein; Aβ: Beta amyloid peptide; BDNF: Brain-derived neurotrophic factor; CGVs: Cryptic genetic variations; CNV: Copy number variants; eQTLs: Expression quantitative trait loci; ER: Endoplasmic reticulum; FOXO: Forkhead box O protein; FTD: Frontotemporal dementia; GWAS: Genome wide association studies; HD: Huntington's disease; HTT: Huntingtin; IIS: Insulin/insulin-like growth factor; ILs: Introgression lines; JNK: c-Jun N-terminal kinase; LOAD: Late-onset Alzheimer's disease; LRRK2: Leucine-rich repeat kinase 2; MAOA: Monogeamine oxidase *amx-2*; MAPK: Mitogen-activated protein kinases; MLH1: MutL homolog 1; mROS: Mitochondrial reactive oxygen species; mTOR: Mechanistic target of rapamycin; mTORC2: Mammalian TORC2; NGDs: Neurodegenerative diseases; PCD: Programmed cell death; PD: Parkinson's disease; polyQ: Polyglutamine; PrD: Prion diseases; QTLs: Quantitative trait locus; QTNs: Quantitative trait nucleotides; RILs: Recombinant inbred lines; ROS: Reactive oxygen species; S6K: Subunit S6 kinase; SNP: Single nucleotide polymorphism; UPR: Unfolded protein response; UPS: Ubiquitin/proteasome system; WES: Whole-exome sequencing

Acknowledgements

We thank Mark Sterken, Basten Snoek, Jana Stastna, Katharina Jovic, Yu Nie, and the members of our laboratories for helpful advice and discussions. We also thank Bei Tian, Jun Qiu, Yongran Ji, Chunyue Wei, and Dianfan Zhou for assistance.

Funding

YW is supported by a PhD Scholarship from Canterbury Christ Church University. JK is supported by the Human Frontiers Science Program. SH is supported by the Leverhulme Trust.

Consent for publication

Not applicable.

Authors' contributions

YW, JK, and SH wrote the manuscript. All authors read and approved the final manuscript.

Competing interests

The authors declare that they have no competing interests.

Author details

[1]Biomolecular Research Group, School of Human and Life Science, Canterbury Christ Church University, Canterbury CT1 1QU, UK. [2]Laboratory of Nematology, Wageningen University, 6708 PB Wageningen, The Netherlands.

References

1. Albert FW, Kruglyak L. The role of regulatory variation in complex traits and disease. Nat Rev Genet. 2015;16:197–212.
2. Andersen EC, Shimko TC, Crissman JR, et al. A powerful new quantitative genetics platform, combining *Caenorhabditis elegans* high-throughput fitness assays with a large collection of recombinant strains. G3 (Bethesda). 2015;5:911–20.
3. Artal-Sanz M, Tavernarakis N. Prohibitin and mitochondrial biology. Trends Endocrinol Metab. 2009;20:394–401.
4. Baleriola J, Walker CA, Jean YY, et al. Axonally synthesized ATF4 transmits a neurodegenerative signal across brain regions. Cell. 2014;158:1159–72.
5. Balla KM, Andersen EC, Kruglyak L, Troemel ER. A wild *C. elegans* strain has enhanced epithelial immunity to a natural microsporidian parasite. PLoS Pathog. 2015;11:1–21.
6. Becanovic K, Norremolle A, Neal SJ, et al. A SNP in the HTT promoter alters NF-[kappa]B binding and is a bidirectional genetic modifier of Huntington disease. Nat Neurosci. 2015;18:807–16.
7. Bendesky A, Pitts J, Rockman M V, et al. Long-range regulatory polymorphisms affecting a GABA receptor constitute a quantitative trait locus (QTL) for social behavior in *Caenorhabditis elegans*. PLoS Genet. 2012; 8(12):e1003157.
8. Bertram L, Bertram L, Tanzi RE, Tanzi RE. The genetic epidemiology of neurodegenerative disease. J Clin Invest. 2005;115:1449–57.
9. Bressler J, Mosley TH, Penman A, et al. Genetic variants associated with risk of Alzheimer's disease contribute to cognitive change in midlife: the Atherosclerosis

Risk in Communities Study. Am J Med Genet Part B Neuropsychiatr Genet. 2017; 174:269–82.

10. Chakraborty S, Bornhorst J, Nguyen TT, Aschner M. Oxidative stress mechanisms underlying Parkinson's disease-associated neurodegeneration in *C. elegans*. Int J Mol Sci. 2013;14:23103–28.
11. Chaudhry M, Wang X, Bamne MN, et al. Genetic variation in imprinted genes is associated with risk of late-onset Alzheimer's disease. J Alzheimers Dis. 2015;44:989–94.
12. Choe KP, Przybysz AJ, Strange K. The WD40 repeat protein WDR-23 functions with the CUL4/DDB1 ubiquitin ligase to regulate nuclear abundance and activity of SKN-1 in *Caenorhabditis elegans*. Mol Cell Biol. 2009;29:2704–15.
13. Choi AMK, Ryter SW, Levine B. Autophagy in human health and disease. N Engl J Med. 2013;368:651–62.
14. Chow CY, Wolfner MF, Clark AG. Using natural variation in Drosophila to discover previously unknown endoplasmic reticulum stress genes. Proc Natl Acad Sci. 2013;110:9013–8.
15. Ciechanover A, Kwon YT. Degradation of misfolded proteins in neurodegenerative diseases: therapeutic targets and strategies. Exp Mol Med. 2015;47:e147.
16. Collier TJ, Redmond DE, Steece-Collier K, et al. Is alpha-synuclein loss-of-function a contributor to parkinsonian pathology? Evidence from non-human primates. Front Neurosci. 2016;10:1–7.
17. Cooper JF, Dues DJ, Spielbauer KK, et al. Delaying aging is neuroprotective in Parkinson's disease: a genetic analysis in *C. elegans* models. npj Park Dis. 2015;1:15022.
18. Cutter AD. Nucleotide polymorphism and linkage disequilibrium in wild populations of the partial selfer *Caenorhabditis elegans*. Genetics. 2006;172:171.
19. Dawson TM, Ko HS, Dawson VL. Genetic animal models of Parkinson's disease. Neuron. 2010;66:646–61.
20. Del Tredici K, Braak H. Lewy pathology and neurodegeneration in premotor Parkinson's disease. Mov Disord. 2012;27:597–607.
21. Driver JA. Inverse association between cancer and neurodegenerative disease: review of the epidemiologic and biological evidence. Biogerontology. 2014;15:547–57.
22. Faber PW, Voisine C, King DC, et al. Glutamine/proline-rich PQE-1 proteins protect *Caenorhabditis elegans* neurons from huntingtin polyglutamine neurotoxicity. Proc Natl Acad Sci U S A. 2002;99:17131–6.
23. Ferrer I, Goutan E, Marín C, et al. Brain-derived neurotrophic factor in Huntington disease. Brain Res. 2000;866:257–61.
24. Fusakio ME, Willy JA, Wang Y, et al. Transcription factor ATF4 directs basal and stress-induced gene expression in the unfolded protein response and cholesterol metabolism in the liver. Mol Biol Cell. 2016;27:1536–51.
25. Gaertner BE, Parmenter MD, Rockman MV, et al. More than the sum of its parts: a complex epistatic network underlies natural variation in thermal preference behavior in *Caenorhabditis elegans*. Genetics. 2012;192:1533–42.
26. Gaertner BE, Phillips PC. *Caenorhabditis elegans* as a platform for molecular quantitative genetics and the systems biology of natural variation. Genet Res (Camb). 2010;92:331–48.
27. Gidalevitz T, Wang N, Deravaj T, et al. Natural genetic variation determines susceptibility to aggregation or toxicity in a *C. elegans* model for polyglutamine disease. BMC Biol. 2013;11:100.
28. Gil EB, Malone Link E, Liu LX, et al. Regulation of the insulin-like developmental pathway of *Caenorhabditis elegans* by a homolog of the PTEN tumor suppressor gene. Proc Natl Acad Sci. 1999;96:2925–30.
29. Gutteling EW, Riksen J a G, Bakker J, Kammenga JE. Mapping phenotypic plasticity and genotype-environment interactions affecting life-history traits in *Caenorhabditis elegans*. Heredity (Edinb). 2007;98:28–37.
30. Harrington AJ, Hamamichi S, Caldwell GA, Caldwell KA. *C. elegans* as a model organism to investigate molecular pathways involved with Parkinson's disease. Dev Dyn. 2010;239:1282–95.
31. Hassan WM, Dostal V, Huemann BN, et al. Identifying Aβ-specific pathogenic mechanisms using a nematode model of Alzheimer's disease. Neurobiol Aging. 2015;36:857–66.
32. Hesp K, Smant G, Kammenga JE. *Caenorhabditis elegans* DAF-16/FOXO transcription factor and its mammalian homologs associate with age-related disease. Exp Gerontol. 2015;72:1–7.
33. Hodgkin J, Doniach T. Natural variation and copulatory plug formation in *Caenorhabditis elegans*. Genetics. 1997;146:149–64.
34. Hohman TJ, Koran MEI, Thornton-Wells TA. Genetic variation modifies risk for neurodegeneration based on biomarker status. Front Aging Neurosci. 2014;6:1–11.
35. Hooli BV, Kovacs-Vajna ZM, Mullin K, et al. Rare autosomal copy number variations in early-onset familial Alzheimer's disease. Mol Psychiatry. 2013;19:676–81.
36. HUANG J, MANNING BD. The TSC1–TSC2 complex: a molecular switchboard controlling cell growth. Biochem J. 2008;412:179–90.
37. Imarisio S, Carmichael J, Korolchuk V, et al. Huntington's disease: from pathology and genetics to potential therapies. Biochem J. 2008;412:191–209.
38. Jack CR, Knopman DS, Jagust WJ, et al. Tracking pathophysiological processes in Alzheimer's disease: an updated hypothetical model of dynamic biomarkers. Lancet Neurol. 2013;12:207–16.
39. Jeon J, Kim W, Jang J, et al. Gene therapy by proteasome activator, PA28γ, improves motor coordination and proteasome function in Huntington's disease YAC128 mice. Neuroscience. 2016;324:20–8.
40. Kamkina P, Snoek LB, Grossmann J, et al. Natural genetic variation differentially affects the proteome and transcriptome in *C. elegans*. Mol Cell Proteomics. 2016;15(5):1670–80.
41. Kammenga JE. The background puzzle: how identical mutations in the same gene lead to different disease symptoms. The FEBS Journal. 2017. doi:10.1111/febs.14080.
42. Kammenga JE, Phillips PC, De Bono M, Doroszuk A. Beyond induced mutants: using worms to study natural variation in genetic pathways. Trends Genet. 2008;24:178–85.
43. Kim EK, Choi E-J. Pathological roles of MAPK signaling pathways in human diseases. Biochim Biophys Acta - Mol Basis Dis. 2010;1802:396–405.
44. Kimura KD, Tissenbaum HA, Liu Y, Ruvkun G. *daf-2*, an insulin receptor-like gene that regulates longevity and diapause in *Caenorhabditis elegans*. Science. 1997;277:942–6.
45. Lambert J-C, Ibrahim-Verbaas CA, Harold D, et al. Meta-analysis of 74,046 individuals identifies 11 new susceptibility loci for Alzheimer's disease. Nat Genet. 2013;45:1452–8.
46. Lapierre LR, Meléndez A, Hansen M. Autophagy links lipid metabolism to longevity in *C. elegans*. Autophagy. 2012;8:144–6.
47. Laplante M, Sabatini DM. Regulation of mTORC1 and its impact on gene expression at a glance. J Cell Sci. 2013;126:1713–9.
48. Lejeune FX, Mesrob L, Parmentier F, Bicep C, Vazquez-Manrique RP, Parker JA, Vert JP, Tourette C, Neri C. Large-scale functional RNAi screen in *C. elegans* identifies genes that regulate the dysfunction of mutant polyglutamine neurons. BMC genomics. 2012;13(1):91.
49. Li S, Armstrong CM, Bertin N, et al. A map of the interactome network of the metazoan *C. elegans*. Science. 2004;303:540–3.
50. Link CD. *C. elegans* models of age-associated neurodegenerative diseases: lessons from transgenic worm models of Alzheimer's disease. Exp Gerontol. 2006;41:1007–13.
51. Lu J-X, Qiang W, Yau W-M, et al. Molecular structure of β-amyloid fibrils in Alzheimer's disease brain tissue. Cell. 2013;154:1257–68.
52. Ma XM, Blenis J. Molecular mechanisms of mTOR-mediated translational control. Nat Rev Mol Cell Biol. 2009;10:307–18.
53. McColl G, Roberts BR, Pukala TL, et al. Utility of an improved model of amyloid-beta (Aβ1-42) toxicity in *Caenorhabditis elegans* for drug screening for Alzheimer's disease. Mol Neurodegener. 2012;7:57.
54. McGrath PT, Xu Y, Ailion M, et al. Parallel evolution of domesticated Caenorhabditis species targets pheromone receptor genes. Nature. 2011;477:321–5.
55. Mercado G, Valdés P, Hetz C. An ERcentric view of Parkinson's disease. Trends Mol Med. 2013;19:165–75.
56. Mihaylova VT, Borland CZ, Manjarrez L, et al. The PTEN tumor suppressor homolog in *Caenorhabditis elegans* regulates longevity and dauer formation in an insulin receptor-like signaling pathway. Proc Natl Acad Sci U S A. 1999;96:7427–32.
57. Monfrini E, Di Fonzo A. In: Rideout HJ, editor. Leucine-rich repeat kinase (LRRK2) genetics and Parkinson's disease BT—leucine-rich repeat kinase 2 (LRRK2). Cham: Springer International Publishing; 2017. p. 3–30.
58. Morris BJ, Willcox DC, Donlon TA, Willcox BJ. FOXO3: a major gene for human longevity—a mini-review. Gerontology. 2015;61:515–25.
59. Mota SI, Costa RO, Ferreira IL, et al. Oxidative stress involving changes in Nrf2 and ER stress in early stages of Alzheimer's disease. Biochim Biophys Acta - Mol Basis Dis. 2015;1852:1428–41.
60. Nalls MA, Pankratz N, Lill CM, et al. Large-scale meta-analysis of genome-wide association data identifies six new risk loci for Parkinson's disease. Nat Genet. 2014;46:989–93.

61. N'Songo A, et al. Comprehensive screening for disease risk variants in early-onset Alzheimer's disease genes in African Americans identifies novel PSEN variants. J Alzheimers Dis. 2017;56(4):1215–22.
62. Ogg S, Ruvkun G. The *C. elegans* PTEN Homolog, DAF-18, acts in the insulin receptor-like metabolic signaling pathway. Mol Cell. 2016;2:887–93.
63. Oh SW, Mukhopadhyay A, Svrzikapa N, et al. JNK regulates lifespan in *Caenorhabditis elegans* by modulating nuclear translocation of forkhead transcription factor/DAF-16. Proc Natl Acad Sci U S A. 2005;102:4494–9.
64. Paaby A, White A, Riccardi D. *C. elegans* harbors pervasive cryptic genetic variation for embryogenesis *C. elegans* harbors pervasive cryptic genetic variation for embryogenesis. 2014. p. 1–31.
65. Palikaras K, Lionaki E, Tavernarakis N. Coupling mitogenesis and mitophagy for longevity. Autophagy. 2015;11:1428–30.
66. Petersen C, Dirksen P, Schulenburg H. Why we need more ecology for genetic models such as *C. elegans*. Trends Genet. 2015;31:120–7.
67. Phillips PC. Epistasis—the essential role of gene interactions in the structure and evolution of genetic systems. Nat Rev Genet. 2008;9:855–67. doi:10.1038/nrg2452.
68. Ramanan VK, Saykin AJ. Pathways to neurodegeneration: mechanistic insights from GWAS in Alzheimer's disease, Parkinson's disease, and related disorders. Am J Neurodegener Dis. 2013;2:145–75.
69. Rittiner JE, Caffall ZF, Hernandez-Martinez R, et al. Functional genomic analyses of mendelian and sporadic disease identify impaired eIF2α signaling as a generalizable mechanism for dystonia. 2016.
70. Rockman MV. The QTN program and the alleles that matter for evolution: all that's gold does not glitter. Evolution (NY). 2012;66:1–17.
71. Rodriguez M, Snoek LB, Riksen JAG, et al. Genetic variation for stress-response hormesis in *C. elegans* lifespan. Exp Gerontol. 2012;47:581–7.
72. Rouault J-P, Kuwabara PE, Sinilnikova OM, et al. Regulation of dauer larva development in *Caenorhabditis elegans* by *daf-18*, a homologue of the tumour suppressor *PTEN*. Curr Biol. 2016;9:329–34.
73. Rual J-F, Venkatesan K, Hao T, et al. Towards a proteome-scale map of the human protein-protein interaction network. Nature. 2005;437:1173–8.
74. Safra M, Ben-Hamo S, Kenyon C, Henis-Korenblit S. The ire-1 ER stress-response pathway is required for normal secretory-protein metabolism in *C. elegans*. J Cell Sci. 2013;126:4136–46.
75. Saito T, Sadoshima J (2016) HHS Public Access. 116:1477–1490.
76. Sanchez-Mut JV, Heyn H, Vidal E, et al. Human DNA methylomes of neurodegenerative diseases show common epigenomic patterns. Transl Psychiatry. 2016;6:e718.
77. Schmid T, Snoek LB, Fröhli E, et al. Systemic regulation of RAS/MAPK signaling by the serotonin metabolite 5-HIAA. PLoS Genet. 2015;11:e1005236.
78. Schneider L, Zhang J. Lysosomal function in macromolecular homeostasis and bioenergetics in Parkinson's disease. Mol Neurodegener. 2010;5:14.
79. Schrag A, Schott JM. Epidemiological, clinical, and genetic characteristics of early-onset parkinsonism. Lancet Neurol. 2006;5:355–63.
80. Schumacher B, Hanazawa M, Lee MH, et al. Translational repression of *C. elegans* p53 by GLD-1 regulates DNA damage-induced apoptosis. Cell. 2005;120:357–68.
81. Selman C, Tullet JMA, Wieser D, et al. Ribosomal protein S6 Kinase 1 signaling regulates mammalian life span. Science. 2009;326:140–4.
82. Serrano-Pozo A, Frosch MP, Masliah E, Hyman BT. Neuropathological alterations in Alzheimer disease. Cold Spring Harb Perspect Med. 2011;1:a006189.
83. Seshadri S, Fitzpatrick AL, Arfan Ikram M, et al. Genome-wide analysis of genetic loci associated with Alzheimer's disease. JAMA. 2010;303:1832–40.
84. Settivari R, VanDuyn N, LeVora J, Nass R. The Nrf2/SKN-1-dependent glutathione S-transferase π homologue GST-1 inhibits dopamine neuron degeneration in a *Caenorhabditis elegans* model of manganism. Neurotoxicology. 2013;38:51–60.
85. Sheaffer KL, Updike DL, Mango SE. The target of rapamycin (TOR) pathway antagonizes pha-4/FoxA to control development and aging. Curr Biol. 2008;18:1355–64.
86. Siitonen A, Nalls MA, Hernández D, et al. Genetics of early-onset Parkinson's disease in Finland: exome sequencing and genome-wide association study. Neurobiol Aging. 2017;53:195.e7–195.e10.
87. Singh KD, Roschitzki B, Snoek LB, *et al.* Natural genetic variation influences protein abundances in *C. elegans* developmental signalling pathways. PLoS One. 2016;11(3):e0149418.
88. Soto C. Protein misfolding in neurodegenerative diseases: the key pending questions. 2013. p. 19–22.
89. Staab TA, Griffen TC, Corcoran C, et al. The conserved SKN-1/Nrf2 stress response pathway regulates synaptic function in *Caenorhabditis elegans*. PLoS Genet. 2013;9(3):e1003354.
90. Stastna JJ, Snoek LB, Kammenga JE, Harvey SC. Genotype-dependent lifespan effects in peptone deprived *Caenorhabditis elegans*. Sci Rep. 2015;5:16259.
91. Stöhr J, Condello C, Watts JC, et al. Distinct synthetic Aβ prion strains producing different amyloid deposits in bigenic mice. Proc Natl Acad Sci. 2014;111:10329–34.
92. Stranger BE, Forrest MS, Dunning M, et al. Relative impact of nucleotide and copy number variation on gene expression phenotypes. Science. 2007;315:848–53.
93. Swertz MA, van der Velde KJ, Tesson BM, et al. XGAP: a uniform and extensible data model and software platform for genotype and phenotype experiments. Genome Biol. 2010;11:R27.
94. Thompson D, Regev A, Roy S. Comparative analysis of gene regulatory networks: from network reconstruction to evolution. Annu Rev Cell Dev Biol. 2015;31:399–428.
95. Tissenbaum HA. Genetics, life span, health span, and the aging process in *Caenorhabditis elegans*. J Gerontol A Biol Sci Med Sci. 2012;67(A):503–10.
96. Triplett JC, Zhang Z, Sultana R, et al. Quantitative expression proteomics and phosphoproteomics profile of brain from PINK1 knockout mice: Insights into mechanisms of familial Parkinson's disease. J Neurochem. 2015;133:750–65.
97. Troulinaki K, Bano D. Mitochondrial deficiency: a double-edged sword for aging and neurodegeneration. Front Genet. 2012;3:1–10.
98. Tullet JMA. DAF-16 target identification in *C. elegans*: past, present and future. Biogerontology. 2015;16:221–34.
99. Ukraintseva SV, Arbeev KG, Akushevich I, et al. Trade-offs between cancer and other diseases: do they exist and influence longevity? Rejuvenation Res. 2010;13:387–96.
100. Van Der Velde KJ, De Haan M, Zych K, et al. WormQTLHD—a web database for linking human disease to natural variation data in *C. elegans*. Nucleic Acids Res. 2014;42:794–801.
101. Van Ham TJ, Thijssen KL, Breitling R, et al. *C. elegans* model identifies genetic modifiers of alpha-synuclein inclusion formation during aging. 2008.
102. Volkers RJM, Snoek LB, van H Hubar CJ, et al. Gene-environment and protein-degradation signatures characterize genomic and phenotypic diversity in wild *Caenorhabditis elegans* populations. BMC Biol. 2013;11:93.
103. Vu V, Verster AJ, Schertzberg M, et al. Natural variation in gene expression modulates the severity of mutant phenotypes. Cell. 2015;162:391–402.
104. Watts JC, Condello C, Stöhr J, et al. Serial propagation of distinct strains of Aβ prions from Alzheimer's disease patients. Proc Natl Acad Sci U S A. 2014; 111:10323–8.
105. Wu RL, Lin M. Opinion: functional mapping—how to map and study the genetic architecture of dynamic complex traits. Nat Rev Genet. 2006;7:229–37.
106. Wullschleger S, Loewith R, Hall MN. TOR signaling in growth and metabolism. Cell. 2006;124:471–84.
107. Yee C, Yang W, Hekimi S. The intrinsic apoptosis pathway mediates the pro-longevity response to mitochondrial ROS in C elegans. Cell. 2014;157:897–909.
108. Zaltieri M, Longhena F, Pizzi M, et al. Mitochondrial dysfunction and alpha-synuclein synaptic pathology in Parkinson's disease: who's on first? Parkinsons Dis. 2015;2015:1–10.
109. Zhang F, Gu W, Hurles ME, Lupski JR. Copy number variation in human health, disease, and evolution. Annu Rev Genomics Hum Genet. 2009;10:451–81.
110. Zhang X, Abels ER, Redzic JS, et al. Potential transfer of polyglutamine and CAG-repeat RNA in extracellular vesicles in Huntington's disease: background and evaluation in cell culture. Cell Mol Neurobiol. 2016;36:459–70.
111. Zinzalla V, Stracka D, Oppliger W, Hall MN. Activation of mTORC2 by association with the ribosome. Cell. 2011;144:757–68.
112. Zoncu R, Efeyan A, Sabatini DM. mTOR: from growth signal integration to cancer, diabetes and ageing. Nat Rev Mol Cell Biol. 2011;12:21–35.

8

A novel RLBP1 gene geographical area-related mutation present in a young patient with retinitis punctata albescens

Concetta Scimone[1,2], Luigi Donato[1,2], Teresa Esposito[3], Carmela Rinaldi[1*], Rosalia D'Angelo[1] and Antonina Sidoti[1,2]

Abstract

Background: Autosomal recessive forms of retinitis punctata albescens (RPA) have been described. RPA is characterized by progressive retinal degeneration due to alteration in visual cycle and consequent deposit of photopigments in retinal pigment epithelium. Five loci have been linked to RPA onset. Among these, the retinaldehyde-binding protein 1 gene, RLBP1, is the most frequently involved and several founder mutations were reported. We report results of a genetic molecular investigation performed on a large Sicilian family in which appears a young woman with RPA.

Results: The proband is in homozygous condition for a novel *RLBP1* single-pair deletion, and her healthy parents, both heterozygous, are not consanguineous. Thenovelc.398delC (p.P133Qfs*258) involves the exon 6 and leads to a premature stop codon, resulting in a truncated protein entirely missing of CRAL-TRIO lipid-binding domain.
Pedigree analysis showed other non-consanguineous relatives heterozygous for the same mutation in the family. Extension of mutation research in the native town of the proband revealed its presence also in healthy subjects, in a heterozygous condition.

Conclusions: A novel *RLBP1* truncating mutation was detected in a young girl affected by RPA. Although her parents are not consanguineous, the mutation was observed in a homozygous condition. Being them native of the same small Sicilian town of Fiumedinisi, the hypothesis of a geographical area-related mutation was assessed and confirmed.

Keywords: Retinitis punctata albescens, RLBP1, Frameshift mutation, Population study, Geographical-area related mutation, RP mutation spectrum

Background

Retinitis pigmentosa (RP) includes more than 70 different forms of inherited eye disorders characterized by progressive vision loss due to photoreceptors degeneration. This heterogeneity is caused by the high number of genes involved in disease's development. Therefore, each RP form differs from another for causative gene, inheritance pattern, symptomatology, age onset, and clinical features. About genetics, more than 50% of cases are affected by autosomal recessive forms due to mutations in almost 50 loci [1]. Among these, the *retinaldehyde-binding protein 1* (*RLBP1*) mutations are cause of retinitis punctata albescens (RPA, OMIM#136880) [2]. RPA is a progressive retinal degeneration belonging to the group of rod-cone dystrophies and has an early onset characterized by night blindness. At diagnosis, fundus appears with punctate white–yellow deposits. Progressive macular atrophy is the main cause of visual acuity loss, and narrowing of the visual field may occur in late teenage [3]. RPA has an incidence of 1/800000 people worldwide, and mutations at RLBP1 gene were reported only in about 1% of patients affected by autosomal recessive forms [2]. However, RPA cases associated with *Rhodopsin* (*RHO*), *Retinol Dehydrogenase 5* (*RDH5*), *Peripherin 2* (*PRPH2*), and *Lecithin Retinol Acyltransferase* (*LRAT*) genes were also reported [4–6]. About RLBP1 gene, different mutations are linked to a wider spectrum of phenotype as Bothnia retinal dystrophy (BD), Newfoundland rod-cone dystrophy (NFRCD), and fundus albipunctatus (FA). These forms differ for age of

* Correspondence: crinaldi@unime.it
[1]Department of Biomedical and Dental Sciences and Morphofunctional Imaging, Division of Molecular Genetics and Preventive Medicine, University of Messina, via C. Valeria 1, I-98125 Messina, Italy
Full list of author information is available at the end of the article

onset, progression, and for severity [7]. A founder effect was proven for mutation p.R234W, causing Bothnia phenotype [8], as well as for splice-junctions mutations detected in NFRCD cases [9]. So, different clinical manifestations are tightly linked to the effects of mutations on protein's structure. *RLBP1* is expressed in retinal pigment epithelium (RPE) and encodes for the cellular retinaldehyde-binding protein 1, CRALBP, involved in visual cycle. Particularly, CRALBP binds 11-cis-retinol which needs to be oxidized in 11-cis-retinal; 11-cis-retinal is carried into photoreceptors where it combines with opsin, resulting in visual pigments formations [10]. This process is common in both cones and rods.

We describe a novel *RLBP1* frameshift mutation, detected in a homozygous condition in a young 31-years-old woman, affected by RPA, certainly with no consanguineous parents up to the 4th generation ancestor, and native in Fiumedinisi, a small Sicilian town in province of Messina. Fiumedinisi is entered in a valley of Nebrodi mountains, at a height of about 200 above sea level. Homozygous condition for a novel mutation is a rare event in patients whose parents are not consanguineous; however, in this case, parents are both native of Fiumedinisi. This leads us to hypothesize that detected mutation is a geographical area-related variant.

Methods

Case analysis

Case description

Here we present a case of a 31-years-old woman (Fig. 1, IV:7), born in Fiumedinisi, a small Sicilian (Italy) town. Her parents, also native of Fiumedinisi, are not consanguineous. About the proband, at the age of five, myopia was diagnosed; two years later, she began suffering from night vision disturbance. Fundus revealed the typical "salt and pepper" aspect (Fig. 2a), without macula involvement; additionally, optical coherent tomography (OCT) examination showed the presence of numerous well-demarcated homogenous dome-shaped lesions, originating from the RPE layer (Fig. 2b). One year later, fundus analysis confirmed the progressive kind of phenotype, so RPA was diagnosed. Thereafter, she has report continuous worsening with also reduction of the visual field due to "retinal-tapetum" degeneration (Fig. 2c). No evidences of eye diseases were reported neither for her parents nor for younger sister.

Molecular analysis

DNA was extracted from peripheral blood. Exons and exon-intron bourdaries of *RLBP1, PRPH2, RHO, LRAT,* and *RDH5* genes were amplified by polymerase chain reaction (PCR). Primers sequences and PCR conditions are available upon request. Direct sequencing was performed using BigDye Terminator v3.1 chemistry and sequencing was ran on a 3500 Genetic Analyzer (Applied Biosystems). Molecular screening of all genes was performed in the proband; then, detected variants were also searched in the other family members.

In-silico prediction

Effects of the novel mutation on CRALBP tertiary structure were predicted by RaptorX (http://raptorx.uchicago.edu/) prediction tool.

Population study

Samples collection

To test the frequency of novel c.398delC variant, two groups made up by 300 and 600 subjects, respectively, heterogeneous for age and gender, were collected. The first one was recruited in Fiumedinisi and sampling criteria include Fiumedinisi native ancestors by at least

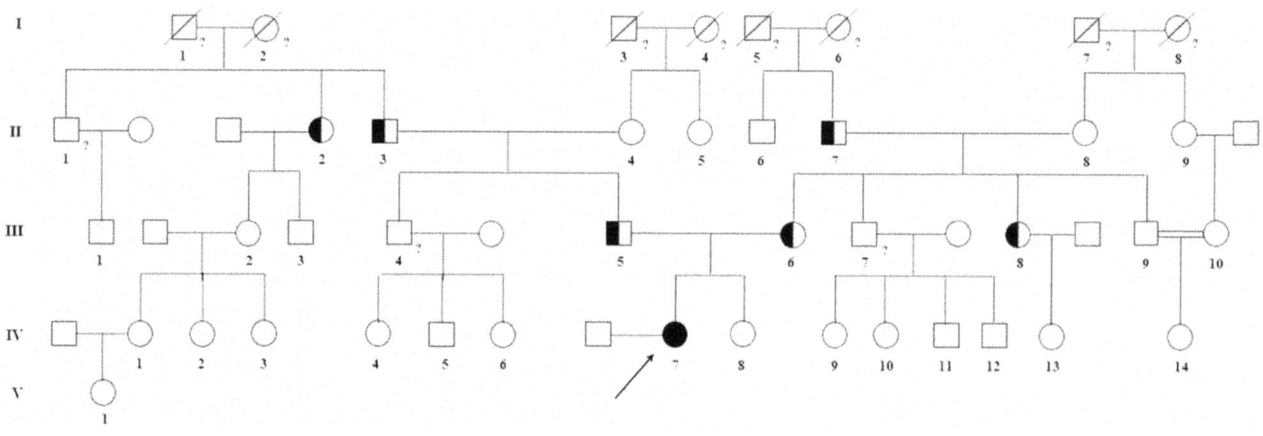

Fig. 1 Familial pedigree. The *arrow* indicates the proband (*IV:7*). *Empty squares* and *circles* symbolize wild-type men and women, respectively. *Half fill* indicates the heterozygous condition for the c.398delC mutation. *Question mark* indicates the unknown genotype due to unavailability of the sample. *Slash* indicates died individuals

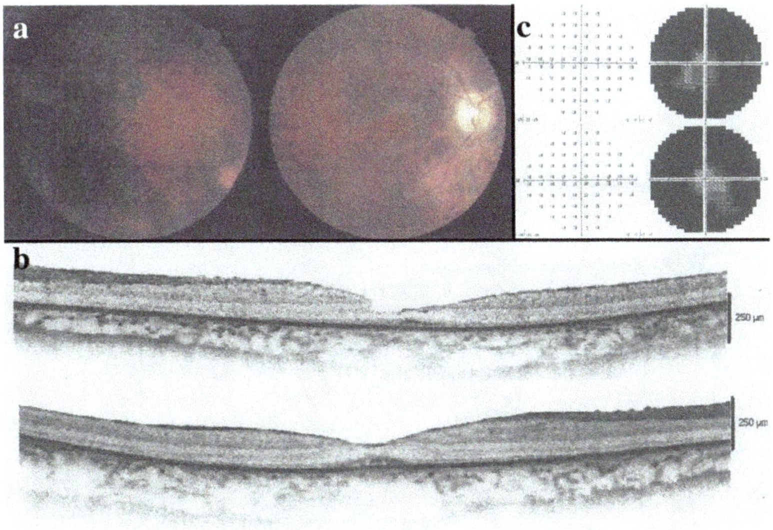

Fig. 2 Clinical investigations. Fundus photograph of proband's both eyes (**a**) shows papillary pallor, fundus albipunctatus, narrowed retinal arteries and temporal displacement of the papilla opaque vessels. A horizontal SD–OCT scan (**b**) shows numerous well-demarcated homogenous dome-shaped lesions originating from the RPE layer and diffused to the IS/OS junction of the photoreceptors, external limiting membrane, and into the outer nuclear layer. Visual field of both eyes (**c**) shows adeep impairment of visual field from tapeto-retinal degeneration; GHT's 30° Threshold Test denotes annular scotom in the explored area with central vision islands

three generations, absence of consanguinity, and absence of inherited disabling ocular diseases. The second group was collected in the Sicilian area: Ionic and Tyrrhenian municipality in province of Messina, like Santa Teresa Riva, Roccalumera, Pace del Mela, were excluded due to their geographic localization, since destinations of the inhabitants of Fiumedinisi who have left their native town. For each population, the c.398delC allele frequency was calculated as $[1 \times (h + 2H)]/2\ N$, where h represents the heterozygous genotype, H thehomozygous genotype, and N the sample size for each population. Deviation from Hardy-Weinberg equilibrium (HWE) was determined using the χ^2test with 2 × 3 contingency tables and 1 degree of freedom. Analysis was performed by the IBM SPSS statistical analysis software.

Molecular analysis

Buccal swabs were used for collecting saliva samples. DNA was purified by QIAamp® DNA Mini kit (Qiagen). Exon 6 of RLBP1 gene was amplified by PCR for all collected samples, then sequenced on 3500 Genetic Analyzer (Applied Biosystems), using BigDyeTermitor v3.1 chemistry.

All patients and controls involved in this study were fully informed and written consent was obtained. For underage subjects, consent was obtained from the parents. The study protocol followed the guidelines of our local ethics committee, and the investigation was conducted with the ethical requirements defined in the Helsinki Declaration.

Results

Case analysis

Direct sequencing of all known RPA causative genes *LRAT, RDH5*, and *RHO* showed no significant variants. Two SNPs rs390659 (c.910C > G, p.Q304E) and rs434102 (c.1013A > G, p.D338G) were detected in *PRPH2*; however, in ClinVar database, these are reported as benign (http://www.ncbi.nlm.nih.gov/clinvar/variation/138904/, http://www.ncbi.nlm.nih.gov/clinvar/variation/138906/). A novel single base-pair frameshift deletion, the c.398delC (p.P133Qfs*258), was detected in exon 6 of RLBP1 gene in a homozygous condition (Fig. 3a). It is not present in mutations and SNPs databases as HGMD, dbSNPs, ClinVar, ExAC, and Ensembl. The variant leads to substitution of proline 133 in glutamine and results in a frameshift with consequent premature termination at the 258th codon, in exon 8, causing the loss of the last 60 amino acids at the C-terminus of the protein. Figure 4 shows the results of in-silico prediction of tertiary structure alteration in mutated protein (Fig. 4b) compared to wild-type one (Fig. 4a). As evident, the c.398delC results in total disruption of functional domain as well as in an altered folding.

About other family's members, the novel mutation was observed in both parents (Fig. 1, III:5, III:6), in heterozygous condition (Fig. 3b) and was absent in the sister (Fig. 1, IV:8; Fig. 3c). Moreover, mutated allele was found both in paternal and in maternal relatives (Fig. 1, II:2, II:3, II:7, III:8).

Population study

As previously described, *RLBP1* mutations are very uncommon and founder mutations were reported. Frequency of

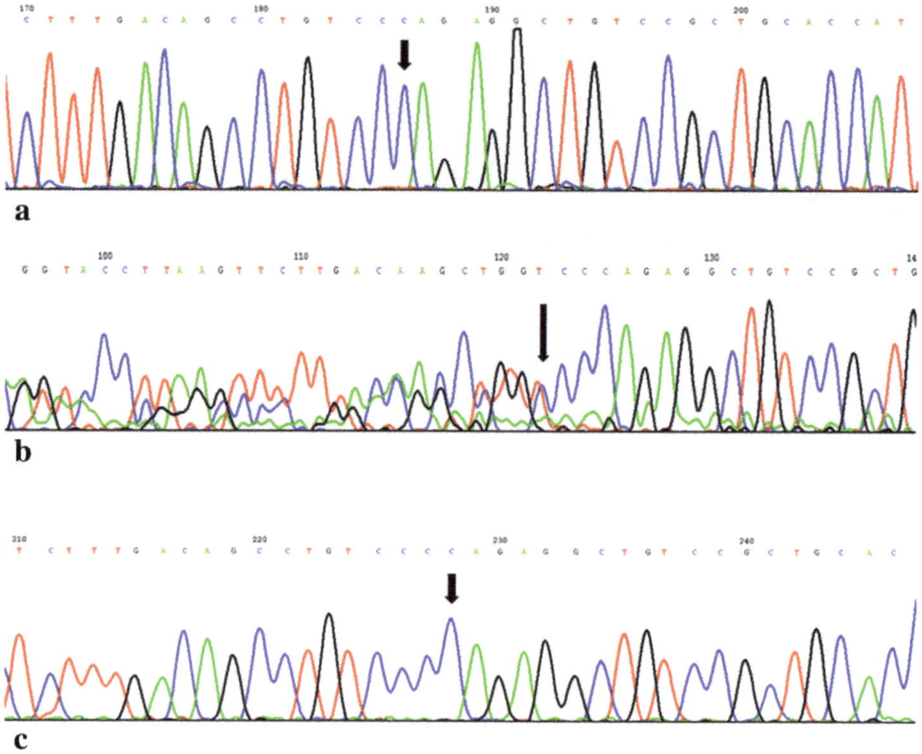

Fig. 3 Partial electropherograms of *RLBP1* exon 6. The *arrows* indicate affected nucleotide. **a** Electropherogram showing the homozygous condition for c.398delC mutation. **b** Electropherogram showing the heterozygous condition, sequence in reverse. **c** Wild-type sequence

c.398delC allele was assessed in both Fiumedinisi and Sicilian population. In the first group, the mutated allele was detected in 0.01%, while none of screened samples of general Sicilian population showed the variant. Table 1 shows the distribution of allele frequencies in Fiumedinisi population; χ^2 test results in a p value = 0.9306 showing the condition of HWE (HWE not consistent for $p < 0.05$).

Discussion

This study reports the results obtained by a molecular analysis performed on a large Sicilian family in which a case of RPA was diagnosed. Molecular screening of all known RPA causative genes showed a novel single base-pair frameshift deletion, c.398delC, in exon 6 of *RLBP1*. It leads to a truncated 258 amino acid protein. Premature

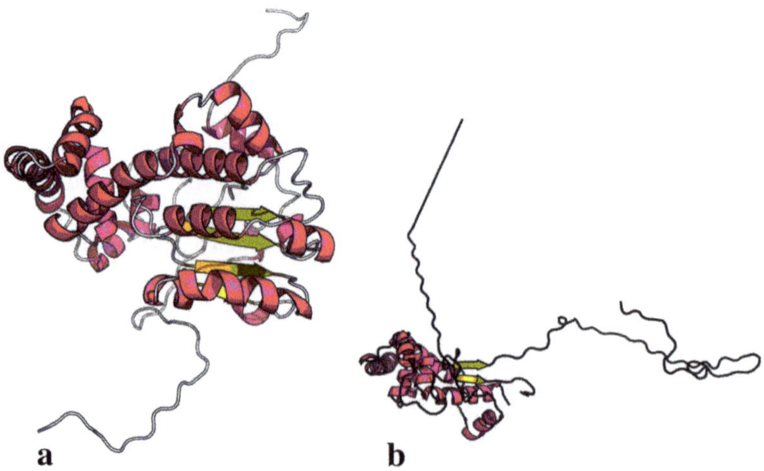

Fig. 4 CRALBP Tertiary structure alteration prediction. Structural models were generated by RaptorX tool. **a** Tertiary structure of wild-type protein. **b** Tertiary structure of p.P133Qfs*258 affected protein

Table 1 Wild-type (+) and c.398delC (−) allele frequencies in Fiumedinisi population

	Samples	Allele +	Allele −	Observed frequency	%	HW-expected frequency	%	Chi-square	*p* value
+/+	297	594	0	297	99.00	297.01	99.00	0.000	
+/−	3	3	3	3	1.00	2.99	1.00	0.000	
−/−	0	0	0	1	0.00	0.01	0.00	0.008	
Total	300	597	3	300	100.00	300.00	100.00	0.008	*0.9306*
	600	*1.00*	*0.01*	600				(*p* value) Chisq w 1 df	

A total of 300 samples were genotyped. Only 0.01% of screened population carries the mutated allele. Chi-square test exhibits a *p* value = 0.9306 suggesting the presence of the equilibrium's condition between wild-type and mutated allele in Fiumedinisi population

stop codon falls within exon 8. CRALBP protein contains one CRAL-TRIO lipid-binding domain, of 162 aminoacids (136-297). As predicted by in-silico analysis, the c.398delC affects proline 133 and causes total disruption of CRAL-TRIO domain, resulting in a loss of ability to bind its ligands. CRAL-TRIO motifs usually bind small hydrophobic molecules. In CRALBP, aminoacids 136-297 form CRAL-TRIO domain and, 12 of these, form the retinoid binding pocket [11]. Therefore, c.398delC has deleterious effects on protein's functionality.

Presence of variant c.398delC was assessed also in the other family's members; despite, they show no pathological phenotypes. Of these, the healthy father and other relatives showed this variant, in a heterozygous condition. This confirms autosomal recessive model of inheritance of RPA. The proband carries the novel c.398delC mutation in a homozygous condition. We proceeded to investigate frequency of variant in Fiumedinisi's population. Fiumedinisi, a small town in province of Messina, is located on the Ionian side, at the height of 190 above sea level (Fig. 5). It has an area of 36 km^2 and extends along Nisi's valley, at Mt. Belvedere's slopes, behind Valle del Mela. The city was founded in VII century B.C., by a group of Calcidesi Greek colonists, attracted by mineral deposits, who inhabited the high area of Mt. Belvedere. Later, Roman, Arab, Norman, Angevin, and Spanish dominions went on. During these centuries, the town was moved to the foot of mountain (where it remains today) and renamed as FlumenDionisyi.

Fig. 5 Focus of geographical area. On the leftmost are showed Italy and Sicily (*pink*); province of Messina is *red* highlighted and enlarged on the right of the picture. Red municipality is Fiumedinisi. Delimited areas are "Valle del Mela" and Nizza, Santa Teresa Riva and Roccalumera, respectively, on the Tyrrhenian and Ionian sides. The *upper picture* is an aerial view of Fiumedinisi

Still today, it results in not being easy to get to. Two big disasters hit the city: an epidemic plague in 1743 and a big flood in 1855. A period of decline, then, started due to high immigration's phenomenon. To date, Fiumedinisi's population counts about 1400 inhabitants, most of which are over the age of 50 years. For our aim, we screened about 25% of population and c.398delC was found in 0.01% and it is in HWE. No other RPA cases were detected in the town. Although there is a high rate of consanguinity in Fiumedinisi's population, its low frequency may be attributed to several dominations, repeated drastic reductions in population, the probable selective disadvantage, and a recent founder effect. This last hypothesis can be demonstrated by several statistic approaches. Among these, the method proposed by Slatkin and Rannala [12, 13] based on allele frequency and coalescent process [14] is what we think to apply to estimate the age of c.398delC mutation, as next development of this paper.

Conclusions

We report a case with RPA caused by the novel mutation, the c.398delC, in RLBP1 gene in homozygous condition necessary for developing the disease. The new mutation enriches the already wide spectrum of RP causative mutations. Hypothesis of a geographic area-related mutation was induced by the detection of this allele in several non-consanguineous inhabitants and by its absence in Sicilian population. However, due to the succession of more dominations, it is very difficult to establish by which ancestral population it was introduced.

Abbreviations

CRALBP: Cellular retinaldehyde-binding protein; FA: Fundus albipunctatus; HWE: Hardy-Weinberg Equilibrium; NFRCD: Newfoundland rod-cone dystrophy; OCT: Optical coherent tomography; RLBP1: Retinaldehyde binding protein 1; RP: Retinitis pigmentosa; RPA: Retinitis punctata albescens; RPE: Retinal pigment epithelium

Acknowledgments

We thank Fiumedinisi's population for disposability, particularly Dr.Bertani Gilda for her precious collaboration in controls collection.

Funding

Authors received no financial support to perform this study.

Authors' contributions

CS wrote the manuscript; LD and TE collected samples and performed molecular analysis; CR supervised the study; RD'A designed the study; AS coordinated the research team. All authors read and approved the final manuscript.

Ethics approval and consent to participate

The study involves human participants. Informed consent was fully obtained for all participants to the study. For underage subjects, consent was obtained from the parents. The study was conducted in accordance with Declaration of Helsinki and it was approved on 04/04/2017 by our local Ethics Committee "Azienda Policlinico Univaersitario G. Martino," protocol number 2317bis.
The paper, subject of this submission, has not been published or submitted in total or in part in any language as contribution either to a journal, as book chapter, or as abstract. All authors read the final version of the manuscript and they agree for publication.

Consent for publication

Subjects involved in this study were fully informed about its aim. They agree with the use of clinical information and consent their publication, in accordance with the current privacy policy.

Competing interests

The authors declare that they have no competing interests.

Author details

[1]Department of Biomedical and Dental Sciences and Morphofunctional Imaging, Division of Molecular Genetics and Preventive Medicine, University of Messina, via C. Valeria 1, I-98125 Messina, Italy. [2]Department of Cutting-Edge Medicine and Therapies, Biomolecular Strategies and Neuroscience, Section of Neuroscience-applied Molecular Genetics and Predictive Medicine, I. E. ME. S. T, via Michele Miraglia 20, I-90139 Palermo, Italy. [3]Department of Experimental Medicine, Division of Human Physiology and Integrate Biological Functions "F. Bottazzi", University of Campania Luigi Vanvitelli, ex II University of Naples, via Santa Maria di Costantinipoli 16, I-80138 Naples, Italy.

References

1. Narayan DS, Wood JP, Chidlow G, Casson RJ. A review of the mechanisms of cone degeneration in retinitis pigmentosa. Acta Ophthalmol. 2016;94:748–54.
2. Morimura H, Berson EL, Dryja TP. Recessive mutations in the RLBP1 gene encoding cellular retinaldehyde-binding protein in a form of retinitis punctata albescens. Invest Ophthalmol Vis Sci. 1999;40:1000–4.
3. Golding AM. Retinitis punctata albescens with pigmentation. Br J Ophthalmol. 1956;40:242–4.
4. Souied E, Soubrane G, Benlian P, Coscas GJ, Gerber S, Munnich A, Kaplan J. Retinitis punctata albescens associated with the Arg135Trp mutation in the rhodopsin gene. Am J Ophthalmol. 1996;121:19–25.
5. Kajiwara K, Sandberg MA, Berson EL, Dryja TP. A null mutation in the human peripherin/RDS gene in a family with autosomal dominant retinitis punctata albescens. Nat Genet. 1993;3:208–12.
6. Littink KW, Van Genderen MM, van Schooneveld MJ, Visser L, Riemslag FC, Keunen JE, et al. A homozygous frameshift mutation in LRAT causes retinitis punctata albescens. Ophthalmology. 2012;119:1899–906.
7. Hipp S, Zobor G, Glöckle N, Mohr J, Kohl S, Zrenner E, et al. Phenotype variations of retinal dystrophies caused by mutations in the RLBP1 gene. Acta Ophthalmol. 2015;93:281–6.
8. Burstedt MS, Sandgren O, Holmgren G, Forsman-Semb K. Bothnia dystrophy caused by mutations in the cellular retinaldehyde-binding protein gene (RLBP1) on chromosome 15q26. Invest Ophthalmol Vis Sci. 1999;40:995–1000.
9. Eichers ER, Green JS, Stockton DW, Jackman CS, Whelan J, McNamara JA, et al. Newfoundland rod-cone dystrophy, an early-onset retinal dystrophy, is caused by splice-junction mutations in RLBP1. Am J Hum Genet. 2002;70:955–64.
10. Kennedy BN, Li C, Ortego J, Coca-Prados M, Sarthy VP, Crabb JW. CRALBP transcriptional regulation in ciliary epithelial, retinal Müller and retinal pigment epithelial cells. Exp Eye Res. 2003;76:257–60.
11. Wu Z, Hasan A, Liu T, Teller DC, Crabb JW. Identification of CRALBP ligand interactions by photoaffinity labeling, hydrogen/deuterium exchange, and structural modeling. J Biol Chem. 2004;279:27357–64.
12. Slatkin M, Rannala B. Estimating the age of alleles by use of intraallelic variability. Am J Hum Genet. 1997;60:447–58.
13. Slatkin M, Rannala B. Estimating allele age. Annu Rev Genomics Hum Genet. 2000;01:225–49.
14. Goldstein DB, Reich DE, Bradman N, Usher S, Seligsohn U, Peretz H. Age estimates of two common mutations causing Factor XI deficiency: recent genetic drft is not necessary for elevated disease incidence among Ashkenazi Jews. Am J Hum Genet. 1999;64:1071–5.

9

Identification of a novel genetic locus underlying tremor and dystonia

Dorota Monies[1,2*†], Hussam Abou Al-Shaar[3], Ewa A. Goljan[1,2], Banan Al-Younes[1,2], Muna Monther Abdullah Al-Breacan[1], Maher Mohammed Al-Saif[4], Salma M. Wakil[1,2], Brian F. Meyer[1,2], Khalid S. A. Khabar[4] and Saeed Bohlega[2,3*†]

Abstract

Background: Five affected individuals with syndromic tremulous dystonia, spasticity, and white matter disease from a consanguineous extended family covering a period of over 24 years are presented. A positional cloning approach utilizing genome-wide linkage, homozygozity mapping and whole exome sequencing was used for genetic characterization. The impact of a calmodulin-binding transcription activator 2, (*CAMTA2*) isoform 2, hypomorphic mutation on mRNA and protein abundance was studied using fluorescent reporter expression cassettes. Human brain sub-region cDNA libraries were used to study the expression pattern of *CAMTA2* transcript variants.

Results: Linkage analysis and homozygozity mapping localized the disease allele to a 2.1 Mb interval on chromosome 17 with a LOD score of 4.58. Whole exome sequencing identified a G>A change in the transcript variant 2 5'UTR of *CAMTA2* that was only 6 bases upstream of the translation start site (c.-6G > A) (NM_001171166.1) and segregated with disease in an autosomal recessive manner. Transfection of wild type and mutant 5'UTR-linked fluorescent reporters showed no impact upon mRNA levels but a significant reduction in the protein fluorescent activity implying translation inhibition.

Conclusions: Mutation of *CAMTA2* resulting in post-transcriptional inhibition of its own gene activity likely underlies a novel syndromic tremulous dystonia.

Keywords: Dystonia, Tremor, Familial, Syndromic

Background

Currently, classification of dystonia relies upon phenomenology rather than etiology. A natural bifurcation is made based upon whether, with the exception of tremor, it is the sole motor feature (isolated dystonia) or is accompanied by other movement disorders including myoclonus (combined dystonia) [1]. Isolated dystonia may be subdivided into generalized or focal/segmental forms, with combined dystonia further categorized based upon the presence of myoclonus or Parkinsonism [2]. The advent of next generation sequencing has led to rapid expansion of phenotypic and genotypic subsets of syndromes primarily characterized as isolated or combined dystonias with more than 20 loci (DYT1-DYT27; some are redundant) identified to date [2, 3].

Familial-isolated dystonias are predominantly inherited as incompletely penetrant autosomal dominant traits most frequently related to the DYT1 or DYT6 loci with mutations in *TOR1A* [4, 5] or *THAP1* [6, 7], respectively. Additional isolated dystonia loci include DYT23 (*CIZ1*) [8], DYT24 (*ANO3*) [9], and DYT25 (*GNAL*) [10, 11], all of which have recently been extensively reviewed [2]. DYT4 (*TUBB4A*) may present as an autosomal dominant disorder in adolescence or early adulthood with prominent laryngeal dysphonia and craniocervical, segmental, or generalized dystonia [12, 13]. A rare recessively inherited generalized isolated dystonia results from mutation of *PRKRA* (DYT16) and is characterized by early onset and frequent association with dystonia-Parkinsonism [2, 14, 15]. Two other recessively inherited

* Correspondence: moniesdm@gmail.com; boholega@kfshrc.edu.sa
†Equal contributors
[1]Department of Genetics, King Faisal Specialist Hospital, and Research Centre, PO Box 3354, Riyadh 11211, Saudi Arabia
[2]Saudi Human Genome Program, King Abdulaziz City for Science and Technology, Riyadh, Saudi Arabia

isolated dystonias have also been described arising from mutation of DYT2 (*HPCA*) [16] or DYT27 (*COL6A3*) [15]. Like these, other dystonia loci such as DYT23 (*CIZ1*) and DYT24 (*ANO3*) are yet to be independently confirmed. Dystonic movements are also associated with mutations of *ADCY5*, identified as the likely cause of a novel movement disorder, familial dyskinesia with facial myokimia (FDFM) [17, 18].

We previously reported a consanguineous extended family having a novel autosomal recessive syndromic tremulous dystonia with spasticity and white matter disease [19]. In this study, we further delineate and genetically characterize the syndrome, identifying *CAMTA2* as a novel candidate gene for a syndromic tremulous dystonia, and describe its clinical course and prognosis over a long follow-up period.

Results

The families originated from the Eastern part of the Arabian Peninsula. Five patients were studied (Fig 1a); their current ages ranging from 33 to 46 years. The detailed clinical description was previously reported [19]. In summary, the disease onset was at 7–8 years of age with tremulous movement in fingers and arms that progressed to affect the face, neck, and trunk. Speech became dysarthric and tremulous. The volitional movement was exacerbated by posture and action. Initially, gait was not impaired. Intelligence, personality, and memory were not affected apart from paranoia and depression in two individuals. Patients exhibited a coarse generalized tremor of 4–5 Hz, present at rest, waxing, and waning in amplitude with side-to-side ("no no" type tremor). In addition, there was variable focal or segmental dystonia with retrocollis shoulder elevation and hyperextension of the fingers. There was generalized hyperreflexia with extensor planter response noted around the age of 15 or 16 years. Sensory examination was intact and there was no truncal ataxia. Patients were followed for up to 24 years with no noticeable increase in their movement disorder. The patients were cognitively intact, they were able to finish college, hold independent jobs, get married, and have children. However, all of them exhibited variable degrees of spasticity affecting their legs more than arms (Additional file 1) and two of them required walking aids in their 30's. Biochemical, organic acid, and lysosomal enzyme studies were normal. Visual and somatosensory evoke potentials showed evidence of prolonged central latencies. Accelerometric recording from the outstretched hand showed 4-Hz tremors. MRI of the brain showed similar abnormalities in all. There was mild symmetrical and confluent white-matter abnormalities sparing the U-fibers and involving the white matter of centrum semiovale, internal capsule, and ventral pons. These findings are not characteristic in any of the classic leukodystrophies. No change was observed in subsequent MRI scans with more than 20 years of follow-up. MR spectroscopy showed no lactate peak or other abnormalities. Patients failed to respond to therapeutic trials of ethanol or anticholinergics. However, some symptomatic benefits were noted with high-dose propranolol, gabapentin, and clonazepam. Patient video. Patients exhibited a coarse generalized tremor of 4–5 Hz, present at rest, waxing and waning in amplitude with side-to-side ("no no" type tremor). In addition, there was variable focal or segmental dystonia with retrocollis shoulder elevation and hyperextension of the fingers. Patients exhibited generalized hyperreflexia with an extensor planter response being noted. (MOV 39038 kb).

Linkage analysis of the nuclear families with three and two affected individuals (Fig. 1a) resulted in independent LOD scores of 2.53 and 2.05, respectively, at rs11651665. Combined analysis (Fig. 1b) localized the disease to a region on chromosome 17 defined proximally by rs743646 and distally by rs34811366 (chr17:4,536,241-6,670,705) with a maximum LOD score of 4.58. Based upon runs of homozygosity shared by all the affected and absent in unaffected individuals (Fig. 1c; red rectangle), we refined the critical interval to a 2.1 Mb region (chr17:4,231,133-6,018,243) containing approximately 60 RefSeq genes (hg19).

Whole exome sequencing of individual IV:7 (Fig. 1a) identified 25,926 variants relative to the hg19 reference sequence, with 1498 variants on chr 17. These were further filtered to exclude all variants present outside our defined ROH (chr 17: 4,231,133-6,018,243) within which we identified 60 variants (Additional file 2). Within the critical interval (chr 17: 4,231,133-6,018,243), 99.42% of bases in the CDS and flanking regions of genes were covered at 30× or better with an average coverage of 425×. By excluding previously reported variants (present in dbSNP with incidence > 0.02; present in 1000 genomes with incidence > 0.02) and retaining those present at a frequency < 2% in more than 1000 Arab exomes held in an in-house database, the list was narrowed to 8. By only focusing on homozygous changes annotated as non-synonymous, splicing variants, frameshift insertions or deletions, and nonsense variants, we decreased the number of candidate variants to 1. It was a G>A change in the 5′UTR of transcript variant 2 of *CAMTA2* (c.-6G>A) (NM_001171166.1) (Fig. 2a). This homozygous variant was identified in 56 reads, was confirmed by Sanger sequencing (Fig. 2b), and segregated with disease in the extended family. *CAMTA2* has many transcript variants only six (1, 2, 3, 4, 7, and 10) of which are protein coding (https://www.proteinatlas.org/ENSG00000108509-CAMTA2/tissue). The aligned cDNA sequence of the translation start site and adjacent 5′ UTR of these six variants show considerable variation (Fig. 3a). The resulting amino terminus amino acid sequences of these transcript variants are aligned and show

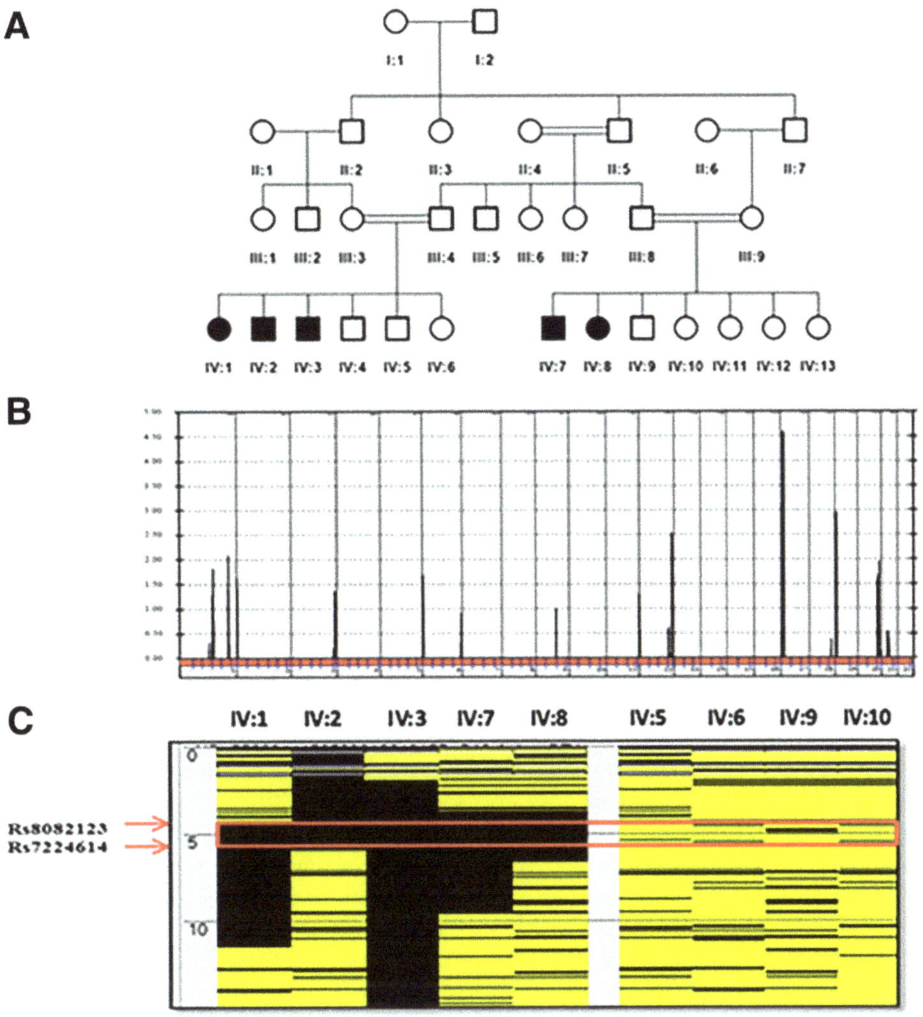

Fig. 1 Identification of a dystonia-associated locus on chromosome 17. **a** Pedigree of an extended family with novel syndromic familial tremulous dystonia. **b** Genome-wide linkage analysis revealed a maximal peak with a LOD score of 4.58 on chromosome 17. **c** AutoSNPa output for chromosome 17 reveals an ROH (boxed in red) shared among affected members (IV:1, IV:2, IV:3, IV:7, and IV:8) and not present in unaffected individuals (IV:4, IV:5, IV:6, and IV:9)

that the first 12 amino acids of transcript variant 2 are not shared by the other protein coding transcript variants (Fig. 3b). Despite the absence of linkage or homozygozity mapping data to suggest involvement of any known dystonia loci, WES of all affected individuals, parents, and unaffected siblings from both nuclear families was undertaken. There were no variants in any of the known dystonia genes (*TOR1A*, *TUBB4A*, *THAP1*, *CIZ1*, *ANO3*, *GNAL*. *SCGE*, *GCH1*, *TH*, *TITF1*, *TAF1*, *PRKRA*, *ATP1A3*, *SLC63*, *ADCY5*) that survived filtration that removed alleles present in public databases at a frequency > 2%, in any affected individual.

The wild-type *CAMTA2* 5′UTR 35 nt translation initiation region was subjected to analysis by a RNAdraw secondary structure program [20]. The structure obtained is a hairpin with three stems that is weakly stable (ΔG=11.1 kcal). The G>A variant 6 bases upstream of the *CAMTA2* transcript variant 2 initiation AUG changed this structure to a more relaxed form (ΔG=9.5 kcal) compared to the wild type. More importantly, the mutant variant secondary structure has only two stems and a wider loop as opposed to the three stem-tight loop of the wild type (Fig. 4a).

In order to evaluate the effects of the *CAMTA2* transcript variant 2 mutation on 5′UTR-linked activities such as transcription and translation, we constructed a fluorescence reporter cassette that contained a 25 nt initiation region upstream of SGFP (Fig. 4b). We subsequently transfected HEK293 cells with the wild type or mutant *CAMTA2* transcript variant 2 5′UTR initiation region containing expression cassettes and measured both the mRNA (qRT-PCR) and protein levels (fluorescence). We observed significant reduction in protein

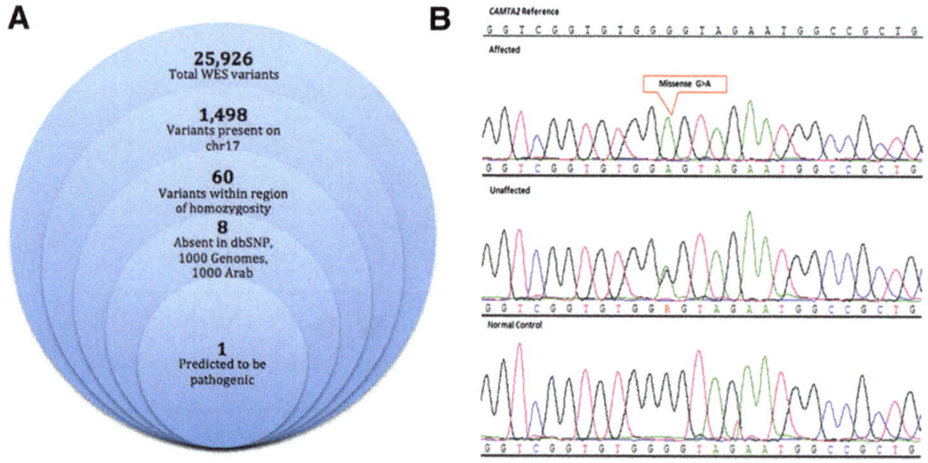

Fig. 2 **a** Filtering strategy used for identification of a causative mutation in the syndromic familial tremulous dystonia patient. **b** DNA electrophoregram with the G>A change in 5'UTR (chr17:4890930) of the *CAMTA2* gene (reverse strand)

(fluorescence levels; (Fig. 4c)) but not mRNA levels of the mutant *CAMTA2* transcript variant 2 expression cassette relative to its wild-type counterpart (Fig. 4d).

PCR amplification of each protein coding isoform of *CAMTA2* indicated differential expression of isoform 2 which was only detected in the vermis cerebellum, left and right cerebellum, cerebral cortex, olfactory lobe, and cerebellar peduncles. Expression of this variant was not detected in the corpus callosum, hippocampus, occipital, or frontal lobes (Fig. 5a). GAPDH expression was used as an internal control for PCR amplification and specificity for each isoform confirmed by Sanger sequencing of amplicons as shown for transcript variant 2 (Fig. 5b). Other protein coding transcript variants of CAMTA2 (1, 3, 4, 7, and 10) were ubiquitously expressed in most sub-regions of the brain tested (Additional file 3).

Discussion

We have described the clinical and genetic characteristics of a novel autosomal recessive familial tremulous dystonia syndrome, characterized by juvenile onset tremulous dystonia with spasticity and white matter disease. Phenotypically, the syndrome was heterogeneous within the families studied with all affected individuals displaying variable forms of both focal or segmental dystonia and coarse, asymmetrical, generalized tremor involving the head, neck and extremities [19]. Isolated dystonia is a classification reserved for exclusive occurrence of dystonia with the exception of tremor, whereas combined dystonia includes additional symptoms such as myoclonus or Parkinsonism by which it is further sub-classified [2].

Disease in the families we report is best characterized as a novel syndromic tremulous dystonia. The phenotype that combines dystonia, tremor, spasticity, and white matter disease clearly defines a novel clinical syndrome [19]. The novel nature of this syndrome is highlighted if not confirmed by identification of a previously undescribed single gene disorder segregating in an autosomal recessive manner within the families studied (LOD score 4.58; chr17:4,536,241-6,670,705). Whole exome and Sanger sequencing identified a homoallelic hypomorphic mutation (c.-6G>A) (NM_001171166.1) just upstream of the translation initiation site in the 5′UTR of *CAMTA2* transcript variant 2, with which disease segregated in the family studied. However, given the nature of WES where 100% coverage of the target region is almost never achieved and the coverage of only flanking regions (10–

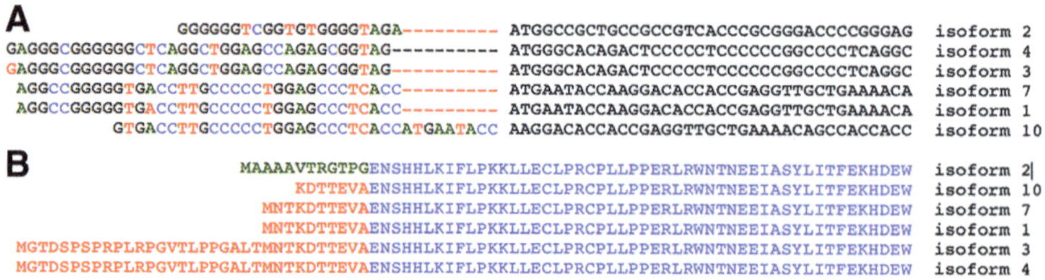

Fig. 3 **a** Alignment of the nucleotide sequences of the six protein coding transcript variants of *CAMTA2*. **b** Amino acid alignment of the six isoforms of CAMTA2

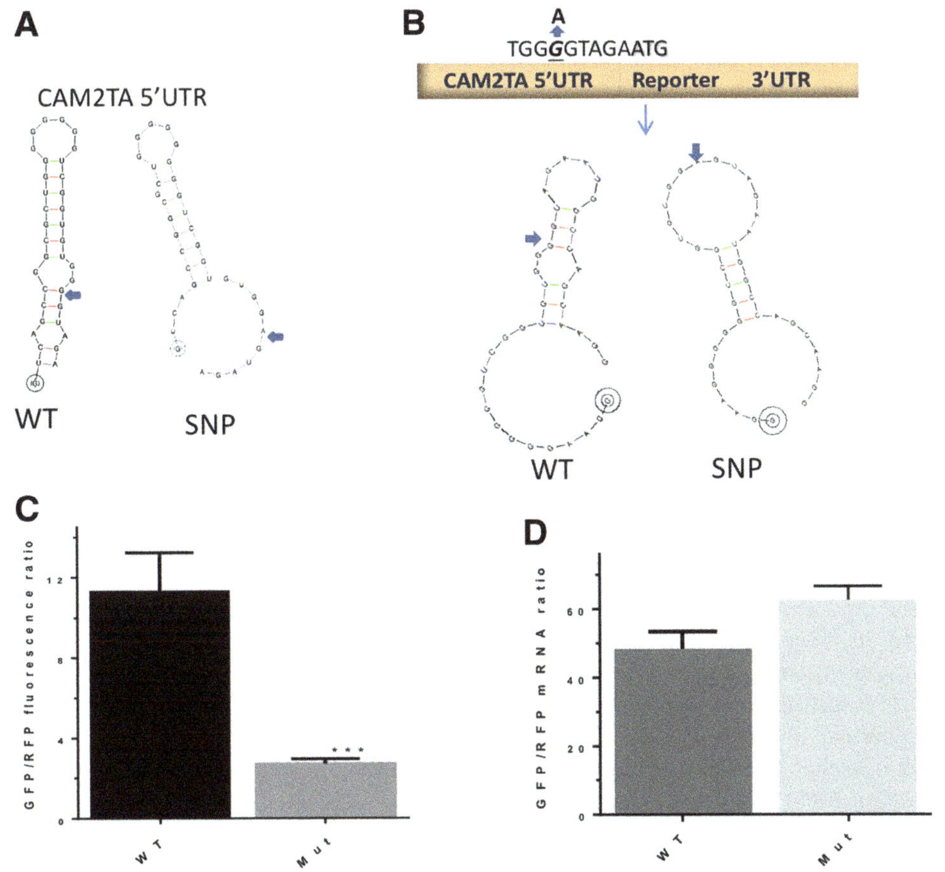

Fig. 4 SNPs nucleotide sequence variation-mediated translational inhibition of CAMTA2. **a** Predicted secondary structures for wild type and mutant initiation region of CAMTA2. **b** Schematic diagram showing the wild type and mutant containing 5′UTR reporter constructs. **c** HEK293 cells (3×10^4 cells/well) in 96-well plates were transfected with 75 ng of purified wild-type and mutant CAMTA2 5′UTR SGFP-expressing constructs along with control RFP expressing constructs (30 ng). After 20 h, fluorescence was quantified using automated image segmentation and quantification as described in the "Methods" section. The GFP/RFP expression ratio of WT and Mut of 5′UTR are shown as mean ± SD ($n = 6$). ***$p < 0.0001$ (Student's t test). **d** Real-time qRT-PCR for reporter mRNA levels. HEK293 cells were transfected with wild type and mutant *CAMTA2* 5′UTR reporters for 24 h. Total RNA was extracted, and qRT-PCR was performed using specific primers for SGFP and RFP as described in the "Methods" section. Data are presented as the mean ± SD of two independent experiments. **$p = 0.0014$ (Student's t test)

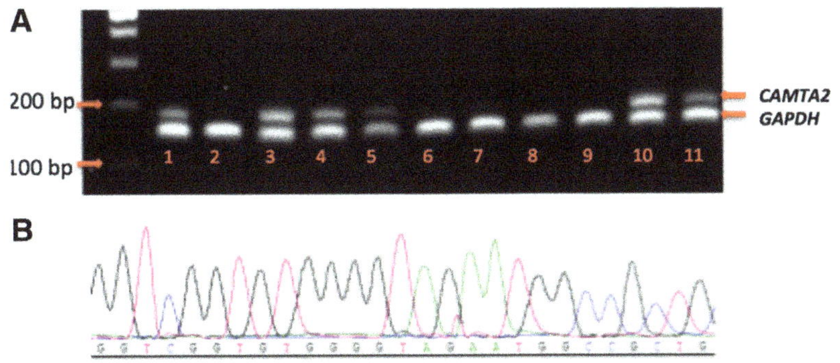

Fig. 5 Expression of *CAMTA2* transcript variant 2 in sub-regions of human brain. **a** PCR amplification of *CAMTA2* transcript variant 2 and *GAPDH* (housekeeping gene) in (1) vermis cerebella, (2) amygdala, (3) cerebellum left, (4) cerebellum right, (5) cerebral cortex, (6) corpus callosum, (7) hippocampus, (8) occipital lobe, (9) frontal lobe, (10) olfactory lobe, and (11) cerebellar peduncles. **b**. Sanger sequencing electrophrogram showing cDNA sequence uniquely identifying *CAMTA2* transcript variant 2

20 bp) of coding sequence in this study, and despite supporting functional evidence, we cannot exclude that a coding point mutation, CNV, deep intronic, or other mutation in the critical interval, in tight linkage disequilibrium with the reported variant, is the true causative mutation.

CAMTA proteins include multiple functional amino acid domains that are evolutionarily conserved including nuclear localization signals, nuclear export signals, a unique DNA binding domain CG-1, TIG-1 involved in protein dimerization, and ANK(ankyrin) repeats involved in protein-protein interactions [21–24]. CAMTA proteins also include a variable number of IQ motifs consisting of a repetitive sequence IQXXXRGXXX known to be associated with binding of Calmodulin (CaM) and CaM-like proteins. Transcription activation domains (TADS) have been reported in Arabadopsis AtCAMTA1, human HsCAMTA2, and functional support derived from reporter genes downstream of TAD's in plant protoplasts or mammalian cell cultures [25]. Cardiac hypertrophy as a consequence transcriptional reprogramming of cardiac gene expression by CAMTA2 interaction with class II histone deacetylase has been reported [26]. In drosophila, CAMTA is highly expressed in retinal cells and has been shown to induce dFbx14 that deactivates the G-protein coupled receptor (GPCR) rhodopsin. This is essential to maintaining Ca^{2+} homeostasis of the photo-transduction pathway, a central mechanism of which is light induced Ca2+ influx and efflux regulated through CaM and dmCAMTA [27]. It is not surprising that *CAMTA2* is strongly expressed in both developing and adult murine retina; however, its function in this context remains unknown. Given the diverse expression and activity of GPCRs, it is reasonable to speculate that regulation of this class of molecule by CAMTA may modulate cellular growth and differentiation as evidenced in cardiac hypertrophy [26]. Both CAMTA1 and CAMTA2 are highly expressed in the heart and brain [27]. CAMTA2 has six protein coding isoforms (CAMTA2 ENSG00000108509) ranging from 178 to 1241 amino acids. The expression pattern and function of each isoform remains unknown.

In this study, we identified a G>A change in the 5′ UTR of *CAMTA2* transcript variant 2. The position of the variant, near a splice site (+ 30) and adjacent to the translation initiation site (− 6) suggested that it might impact *CAMTA2* transcript variant 2 expression or translation. Our results clearly indicated that the *CAMTA2* mutant affected processes downstream of mRNA transcription, most likely translation as protein levels were dramatically reduced while there was no effect on mRNA levels (Fig. 4c–d). As protein expression was severely impacted without any alteration in the coding region (i.e, no amino acid change), the hypomorphic nature of the underlying mutation is clearly evident. It is known that secondary structures in the 5′UTR with stable high-energy forms ($\Delta G<-50$ kcal/mol) inhibit translation if near the mRNA methyl G cap [28]. Hairpin structures of weaker energy coupled with positions far away from the 5'end cap such as in our case, in contrast, enhance translation [28]. The wild-type *CAMTA2* 5′ UTR-predicted structure is near the initiation codon and of weakly stable structure, thus may facilitate translation machinery entry and therefore enhances translation. These activities are lost in the case of the mutant variant. In general, mutations near initiation regions can dramatically affect translation. Context sequence upstream of the initiation codon can modulate the scanning activity of the 40S ribosomal subunit affecting translation; mutations in this region, such as those in the Kozak consensus sequence that direct ribosomes to initiate protein synthesis, negatively affect translation initiation [29].

Further support for the pathogenicity or association of the *CAMTA2* variant with the phenotype described (for which we propose the name Bohlega Syndrome) in affected members of the family studied comes from the expression pattern of *CAMTA2* transcript variant 2 in human brain sub-regions where it is expressed in the vermis cerebellum, left and right cerebellum, cerebral cortex, olfactory lobe, and cerebellar peduncles (Fig. 5a). Expression of this isoform was not detected in the corpus callosum, hippocampus, and occipital or frontal lobes (Fig. 5a). Other CAMTA2 protein coding transcripts were ubiquitously expressed in all brain subregions tested. In mouse brain, CAMTA2 is expressed in Purkinje cells and granule cells of the dentate gyrus, pyramidal layer, and cerebellum (Additional file 4), but the distribution of each isoform is not known. Dystonia and Parkinsonism, including tremor, have largely been attributed to dysfunction of the basal ganglia. However, more recently, the role of Purkinje cells, the cerebellum, and their links to the basal ganglia in dystonia and Parkinson's disease have been questioned [30–32]. Anatomical studies have identified connections from the basal ganglia and cerebellum via the thalamus to the cortex [33]. Of note, the ability of CAMTA2 knockout mice to mount a hypertrophic response to cardiac stress such as aortic banding is severely compromised [27]. There was no report of any tremor or dystonia in the CAMTA2 knockout mouse. In the affected individuals from our study, absence of any clinical cardiac symptoms extends the apparent discordance between the human and mouse phenotypes, albeit the mouse phenotype was induced by severe stress (aortic banding) [27]. However, the mouse ENCODE consortium mapped transcription factor binding and other regulatory domains across the mouse genome and compared them with the human genome. While they identified substantial conservation, they also identified a large degree of divergence between mouse and man in transcription and transcription regulation [34]. Therefore, differences in mouse and human phenotypes associated with genes such as *CAMTA2* involving transcription or transcription regulation should not be unexpected.

Conclusions

Our study identifies a novel dystonia locus *CAMTA2* that opens a new line of investigation associated with the role of CAMTA2 in the growth, differentiation, survival, and function of cells and their cellular mechanisms within the brain as related to movement disorders. It also highlights the value of rare inherited disorders in providing insight into more common disease states such as dystonia and tremor.

Methods

Subjects and nucleic acid extraction

Five patients from a consanguineous extended family (two nuclear families) (Fig. 1a) were examined in the Department of Neurosciences, King Faisal Specialist Hospital and Research Center (KFSHRC). They were diagnosed with familial tremulous dystonia. All subjects were enrolled under a KFSHRC IRB-approved (RAC# 2090011) protocol with full-written informed consent.

Genomic DNA extraction

Genomic DNA was extracted from peripheral blood samples of patients and family members using a Gentra Puregene Kit (QIAGEN GmbH, Hilden, Germany). Purity and quantity of DNA samples were assessed spectrophotometrically and stored at – 20 °C until required.

Clinical data and imaging

Patients were seen and examined in the neurology clinic of King Faisal Specialist Hospital and Research Centre (KFSHRC) by certified neurologists (H.A. and S.B.). Briefly, the clinical data was obtained from the charts of patients at the time of the study and during the 20 years of follow-up. Imaging was performed by magnetic resonance imaging (MRI) 3 T using the following sequences: T1, T2, FLAIR, DWI, ADC, and MR spectroscopy. The other clinical, neurophysiological, and laboratory methodology may be reviewed from the previously published clinical paper [19].

Linkage and Homozygosity mapping

All participating individuals (affected and unaffected) were genotyped using an Affymetrix Axiom Genome-Wide CEU array containing 587,353 markers (Affymetrix, Santa Clara, CA, USA) following the manufacturer's protocol (http://www.affymetrix.com/support/technical/manuals.affx). Resulting genotypes were analyzed for shared runs of homozygosity (ROH) using autoSNPa (http://dna.leeds.ac.uk/autosnpa/). Linkage analysis was performed using the Allegro module of Easy Linkage assuming autosomal recessive inheritance and 100% penetrance (http://genetik.charite.de/hoffmann/easyLINKAGE/index.html). For linkage analysis, the extended family was broken down to reflect the two consanguineous nuclear families, with both independent LOD scores and a combined total LOD score being calculated, in a single multipoint analysis of both nuclear families.

Whole exome sequencing and analysis

DNA was amplified to obtain a whole exome Ion Proton AmpliSeq library which was further used for emulsion PCR on an Ion One Touch System followed by an enrichment process using an Ion OneTouch ES, both procedures following the manufacturer's instructions (Life Technologies, Carlsbad, CA, USA). The template-positive Ion PI Ion Sphere particles were processed for sequencing on the Ion Proton instrument (Life Technologies, Carlsbad, CA, USA). Approximately 15–17 Gb of sequence was generated per sequencing run. Reads were mapped to UCSC hg19 (http://genome.ucsc.edu/) and variants identified using the Ion Torrent pipeline (Life Technologies, Carlsbad, CA, USA). Based upon the consanguineous inbred pedigree and a clear autosomal recessive inheritance pattern in the family studied, the resultant variant caller file (vcf) was filtered to include only homozygous variants in the critical interval on chromosome identified by linkage and homozygosity mapping, within the pre-determined ROH shared by all affected individuals only. These were further filtered to only retain variants that were present in public databases including NHLBI Exome Sequencing Project, 1000 genomes project (phase 3), Exome Aggregation Consortium (ExAC) (version 0.3.1), dbSNP (build 141), and Kaviar (Known VARiants) (September 2015) databases at an incidence of < 2%. Included variants were further selected based upon pathogenicity predicted by Polyphen2 (http://genetics.bwh.harvard.edu/pph2/), SIFT (http://sift.jcvi.org/), and CADD (http://cadd.gs.washington.edu/info). Potential causative variant(s) were validated by Sanger sequencing and further vetted for familial segregation based upon autosomal recessive inheritance. For validation, an amplicon incorporating the variant of interest was amplified and sequenced using a BigDye Terminator kit (Applied Biosystems, Foster City, CA) and run on an ABI 3730xl automated sequencer (Applied Biosystems, Foster City, CA). SeqScape v.2.6 software (Applied Biosystems, Foster City, CA) was used to align sequence data and confirm variants. *CAMTA2* transcript variant 2 is a human isoform and aligns with the human reference sequence of this gene.

Expression constructs

The *CAM2TA* 5′UTR reporter expression constructs were generated by using a cloning-free PCR method [35]. The *CAM2TA* 5′UTR wild type and mutant sequences (underlined) were incorporated in the forward primer: CCTATCAGTGATAGGGCGGCTGGG*TATATAA*TGGAAGGGG GGTCGGTGTGGG/AGTAGAATGGCCAGCAAGG. The sequence also contained a minimal promoter (TATA box in italics), 15 bases of complementary sequence to an expression vector containing strong modified EGFP, *UWr*-EGFP (SGFP), and a 3′UTR [36]. The reverse primer is complementary to a downstream region of the poly(A) site on the vector. Briefly, PCR was performed with the following conditions: 2.5 units of HotStart Taq(Qiagen) and 0.2 units of Pfx polymerase (Invitrogen, Carlsbad, CA) mix, 100 ng, template and cycle conditions: 95 °C for 12 min, 31 cycles of: 94 °C, 1 min., 55 °C, 1 min., 72 °C, 4 min., and final extension at 72 °C for 7 min. The PCR products were purified using Qiagen PCR purification columns and eluted in sterile water. The PCR products were run on a 1.2% agarose gel and visualized by ethidium bromide under UV light to verify size and quality. The resultant purified constructs contained a minimal promoter, *CAM2TA2* 5′UTR, SGFP reporter, and 3′UTR, and were used for transfection.

Cells and transfection

HEK293 was obtained from the ATCC (Rockville, MD) and cultured in DMEM medium (Invitrogen, Carlsbad, CA) supplemented with 10% FBS and antibiotics at 37 ° C in 5% CO_2. Cells were seeded at 3×10^4 cells per well in 96-well clear-bottom black plates (Matrix technologies, Hudson, NH) and incubated overnight. Cells were transfected with purified reporter constructs. Transfections with *CAM2TA2 5′UTR*-SGFP reporter along with a normalization control RFP expression plasmid driven by constitutive yet non-inducible RPS30 promoter [37] were performed in serum-free medium for 4 h, using Lipofectinamine 2000 (Invitrogen) according to the manufacturer's instructions. Subsequently, the cell culture were replaced with completed DMEM medium and incubated further for 16 h. At these conditions, no effect on cell viability was observed. All transfections were performed in several replicates as indicated in the text. Fluorescence were acquired by the BD imaging pathway and quantified using the Proxcell imaging segmentation and quantification program [38]. The variance in GFP fluorescence among replicate microwells was < 10%. Data are presented as mean values ± standard deviation (SD) of total fluorescence intensity in each well, with at least six replicate readings.

Real-time RT-PCR

Quantitative real-time (qRT-PCR) was performed for SGFP and RFP reporter mRNA using SybrGreen Real-Time PCR. Briefly, total RNAs were treated with DNase I, and cDNAs were used with Lightcycler Faststart DNA master SYBR Green I, (Roche, USA). The following primers were used: SGFP forward: GCCAGGTCTTTTCTGCAGTCACCG; reverse: GTAGTGCAGATGAACTTCAGG); human GAPDH forward: GGCAAATTCAACGGCACAGT; reverse: GATGGTGATGGGCTTCCC. In all cases, relative quantification of expression was determined using the standard curve method and the values obtained were within both detectable and linear range. Average concentrations were normalized to the endogenous GAPDH gene. The final results were converted to ratio ± SD of the specific mRNA levels to housekeeping mRNA levels. qRT-PCR was performed in multiplex using the Chroma 4DNA Engine thermocycler (Bio-Rad).

Amplification of *CAMTA2* protein coding isoforms from human cDNA libraries

We used cDNA libraries prepared from total RNA isolated from normal human adult brain tissues: vermis cerebelli, amygdala, cerebellum left, cerebellum right, cerebral cortex, corpus callosum, hippocampus, occipital lobe, frontal lobe, olfactory, and cerebellar peduncles (BioChain Institute, Newark, CA) to determine the expression pattern of CAMTA2 isoforms in human brain sub-regions, skeletal muscle, and cardiac muscle. The cDNA libraries were used as template for PCR amplification of protein coding *CAMTA2* isoforms (isoform 1, isoform 2, isoform 3, isoform 4, isoform 7, and isoform 10). Briefly, PCR mixtures containing 2.5 ng cDNA, 1× PCR buffer, 2.5 mM each dNTP, 0.5 μM each primer, and 0.25 U HotStar Taq polymerase (Qiagen) were cycled 30 times at 95 °C for 15 min, 95 °C for 45 s, 64 °C for 45 s, and 72 °C for 60 s. *GAPDH* was amplified in a multiplex reaction as an internal PCR control. Primer information is available on request. The resulting amplicons were evaluated by electrophoresis on a 2% agarose gel. Representative bands were sequenced to confirm amplicon specificity.

Additional files

Additional file 1: Patient video. Patients exhibited a coarse generalized tremor of 4–5 Hz, present at rest, waxing and waning in amplitude with side-to-side ("no no" type tremor). In addition, there was variable focal or segmental dystonia with retrocollis shoulder elevation and hyperextension of the fingers. Patients exhibited generalized hyperreflexia with an extensor planter response being noted.

Additional file 2: Variants in candidate interval.

Additional file 3: Expression of CAMTA2 isoforms, 1, 3, 4, 7 and 10 in sub-regions of the brain.

Additional file 4: CAMTA2 expression in brain.

Acknowledgements
The authors would like to thank the family members for their participation in this study. We thank the Genotyping and Sequencing Core Facilities of King Faisal Specialist Hospital & Research Centre (KFSHRC) for their contributions. Work described was supported by grants from KFSHRC (RAC#2090011) and King Abdulaziz City for Science and Technology (KACST) (KACST-12-BIO2943-20). We would also like to acknowledge the support of KACST through the Saudi Human Genome Project (SHGP).

Funding
This work was funded through grants from King Faisal Specialist Hospital & Research Centre (RAC#2090011) and King Abdulaziz City for Science and Technology (KACST-12-BIO2943-20).

Authors' contributions
DM contributed to the conception and design, acquisition and analysis of data, and drafting of the manuscript and figures. HA participated in the acquisition and analysis of data, and drafting of the manuscript and figures. EN participated in the acquisition and analysis of data and drafting of the manuscript and figures. BA contributed to the acquisition and analysis of data and drafting of the manuscript and figures. MAB contributed to the acquisition and analysis of data. MAS participated in the acquisition and analysis of data. SW contributed to the acquisition and analysis of data and drafting of the manuscript and figures. BM participated in the conception and design, acquisition and analysis of data and drafting of the manuscript and figures. KK participated in the conception and design, acquisition and analysis of data and drafting of the manuscript and figures. SB contributed to the conception and design, acquisition and analysis of data and drafting of the manuscript and figures. All authors read and approved the final manuscript.

Consent for publication
Written informed consent was obtained to publish the video included in the supplement to this article.

Competing interests
The authors declare no competing interest regarding the production of this article. The authors have no personal financial or institutional interest in any of the drugs, materials, or devices described in this article.

Author details
[1]Department of Genetics, King Faisal Specialist Hospital, and Research Centre, PO Box 3354, Riyadh 11211, Saudi Arabia. [2]Saudi Human Genome Program, King Abdulaziz City for Science and Technology, Riyadh, Saudi Arabia. [3]Department of Neurosciences, King Faisal Specialist Hospital and Research Centre, PO Box 3354, Riyadh 11211, Saudi Arabia. [4]Biomolecular Medicine, Research Centre, King Faisal Specialist Hospital and Research Centre, Riyadh, Saudi Arabia.

References

1. Albanese A, Bhatia K, Bressman SB, et al. Phenomenology and classification of dystonia: a consensus update. Movement disorders : official journal of the Movement Disorder Society. 2013;28(7):863–73.
2. Balint B, Bhatia KP. Isolated and combined dystonia syndromes—an update on new genes and their phenotypes. European journal of neurology : the official journal of the European Federation of Neurological Societies. 2015;22(4):610–7.
3. Charlesworth G, Bhatia KP, Wood NW. The genetics of dystonia: new twists in an old tale. Brain : a journal of neurology. 2013;136(Pt 7):2017–37.
4. Ozelius LJ, Hewett J, Kramer P, et al. Fine localization of the torsion dystonia gene (DYT1) on human chromosome 9q34: YAC map and linkage disequilibrium. Genome Res. 1997;7(5):483–94.
5. Zirn B, Grundmann K, Huppke P, et al. Novel TOR1A mutation p.Arg288Gln in early-onset dystonia (DYT1). J Neurol Neurosurg Psychiatry. 2008;79(12):1327–30.
6. Clot F, Grabli D, Burbaud P, et al. Screening of the THAP1 gene in patients with early-onset dystonia: myoclonic jerks are part of the dystonia 6 phenotype. Neurogenetics. 2011;12(1):87–9.
7. Fuchs T, Gavarini S, Saunders-Pullman R, et al. Mutations in the THAP1 gene are responsible for DYT6 primary torsion dystonia. Nat Genet. 2009;41(3):286–8.
8. Xiao J, Uitti RJ, Zhao Y, et al. Mutations in CIZ1 cause adult onset primary cervical dystonia. Ann Neurol. 2012;71(4):458–69.
9. Charlesworth G, Plagnol V, Holmstrom KM, et al. Mutations in ANO3 cause dominant craniocervical dystonia: ion channel implicated in pathogenesis. Am J Hum Genet. 2012;91(6):1041–50.
10. Fuchs T, Saunders-Pullman R, Masuho I, et al. Mutations in GNAL cause primary torsion dystonia. Nat Genet. 2013;45(1):88–92.
11. Saunders-Pullman R, Fuchs T, San Luciano M, et al. Heterogeneity in primary dystonia: lessons from THAP1, GNAL, and TOR1A in Amish-Mennonites. Movement disorders : official journal of the Movement Disorder Society. 2014;29(6):812–8.
12. Hersheson J, Mencacci NE, Davis M, et al. Mutations in the autoregulatory domain of beta-tubulin 4a cause hereditary dystonia. Ann Neurol. 2013; 73(4):546–53.
13. Wilcox RA, Winkler S, Lohmann K, Klein C. Whispering dysphonia in an Australian family (DYT4): a clinical and genetic reappraisal. Movement disorders : official journal of the Movement Disorder Society. 2011;26(13):2404–8.
14. Camargos S, Scholz S, Simon-Sanchez J, et al. DYT16, a novel young-onset dystonia-parkinsonism disorder: identification of a segregating mutation in the stress-response protein PRKRA. Lancet Neurol. 2008;7(3):207–15.
15. Zech M, Lam DD, Francescatto L, et al. Recessive mutations in the alpha3 (VI) collagen gene COL6A3 cause early-onset isolated dystonia. Am J Hum Genet. 2015;96(6):883–93.
16. Charlesworth G, Angelova PR, Bartolome-Robledo F, et al. Mutations in HPCA cause autosomal-recessive primary isolated dystonia. Am J Hum Genet. 2015;96(4):657–65.
17. Chen YZ, Friedman JR, Chen DH, et al. Gain-of-function ADCY5 mutations in familial dyskinesia with facial myokymia. Ann Neurol. 2014;75(4):542–9.
18. Chen YZ, Matsushita MM, Robertson P, et al. Autosomal dominant familial dyskinesia and facial myokymia: single exome sequencing identifies a mutation in adenylyl cyclase 5. Arch Neurol. 2012;69(5):630–5.
19. Bohlega S, Stigsby B, al-Kawi MZ, et al. Familial tremulous and myoclonic dystonia with white matter changes in brain magnetic resonance imaging. Movement disorders : official journal of the Movement Disorder Society. 1995;10(4):513–7.
20. Matzura O, Wennborg A. RNAdraw: an integrated program for RNA secondary structure calculation and analysis under 32-bit Microsoft Windows. Computer applications in the biosciences : CABIOS. 1996;12(3):247–9.
21. Aravind L, Koonin EV. Gleaning non-trivial structural, functional and evolutionary information about proteins by iterative database searches. J Mol Biol. 1999;287(5):1023–40.
22. Bouche N, Scharlat A, Snedden W, Bouchez D, Fromm H. A novel family of calmodulin-binding transcription activators in multicellular organisms. J Biol Chem. 2002;277(24):21851–61.
23. Muller CW, Rey FA, Sodeoka M, Verdine GL, Harrison SC. Structure of the NF-kappa B p50 homodimer bound to DNA. Nature. 1995;373(6512):311–7.
24. Sedgwick SG, Smerdon SJ. The ankyrin repeat: a diversity of interactions on a common structural framework. Trends Biochem Sci. 1999;24(8):311–6.
25. Finkler A, Ashery-Padan R, Fromm H. CAMTAs: calmodulin-binding transcription activators from plants to human. FEBS Lett. 2007;581(21):3893–8.
26. Schwartz RJ, Schneider MD. CAMTA in cardiac hypertrophy. Cell. 2006; 125(3):427–9.
27. Song K, Backs J, McAnally J, et al. The transcriptional coactivator CAMTA2 stimulates cardiac growth by opposing class II histone deacetylases. Cell. 2006;125(3):453–66.
28. Babendure JR, Babendure JL, Ding JH, Tsien RY. Control of mammalian translation by mRNA structure near caps. RNA. 2006;12(5):851–61.
29. Kozak M. Regulation of translation via mRNA structure in prokaryotes and eukaryotes. Gene. 2005;361:13–37.
30. Babij R, Lee M, Cortes E, Vonsattel JP, Faust PL, Louis ED. Purkinje cell axonal anatomy: quantifying morphometric changes in essential tremor versus control brains. Brain : a journal of neurology. 2013;136(Pt 10):3051–61.

31. Lewis MM, Galley S, Johnson S, Stevenson J, Huang X, McKeown MJ. The role of the cerebellum in the pathophysiology of Parkinson's disease. The Canadian journal of neurological sciences. J Can Sci Neurol. 2013;40(3):299–306.
32. Prudente CN, Hess EJ, Jinnah HA. Dystonia as a network disorder: what is the role of the cerebellum? Neuroscience. 2014;260:23–35.
33. Middleton FA, Strick PL. Basal ganglia and cerebellar loops: motor and cognitive circuits. Brain Res Brain Res Rev. 2000;31(2–3):236–50.
34. Yue F, Cheng Y, Breschi A, et al. A comparative encyclopedia of DNA elements in the mouse genome. Nature. 2014;515(7527):355–64.
35. al-Haj L, Al-Ahmadi W, Al-Saif M, Demirkaya O, Khabar KS. Cloning-free regulated monitoring of reporter and gene expression. BMC Mol Biol. 2009;10:20.
36. Al-Saif M, Khabar KS. UU/UA dinucleotide frequency reduction in coding regions results in increased mRNA stability and protein expression. Molecular therapy : the journal of the American Society of Gene Therapy. 2012;20(5):954–9.
37. Hitti E, Al-Yahya S, Al-Saif M, et al. A versatile ribosomal protein promoter-based reporter system for selective assessment of RNA stability and post-transcriptional control. RNA. 2010;16(6):1245–55.
38. Mahmoud L, Al-Saif M, Amer HM, Sheikh M, Almajhdi FN, Khabar KS. Green fluorescent protein reporter system with transcriptional sequence heterogeneity for monitoring the interferon response. J Virol. 2011;85(18): 9268–75.

10

Recessive *VARS2* mutation underlies a novel syndrome with epilepsy, mental retardation, short stature, growth hormone deficiency, and hypogonadism

Abdulaziz Alsemari[1,3], Banan Al-Younes[2,3], Ewa Goljan[2,3], Dyala Jaroudi[2,3], Faisal BinHumaid[2,3], Brian F. Meyer[2,3], Stefan T. Arold[4] and Dorota Monies[2,3*]

Abstract

Background: Most mitochondrial and cytoplasmic aminoacyl-tRNA synthetases (aaRSs) are encoded by nuclear genes. Syndromic disorders resulting from mutation of aaRSs genes display significant phenotypic heterogeneity. We expand aaRSs-related phenotypes through characterization of the clinical and molecular basis of a novel autosomal-recessive syndrome manifesting severe mental retardation, ataxia, speech impairment, epilepsy, short stature, microcephaly, hypogonadism, and growth hormone deficiency.

Results: A G>A variant in exon 29 of *VARS2* (c.3650G>A) (NM_006295) was identified in the index case. This homozygous variant was confirmed by Sanger sequencing and segregated with disease in the family studied. The c.3650G>A change results in alteration of arginine to histidine at residue 1217 (R1217H) of the mature protein and is predicted to be pathogenic.

Conclusions: These findings contribute to a growing list of aaRSs disorders, broadens the spectrum of phenotypes attributable to *VARS2* mutations, and provides new insight into genotype-phenotype correlations among the mitochondrial synthetase genes.

Keywords: Syndromic, Angelman, Mitochondrial, Dysmorphism, Hypogonadism, Vitamin D deficiency

Background

Mental retardation (MR) affects approximately 1–3% of the human population. Both genetic and environmental factors are responsible for causing MR, but in almost 60% of the cases, the actual cause cannot be identified precisely [1, 2]. In developed countries, most severe forms of intellectual disability are thought to have a genetic cause [3]. Mutations in more than 400 genes have been linked to intellectual disability, but most of these mutations have a very low prevalence and their phenotypes are often indistinguishable [4].

* Correspondence: moniesdm@gmail.com
[2]Department of Genetics, King Faisal Specialist Hospital and Research Centre, MBC 03, PO Box 3354, Riyadh 11211, Saudi Arabia
[3]Saudi Human Genome Project, King Abdulaziz City for Science and Technology, Riyadh, Saudi Arabia

A number of single-gene disorders result in mental retardation. Many of these are associated with atypical or dysmorphic physical characteristics. Such conditions include fragile X syndrome, neurofibromatosis, tuberous sclerosis, Noonan's syndrome, Cornelia de Lange's syndrome, and Angelman syndrome(AS) [5]. AS is usually considered as a diagnosis in individuals with severe developmental delay, in combination with seizures, ataxia, hypermotoric behaviors, and absence of speech. Ninety percent of these individuals have an identifiable molecular defect that results in the loss of expression of maternally inherited *UBE3A* (encoding E6AP) in the brain [6]. Over the last few years, many novel syndromes that mimic AS have been delineated and it is now recognized that some individuals who were clinically diagnosed with "test-negative AS" have one of these AS-like syndromes instead. Their phenotypes include severe intellectual disability, motor delay, variable degrees of speech delay ranging from being completely non-verbal to being able to speak in short sentences, autistic and maladaptive behaviors, short attention span, and seizures [7].

We report a novel autosomal recessive syndromic disorder with clinical resemblance to AS. The clinical spectrum includes severe mental retardation, ataxia, speech impairment, epilepsy, short stature, microcephaly, flap occiput, protruding tongue, prognathia, tremulous movement of limbs, apparent happy demeanor, easily excitable personality, and excessive chewing mouth behaviors. In addition, there were severe growth hormone deficiency, hypogonadism, and severe osteomalacia. Whole exome sequencing of the affected patients identified a homozygous mutation in *VARS2*.

Methods

Subjects and nucleic acid extraction

Four patients from a consanguineous extended family from the same region of Saudi Arabia were examined by a neurologist (AAS), in the department of Neurosciences, King Faisal Specialist Hospital and Research Center (KFSHRC). They were diagnosed with an Angelman-like syndrome. All subjects were enrolled under a KFSHRC IRB-approved (RAC# 13-MED2056-20) protocol with full-written informed consent. Genomic DNA was extracted from peripheral blood samples using standard procedures (Flexi Gene DNA Handbook, Qiagen). Samples were quantitated spectrophotometrically and stored at – 20 °C.

Linkage and homozygosity mapping

All participating individuals (affected and unaffected) were genotyped using an Affymetrix Axiom array (Affymetrix, Santa Clara, CA, USA) following the manufacturer's protocol (http://www.affymetrix.com/support/technical/manuals.affx). Resulting genotypes were analyzed for shared runs of homozygosity (ROH) using autoSNPa (http://dna.leeds.ac.uk/autosnpa/). Linkage analysis was performed using the Allegro module of Easy Linkage assuming autosomal recessive inheritance and 100% penetrance (http://genetik.charite.de/hoffmann/easyLINKAGE/index.html).

Whole exome sequencing and analysis

DNA was amplified to obtain a whole exome Ion Proton AmpliSeq library which was further used for emulsion PCR on an Ion OneTouch System followed by an enrichment process using an Ion OneTouch ES, both procedures following the manufacturer's instructions (Life Technologies, Carlsbad, CA, USA). The template-positive Ion PI Ion Sphere particles were processed for sequencing on the Ion Proton sequencer (Life Technologies, Carlsbad, CA, USA). Approximately 15–17 Gb of sequence was generated per sequencing run. Reads were mapped to UCSC hg19 (http://genome.ucsc.edu/) and variants identified using the Ion Torrent pipeline (Life Technologies, Carlsbad, CA, USA). The resultant variant caller file (vcf) was filtered to include only homozygous variants within the pre-determined ROH (shared by affecteds only) that were absent from dbSNP, 1000 genomes, approximately 2000 Saudi exomes. Included variants were further selected based upon pathogenicity predicted by Polyphen2 (http://genetics.bwh.harvard.edu/pph2/) and SIFT (http://sift.jcvi.org/). Potential causative variants were confirmed by Sanger sequencing and screened further for familial segregation based upon autosomal recessive inheritance. For confirmation, an amplicon incorporating the candidate variant was amplified and sequenced using a BigDye Terminator kit (Applied Biosystems, Foster City, CA) and run on an ABI 3730xl automated sequencer (Applied Biosystems, Foster City, CA). SeqScape v.2.6 software (Applied Biosystems, Foster City, CA) was used to align sequence data and identify variants.

Computational structural analysis of mutants

Sequences were retrieved from the Uniprot database. BLAST and SwissModel [8] were used to search for suitable structural templates in the Protein Data Bank (PDB). SwissModel, RaptorX [9], and I-Tasser [10] were used to produce homology models. Models were manually inspected, and mutations evaluated, using the Pymol program (pymol.org). Disorder and secondary structure elements were predicted using RaptorX. Transmembrane helices were predicted using Phobius [11]. Functional information was compiled from various resources, including Uniprot [12], InterPro [13], and publications associated with the model templates used.

Results

Clinical profile

The index case (V:3) (Fig. 1a), a 23-year-old male was a product of full-term pregnancy and normal delivery. He was well until the age of 4 months when he started to have generalized tonic clonic seizures. At the age of 1 year, the family noticed that the patient had delayed development. He had delayed sitting, delayed standing, and only started to crawl at the age of 2 years. The patient eventually exhibited severe developmental delay and severe speech impairment. The speech impairment was more receptive than expressive. Upon reviewing his family history, there were three other similarly affected siblings (one brother and two sisters). He also has four similarly affected cousins (two males and two females) (Fig. 1a).

Affected individuals usually have incoordination of movement until adolescence at which point they stop walking,

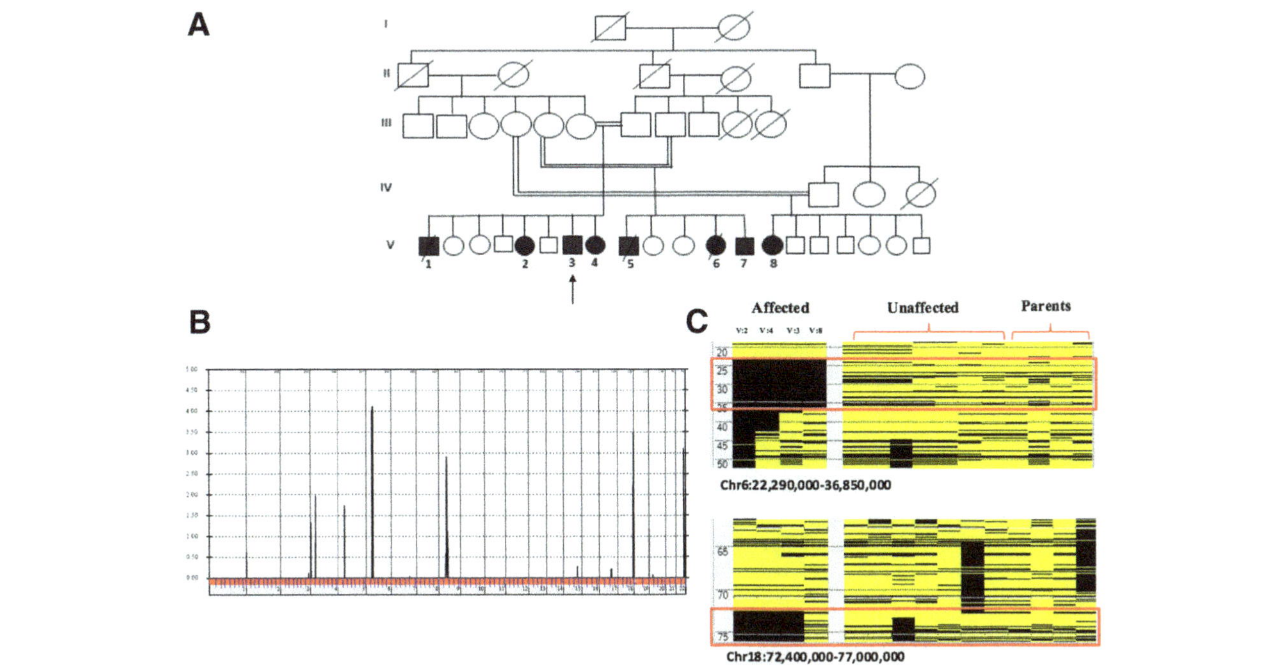

Fig. 1 Identification of disease locus on chromosome 6: **a** Pedigree of extended family with a novel syndromic disorder. **b** Genome-wide linkage analysis revealed a peak with a maximum LOD score of 4.11 on chromosome 6. **c** AutoSNPa output for chromosome 6 reveals an ROH (boxed in red) shared among affected members (V:2, V:3, V:4, and V:8) and not present in unaffected individuals or parents. Output for chromosome 18 excludes this locus because the ROH is not shared with affected individual V:10

consequent to flexed joint deformities of the lower limbs. Their behavioral pattern is best described as frequent episodic smiling and an apparent happy demeanor (Fig. 2a), with an easily excitable personality and habitual hand flapping movement. They also have a history of hyperphagia and usually eat a lot more than their normal siblings. Their examination showed severe mental retardation with expressive and receptive aphasia. Their speech and cognition deteriorates to a degree at which they became unable to produce words or to understand commands. There is no facial, axillae, or pubic hair. The affected male had small testicles. They have flat occiput and constant protruding tongue with tongue thrusting. They have microcephaly, prognathia, and wide-spaced teeth (Fig. 2b). During movement of the upper limbs, there are usually flexed arm positions (Fig. 2c). They are of short stature and increasingly become chair bound (Fig. 2d). They typically show an excessive chewing mouth behavior. Their examination showed brisk reflex of the upper and lower limbs. The older affected son (V:1) expired at the age of 23 because of a chest infection.

Laboratory investigations of the index case showed that he had a normal blood count, normal liver function test, and normal electrolytes and renal function test. His peripheral smear morphology was normal. His calcium was low at 2.3 mmol/l, while his phosphate amylase was normal. He had high alkaline phosphatase at 288 U/L (normal range is 40–135). The patient had normal synacthen test and a low IGF level (26 μg/l) (normal range between 116 and 358). IGF binding protein-3 was 1.8 mg/l, which is extremely low (normal range is 3.4–7.8 mg/l). His FSH was 1.8 IU/L and LH was 0.5 IU/L, both in the lower range. The prolactin level and TSH was normal. Testosterone was extremely low < 0.1 nmol/l (normal range is 9.9–27.8). Free androgen index is low < 0.1% (normal range is 14.8–94.8). The 25-hydroxy vitamin D level is undetectable in blood.

Twenty-four-hour urine collection for oligosaccharides was normal, and mucopolysaccharide was negative in the urine. Chromosomal analysis was 46,XY and attributed as no apparent chromosomal abnormalities. FISH analysis to test *SNRPN* deletion within the 15q11.2 locus for AS was negative. The patient also had a growth hormone stimulation test, and the result showed a flat response with a fasting growth hormone of 0.6 m U/L. After 10 min, it was 0.6 m U/L and after 1 h was 1.3 m U/L. This indicated severe growth hormone failure.

A skeletal survey was also done for the patient, and the result was in keeping with metabolic bone disease with evidence of severe osteomalacia and multiple looser zones and fractures, microcephaly, scoliosis, and kyphosis. Serum homocysteine was normal. MRI of the brain was performed and concluded that there were relative atrophic changes seen involving the cerebellum; otherwise, there was no evidence of white matter disease, cortical matter disease, or cortex abnormality.

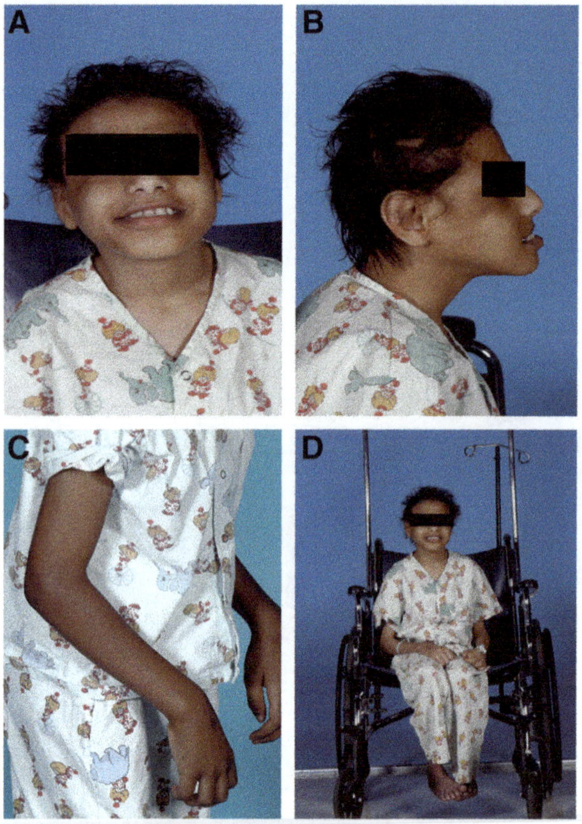

Fig. 2 Clinical features: **a** Apparent happy demeanor with easily excitable personality. **b** Microcephaly, prognathia, and wide spaced teeth. **c** There are usually flexed arm positions particularly during movement. **d** Short stature and increasingly chair bound

A muscle biopsy was studied for the index patient. It was interpreted as having type II fiber atrophy with glycogen excess associated with a mild increase in mitochondria. Similar findings were observed in his brother and sister. There were no deficiencies in mitochondrial electron transport chain enzymes. Citrate synthase activity was increased, suggesting increased mitochondrial proliferation.

Linkage and homozygosity mapping

Linkage analysis (Fig. 1b) localized the disease to a region on chromosome 6 defined proximally by rs743646 and distally by rs34811366 (chr6:4,536,241-6,670,705) with a maximum LOD score of 4.20. Based upon runs of homozygosity shared by all affecteds and absent in all unaffected individuals (Fig. 1c; red rectangle), we refined the critical interval to a 2.1 Mb region (chr6:4,231,133-6,018,243) containing approximately 60 RefSeq genes (hg19).

Whole exome sequencing and analysis

Whole exome sequencing of the index case (V:3) identified 25,168 variants relative to hg19, with 1342 variants on chr 6. These were further filtered to exclude all variants present outside our defined ROH (chr 6: 4,231,133-6,018,243) within which we identified 461 variants. By excluding previously reported variants (present in dbSNP, 1000 genomes and 2000 Arab exomes), the list was narrowed to 8. By only focusing on homozygous changes that were non-synonymous, splicing variants, frameshift insertions or deletions, and nonsense variants, we decreased the number of candidate variants to 1. It was a G>A change in exon 29 of *VARS2* (NM_006295:c.3650G>A:R1217H) (Fig. 3a). This homozygous variant was confirmed by Sanger sequencing, and the mutation segregated recessively with disease in the families studied (Fig. 3b).

Structure and function

VARS2 belongs to the class-I aminoacyl-tRNA synthetase family. The aminoacyl-tRNA synthetase (also known as aminoacyl-tRNA ligase) catalyzes the attachment of an amino acid to its cognate transfer RNA molecule in a highly specific two-step reaction. Class I aminoacyl-tRNA synthetases contain a characteristic Rossman fold catalytic domain and are mostly monomeric. tRNA binding involves an alpha-helical structure that is conserved between class I and class II synthetases. R1217 is located in the C-terminal coiled-coil domain of the tRNA synthetase domain of VARS2 (Fig. 4a). R1217 directly binds to the phosphate backbone of the tRNA D- loop. R1217 also forms a hydrogen bond with the ribose moiety of an adenosine that is flipped out and points toward the coiled-coil domain (Fig. 4b). R1217 has therefore an important function in the specific recognition of tRNA. The mutation of R1217 into a histidine abolishes this function. Hence, the R1217H mutant is expected to have compromised abilities to recognize the correct tRNA. This is consistent with the prediction of pathogenicity of this variant by models including Polyphen, SIFT, and CADD.

Discussion

Aminoacyl-tRNA synthetase enzymes (aaRSs) are involved in translation at the second step in creating a protein from a gene's instructions. The mRNA interacts with the ribosomal complex which reads the sequence of mRNA bases. Each sequence codes for one particular amino acid, and tRNA assembles the protein, one amino acid at a time. Subsequently, aaRSs help attach the amino acid to the tRNA [14].

A typical animal cell contains 37 cytoplasmic and mitochondrial synthetase genes, all of which are encoded on nuclear chromosomes. Mitochondrial synthetases are given the same name as cytosolic synthetases, appended with the numeral 2 (e.g., HARS and HARS2) [15]. Mitochondrial and cytoplasmic aaRSs are encoded by distinct nuclear genes, with the exception of GARS and KARS, which are present in both cellular compartments [16]. Mt-tRNA genes are located in three transcription units: the short H-strand unit (with the rRNA region and two

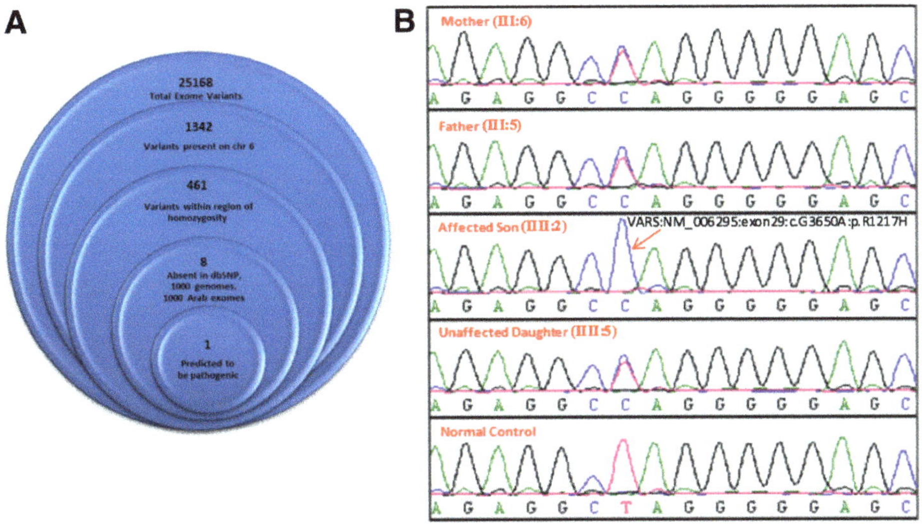

Fig. 3 *VARS2* mutation: **a** Filtering strategy used for identification of a causative mutation. **b** DNA electrophoregram with the G>A change in *VARS2* (NM_006295:exon29:c.G3650A:p.R1217H)

tRNA genes) is transcribed twice as fast as the L-strand that contains eight tRNA genes and more frequently the entire H-strand unit that produces 14 tRNAs [17]. Currently, pathological mutations that cause human diseases have been identified in many genes that encode cytoplasmic and mitochondrial tRNA-synthetases. The first cytoplasmic aaRSs mutation associated with a human

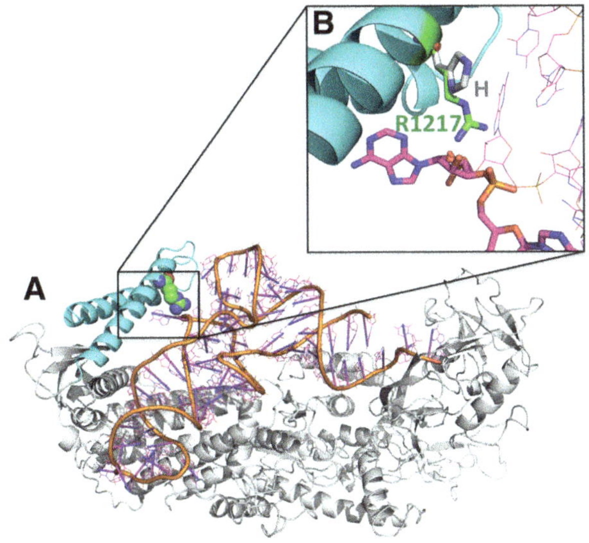

Fig. 4 Computational structural analysis of mutant: Homology model of the VARS2 tRNA synthetase domain, based on the PDB id 1gax (36% sequence id, RaptorX *p* value 1.12e-19). **a** The C-terminal coiled-coild domain is shown in cyan. R1217 is highlighted with carbon atoms shown in green. tRNA is shown with orange backbone. **b** Inlay: Zoom of the indicated region. H1217 is shown with gray carbon atoms. The tRNA D loop is shown with carbon atoms in magenta

disease, Charcot-Marie-Tooth, was discovered in glycyl-tRNA synthetase (GARS) [18]. Typically, defects in each mtARS have been identified in one tissue-specific disease (most commonly affecting the brain) or in one syndrome [19].

VARS2 encodes mitochondrial valyl-tRNA (Val-tRNA) synthetase, which is involved in mitochondrial translation [16]. The deduced 993-amino acid protein has an N-terminal mitochondrial targeting signal with a predicted cleavage site after residue 29. VARS2 has characteristics of a class I mitochondrial aminoacyl-tRNA synthetase, including a classical Rossmann fold [20].

A homozygous mutation of *VARS2* was reported in a sporadic case with microcephaly and epilepsy [16]. It was associated with isolated deficiency of the mitochondrial respiratory chain (MRC) complex I. The disorder was secondary to a homozygous c.1100C>T transition in *VARS2*, resulting in a threonine 367 to isoleucine (T367I) substitution at a highly conserved residue. The affected patient was an 8-year-old with psychomotor delay, facial dysmorphisms, and microcephaly. At 4 years of age, there was onset of partial seizures. His brain MRI displayed hyperintense lesions in the periventricular regions, the insula, and the right fronto-temporal cortex. Brain proton magnetic resonance spectroscopy revealed a lactate peak in the frontal white matter. Muscle computerized tomography showed bilateral quadriceps hypotrophy, but no evident histological alteration was observed in a muscle biopsy [16].

In a British boy with combined oxidative phosphorylation deficiency-20, a compound heterozygous mutation of *VARS2*: a c.1135G>A transition, resulting in an alanine 379 to threonine (A379T) substitution, and a c.1877C-A transversion, resulting in an alanine 626 to

Table 1 Genetic and phenotypic heterogeneity

Clinical features	Diodato et al.	Taylor et al.	This study
	NM_001167734.1: c.1100C>T, p.Thr367Ile (Hom)	NM_001167734.1: c.1135G>A,p.Ala379Thr (Het) c.1877C>A, p.Ala626Asp (Het)	NM_006295: c.A3253G; p.S1085G
Mental retardation	X		X
Facial dysmorphisms	X		X
Ataxia		X	X
Combined oxidative phosphorylation deficiencies		X	
Microcephaly	X		X
Progressive external ophthalmoplegia		X	
Epilepsy	X		X
Short stature			X
Hyperintense lesions in the periventricular regions, the insula, and the right frontotemporal cortex	X		
Evidence of severe growth hormone deficiency			X
Hypogonadism			X
Severe osteomalacia			X

X : affected

aspartic acid substitution (A626D). He presented with progressive external ophthalmoplegia, ataxia, and combined oxidative phosphorylation deficiencies [21].

Our family share with the reported cases, the findings of microcephaly, epilepsy, and the progressive cerebral encephalopathy; however, several dysmorphic features were noticeable in clinical assessment of the affected members from our family. The hypogonadism, vitamin D resistant osteomalacia, and the growth hormone deficiency were significant and novel associations with the *VARS2* mutation in the family from this study (Table 1).

In our *VARS2* family, there were no apparent deficiencies in mitochondrial electron transport chain enzymes. However, citrate synthase activity was increased, suggesting increased mitochondrial proliferation. We identified a homozygous change *VARS2:*NM_006295:c.3650G>A:R1217H, segregating with disease in our family. This mutation is in the anticodon-binding domain of tRNA synthetase, while the reported isolated *VARS2* homozygous mutation [16] is located in the catalytic core domain of valyl-tRNA synthetase.

The lack of clear biochemical phenotypes of OXPHOS or mitochondrial protein synthesis defects in skin fibroblasts and myoblasts from most of mutant aaRSs patients prevented the use of cultured cells for functional studies. In a few RARS2 patients, fibroblasts showed OXPHOS defects, with RC enzymes variably affected; YARS2 patients presented deficient mitochondrial protein synthesis in myotubes but not in cultured fibroblasts [22].

Conclusions

In summary, we describe a novel autosomal recessive syndromic disorder associated with a mutation in *VARS2* (mitochondrial valyl tRNA synthetase). Clinical features included severe mental retardation, ataxia, speech impairment, epilepsy, short stature, microcephaly, dysmorphia, an easily excitable personality, and excessive chewing mouth behaviors. In addition, there was clinical and chemical evidence of severe growth hormone deficiency, hypogonadism, and severe osteomalacia. Our findings further expand the phenotypic spectrum of Aminoacyl-tRNA synthetase enzymes.

Acknowledgements

The authors would like to thank the family members for their participation in this study. We thank the Genotyping and Sequencing Core Facilities of King Faisal Specialist Hospital and Research Centre for their contributions. We would also like to acknowledge the support of King Abdulaziz City for Science and Technology (KACST) through the Saudi Human Genome Project (SHGP).

Funding

This work was funded through grants from King Faisal Specialist Hospital and Research Centre (RAC 2140029) and King Abdulaziz City for Science and Technology (KACST#13-MED2056-20).

Author's contributions

AA contributed to the conception and design, acquisition and analysis of data, and drafting of the manuscript. BA contributed to the acquisition and analysis of data and preparation of figures. EG contributed to the acquisition and analysis of data and preparation of figures. DJ contributed to the acquisition and analysis of data and preparation of figures. FB contributed to the acquisition and analysis of data. BFM contributed to the conception and design, acquisition and analysis of data, and drafting of the manuscript and figures. STA contributed to the acquisition and analysis of data. DM contributed to the conception and design, acquisition and analysis of data, and drafting of the manuscript and figures. All authors read and approved the final manuscript.

Consent for publication

Written informed consent was obtained to publish the video included in the supplement.

Competing interests

The authors declare that they have no competing interests.

Author details

[1]Department of Neurosciences, King Faisal Specialist Hospital and Research Centre, Riyadh, Saudi Arabia. [2]Department of Genetics, King Faisal Specialist Hospital and Research Centre, MBC 03, PO Box 3354, Riyadh 11211, Saudi Arabia. [3]Saudi Human Genome Project, King Abdulaziz City for Science and Technology, Riyadh, Saudi Arabia. [4]King Abdullah University of Science and Technology (KAUST), Computational Bioscience Research Center (CBRC), Division of Biological and Environmental Sciences and Engineering (BESE), Thuwal 23955-6900, Saudi Arabia.

References

1. Al-Mujadi H, AR AR, Katzarov MG, Dehrab NA, Batra YK, Al-Qattan AR. Preemptive gabapentin reduces postoperative pain and opioid demand following thyroid surgery. Can J Anaesth. 2006;53:268–73.
2. Fernell E. Aetiological factors and prevalence of severe mental retardation in children in a Swedish municipality: the possible role of consanguinity. Dev Med Child Neurol. 1998;40:608–11.
3. Ropers HH. Genetics of early onset cognitive impairment. Annu Rev Genomics Hum Genet. 2010;11:161–87.
4. de Ligt J, Willemsen MH, van Bon BW, Kleefstra T, Yntema HG, Kroes T, Vulto-van Silfhout AT, Koolen DA, de Vries P, Gilissen C, et al. Diagnostic exome sequencing in persons with severe intellectual disability. N Engl J Med. 2012;367:1921–9.
5. Fryns JP, de Ravel TJ. London Dysmorphology Database, London Neurogenetics Database and Dysmorphology Photo Library on CD-ROM [Version 3] 2001R. M. Winter, M. Baraitser, Oxford University Press, ISBN 019851-780, pound sterling 1595. Hum Genet. 2002;111:113.
6. Ramsden SC, Clayton-Smith J, Birch R, Buiting K. Practice guidelines for the molecular analysis of Prader-Willi and Angelman syndromes. BMC Med Genet. 2010;11:70.
7. Tan WH, Bird LM, Thibert RL, Williams CA. If not Angelman, what is it? A review of Angelman-like syndromes. Am J Med Genet A. 2014;164A:975–92.
8. Arnold K, Bordoli L, Kopp J, Schwede T. The SWISS-MODEL workspace: a web-based environment for protein structure homology modelling. Bioinformatics. 2006;22:195–201.
9. Kallberg M, Margaryan G, Wang S, Ma J, Xu J. RaptorX server: a resource for template-based protein structure modeling. Methods Mol Biol. 2014;1137:17–27.
10. Yang J, Yan R, Roy A, Xu D, Poisson J, Zhang Y. The I-TASSER Suite: protein structure and function prediction. Nat Methods. 2015;12:7–8.
11. Kall L, Krogh A, Sonnhammer EL. A combined transmembrane topology and signal peptide prediction method. J Mol Biol. 2004;338:1027–36.
12. UniProt C. UniProt: a hub for protein information. Nucleic Acids Res. 2015; 43:D204–12.
13. Hunter S, Jones P, Mitchell A, Apweiler R, Attwood TK, Bateman A, Bernard T, Binns D, Bork P, Burge S, et al. InterPro in 2011: new developments in the family and domain prediction database. Nucleic Acids Res. 2012;40:D306–12.
14. Francklyn C, Musier-Forsyth K, Martinis SA. Aminoacyl-tRNA synthetases in biology and disease: new evidence for structural and functional diversity in an ancient family of enzymes. RNA. 1997;3:954–60.
15. Abbott JA, Francklyn CS, Robey-Bond SM. Transfer RNA and human disease. Front Genet. 2014;5:158.
16. Diodato D, Melchionda L, Haack TB, Dallabona C, Baruffini E, Donnini C, Granata T, Ragona F, Balestri P, Margollicci M, et al. VARS2 and TARS2 mutations in patients with mitochondrial encephalomyopathies. Hum Mutat. 2014;35:983–9.
17. King MP, Attardi G. Post-transcriptional regulation of the steady-state levels of mitochondrial tRNAs in HeLa cells. J Biol Chem. 1993;268:10228–37.
18. Antonellis A, Ellsworth RE, Sambuughin N, Puls I, Abel A, Lee-Lin SQ, Jordanova A, Kremensky I, Christodoulou K, Middleton LT, et al. Glycyl tRNA synthetase mutations in Charcot-Marie-Tooth disease type 2D and distal spinal muscular atrophy type V. Am J Hum Genet. 2003;72:1293–9.
19. Euro L, Konovalova S, Asin-Cayuela J, Tulinius M, Griffin H, Horvath R, Taylor RW, Chinnery PF, Schara U, Thorburn DR, et al. Structural modeling of tissue-specific mitochondrial alanyl-tRNA synthetase (AARS2) defects predicts differential effects on aminoacylation. Front Genet. 2015;6:21.
20. Bonnefond L, Fender A, Rudinger-Thirion J, Giege R, Florentz C, Sissler M. Toward the full set of human mitochondrial aminoacyl-tRNA synthetases: characterization of AspRS and TyrRS. Biochemistry. 2005;44:4805–16.
21. Taylor RW, Pyle A, Griffin H, Blakely EL, Duff J, He L, Smertenko T, Alston CL, Neeve VC, Best A, et al. Use of whole-exome sequencing to determine the genetic basis of multiple mitochondrial respiratory chain complex deficiencies. JAMA. 2014;312:68–77.
22. Diodato D, Ghezzi D, Tiranti V. The mitochondrial aminoacyl tRNA synthetases: genes and syndromes. Int J Cell Biol. 2014;2014:787956.

Overexpressed HSF1 cancer signature genes cluster in human chromosome 8q

Christopher Q. Zhang[1,4], Heinric Williams[3,4], Thomas L. Prince[3,4†] and Eric S. Ho[1,2*†]

Abstract

Background: HSF1 (heat shock factor 1) is a transcription factor that is found to facilitate malignant cancer development and proliferation. In cancer cells, HSF1 mediates a set of genes distinct from heat shock that contributes to malignancy. This set of genes is known as the HSF1 Cancer Signature genes or simply HSF1-CanSig genes. HSF1-CanSig genes function and operate differently than typical cancer-causing genes, yet it is involved in fundamental oncogenic processes.

Results: By utilizing expression data from 9241 cancer patients, we identified that human chromosome 8q21-24 is a location hotspot for the most frequently overexpressed HSF1-CanSig genes. Intriguingly, the strength of the HSF1 cancer program correlates with the number of overexpressed HSF1-CanSig genes in 8q, illuminating the essential role of HSF1 in mediating gene expression in different cancers. Chromosome 8q21-24 is found under selective pressure in preserving gene order as it exhibits strong synteny among human, mouse, rat, and bovine, although the biological significance remains unknown. Statistical modeling, hierarchical clustering, and gene ontology-based pathway analyses indicate crosstalk between HSF1-mediated responses and pre-mRNA 3′ processing in cancers.

Conclusions: Our results confirm the unique role of chromosome 8q mediated by the master regulator HSF1 in cancer cases. Additionally, this study highlights the connection between cellular processes triggered by HSF1 and pre-mRNA 3′ processing in cancers.

Keywords: Heat shock factor 1 (HSF1), HSF1 cancer signature (CanSig), Malignancy, Pre-mRNA 3′ processing, Chromosome 8q, Synteny

Introduction

Heat-shock factor 1 (HSF1) is a master transcription factor that initiates the expression of heat shock proteins (HSPs) and other genes in response to cellular stress, thus allowing cells to adapt and prolong survival. Cancer cells, however, hijack this protective mechanism to allow them to continue proliferating in their toxic microenvironments. Several studies have linked increased HSF1 activity to malignant cell growth [1]. This pro-malignant activity is associated with HSF1 binding to the promoters and initiating the expression of certain genes independent of heat shock [1]. Identified by Mendillo, Santagata, and coworkers, the HSF1 Cancer Signature (HSF1-CanSig) is a set of 475 genes overexpressed in a number of highly malignant cancer cells and primary tumors [1].

HSF1 itself is observed to be overexpressed in across different tumor types and to promote proliferation, migration, and invasion [2–5]. Frequently, overexpressed genes across different tumors or cancer types are understood to be likely oncogenic and contribute to tumorigenesis [6]. Consequently, this drives the overexpression of the HSF1-CanSig in tumor cells and possibly the surrounding stroma leading to metastasis and poor clinical outcomes [1, 2]. The biological processes and features that link the HSF1-CanSig to malignant cell growth, however, are still unclear.

To elucidate the role of the HSF1-CanSig in cancer, we mined and analyzed the overexpression in 9241 cancer cases from 27 unique primary tumor sites from The Cancer Genome Atlas (TCGA) hosted in cBioPortal [7]. We found that 27 of the top 100 most frequently overexpressed HSF1-CanSig genes are clustered in a highly syntenic region of chromosome 8q. We observed that

* Correspondence: hoe@lafayette.edu
†Equal contributors
[1]Department of Biology, Lafayette College, Easton, PA 18042, USA
[2]Department of Computer Science, Lafayette College, Easton, PA 18042, USA
Full list of author information is available at the end of the article

this HSF1-driven chromosome 8q gene is set to be overexpressed in a majority of cancer types. Furthermore, our gene ontology analysis indicates a link between HSF1-driven transcription and mRNA processing located on chromosome 8q.

Results

Top 100 high-scoring HSF1-CanSig genes are disproportionately located in chromosome 8q

We devised an overexpression score to quantify the prevalence of HSF1-CanSig genes being overexpressed among cancer cases (see the "Materials and methods" section for details). The overexpression score reflects the prevalence of a gene being overexpressed ≥ 2 standard deviations above the control reference as calculated by cBioPortal for different primary tumor sites [8]. A gene associated with a high overexpression score indicates that the gene is often upregulated in cancer cases. Overexpression scores of all 475 HSF1-CanSig genes from different primary tumor sites were aggregated, ranked, and examined by their percentage distribution per chromosome arm in tiers: top 100, 200, 300, and all HSF1-CanSig genes. If ranking has no effect on distribution, the distribution of genes in each tier should resemble the distribution of all HSF1-CanSig genes, meaning that they distribute like the result of random sampling of HSF1-CanSig genes. Intriguingly, 27% of the top 100 groups comprises of genes encoded in chromosome 8q (Fig. 1a, leftmost light blue bar) given that 8q consists of only 6% (29 of 475) of all HSF1-CanSig (leftmost gray bar). Such percentage declines continuously with from the top 200 group to top 300 group.

To assess the statistical significance of the distribution exhibited in Fig. 1a, we tested the probability of observing certain number of genes per chromosome arm in different tiers by hypergeometric test and Fisher's exact test (Additional file 1: Table S1). The four chromosome arms that encode the highest number of HSF1-CanSig genes are 1p (30), 8q (29), 11q (29), and 17q (30). When genes were examined in tiers by ranking, HSF1-CanSig genes encoded in 8q are included most disproportionately in different tiers but not for genes in the other three chromosome arms, i.e., 1p, 11q, and 17q. The group that attained the smallest p value of Fisher's test was the top 100 group ($5.41e^{-20}$), whereas hypergeometric test found that the top 50 group achieved the lowest p value ($3.13e^{-18}$) followed by the top 100 group ($1.09e^{-17}$). But the former captures only 22 8q genes out of 29 versus 27 8q genes by the latter. To balance statistical rigor and coverage, we decided to focus on the top 100 frequently overexpressed HSF1-CanSig genes. Importantly, regardless of which tier we choose, the most striking finding is that a large proportion of top-ranking genes are encoding in chromosome 8q as supported by Additional file 1: Table S1.

Next, we examine whether the skew distribution of 8q genes in the top 100 group is influenced by the intrinsic distribution of cancer or protein-coding genes. To rule out this possibility, the aggregated overexpression scores of the top 100 high-scoring HSF1-CanSig genes were compared with two null models or references: census cancer gene and all protein-coding genes, which consist of only 2–3% of 8q genes (gray and black bars in Fig. 1b, respectively). A high correlation between the two references based on the percentage of genes per chromosome arm was shown ($R = 0.92$, $p < 5.98e^{-8}$), suggesting that cancer genes exhibit similar distribution as protein coding genes in the human genome. To ascertain the bias of HSF1-CanSig 8q genes quantitatively, a hypergeometric test was used. The probability of observing 27 or more HSF1-CanSig 8q genes (listed in Table 1) in a sample of 100 genes by chance is 1.02×10^{-22}.

Overexpression of HSF1-CanSig 8q genes is primary tumor site-specific

Furthermore, we examined the distribution of HSF1-CanSig 8q genes by individual primary tumor sites. Skewed distribution of HSF1-CanSig genes on chromosome 8q, similar to that in Fig. 1b, was observed in 22 out of 27 primary sites with a variable degree of 8q bias (plots for individual primary site can be found in Additional file 2). The five primary tumor sites that showed no bias are the adrenal gland, kidney, nervous system, thymus, and thyroid. An exponential distribution ($p_i = \lambda e^{-\lambda i}$ or in linear form: $\ln p_i = -\lambda i + \ln \lambda$) was used to quantify the extent of skewness: p_i is the probability or proportion of the ith-ranked chromosome, λ is a coefficient, and i is the rank in the range 1 to 10. A large magnitude of λ indicates a high degree of skewness and vice versa. The λ values of primary sites span a wide spectrum of chromosome 8q bias as shown in Fig. 2, where higher 8q bias are observed in breast and liver cancers than in lymph nodes and soft tissue, for instance. Primary sites lacking 8q bias primary tumor sites tend to be cornered at a region with small λ value (labeled in gray). This result reveals that the overexpression of these HSF1-CanSig 8q genes is primary tumor site-specific.

HSF1-CanSig genes clustered at a 44-Mbp region near the end of chromosome 8q

The 27 HSF1-CanSig 8q genes ranked in the top 100 high-scoring HSF1-CanSig genes are tabulated in Table 1. Remarkably, all HSF1-CanSig 8q genes including HSF1 itself occupy the last three cytogenetic bands of chromosome 8q, i.e., 8q21.3 to 8q.24.3 (~ 44 Mbps). The 3′-

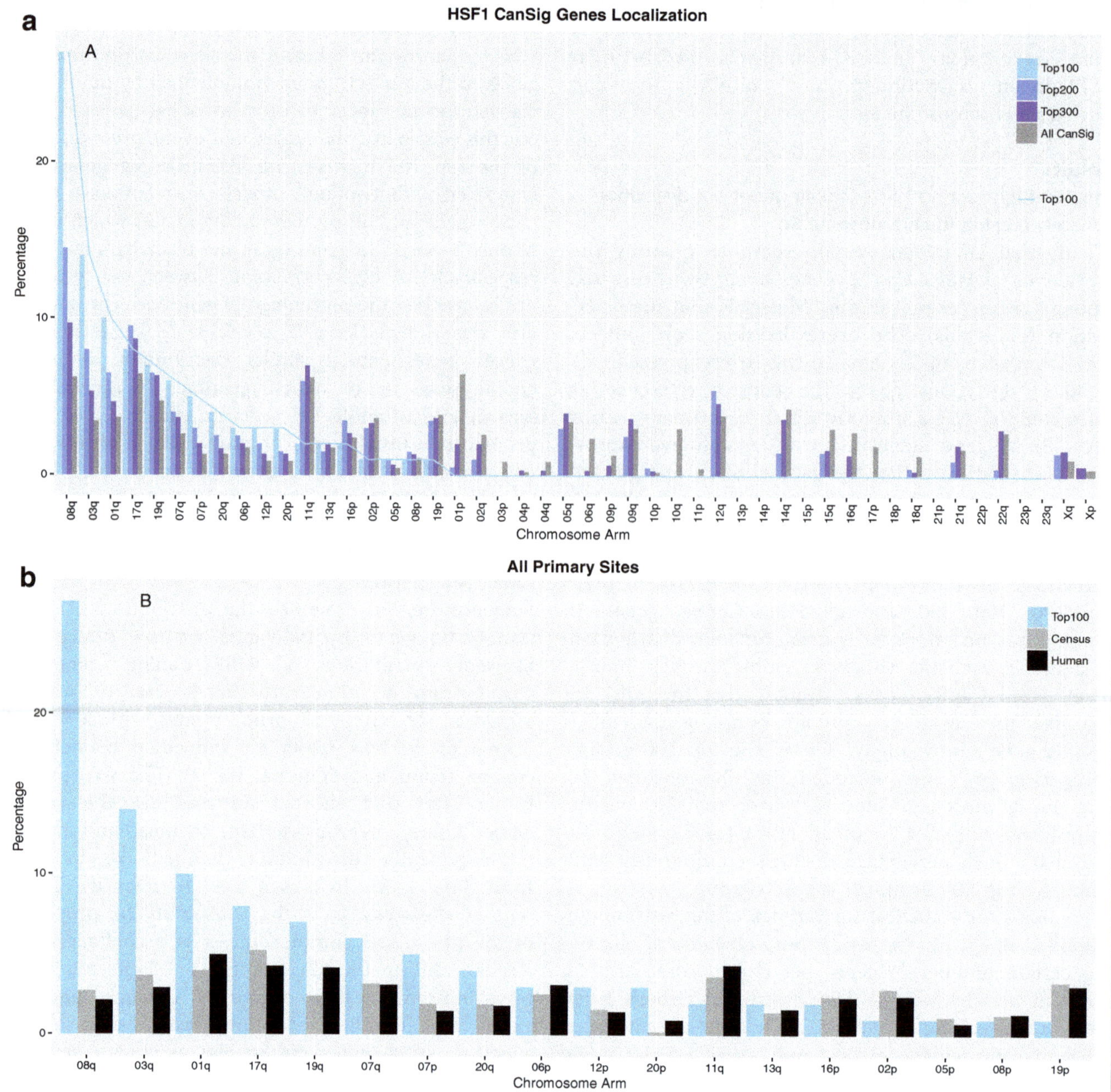

Fig. 1 **a** Percentage distribution of HSF1-CanSig genes by chromosome arms from different tiers: top 100, top 200, top 300, and all HSF1-CanSign genes. **b** Percentage distribution of the top 100 high-scoring HSF1-CanSig genes by chromosome arm. Two references were used for comparison: 633 cancer census genes (labeled census) from COSMIC [49] and 19,073 human protein-coding genes (labeled all) based on HuGO [48]

most gene ZNF250 is ~237 Kbps from the chromosome's 3′ end with only three genes encoded further downstream.

Frequently overexpressed HSF1 associates with the overexpression of HSF1-CanSig genes in 8q

Cancer is heterogeneous both within and between cancer types. Hence, we sought to rank individual HSF1-CanSig 8q genes in different primary tumor sites (Table 2). To cover the entire chromosomal segment encoding all 29 HSF1-CanSig 8q genes, two 8q genes, GPT and KLF10, were included even though they were not ranked in the top 100. The one-sample two-sided Student's t test was used to assess the ranking distribution of them (the second column from the right in Table 2) for each primary tumor site where the p values of the t test are shown in the last column of Table 2. At 95% confidence level, ranks of 8q genes in 20 primary tumor sites (from the ovary to pleura in Table 2) are detected with statistically significant deviation from the

Table 1 Twenty-seven out of the top 100 high-scoring HSF1-CanSig genes are clustered near the end of chromosome 8q

Gene symbol	Name	Location	Rank
PUF60	Poly(U) binding splicing factor 60	08q24.3	1
HSF1	Heat shock transcription factor 1	08q24.3	2
NUDCD1	Nudc domain containing 1	08q23.1	3
MRPL13	Mitochondrial ribosomal protein L13	08q24.12	4
ENY2	ENY2, transcription and export complex 2 subunit	08q23.1	5
CPSF1	Cleavage and polyadenylation specific factor 1	08q24.3	7
SHARPIN	SHANK associated RH domain interactor	08q24.3	8
AZIN1	Antizyme inhibitor 1	08q22.3	9
MAF1	MAF1 homolog, negative regulator of RNA polymerase III	08q24.3	10
PABPC1	Poly(A) binding protein cytoplasmic 1	08q22.3	11
TRAPPC9	Trafficking protein particle complex 9	08q24.3	12
ZNF250	Zinc finger protein 250	08q24.3	13
ZNF34	Zinc finger protein 34	08q24.3	15
NBN	Nibrin	08q21.3	16
JRK	Jrk helix-turn-helix protein	08q24.3	19
DPY19L4	Dpy-19 Like 4 (*C. elegans*)	08q22.1	22
CYHR1	Cysteine and histidine rich 1	08q24.3	23
TPD52	Tumor protein D52	08q21.13	28
PLEC	Plectin	08q24.3	32
MFSD3	Major facilitator superfamily domain containing 3	08q24.3	35
KIFC2	Kinesin family member C2	08q24.3	40
C8ORF37	Chromosome 8 open reading frame 37	08q22.1	49
NDRG1	N-Myc downstream regulated 1	08q24.22	54
SLC45A4	Solute carrier family 45 member 4	08q24.3	56
MROH6	Maestro heat like repeat family member 6	08q24.3	59
LY6K	Lymphocyte antigen 6 family member K	08q24.3	66
LRP12	LDL receptor related protein 12	08q22.3	69

Rank is based on the aggregated overexpression score of all primary tumor sites

mean rank of 233 (see the "Materials and methods" section for the determination of the mean rank). No single primary tumor site had all 29 HSF1-CanSig 8q genes overexpressed in the top 100 HSF1-CanSig genes. Surprisingly there were no primary tumor sites with significant low-average rank (the lowest two are the thymus, $p = 0.896$, and the thyroid, $p = 0.283$, in Table 2), suggesting the absence of a strong suppression among HSF1-CanSig 8q genes in those primary tumor sites.

While no single HSF1-CanSig 8q gene is ranked in the top 100 across all primary sites, HSF1 is frequently overexpressed in most primary tumor sites with kidney as the only exception (26 out of 27, shown at the bottom of HSF1 column in Table 2). HSF1 is also the gene with the highest average rank (33) followed by CPSF1 (52). Additionally, the rank of HSF1 correlates with the average rank of primary sites ($R = 0.71$), but the most correlated 8q gene is CYHR1, not HSF1. Importantly, HSF1 is the only gene in which its rank is always lower than the average rank (Additional file 3: Figure S1).

Lastly, primary tumor sites associated with a small number of high-scoring HSF1-CanSig 8q genes, such as the lymph nodes, soft tissue, thymus, and thyroid, tend to exhibit lower skewness (see Fig. 2). These results suggest a minor role of HSF1-mediated activities in those primary tumor sites.

Primary tumor sites are segregated by ranks

We next asked if the ranks of HSF1-CanSig 8q genes in different primary tumor sites share similar patterns. By using hierarchical clustering with two distinct distance methods, primary sites were clustered using the ranks of HSF1-CanSig 8q genes as shown in Fig. 3. Both panels in Fig. 3 show that primary tumor sites can be reliably split into two groups by the average rank, namely, the top-ranking and the low-ranking groups. The medians

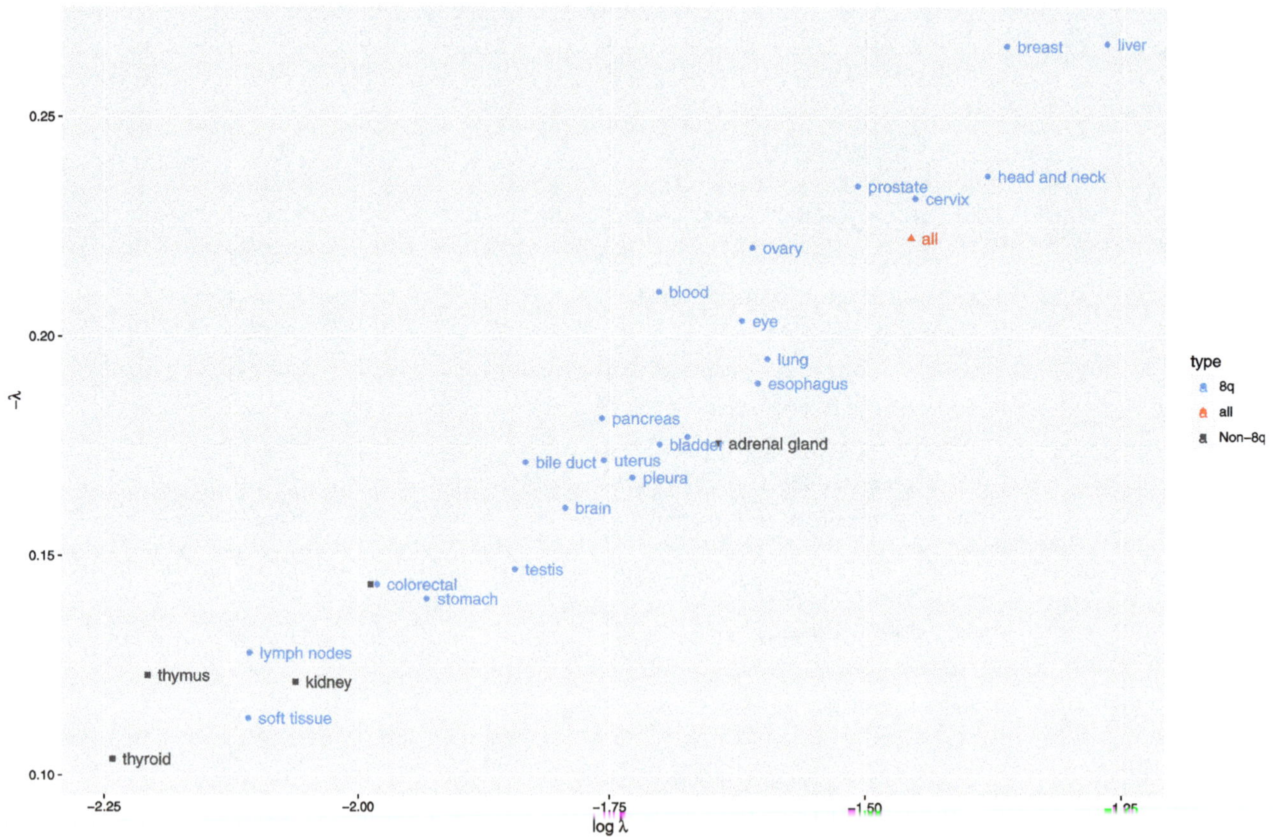

Fig. 2 Quantification of 8q skewness. The percentages of the top 10 chromosomes were fitted to an exponential model. The model outputs (λ = slope, $\ln(\lambda)$ = intercept) are plotted. Primary tumor sites with 8q and without 8q clustering are in blue and gray, respectively. The red triangle represents the fitting of data when 8q genes from all primary sites are ranked together

of the top-ranking groups produced by the two distance methods reflect higher consistency as they vary within a narrow range of 65–68. But the medians of the low-ranking groups show a larger difference from 148 to 198.

Regarding the range of average rank within each group, if the outlier, i.e., soft tissue with average rank 213, is taken out from the top-ranking correlation-distance group in Fig. 3b, the range of average ranks in both becomes identical, 34 to 102. A cutoff value of 102 seems to divide the two groups. Moreover, top-ranking groups created by the two distance methods share 13 out of 14 to 17 primary sites (marked by asterisks in Fig. 3). The low-ranking groups however display a larger variation in rank medians 146–198, but 9 out of 10 to 13 primary sites are clustered together by both methods (marked by plus signs in Fig. 3). This result suggests the feasibility of stratifying primary tumor sites by the over-expression of HSF1-CanSig 8q genes.

HSF1-CanSig 8q genes are highly syntenic in mammals

Since the proximity of HSF1-CanSig 8q genes seems to be a contributing factor for their cohesive expression pattern, we asked whether their proximity reveals biological significance. We sought answers in the light of evolution by examining the synteny of the genomic region spanning the 5′- and 3′-most HSF1-CanSig 8q genes, i.e., TPD53 and ZNF250, respectively. It means that other non-HSF1-CanSig genes were also included in the syntenic analysis. This region consists of 267 genes with known genomic coordinates in human where 29 of them are HSF1-CanSig genes. We aligned these human genes with homologs from bovine, mouse, and rat. Their pairwise percentage of synteny can be found in Table 3A, which varies from the lowest 50% (mouse and bovine) to the highest 84% (human and mouse). As our focus is on the HSF1-CanSig genes, we highlighted their genomic order in Table 3B. Syntenic information of all 267 genes can be found in Additional file 1: Table S5.

Twenty-three out of 29 human HSF1-CanSig 8q genes are syntenic to mouse and rat. Although three non-syntenic genes (Nbn, Dpy19l4, and 2610301B20Rik in mouse and Nbn, Dpy19l4, and MGC94199 in rat) reside in other chromosomes, the order of their locations still agrees with that in humans. If these three genes are included in calculating the percentage of synteny, human-

Table 2 Ranks of HSF1-CanSig 8q genes by primary tumor site. The bracketed number next to the name of a primary site denotes the number of cases recruited in the study. Genes displayed in the table heading are arranged in a syntenic order. The number inside a cell denotes the rank of the gene among HSF1-CanSig genes with the most frequently overexpressed genes ranked 1. Ranks ≤ 100 are highlighted in solid black. Some genes may share the same rank if they have the same overexpression score. Count includes only genes with rank ≤ 100. Average rank is the arithmetic mean of ranks. HSF1-CanSig genes GPT and LRP12 are not part of the top 100 high-scoring genes. *p* value is determined by one-sample two-sided Student's *t* test. The table is sorted by average rank and count in ascending and descending order, respectively

Primary Site	TPD52	NBN	DPY19L4	C8ORF37	PABPC1	KLF10	AZIN1	LRP12	NUDCD1	ENY2	MRPL13	NDRG1	TRAPPC9	SLC45A4	JRK	LY6K	MROH6	PUF60	PLEC	SHARPIN	MAF1	HSF1	CPSF1	CYHR1	KIFC2	GPT	MFSD3	ZNF34	ZNF250	Count	Average Rank	p-value
ovary (558)	30	18	49	85	35	60	11	40	23	25	28	40	6	32	33	136	70	2	9	1	5	3	29	16	50	120	15	4	7	27	34	1.47E-23
eye (80)	29	10	23	27	332	245	7	21	17	2	4	25	16	12	6	42	42	3	23	8	14	21	6	9	16	37	2	11	21	27	36	9.29E-15
prostate (498)	2	11	28	45	1	120	5	30	5	3	8	154	16	87	22	135	59	11	28	15	9	6	18	28	13	135	35	30	7	25	37	4.70E-20
breast (1100)	25	34	23	43	9	187	7	58	2	3	1	46	12	92	35	27	64	4	56	22	17	5	29	104	40	163	59	13	10	26	41	1.44E-19
colorectal (382)	55	35	8	62	13	34	7	60	12	17	30	104	22	113	41	177	60	9	132	18	12	20	23	24	120	342	42	19	22	23	56	7.09E-14
skin (472)	129	63	32	93	25	250	32	164	65	17	16	129	27	38	15	173	74	5	108	53	39	15	6	3	53	59	42	11	38	23	61	2.26E-15
bladder (408)	41	16	16	29	3	125	5	125	11	4	9	83	32	117	67	37	63	6	146	44	12	1	27	83	202	188	117	146	21	21	61	2.92E-15
liver (373)	29	19	25	113	2	376	14	113	3	6	1	74	13	72	20	415	21	4	30	5	7	11	9	17	33	356	23	15	10	24	63	1.08E-08
lung (517)	26	29	32	59	7	257	19	111	6	13	48	43	20	304	39	39	59	1	155	22	5	4	11	18	39	461	48	9	15	24	65	2.23E-09
head and neck (522)	209	45	70	70	21	26	31	75	16	21	51	16	84	280	112	23	152	3	23	14	19	2	10	33	28	447	84	29	53	24	71	7.05E-10
esophagus (185)	50	69	32	91	75	270	26	188	14	31	9	38	34	75	50	20	170	4	22	19	15	2	26	19	57	299	188	69	99	24	71	7.89E-12
pancreas (179)	39	97	31	153	21	383	4	97	2	3	1	465	50	12	12	60	39	21	9	18	50	24	9	78	60	153	18	153	19	24	72	9.73E-09
stomach (415)	13	25	22	97	12	450	4	161	9	10	7	31	33	180	92	128	55	3	142	45	59	8	14	17	59	148	31	274	110	21	77	2.62E-09
uterus (177)	83	223	16	104	12	386	27	262	4	23	56	27	71	18	35	188	43	56	35	27	7	7	10	47	65	310	65	23	16	23	77	3.21E-09
cervix (306)	167	53	67	167	16	149	34	118	34	41	19	96	149	135	96	41	268	18	268	36	21	20	28	46	53	393	82	62	72	20	95	5.10E-09
testis (156)	51	5	152	17	8	101	16	127	183	183	51	86	26	101	21	338	40	51	9	56	152	35	14	295	56	127	463	6	127	17	100	4.14E-07
brain (530)	241	22	114	262	15	319	319	202	28	13	23	163	68	114	18	35	449	22	42	42	10	8	13	35	96	37	163	28	30	19	101	1.46E-06
blood (179)	20	33	177	25	25	128	15	100	37	249	33	100	9	33	2	462	3	79	47	100	79	79	177	47	289	464	58	79	9	22	102	3.56E-06
bile duct (36)	9	14	41	74	414	144	414	414	74	144	41	414	6	74	9	22	74	74	269	74	144	41	41	41	144	269	269	41	22	18	131	4.44E-04
pleura (87)	230	230	230	300	58	386	91	230	58	18	91	230	58	58	143	300	386	143	143	91	58	5	91	143	18	29	91	91	230	15	146	1.71E-04
adrenal gland (79)	138	82	69	82	232	329	393	138	23	82	113	232	138	329	100	466	466	100	82	69	159	46	35	192	232	138	393	439	82	12	185	7.45E-02
nervous system (528)	245	145	272	272	145	190	272	26	166	99	190	409	125	215	112	442	332	99	215	59	59	22	59	442	426	190	190	39	22	9	189	7.60E-02
lymph nodes (48)	106	56	358	56	106	211	358	358	211	106	106	358	106	211	11	56	358	358	358	358	211	56	26	211	458	211	211	358	56	7	207	3.06E-01
soft tissue (263)	334	38	110	285	72	257	68	215	4	22	2	215	285	309	196	426	387	72	110	285	232	93	82	410	447	460	447	126	196	9	213	4.75E-01
kidney (534)	70	196	437	344	63	131	80	175	217	30	429	241	344	450	241	450	48	70	149	48	58	196	63	407	217	437	283	283	379	9	225	7.80E-01
thymus (120)	29	90	133	350	213	350	213	281	133	406	281	213	90	281	350	406	52	90	350	52	281	52	213	281	281	281	461	350	281	7	236	8.96E-01
thyroid (509)	443	452	421	207	352	278	421	124	400	57	180	352	96	452	400	421	57	278	96	36	43	96	324	443	207	207	180	352	236	7	262	2.83E-01
Count	17	22	17	16	20	3	20	9	21	22	21	13	21	12	20	11	18	24	14	25	21	26	24	17	16	4	15	18	20	17\19		
Average Rank	105	78	111	130	85	227	107	149	65	60	68	162	72	155	84	202	144	59	113	60	66	33	52	129	139	239	150	113	81		112	

rat and human-mouse share 72 and 84% of synteny, respectively. In comparison between bovine and human, all 8q homologs in bovine are found in a single chromosome, i.e., chromosome 14, but in the complementary strand of the bovine's reference genome. Twenty-five of the 29 8q genes are syntenic. If the entire 8q region is considered, bovine's genes share only 57% of synteny with human versus 61–71% in rat and mouse, indicating bovine genome has accumulated a larger scale of genome rearrangement than human, mouse, and rat since divergence. Regardless, the gene order is largely preserved between human and bovine. This finding is congruent with the estimated divergence time among these four mammals [9]. According to the report, bovine and human and mouse and human diverge in about 71–113 million years ago (MYA) and 62–101 MYA, respectively. Furthermore, our findings concur with the synteny analyses from the genome sequencing project of human (Figure 46 of [10]), Brown Norway rat (Figure 4 of [11]), and Taurine cattle (Figure S12 of [12]). To further investigate the synteny of human 8q in mouse, rat, and bovine, the tool Cinteny [13] was used to visualize the extent of synteny and they can be found in Additional file 3: Figure S2.

There are two possible explanations to account for the observed syntenic conservation among these genes: (i) the divergence time between taxa is insufficient to observe significant rearrangement or (ii) the proximity of genes is essential for biological functions. Classic examples are the Hox and globin gene clusters. As the estimation from [9] is adequate to disregard the former, we were intrigued to explore the latter, i.e., the biological function shared among these genes.

Synteny does not explain coexpression

To determine whether or not the overexpression of HSF1-CanSig 8q genes in cancers is related to synteny, we analyzed the expression pattern of genes flanking HSF1 (genomic coordinates of HSF1's neighboring genes can be found in Additional file 1: Table S7) in each cancer case to shed light on the clustering of HSF1-CanSig 8q genes. We reasoned that if the expression level of HSF1's neighboring genes correlates with their distance from HSF1, synteny may be a factor accounting for the clustering of CanSig 8q genes; otherwise reasons other than synteny may shape the clustering.

We selected 1682 cancer cases with HSF1 overexpressed two or more standard deviations above the reference. As shown in Fig. 4a, expression of HSF1 and its neighboring genes do not correlate with the distance between them. Genes close to HSF1 such as MROH1, BOP1, and DGAT1 do not show higher correlation (in terms of overexpression with HSF1) than HSF1-CanSig 8q genes that are farther from them. For example,

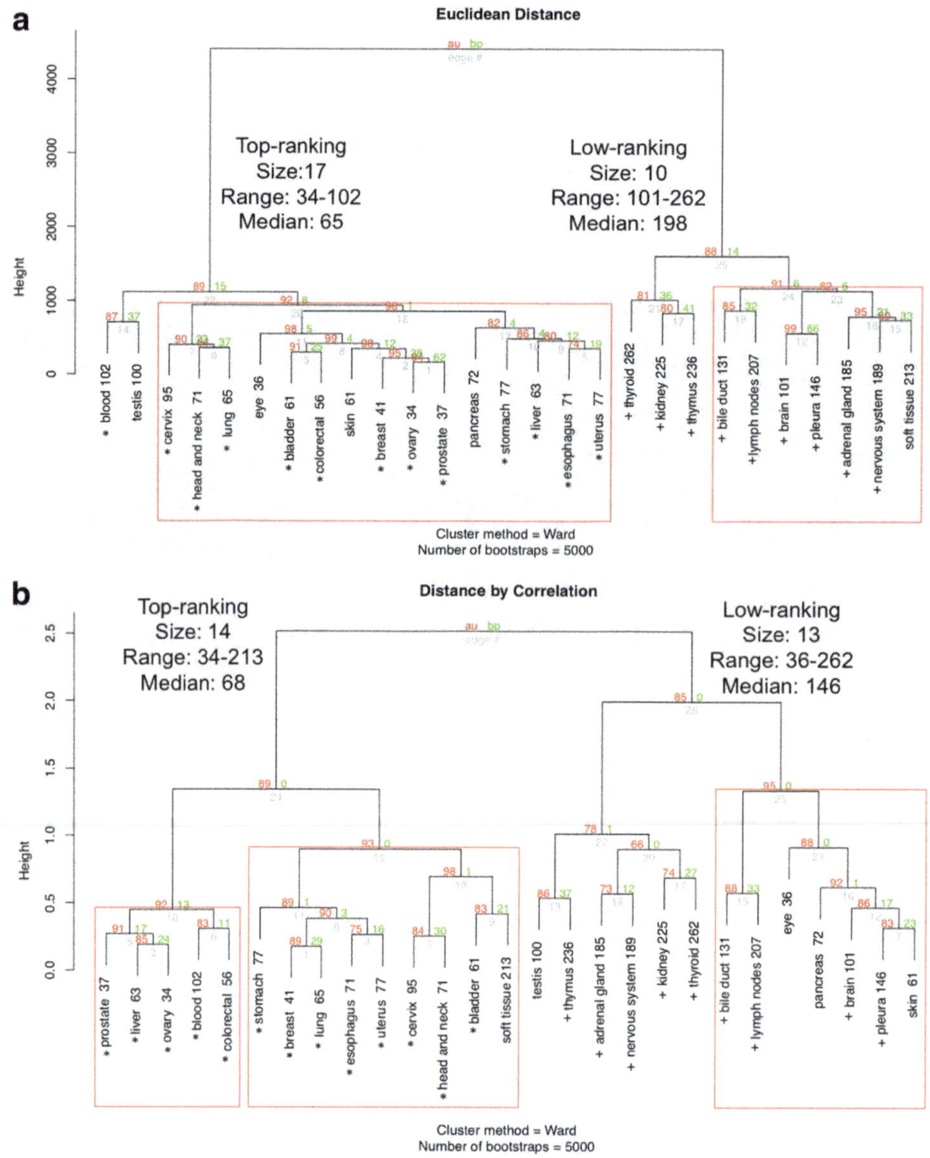

Fig. 3 Rank analysis by hierarchical clustering with bootstrapping. The number beside each primary tumor site is the average rank. Subtrees with high bootstrap value (≥ 90) are highlighted in red boxes. Each dendrogram was built after bootstrapping 5000 times. The red and green numbers at the branch point represent approximately unbias (AU) *p* value and bootstrap probability (BP), respectively. AU is recommended as a better parameter than BP to assess the clustering reliability at each branch point [54]. Cluster subtree highlighted by red box indicates AU ≥ 90. For the top-ranking groups, primary tumor sites shared between two distance methods are marked by asterisks. Similarly, for the low-ranking groups, primary sites are marked by plus signs. **a** Cluster dendrogram based on Euclidean distance. **b** Same input data as panel **a** but correlation was used as the distance calculation method

although 8q genes SHARPIN, MAF1, and CPSF1 are far apart from HSF1 than MROH1, BOP1, and DGAT1, they exhibit higher correlation with HSF1's expression. Similar variability is also observed when correlations are examined by individual primary tumor site (Additional file 4). Thus, our data suggests that synteny is not a contributing factor for the co-overexpression of HSF1-CanSig genes clustered in chromosome 8q21-24. It is however intriguing to see that several overexpressed non-HSF1-CanSig genes are correlated with HSF1 such as MAF1 ($R = 0.7$), CYC1 ($R = 0.6$), and SLC52A2 ($R = 0.6$) (see Fig. 4b).

Network analysis links HSF1-mediated responses to pre-mRNA 3′ processing

Next, we took a systems approach to determine if HSF1-CanSig 8q genes work cooperatively in biological networks. WebGestalt gene network (or pathway) analysis was performed to determine whether shared biological processes are common to these genes. The basic

Table 3 Synteny analysis of HSF1-CanSig 8q genes. A. Percentage of syntenic genes between species. Percentages are based on the number of genes in the column's species. The percentages in lower triangle excluded the separated chr4q in mouse and chr5q in rat. Whereas, the percentages in the upper triangle (shaded) include the separated chr4q in mouse and chr5q in rat. B. Order of human genes is used as the reference. Genes are ordered in 5'-to-3' direction according to respective reference genomes from top to bottom except for bovine, which is in reverse order. Non-syntenic genes are marked in boxes. Superscript + indicates the chromosomal location of the gene was determined by BLAT. Exact genomic coordinates of these genes can be found in Additional file 1: Table S6

A	Human	Mouse	Rat	Bovine
Human		84	72	57
Mouse	71		82	59
Rat	61	69		65
Bovine	57	50	55	

B			
Bovine (bosTau8) Chr14	Human (hg38) Chr8q	Mouse (mm10) Chr15q	Rat (rn6) Chr7q
MRPL13			
	TPD52	*Chr4q*$^{\#}$	*Chr5q*$^{\#}$
NBN	NBN	*Nbn*$^{\#}$	*Nbn*$^{\#}$
DPY19L4^{+}	DPY19L4	*Dpy19l4*$^{\#}$	*Dpy19l4*$^{\#}$
C14H8orf37^{+}	C8orf37	*2610301B20Rik*$^{\#}$	*MGC94199*$^{\#}$
PABPC1	PABPC1	Pabpc1	Pabpc1
KLF10	KLF10	Klf10	Klf10
AZIN1	AZIN1	Azin1	Azin1
LRP12	LRP12	Lrp12	Lrp12
NUDCD1	NUDCD1	Nudcd1	Nudcd1
ENY2	ENY2	Eny2	Eny2
	MRPL13	Mrpl13	Mrpl13
TPD52			
SLC45A4^{+}			
NDRG1	NDRG1	Ndrg1	Ndrg1
TRAPPC9^{+}	TRAPPC9	Trappc9	Trappc9
	SLC45A4	Slc45a4	Slc45a4^{+}
JRK^{+}	JRK	Jrk	Jrk
N.A.	LY6K	N.A.	N.A.
MROH6^{+}	MROH6	Mroh6	Mroh6^{+}
PUF60	PUF60	Puf60	Puf60
PLEC^{+}	PLEC	Plec	Plec
SHARPIN	SHARPIN	Sharpin	Sharpin
MAF1	MAF1	Maf1	Maf1^{+}
HSF1	HSF1	Hsf1	Hsf1
CPSF1	CPSF1	Cpsf1	Cpsf1
CYHR1	CYHR1	Cyhr1	Cyhr1
KIFC2	KIFC2	Kifc2	Kifc2
GPT	GPT	Gpt	Gpt
MFSD3^{+}	MFSD3	Mfsd3	Mfsd3
ZNF34	ZNF34	N.A.	N.A.
ZNF250^{+}	ZNF250	Zfp647^{+}	Zfp647^{+}

mechanism of WebGestalt is to detect the enrichment of Gene Ontology (GO) terms associated with a set of genes with statistical support. We used 29 HSF1-CanSig 8q genes to query 33 TCGA RNA-Seq studies in various primary tumor sites (the list of studies queried can be found in Additional file 1: Table S8). A sample of result generated by WebGestalt can be found in Additional file 3.

Out of the 33 TCGA datasets, eight studies were found to be enriched by HSF1-CanSig 8q genes: adrenocortical carcinoma, colon adenocarcinoma, esophageal carcinoma, kidney chromophobe, hepatocellular carcinoma,

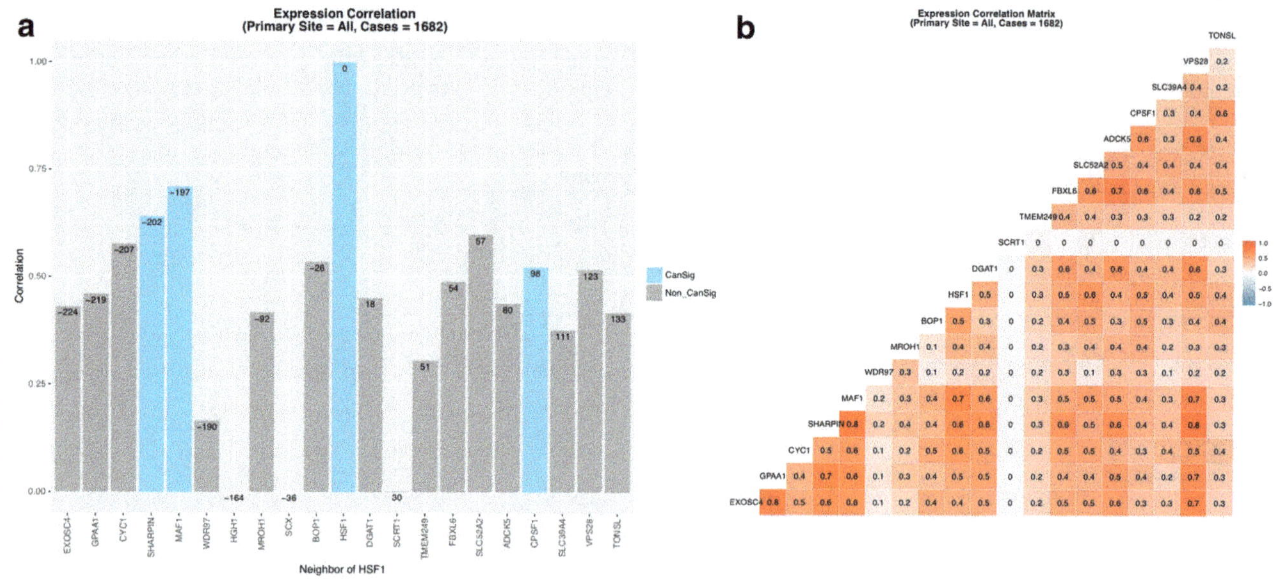

Fig. 4 Expression correlation between genes in the neighborhood of HSF1. **a** HSF1 is set as the focal gene. No expression data can be obtained for HGH1 and SCX in cBioPortal; therefore, their correlations were set to zero. The numbers on the top of the bars represent the distance of the gene from HSF1 in Kbps. Negative signs indicate genes are located upstream of HSF1. **b** Correlation matrix of all 21 genes. Blank boxes represent insufficient number of cases to determine the correlation

lung adenocarcinoma, ovarian serous cystadenocarcinoma, and stomach adenocarcinoma. Intriguingly, all results pointed to the enrichment of RNA polyadenylation (GO:0043631, *p* values 2.40e-3 to 4.34e-2), mRNA polyadenylation (GO:0006378, *p* values 2.40e-3 to 4.34e-2), and 3′ end processing (GO:0031124, *p* values 3.03–4.30e-2). With no exception, the CanSig 8q genes that contributed to these hits were HSF1, PABPC1, and CPSF1.

To corroborate WebGestalt's results, we repeated the hierarchical clustering analysis of rank data (Table 2) but using only HSF1 and three pre-mRNA 3′ processing genes: PABPC1, CPSF1, and PUF60. (Note that the clustering method requires a minimum of four genes; therefore, PUF60, a RNA processing factor, was included here even though WebGestalt did not highlight it.) If these four genes dominate the average rank of HSF1-CanSig 8q genes in different primary tumor sites, the clustering results produced by them should highly resemble to the results in Fig. 3. Figure 5a, b shows two cluster dendrograms produced by the two distance methods which share high similarity with Fig. 3a, b, respectively. When the two cluster dendrograms generated by the Euclidean distance (Figs. 3a and 5a) are compared, the top-ranking and low-ranking groups are found to share 15 and 9 common primary sites, respectively, which means 24 out of 27 primary sites. Similarly, the two dendrograms using correlation distance method (Figs. 3b and 5b) share 12 and 9 common primary sites, respectively. It means that these four genes are representative for all HSF1-CanSig 8q genes, indicating influential role played by them. It also suggests that HSF1 mediates pre-mRNA 3′ processing in cancer development.

Discussion

We have undertaken a cancer bioinformatics approach to study the clustering of HSF1-CanSig genes in chromosome 8q. Based on the TCGA expression data obtained from cancer cases, we found that HSF1-CanSig genes are clustered at chromosome 8q substantiated by compelling statistical evidence under two null models: all protein-coding genes and all HSF1-CanSig genes. λ value analysis further reveals variation of 8q prevalence in 27 primary tumor sites (Fig. 2). This result may support the use of λ to characterize the strength of HSF1 cancer program in various cancers: bone [14], breast [1], colon [1], esophagus [3], head and neck [15], liver [16], lung [1], ovarian [5], pancreas [17], prostate [18], uterine [4], and skin cancer [19]. Crucially, this perspective has been confirmed by the hierarchical clustering analysis as all of the above cancers, except bone, are members of the top-ranking group (Figs. 3 and 5).

The rank analysis discovers the association between HSF1 and HSF1-CanSig genes encoded in chromosome 8q in 27 primary tumor sites. HSF1's rank correlates with and always falls below the average rank across primary sites (Additional file 3: Figure S1). The ranks of 20 out of 27 primary tumor sites are asymmetrically skewed to high ranks with statistical significance, suggesting the overexpression of HSF1-CanSig 8q genes is coordinated

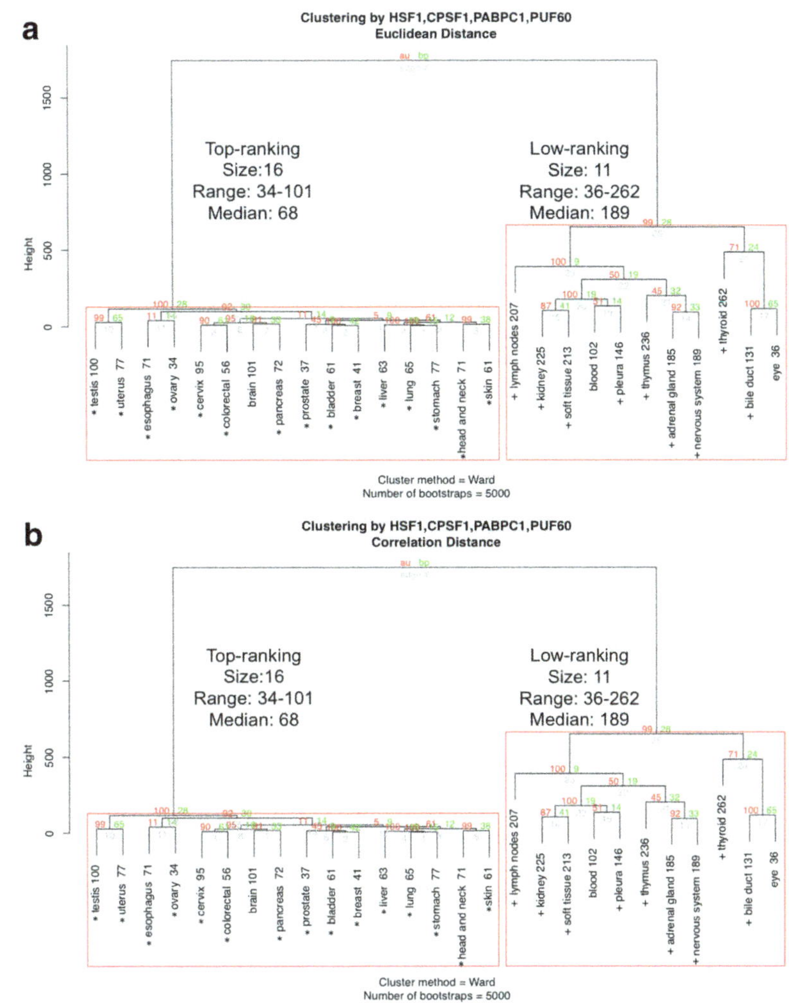

Fig. 5 a Hierarchical clustering of rank data (Table 2) by HSF1 and three pre-mRNA 3′ processing factors: PABPC1, CPSF1, and PUF60. Distance method is based on the Euclidean distance. In the top-ranking group, the asterisk denotes the primary tumor site shared with Fig. 3a. In the low-ranking group, the plus sign indicates the primary site shared with Fig. 3a. **b** Clustering by correlation distance

in those tumor sites. In contrast, when the rank of HSF1 is greater than the cutoff rank ~ 102, more diverse overexpression patterns are observed among HSF1-CanSig 8q genes, implying the diminishing control of HSF1 in those cancers. These observations engender HSF1 to be an important target for the development of cancer treatment.

Clustered HSF1-CanSig 8q genes co-express at a high level in several primary tumor sites. Is the clustering of HSF1-CanSig 8q genes a contributing factor for co-expression biologically? As "Nothing in biology makes sense except in the light of evolution" [20], we sought answers from synteny. Our results (Table 3B) indicate the majority (22–23) of HSF1-CanSig 8q genes are syntenic among human, mouse, rat, and bovine. Despite the neighboring gene model suggesting an evolutionary explanation of this co-expression [21, 22], our results do not align with such a model. When the expressions of genes flanking HSF1 were examined, our results demonstrate incoherent overexpression levels attributed to distance. Thus, other forces such as the transcriptional activation by HSF1 or its downstream products may be the cause of the co-expression of HSF1-CanSig 8q genes in tumors. It still remains as a question whether this is common for other cancer-related transcription factors. Related to this analysis is the finding that a few non-HSF1-CanSig genes encoded in chromosome 8q whose expressions correlate with HSF1: GPAA1 ($R \sim 0.8$) and EXOSC4 ($R \sim 0.75$) in the ovary and FBXL6 ($R \sim 0.8$) in the prostate (Additional file 4). These findings concur with previous studies that genes in regions 8q21, 8q22, and 8q24 are notable for cancers [23–25].

Hierarchical clustering and network analysis have fostered the connection between HSF1 and pre-mRNA 3′ processing. Three essential pre-mRNA processing factors are discovered in relation to the overexpression of

HSF1-CanSig genes through network analysis: PABPC1, CPSF1, and PUF60. CPSF1 is an essential polyadenylation factor. To date, no biochemical evidence supports any association between HSF1 and CPSF1. However, the interaction between the two processes, heat shock and polyadenylation, has been confirmed by two previous studies [26, 27], suggesting intermediaries may be involved in the crosstalk between these two processes. The poly(A) tail of a mRNA serves as a checkpoint for mRNA nuclear export. Intriguingly, HSF1-TRP interaction had been reported to facilitate the export of HSP70 mRNA under stress [28]. HSF1 had also been shown to form complexes with two core polyadenylation factors: symplekin and CstF64 [29, 30]. Alternative polyadenylation (APA) is ubiquitous and tissue-specific in mammals [31–33]. Importantly, it has been confirmed to play an essential role in many tumor types [34–43]. In rare cases, APA interferes with the protein encoded by the gene. Two classic, non-cancer examples are calcitonin-related polypeptide-alpha gene (CALCA) [44] and immunoglobulin M (IgM) [45]. But more often, APA alters the 3′ untranslated regions (UTRs) by either shortening or lengthening the constitutive 3′ UTRs. 3′ UTR is known to harbor binding sites of regulatory proteins, and miRNAs, in addition to secondary structures [46]. Their binding can either attenuate or bolster mRNA stability, affecting downstream protein synthesis and subsequently the proteome of cells.

It is noteworthy that two other HSF1-CanSig 8q genes are associated with pre-mRNA 3′-end processing: PUF60 and PABPC1. PUF60 is a member of the highly conserved nucleic acid-binding protein family: pumilio and FBF homology protein (PUF). PUF60 involves in pre-mRNA splicing and transcriptional regulation. Notably, PABPC1 binds to poly(A) tail of eukaryotic mRNA, promoting mRNA translatability. A proteomic study had identified the presence of PUF60 and PABPC1 in the polyadenylation complex [47], ascertaining the linkage between heat shock and polyadenylation.

Conclusion

The role of HSF1 played in cancer proliferation and malignancy is critical and well established. The cancer bioinformatics approach taken by us provides new information about the overexpression of HSF1-CanSig 8q genes mediated by HSF1 in different tumor types, illuminating the connection between malignancy progression driven by HSF1 and pre-mRNA 3′ processing. However, the true underlying biological mechanisms that drives HSF1 relationship with chromosome 8q is multifaceted and largely unknown but may hold the key to understanding tumor development in certain tissues and organs. As the activation of the master regulator HSF1 varies among different primary tumor sites, HSF1-CanSig 8q genes may be developed as prognosis biomarkers for improving clinical outcomes.

Materials and methods

HSF1-CanSig genes and cancer expression data

The list of 456 HSF1 CanSig genes was downloaded from Table S5 of [1]. To facilitate searching by gene symbols, they were adhered to the standard HUGO Gene Nomenclature [48]. As a result, 87 non-HuGO gene symbols from [1] were replaced by the official HuGO gene symbols. Six of them were converted to multiple gene symbols, expanding the original list to 475 HSF1-CanSig genes. The complete list of HSF1-CanSig genes and their full names and chromosome locations can be found in Additional file 1: Table S2.

mRNA expression (in fold change) of the HSF1-CanSig genes in different cancer primary sites were retrieved from cBioPortal via its Web API [7]. At the time of writing, cBioPortal contains 159 cancer studies conducted in 29 primary tumor sites (Additional file 1: Table S3). Our goal is to include one RNA-Seq study per primary site. The following criteria were used to select expression data for our analysis:

1. The study is focused on a specific primary tumor.
2. Transcriptome data is generated by the study.
3. Only one study is selected per primary tumor site. If multiple transcriptome datasets are found for a primary site, the study recruited the largest number of cancer cases is chosen.
4. Only expressions of HSF1-CanSig genes are considered.

Transcriptome data were found in the majority, but not all, of cancer studies. In this report, 27 studies were selected according to the above criteria and they are tabulated in Additional file 1: Table S9.

Overexpression score of HSF1-CanSig genes

A cancer study consists of a group of cancer cases or patients, and each cancer case is associated with a set of mRNA expression *z*-scores of genes. But in this report, our focus is solely on the 475 HSF1-CanSig genes mentioned above. cBioPortal pre-computes a *z*-score for each gene, representing the number of standard deviations of its expression deviates from the reference gene population [8]. For example, a *z*-score value 2.8318 for gene ANAT in the case TCGA-OR-A5J1-01 from the Adrenocortical Carcinoma study indicates that the gene ANAT is overexpressed by 2.8318 standard deviations above the reference.

To quantify the prevalence of overexpression of HSF1-CanSig genes among cancer cases, we devised the following scoring scheme to assess the magnitude of a gene's overexpression as:

$$Overexpression\ Score = Number\ of\ overexpressed\ cases / Total\ number\ of\ cases$$

A gene is considered overexpressed if its cBioPortal's expression *z*-score is ≥ 2, the same threshold used in cBioPortal website. For instance, the Adrenocortical Carcinoma study consists of 79 cancer cases (see Table 3), UBE2B gene was detected to overexpress in 33 cases. The overexpression score of UBE2B is 33/79 = .4177. The full list of genes' overexpression scores by primary site can be found in Additional file 1: Table S4.

Chromosome localization of top 100 HSF1-CanSig genes

When expression data of different primary tumor sites was combined for analysis, the overexpression score of an HSF1-CanSig gene was calculated by adding the overexpression scores of the gene in all primary sites. The top 100 highest scoring genes were used for the chromosome localization analysis. The complete list of the top 100 HSF1-CanSig genes can be found in Additional file 1: Table S5. Two sets of genes were chosen as the reference or null model: all human protein-coding genes based on HuGO [48] and genes implicated in cancer when mutated according to the COSMIC Cancer Gene Census [49]. In total, 19,073 and 633 of protein-coding genes and cancer census genes were used as references, respectively.

A hypergeometric model was used to assess the probability of localization. $P(i \geq k) = 1 - \Sigma_{i=0}^{i=k-1} \binom{K}{i}\binom{N-K}{n-i} / \binom{N}{n}$, where K is the number of genes encoded in the chromosome 8q, k is the number of 8q genes among the top 100 high-scoring HSF1-CanSig genes, n is the sample size, and N is the total number of protein-coding genes. For $K = 401$, $k = 27$, $n = 100$, and $N = 19{,}073$, the probability of observing 27 or more HSF1-CanSig 8q genes in a random sample of 100 genes by chance is 1.02×10^{-22}.

There are 29 HSF1-CanSig genes in 8q. The probability of finding 27 out of 29 HSF1-CanSig 8q genes in a sample of 100 HSF1-CanSig genes can be determined by hypergeometric model. Using the same formula above: K (= 29) is the number of HSF1-CanSig genes encoded in 8q, k (= 27) is the number of 8q genes in the top 100, n (= 100) is the sample size, and N (= 475) is the total number of HSF1-CanSig genes. The probability is 6.85×10^{-18}.

λ value analysis was performed by fitting of the linear form of the exponential distribution, i.e., $\ln p_i = -\lambda i + \ln \lambda$, to the rank data in Additional file 1: Table S5. Only the percentages of the top 10 high-ranking chromosome arms were used for fitting. Fitting was done by lm() function in *R*.

Synteny analysis

The genomic segment encoding human HSF1-CanSig 8q genes were analyzed for syntenic conservation. This segment spans from 80,034,869 (TPD52) to 144,901,461 (ZNF250) in chromosome 8 of human. By using the homology information from NCBI HomoloGene database (build 68) [50], 267, 165, 237, and 199 homologous genes were identified with genomic coordinates in human (hg38), bovine (bosTau8), mouse (mm10), and rat (rn6), respectively. Their genomic locations were retrieved automatically from the UCSC Genome Browser [51] through the Python package CruzDb [52]. For genomic locations that cannot be obtained by this approach, they were either determined manually by aligning the cDNA sequences on the genome by BLAT [53] or discarded. Genomic coordinates of the genes can be found in Additional file 1: Table S6.

Coexpression analysis

Ten genes immediately flanking the upstream and downstream regions of HSF1 were identified in the human genome (hg38) using the UCSC Genome Browser [51]. This region spans about 365 kbps. The ten immediate upstream genes of HSF1 are EXOSC4, GPAA1, CYC1, SHARPIN*, MAF1*, WDR97, HGH1, MROH1, BOP1, and SCX. The ten downstream genes are DGAT1, SCRT1, TMEM249, FBXL6, SLC52A2, ADCK5, CPSF1*, SLC39A4, VPS28, and TONSL (HSF1-CanSig genes are suffixed by an asterisk). Genomic coordinates of them can be found in Additional file 1: Table S7. Expressions (in fold changes) of these 21 genes, including HSF1, were retrieved from cBioPortal if HSF1's expression of a cancer case is ≥ 2 standard deviations higher than the reference. Correlation of expression fold change was computed between each of the 20 genes and HSF1 using the function cor.test in *R* where the method was Pearson.

Student's *t* test and hierarchical clustering

Our goal is to assess the statistical significance of the average rank of HSF1-CanSig 8q genes per primary site (listed in the second column from the right in Table 2). By assuming ranks are independent and uniformly distributed, the average rank of a sample consisting of 29 ranks randomly drawn from 1 to 466 without replacement does distribute normally with mean 233. (Note that the largest rank is 466 instead of 475 because of ties.) Thus, *t* test is suitable for this analysis. *R*'s t.test() function was used with mean 233 (null hypothesis). The mean was 233 as the range of rank is 1 to 466. *p* values of the *t* test can be found in the rightmost column in Table 2.

Hierarchical clustering was performed using *R*'s package pvclust [54]. One distinct feature offered by pvclust is the assessment of cluster subtrees by resampling bootstrap. The input to pvclust was the gene ranks in Table 2. The cluster method and the number of bootstrapping were "Ward" and 5000, respectively. Two distance calculation methods were used for cross-validation purpose: Euclidean distance and correlation.

WebGestalt analysis

We used WebGestalt to uncover possible gene networks commonly perturbed by HSF1-CanSig 8q genes [55]. At WebGestalt website [56], we selected "hsapiens" as the organism, "Network Topology-based Analysis (NTA)" as the method of interest, "network" as the functional database, and "genesymbol" as gene ID type. Based on these parameters, WebGestalt automatically shortlisted 33 cancer studies from The Cancer Genome Atlas (TCGA) that were available for searching. We used 27 HSF1-CanSig 8q genes, including HSF1, identified from the top 100 most frequently overexpressed genes to query WebGestalt while values of other parameters remained in default settings. We repeated the search for all 33 TCGA cancer studies (Additional file 1: Table S8).

Additional files

Additional file 1: Table S1. Statistical tests of observing HSF1-CanSig genes by chromosome arm. **Table S2.** List of 475 HSF1-CanSig genes. **Table S3.** List of 29 primary tumor sites. **Table S4.** Overexpression scores of HSF1-CanSig genes in each primary site. **Table S5.** Top 100 most frequently overexpressed HSF1-CanSig genes. **Table S6.** Genomic coordinates of syntenic genes. **Table S7.** Genomic coordinates of HSF1's flanking genes. **Table S8.** List of TCGA studies queried in WebGestalt website.

Additional file 2: Percentage distribution of HSF1-CanSig genes by chromosome arm for each primary tumor site. For each primary site, two plots are included, one using all protein-coding genes and cancer census genes as references. The other uses HSF1-CanSig genes as the reference.

Additional file 3: Rank of HSF1 among primary sites. Syntenty visualization by Cinteny. An example showing the output web page of WebGestalt

Additional file 4: Expression correlation analysis of syntenic genes for each primary tumor sites.

Abbreviations

GO: Gene Ontology; HSF1-CanSig Gene: Heat Shock Factor 1 Cancer Signature Gene; HuGO: Human Genome Organization; MYA: Million years ago; TCGA: The Cancer Genome Atlas; UTR: Untranslated region

Acknowledgements

The authors thank the patient participants in TCGA and the cBioPortal support team at the Memorial Sloan Kettering Cancer Center.

Funding

ESH is kindly supposed by the Biology Department at Lafayette College. CQZ is supported by the Weis Center for Research. TLP and HW are supported by Mowad Endowment for New Discoveries, Department of Urology, Geisinger Clinic.

Authors' contributions

All authors conceived the study. CQZ, TLP, and ESH conducted the data analysis. ESH performed the statistical analysis and programming. CQZ, HW, TLP, and ESH wrote the manuscript. All authors read and approved the final manuscript.

Consent for publication

All authors read and approved the final manuscript.

Competing interests

The authors declare that they have no competing interests.

Author details

[1]Department of Biology, Lafayette College, Easton, PA 18042, USA.
[2]Department of Computer Science, Lafayette College, Easton, PA 18042, USA.
[3]Urology Department, Geisinger Medical Center, Danville, PA 17822, USA.
[4]Weis Research Center, Geisinger Medical Center, Danville, PA 17822, USA.

References

1. Mendillo ML, Santagata S, Koeva M, Bell GW, Hu R, Tamimi RM, Fraenkel E, Ince TA, Whitesell L, Lindquist S. HSF1 drives a transcriptional program distinct from heat shock to support highly malignant human cancers. Cell. 2012;150(3):549–62.
2. Scherz-Shouval R, Santagata S, Mendillo ML, Sholl LM, Ben-Aharon I, Beck AH, Dias-Santagata D, Koeva M, Stemmer SM, Whitesell L, et al. The reprogramming of tumor stroma by HSF1 is a potent enabler of malignancy. Cell. 2014;158(3):564–78.
3. Liao Y, Xue Y, Zhang L, Feng X, Liu W, Zhang G. Higher heat shock factor 1 expression in tumor stroma predicts poor prognosis in esophageal squamous cell carcinoma patients. J Transl Med. 2015;13:338.
4. Engerud H, Tangen IL, Berg A, Kusonmano K, Halle MK, Oyan AM, Kalland KH, Stefansson I, Trovik J, Salvesen HB, et al. High level of HSF1 associates with aggressive endometrial carcinoma and suggests potential for HSP90 inhibitors. Brit J Cancer. 2014;111(1):78–84.
5. Powell CD, Paullin TR, Aoisa C, Menzie CJ, Ubaldini A, Westerheide SD. The heat shock transcription factor HSF1 induces ovarian cancer epithelial-mesenchymal transition in a 3D spheroid growth model. PLoS One. 2016;11(12)
6. Santarius T, Shipley J, Brewer D, Stratton MR, Cooper CS. EPIGENETICS AND GENETICS: a census of amplified and overexpressed human cancer genes. Nat Rev Cancer. 2010;10(1):59–64.
7. cBioPortal Web API. http://www.cbioportal.org/web_api.jsp. Accessed 1 May 2017.
8. Gao J, Aksoy BA, Dogrusoz U, Dresdner G, Gross B, Sumer SO, Sun Y, Jacobsen A, Sinha R, Larsson E, et al. Integrative analysis of complex cancer genomics and clinical profiles using the cBioPortal. Sci Signal. 2013;6(269):pl1.
9. Donoghue PC, Benton MJ. Rocks and clocks: calibrating the tree of life using fossils and molecules. Trends Ecol Evol. 2007;22(8):424–31.
10. Lander ES, Consortium IHGS, Linton LM, Birren B, Nusbaum C, Zody MC, Baldwin J, Devon K, Dewar K, Doyle M, et al. Initial sequencing and analysis of the human genome. Nature. 2001;409(6822):860–921.
11. Gibbs RA, Weinstock GM, Metzker ML, Muzny DM, Sodergren EJ, Scherer S, Scott G, Steffen D, Worley KC, Burch PE, et al. Genome sequence of the Brown Norway rat yields insights into mammalian evolution. Nature. 2004; 428(6982):493–521.

12. Elsik CG, Tellam RL, Worley KC, Gibbs RA, Abatepaulo ARR, Abbey CA, Adelson DL, Aerts J, Ahola V, Alexander L, et al. The genome sequence of taurine cattle: a window to ruminant biology and evolution. Science. 2009; 324(5926):522–8.
13. Sinha AU, Meller J. Cinteny: flexible analysis and visualization of synteny and genome rearrangements in multiple organisms. BMC Bioinf. 2007;8
14. Zhou Z, Li Y, Jia Q, Wang Z, Wang X, Hu J, Xiao J. Heat shock transcription factor 1 promotes the proliferation, migration and invasion of osteosarcoma cells. Cell Prolif. 2017;
15. Ishiwata J, Kasamatsu A, Sakuma K, Iyoda M, Yamatoji M, Usukura K, Ishige S, Shimizu T, Yamano Y, Ogawara K, et al. State of heat shock factor 1 expression as a putative diagnostic marker for oral squamous cell carcinoma. Int J Oncol. 2012;40(1):47–52.
16. Fang F, Chang RM, Yang LY. Heat shock factor 1 promotes invasion and metastasis of hepatocellular carcinoma in vitro and in vivo. Cancer-Am Cancer Soc. 2012;118(7):1782–94.
17. Dudeja V, Chugh RK, Sangwan V, Skube SJ, Mujumdar NR, Antonoff MB, Dawra RK, Vickers SM, Saluja AK. Prosurvival role of heat shock factor 1 in the pathogenesis of pancreatobiliary tumors. Am J Physiol-Gastr L. 2011; 300(6):G948–55.
18. El Gammal AT, Bruchmann M, Zustin J, Isbarn H, Hellwinkel OJC, Kollermann J, Sauter G, Simon R, Wilczak W, Schwarz J, et al. Chromosome 8p deletions and 8q gains are associated with tumor progression and poor prognosis in prostate cancer. Clin Cancer Res. 2010;16(1):56–64.
19. Toma-Jonik A, Widlak W, Korfanty J, Cichon T, Smolarczyk R, Gogler-Piglowska A, Widlak P, Vydra N. Active heat shock transcription factor 1 supports migration of the melanoma cells via vinculin down-regulation. Cell Signal. 2015;27(2):394–401.
20. Dobzhansky T. Nothing in biology makes sense except in light of evolution. Am Biol Teach. 1973;35(3):125–9.
21. Ghanbarian AT, Hurst LD. Neighboring genes show correlated evolution in gene expression. Mol Biol Evol. 2015;32(7):1748–66.
22. Liao BY, Zhang JZ. Coexpression of linked genes in mammalian genomes is generally disadvantageous. Mol Biol Evol. 2008;25(8):1555–65.
23. van Duin M, van Marion R, Vissers K, Watson JEV, van Weerden WM, Schroder FH, Hop WCJ, van der Kwast TH, Collins C, van Dekken H. High-resolution array comparative genomic hybridization of chromosome arm 8q: evaluation of genetic progression markers for prostate cancer. Genes Chromosomes Cancer. 2005;44(4):438–49.
24. Salinas CA, Kwon E, Carlson CS, Koopmeiners JS, Feng Z, Karyadi DM, Ostrander EA, Stanford JL. Multiple independent genetic variants in the 8q24 region are associated with prostate cancer risk. Cancer Epidemiol Biomarkers. 2008;17(5):1203–13.
25. Haiman CA, Patterson N, Freedman ML, Myers SR, Pike MC, Waliszewska A, Neubauer J, Tandon A, Schirmer C, McDonald GJ, et al. Multiple regions within 8q24 independently affect risk for prostate cancer. Nat Genet. 2007; 39(5):638–44.
26. Di Giammartino DC, Shi YS, Manley JL. PARP1 represses PAP and inhibits polyadenylation during heat shock. Mol Cell. 2013;49(1):7–17.
27. Hollerer I, Curk T, Haase B, Benes V, Hauer C, Neu-Yilik G, Bhuvanagiri M, Hentze MW, Kulozik AE. The differential expression of alternatively polyadenylated transcripts is a common stress-induced response mechanism that modulates mammalian mRNA expression in a quantitative and qualitative fashion. RNA. 2016;22(9):1441–53.
28. Skaggs HS, Xing H, Wilkerson DC, Murphy LA, Hong Y, Mayhew CN, Sarge KD. HSF1-TPR interaction facilitates export of stress-induced HSP70 mRNA. J Biol Chem. 2007;282(47):33902–7.
29. Nagaike T, Manley JL. Transcriptional activators enhance polyadenylation of mRNA precursors. RNA Biol. 2011;8(6):964–7.
30. Xing HY, Mayhew CN, Cullen KE, Park-Sarge OK, Sarge KD. HSF1 modulation of hsp70 mRNA polyadenylation via interaction with symplekin. J Biol Chem. 2004;279(11):10551–5.
31. Beaudoing E, Gautheret D. Identification of alternate polyadenylation sites and analysis of their tissue distribution using EST data. Genome Res. 2001; 11(9):1520–6.
32. Zhang HB, Lee JY, Tian B. Biased alternative polyadenylation in human tissues. Genome Biol. 2005;6(12)
33. Hoque M, Ji Z, Zheng DH, Luo WT, Li WC, You B, Park JY, Yehia G, Tian B. Analysis of alternative cleavage and polyadenylation by 3′ region extraction and deep sequencing. Nat Methods. 2013;10(2):133–9.
34. Xia Z, Donehower LA, Cooper TA, Neilson JR, Wheeler DA, Wagner EJ, Li W. Dynamic analyses of alternative polyadenylation from RNA-seq reveal a 3′-UTR landscape across seven tumour types. Nat Commun. 2014;5
35. Singh P, Alley TL, Wright SM, Kamdar S, Schott W, Wilpan RY, Mills KD, Graber JH. Global changes in processing of mRNA 3′ untranslated regions characterize clinically distinct cancer subtypes. Cancer Res. 2009;69(24): 9422–30.
36. Mayr C, Bartel DP. Widespread shortening of 3′ UTRs by alternative cleavage and polyadenylation activates oncogenes in cancer cells. Cell. 2009;138(4): 673–84.
37. Morris AR, Bos A, Diosdado B, Rooijers K, Elkon R, Bolijn AS, Carvalho B, Meijer GA, Agami R. Alternative cleavage and polyadenylation during colorectal cancer development. Clin Cancer Res. 2012;18(19):5256–66.
38. Masamha CP, Xia Z, Yang JX, Albrecht TR, Li M, Shyu AB, Li W, Wagner EJ. CFIm25 links alternative polyadenylation to glioblastoma tumour suppression. Nature. 2014;510(7505):412-+.
39. YG F, Sun Y, Li YX, Li J, Rao XQ, Chen C, AL X. Differential genome-wide profiling of tandem 3′ UTRs among human breast cancer and normal cells by high-throughput sequencing. Genome Res. 2011;21(5):741–7.
40. Liaw HH, Lin CC, Juan HF, Huang HC. Differential MicroRNA regulation correlates with alternative polyadenylation pattern between breast cancer and normal cells. PLoS One. 2013;8(2)
41. Elkon R, Drost J, van Haaften G, Jenal M, Schrier M, Vrielink JAFO, Agami R. E2F mediates enhanced alternative polyadenylation in proliferation. Genome Biol. 2012;13(7)
42. Sandberg R, Neilson JR, Sarma A, Sharp PA, Burge CB. Proliferating cells express mRNAs with shortened 3′ untranslated regions and fewer microRNA target sites. Science. 2008;320(5883):1643–7.
43. Buchert M, Papin M, Bonnans C, Darido C, Raye WS, Garambois V, Pelegrine A, Bourgaux JF, Pannequin J, Joubert D, et al. Symplekin promotes tumorigenicity by up-regulating claudin-2 expression. Proc Natl Acad Sci U S A. 2010;107(6):2628–33.
44. Amara SG, Jonas V, Rosenfeld MG, Ong ES, Evans RM. Alternative RNA processing in calcitonin gene-expression generates messenger-RNAs encoding different polypeptide products. Nature. 1982;298(5871):240–4.
45. Alt FW, Bothwell ALM, Knapp M, Siden E, Mather E, Koshland M, Baltimore D. Synthesis of secreted and membrane-bound immunoglobulin-mu heavy-chains is directed by messenger-RNAs that differ at their 3′ ends. Cell. 1980; 20(2):293–301.
46. Wu XB, Bartel DP. Widespread influence of 3′-end structures on mammalian mRNA processing and stability. Cell. 2017;169(5):905-+.
47. Shi YS, Di Giammartino DC, Taylor D, Sarkeshik A, Rice WJ, Yates JR, Frank J, Manley JL. Molecular architecture of the human pre-mRNA 3′ processing complex. Mol Cell. 2009;33(3):365–76.
48. HuGO Gene Nomenclature Committee. ftp://ftp.ebi.ac.uk/pub/databases/genenames/new/tsv/hgnc_complete_set.txt. Accessed 26 Apr 2017.
49. COSMIC Cancer Gene Census. http://cancer.sanger.ac.uk/census. Accessed 1 May 2017.
50. NCBI HomoloGene. ftp://ftp.ncbi.nih.gov/pub/HomoloGene/build68/. Accessed 1 May 2017.
51. UCSC Genome Browser. http://genome.ucsc.edu. Accessed 1 May 2017.
52. Pedersen BS, Yang IV, De S. CruzDB: software for annotation of genomic intervals with UCSC genome-browser database. Bioinformatics. 2013;29(23): 3003–6.
53. Kent WJ. BLAT—the BLAST-like alignment tool. Genome Res. 2002;12(4): 656–64.
54. Suzuki R, Shimodaira H. Pvclust: an *R* package for assessing the uncertainty in hierarchical clustering. Bioinformatics. 2006;22(12):1540–2.
55. Wang J, Vasaikar S, Shi Z, Greer M, Zhang B. WebGestalt 2017: a more comprehensive, powerful, flexible and interactive gene set enrichment analysis toolkit. Nucleic Acids Res. 2017;
56. WebGestalt 2017. http://www.webgestalt.org/option.php. Accessed May 2017.

12

The genetic structure of the Belgian population

Jimmy Van den Eynden[1,2*], Tine Descamps[1], Els Delporte[1], Nancy H. C. Roosens[1], Sigrid C. J. De Keersmaecker[1], Vanessa De Wit[1], Joris Robert Vermeesch[3], Els Goetghebeur[4], Jean Tafforeau[1], Stefaan Demarest[1], Marc Van den Bulcke[1] and Herman Van Oyen[1,5*]

Abstract

Background: National and international efforts like the 1000 Genomes Project are leading to increasing insights in the genetic structure of populations worldwide. Variation between different populations necessitates access to population-based genetic reference datasets. These data, which are important not only in clinical settings but also to potentiate future transitions towards a more personalized public health approach, are currently not available for the Belgian population.

Results: To obtain a representative genetic dataset of the Belgian population, participants in the 2013 National Health Interview Survey (NHIS) were invited to donate saliva samples for DNA analysis. DNA was isolated and single nucleotide polymorphisms (SNPs) were determined using a genome-wide SNP array of around 300,000 sites, resulting in a high-quality dataset of 189 samples that was used for further analysis. A principal component analysis demonstrated the typical European genetic constitution of the Belgian population, as compared to other continents. Within Europe, the Belgian population could be clearly distinguished from other European populations. Furthermore, obvious signs from recent migration were found, mainly from Southern Europe and Africa, corresponding with migration trends from the past decades. Within Belgium, a small north-west to south-east gradient in genetic variability was noted, with differences between Flanders and Wallonia.

Conclusions: This is the first study on the genetic structure of the Belgian population and its regional variation. The Belgian genetic structure mirrors its geographic location in Europe with regional differences and clear signs of recent migration.

Keywords: Genetic variability, Population genomics, Public health genomics

Background

After the completion of the Human Genome Project in 2003, international efforts were initiated to map human genetic variation between populations. This variation has been described for 26 populations worldwide via the 1000 Genomes Project [1, 2]. While this is a valuable resource for studying global genetic variation, both the number of samples per population and the total number of populations studied are relatively low. To gain sufficient statistical power and avoid false positives/negatives due to unmatched control populations in genotype-phenotype association studies, population-based genetic reference data are required from more specific and extended populations [3]. To address this, population-based whole genome sequencing initiatives have been performed at the national level throughout Europe [4–7]. From these genetic population studies, it has become clear that there is a strong correlation between the geographical location and the genetic structure of different populations [8, 9], also at the more regional level (e.g., along a north-south axis in the Netherlands) [6, 7, 9, 10]. Currently, this genetic information is not available for the Belgian population.

Therefore, a population-based cross-sectional study, called BelPHG-21 (Belgian Public Health Genomics in the twenty-first century), was organized that aims at describing the genetic variability in the Belgian population and relating

* Correspondence:
jimmy.van.den.eynden@gu.se; herman.vanoyen@wiv-isp.be
[1]Scientific Institute of Public Health, Brussels, Belgium

this variability to several indicators of health and disease to anticipate a future transition towards precision public health. To accomplish this, the BelPHG-21 study was organized in the context of the 2013 Belgian National Health Interview Survey (NHIS). The NHIS is a population-based survey that has been periodically organized since 1997. It has the purpose to assess the health status and its distribution in the population and to describe the association of this health status with its main determinants [11].

Here, we report on the BelPHG-21 results related to the genetic structure of the Belgian population. By incorporating NHIS information about the study subject's residence and parental country of birth, we demonstrate small but clear genetic differences between different regions and illustrate how the population's genetic structure is shaped by recent migration waves.

Results

Factors determining consent for former NHIS participants to participate in the BelPHG-21 population genetic study

To study the genetic structure and variability in the Belgian population, a subset of participants from the most recent NHIS, conducted on 10,829 inhabitants in 2013 [12], was invited to donate saliva samples for DNA analysis (Fig. 1). From the invited subsample of 1468 individuals, 210 (14%) consented and submitted saliva samples for DNA analysis. Using the NHIS information related to demographics, education, employment, income, health behavior, lifestyle characteristics, physical and social environment, and wellbeing characteristics, we applied a predictive weighted binomial regression analysis to identify the factors that determined consent to participate in this study. This resulted in a model containing four variables (smoking behavior, education, age, and region of residence) (Table 1). The odds for consent were higher in non-smokers (and higher in former smokers than never smokers), higher age, and higher educational attainment. Furthermore, the likelihood for consent was also higher in the Flemish Region compared to the other regions.

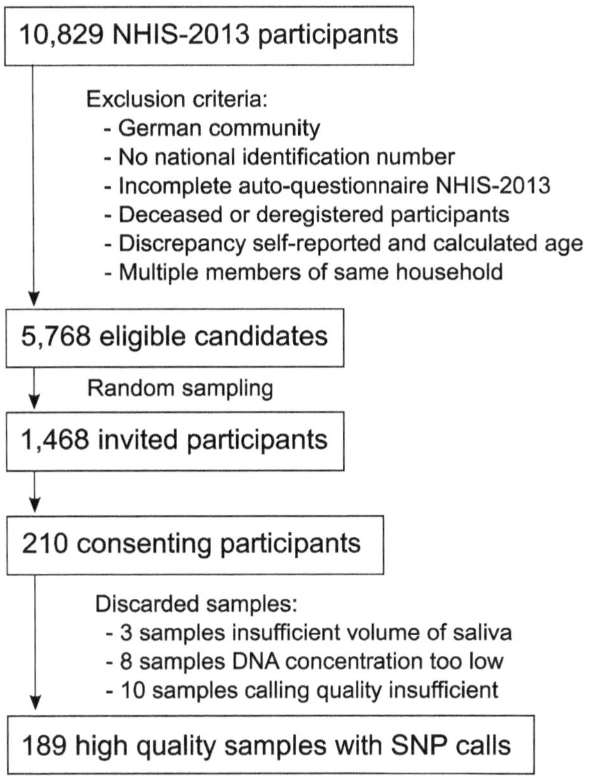

Fig. 1 BelPHG-21 study flow chart, Belgium 2016

DNA sampling from the Belgian population

DNA was extracted from saliva samples obtained from 210 consenting participants. Three of these samples contained insufficient saliva, while the DNA concentration of eight other samples was considered too low for downstream analysis (Fig. 1). From the remaining 199 samples, single nucleotide polymorphisms (SNPs) at ± 300,000 sites were measured using the whole genome scanning Illumina HumanCytoSNP-12 microarray. Microarray results from 10 samples were excluded due to insufficient call rates (i.e., lower than 0.6), resulting in 189 samples that were used for further analysis. These samples were genotyped using the hg19 human genome build as a reference and variant allele frequencies were calculated for a total number of 261,079 SNPs for which genotyping information was available (Additional file 1: Table S1).

Belgium is composed of three different geographical regions: Flanders, Brussels, and Wallonia. The 189 samples that were used for analysis in this study were donated by volunteers in all three regions: 98 from Flanders, 62 from Wallonia, and 29 from Brussels

Table 1 Variables predicting consent, BelPHG-21 study, Belgium 2016

Variable	Parameter	OR	95% CI OR
Smoking habits (ref "Smoker")	Former smoker	2.21	[1.38, 3.62]
	Never smoker	1.38	[0.92, 2.15]
Age	Age (years)	1.02	[1.01, 1.02]
Residence (ref "Flanders")	Brussels' region	0.45	[0.24, 0.77]
	Walloon region	0.58	[0.41, 0.80]
Education (ref "No diploma")	Lower secondary	2.29	[1.38, 7.02]
	Higher secondary	4.13	[2.12, 9.50]
	Higher	6.66	[3.40, 15.44]
	Not applicable	5.31	[2.25, 14.06]

A weighted binomial regression analysis to predict consent contained four variables (smoking habits, age, residence, and education). For each parameter, the odds ratio (OR) and 95% CI interval are given

(Fig. 2). These numbers are proportional to the population size of the respective regions with an overrepresentation from the Brussels region (15.3, 17.3, and 24.7 samples per million inhabitants for Flanders, Wallonia, and Brussels respectively) (Additional file 2: Table S2). This oversampling of the Brussels region is related to the way the NHIS was constructed [11]. Similarly, the distribution of these samples was relatively proportional to the population size from each of 10 provinces in Belgium with a variation from 11.3 to 26.7 samples per million inhabitants (Fig. 2 and Additional file 2: Table S2).

From the samples used for analysis, 35 (18.5%) were donated by an individual with non-Belgian roots (here defined as an individual whose mother and/or father was born outside Belgium, as registered in NHIS). These individuals mainly originated from neighboring countries (16), Southern Europe (7), and Africa (6). Proportionally, the highest numbers of non-Belgian samples were obtained from individuals living in Brussels (11/29 samples; 37.9%), followed by Wallonia (18/62; 29.0%), and Flanders (6/98; 6.1%) (Fig. 2). These numbers are representative for the current structure of Belgian population, with important migration waves in the past 50 years, mainly from Southern Europe and Northern Africa, and with the highest immigration numbers found in the Brussels region [13].

The Belgian population is a typical European population with genetic signals of recent migration from the African continent

We first compared the genetic structure of the Belgian population with other populations worldwide. To restrict this analysis to the most informative and independent SNPs, only SNPs located on autosomes with a minor allele frequency of at least 5%, less than 2% missing values, and a linkage disequilibrium lower than 0.2 were used for structural comparison. These filtering criteria resulted in a selection of 47,802 SNPs. These SNPs were used for a principal component (PC) analysis using continental population data (i.e., African, American, South Asian, East Asian, and European populations) from the 1000 Genomes Project [1, 2, 14]. The main variance was captured by the first four PCs (10.4%, Additional file 3: Figure S1a). Using the first two PCs, which capture 5.2 and 3.2% of the total variance respectively, an expected and clear separation was observed between the different continental populations, with the largest differences between African, East Asian, and European populations (Fig. 3). By applying this PC model on the Belgian SNP genotype data, 184/189 samples were perfectly mapped on the European population. The remaining five samples were mapped towards or on the African population, and for four of these samples, an African origin was indeed confirmed in the NHIS data (Fig. 3 and Additional file 3: Figure S1b). These

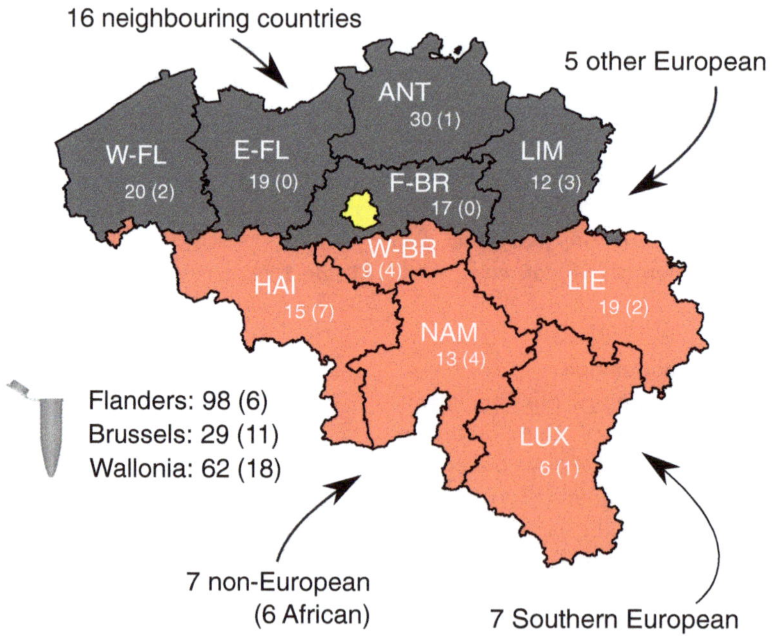

Fig. 2 DNA sampling in the BelPHG-21 study, Belgium 2016. DNA was sampled and analyzed from 189 NHIS-2013 participants. The number of collected samples is indicated for the three regions and 10 provinces. Numbers between brackets indicate the number of samples collected from participants with a migration background. Main origins from migrants are indicated by the arrows. See Additional file 2: Table S2 for more details and province abbreviations

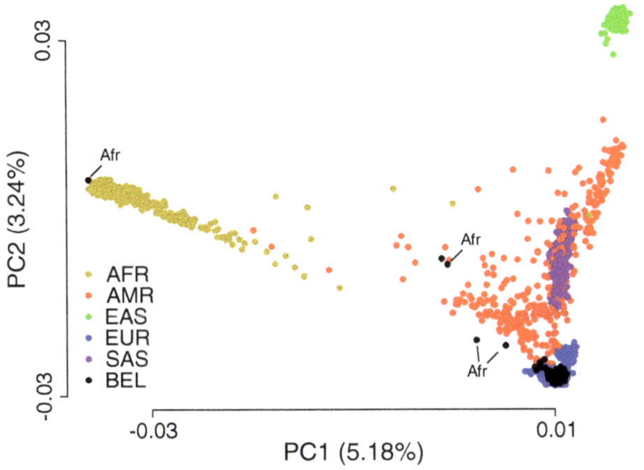

Fig. 3 Genetic variation of the Belgian population related to continental populations. Plot of the first two principal components (PC) based on a PC analysis using 47,802 SNPs from five continental populations as indicated (AFR African, AMR American, EAS East Asian, EUR European, SAS South Asian). Belgian data (BEL) were mapped independently. Total variability captured by each PC is indicated by axis labels. For Belgian samples with a migration background that map outside the European population, the background is indicated. See main text for SNP filtering criteria

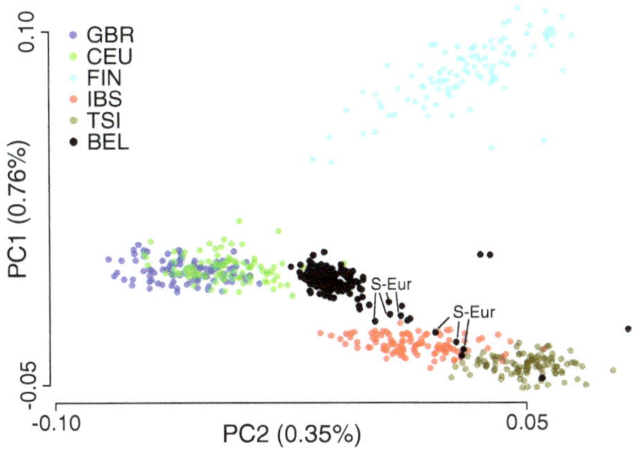

Fig. 4 Genetic variation of the Belgian population related to other European populations. Plot of the first two principal components (PC) based on a PC analysis using 41,083 SNPs from five other European populations as indicated (GBR British, CEU Utah Residents with Northern and Western European Ancestry, FIN Finnish, IBS Iberian population in Spain, TSI Toscani in Italy). Belgian data (BEL) were mapped independently. Total variability captured by each PC is indicated by the axis labels. All Belgian samples with a South European migration background are indicated. See Additional file 4: Figure S2 for more details related to migration backgrounds. See main text for SNP filtering criteria

findings show that the genetic structure of the Belgian population is a typical European population with signals of recent migration from the African continent.

The genetic structure of the Belgian population mirrors its geographical location in Europe

Our next goal was to examine how the Belgian population differs from other European populations. After a similar SNP filtering strategy as described above, the PC analysis was repeated using the European populations from the 1000 Genomes Project. Only the first two PCs were informative for a rather small variance of 0.76 and 0.35% respectively (Additional file 4: Figure S2a). As expected, the PC plot mimics European geography with a clear separation between Finnish, British, and South European populations. In agreement with this geographical orientation, most of the Belgian population mapped between the British and Southern European populations. Few samples mapped closer or within the South European populations, and for seven of these, a South European origin could indeed be confirmed (Fig. 4 and Additional file 4: Figure S2b). These data show the uniqueness of the Belgian genetic structure in Europe with geographically related differences with the other European populations.

Minor genetic differences between Flanders and Wallonia

To examine potential differences between the three main geographic regions in Belgium, we excluded the 35 samples with non-Belgian origins and indicated regional information on the European PC model (Fig. 5). This PC analysis demonstrates a minor but clear genetic substructure in the Belgian population along a north-west to south-east geographic gradient, with Flemish inhabitants mapping in the north-west, Walloon inhabitants mapping in the south-east, and individuals from Brussels in between. A similar analysis based on sample origin information from the 10 different provinces confirmed these geographic-related genetic differences (Additional file 5: Figure S3).

To quantify the genetic differences between the three Belgian regions and compare them with other European

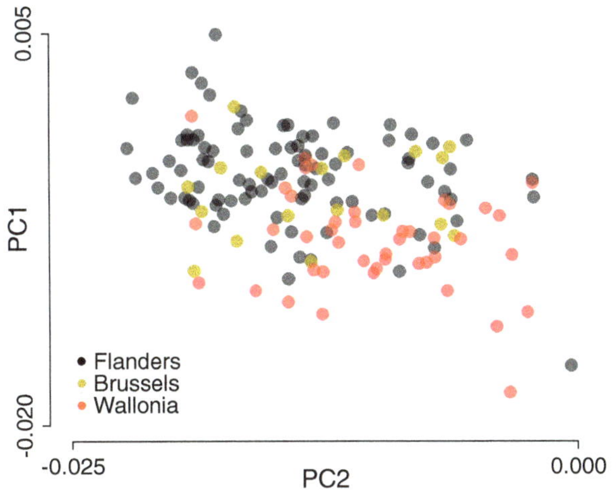

Fig. 5 Regional genetic variation within the Belgian population. Belgian samples were mapped on the European PC model (41,083 SNPs) and colored based on the region of inhabitance as indicated

populations, we calculated fixation indices (i.e., Fst values). Fst is a metric with values between 0 and 1 that is used to measure genetic distance between populations, in which high pairwise values indicate large population differentiation and vice versa. Fst values between the three Belgian regions were very small (between 1.8 and 2.5e−04) and smaller than with any other European population (Fst 3.2e −04 or higher) (Table 2), confirming the results from the PC analyses.

Finally, we correlated the variant allele frequencies of the Belgian population with other European populations. This analysis also allowed us to compare the Belgian population to the geographically and linguistically related population of the Netherlands, for which allele frequency data, but no genotyping data, are publicly available (Genome of the Netherlands project, GoNl [6]). As expected, and similar to the PC and Fst analysis, extremely good correlations were found with all European populations (Pearson correlation coefficients of 0.972 and higher). The best correlation was found with the GoNl data (Pearson correlation coefficient of 0.992; Additional file 6: Figure S4).

In summary, these data show minor but clear genetic differences between the different regions in Belgium with a north-west to south-east geographical correlation. However, these differences were smaller than with any other European population.

Discussion

The BelPHG-21 study aims at developing an approach to estimate genetic variability at the level of the Belgian population and exploring how this genetic information can be associated with information collected via NHIS questionnaires. Here, we reported on the genetic structure of the Belgian population using ± 300,000 genome-wide SNPs, derived from saliva DNA sampled from 189 participants from the latest NHIS (2013). This is the first study on the genetic structure of the Belgian population and its regional differences. The Belgian population was found to be a typical European population with signals of recent migration from the African continent. Within Europe, the Belgian population has its unique properties, which clearly mirrors the geographic orientation and again with signs of recent migration from Southern Europe. While minor genetic differences were observed between the different Belgian regions, these were smaller than with any other population studied.

The linkage possibilities to health, demographic, and related data from the NHIS provide a unique opportunity to study the different factors that determine participation in a genetic study and to better understand public willingness to participate in population genetic research. Consent rates to participate in this study were rather low (14%). Related studies that recruited people with previous participation in health research reported on variable consent rates between 21 and 85% [15–18]. The 3-year time gap between the NHIS and the second contact, inviting participation in the BelPHG-21 study, may be a reason for the relatively low participation rate. Furthermore, subjects were contacted only by an invitation letter (and reminder). We noted regional differences in the consent rate, a higher consent for non-smokers and a positive association with age and education level. The higher likelihood of educated people to give consent is in agreement with other studies [19–21]. The correlation with age is more ambiguous, with reports of both younger age as well as older age being associated with lower consent [15, 20, 22]. The observation of smokers to be less likely to provide consent is in contradiction to other studies [23]. However, in our study, the difference was attributed to a higher consent rate for former smokers than for current smokers, while no significant difference was observed between current smokers and never smokers, indicating a more complex association with smoking status.

The close mirroring of population genetic variation with geography is a well-known phenomenon that has been described at the European level previously [8, 9]. Novembre et al. described a geographical map of Europe that arose naturally as an efficient PCA-based two-dimensional summary of genetic variation in European individuals from different countries (including 43 Belgians) [8]. This has been shown not only on national, but also on regional levels,

Table 2 Fst pairwise genetic distances between Belgian regional and other European populations, BelPHG-21 study, Belgium 2016

	GBR	FIN	IBS	CEU	TSI	Flanders	Wallonia	Brussels
GBR	0							
FIN	6.7e−03	0						
IBS	2.4e−03	1.0e−02	0					
CEU	3.2e−04	6.2e−03	2.2e−03	0				
TSI	3.7e−03	1.1e−02	1.5e−03	3.3e−03	0			
Flanders	6.5e−04	6.7e−03	1.9e−03	3.2e−04	2.7e−03	0		
Wallonia	7.1e−04	7.1e−03	1.4e−03	4.1e−04	2.1e−03	2.5e−04	0	
Brussels	6.5e−04	7.0e−03	1.8e−03	4.3e−04	2.5e−03	1.8e−04	1.8e−04	0

as exemplified by genetic substructures that were observed in Sweden [10, 24], Finland [7, 25], the Netherlands [6], and several other European countries [9]. While this mirroring of the Belgian genetic structure with its geographical orientation in Europe was expected, it is quite remarkable to see the same phenomenon at regional and to a lesser extent even at provincial levels in a small country like Belgium.

Belgium is a country of immigration, with substantial labor-related migration waves after the Second World War [13]. This is reflected in this study, in which 18.5% of the samples originated from individuals with a migration background. Apart from the expected origin from neighboring countries, the most frequent migration backgrounds were South European and African, in agreement with migration data. This non-Belgian origin was clearly reflected in the genetic structure of the Belgian population. Information on migration background was derived from NHIS data and based on to the country of birth of the parents of the study participant. While this study demonstrated the usefulness of this type of NHIS-related information, there are some limitations. First, the information is limited to the first generation only (parents). This likely explains why one individual with a clear non-European genetic background was not found to have a migration background based on our criteria, which is unlikely. Secondly, there is also a risk of false positive migration findings as country of birth does not necessarily imply the individual (or his parents) originated in that country.

The NHIS contains information related to health, lifestyle, environment, etc. [11]. Therefore, linking genetic information to NHIS data has great potential for genome-wide association studies. While the current sample size (189) is rather limited, we will use the experience gathered in this study to set up a larger genetic study linked to the next NHIS (2018) where we aim to sequence the entire genome from a representative sample of ± 1000 individuals from the Belgian population using whole genome sequencing techniques. This genetic epidemiology study will also give a concrete idea about the expected number of harmful variants in the genome of a typical Belgian individual, including the prevalence of several genetic disease carriers. Results from these studies will be an invaluable resource in the transition towards precision public health, focusing on subsets of the population at increased risk, rather than on the entire population.

Conclusions

This is the first study on the genetic structure of the Belgian population and its regional variation. The Belgian population is a typical European population with minor but clear differences between the regions and clear signs of recent migration.

Methods

BelPHG-21 study design

The BelPHG-21 study is a cross-sectional population-based study, conducted in 2016. The study population was selected from the participants of the most recent (2013) NHIS. Persons from the German community or with a missing national identification number were excluded for logistic reasons. Only one member per household was included, and the participants that did not complete the auto-questionnaire part of the NHIS in 2013 (which contains information on health, lifestyle, and environment), that were deceased, deregistered or with a discrepancy between self-reported and calculated age were excluded. From the resulting 5768 eligible candidates, a random subsample of 1468 individuals was invited to participate in the BelPHG-21 study. Individuals were invited via postal mail. They received an invitation letter with all the information on the BelPHG-21 study as well as informed consent forms to return in case of participation. Individuals that consented to participate in the study were subsequently sent a saliva sample collection kit (Oragene® saliva collection device (OG-500, DNA Genotek Inc. Ottawa, Canada)) accompanied by an information letter, a user manual for sample collection, a safety bag, and a pre-paid and UN3373-labeled bubble. Each package complied with the guidance on regulations for the transport of infectious substances. All communication with study participants (consent, sampling kits) was performed by a trusted third party (Belgian Federal Service of Internal Affairs) that was not involved in the downstream research in any way. They submitted the received samples to the BelPHG-21 research team and provided linking information to the NHIS database.

Sample preparation and SNP array

Upon receipt of saliva samples by the BelPHG-21 research team, DNA was manually extracted from the total sample following the manufacturer's instructions (DNA Genotek, PD-PR-015 Issue 10/2015-01). DNA samples were quantified using UV absorbance, and SNPs at ± 300,000 sites were determined using the whole genome scanning 12-sample Illumina HumanCytoSNP-12v2.1 BeadChip according to the manufacturer's instructions. Images were captured on Cytoscan (Illumina), and data were primarily analyzed using Illumina's GenomeStudio software.

NHIS data

Information related to the study participants' residence (region and province), migration background, and a set of variables that could influence the consent rate was obtained from the NHIS-2013 database. A subject was considered to have a migration background when at least one of its parents was born outside of Belgium. The variables that were considered to potentially influence consent rates were the background characteristics (demographic information, education, employment, income), health behavior and lifestyle characteristics (substance abuse, nutritional status, physical activity), physical and social environment characteristics (housing, passive smoking), and health and wellbeing characteristics (chronic diseases, mental health, perceived health). Selection of the variables for the predictive model was conducted by stepwise-weighted binomial regression analysis. Survey weights were calculated based on age, gender, and region of residence.

Demographic and geographic data

Demographic data from the 2013 Belgian population were downloaded from Eurostat (http://ec.europa.eu/eurostat). Belgian geographic data were obtained using the BelgiumMaps.Statbel R package (available at http://www.datatailor.be/rcube/).

SNP data processing

Genotype calls were imported in R and converted in a *sample x SNP* matrix. Samples with genotyping call rates lower than 0.6 were excluded from further analysis. All genotypes were converted so that the reference allele always referred to the positive strand of the human genome build hg19. Genomic coordinates for all SNPs were retrieved from dbSNP Build 144 [26]. SNPs not present in dbSNP were excluded from further analysis. The R SNPRelate package [27] was used for downstream analysis (calculation of allele frequencies, principal component analysis, ...). Therefore, genotyping data were first converted in genomic data structure (gds) format.

Principal component analysis

A principal component (PC) analysis was performed using phase 3 data from 1000 genomes [2]. VCF and sample information data were downloaded from the 1000 genomes ftp site (ftp://ftp.1000genomes.ebi.ac.uk/vol1/ftp/release/20130502/), and genotype information from all SNPs overlapping with the BelPHG-21 study was extracted using the R VariantAnnotation package [28]. Belgian data were mapped independently on the PC model. The PC analysis was performed using the most informative SNPs by applying the following filtering criteria: (1) SNPs measured in all datasets, (2) lower than 2% missing values in both the 1000 genome dataset that was used to perform the PC analysis and the Belgian data, (3) only autosomes, (4) minor allele frequency higher than 5%, and (5) linkage disequilibrium lower than 0.2.

Variant allele frequency correlations

Correlations between variant allele frequencies (VAF) were determined using Pearson's correlation. Correlations were determined for SNPs after filtering using the higher mentioned criteria. Variant allele frequencies from the Genome of the Netherlands (GoNl) project were derived from the release 5 vcf files downloaded at https://molgenis26.target.rug.nl/downloads/gonl_public/variants/release5/.

Additional files

Additional file 1: Table S1. SNP variant allele frequencies of the Belgian population, BelPHG-21 study, Belgium 2016. Variant allele frequencies (VAF) for all SNPs were calculated with reference to the human genome build hg19 on all samples or after exclusion of samples with a foreign origin (indicated by tab names). Columns indicate SNP ids, chromosome, position, reference allele, variant allele, VAF, and frequencies of homozygous reference (AA), heterozygous (AB), and homozygous variant (BB) alleles respectively.

Additional file 2: Table S2. Population details of the samples analyzed in the BelPHG-21 study, Belgium 2016. Indication of the number of samples taken from NHIS-2013 participants in each region and province, total number of inhabitants in 2013, number of samples taken per million inhabitants, number of invited participants, and sampling percentages.

Additional file 3: Figure S1. Genetic variation of the Belgian population related to continental populations. Principal component (PC) analysis using 47,802 SNPs from five continental populations with mapping of the Belgian population. (a) Screeplot showing the variability captured by the first 10 PCs. (b) Pairwise plots of the first four PCs. Populations are indicated below the plot (AFR: African; AMR: American; EAS: East Asian; EUR: European; SAS: South Asian; BEL: Belgian).

Additional file 4: Figure S2. Genetic variation of the Belgian population related to other European populations. Principal component (PC) analysis based on 41,083 SNPs from five different European populations (see Fig. 4 for details). (a) Screeplot showing the variability captured by the first 10 PCs. (b) PC plot of five European populations with mapping of the Belgian population and indication of their migration backgrounds. For visualization purposes, the other European populations are shown in gray.

Additional file 5: Figure S3. Provincial genetic variation within the Belgian population. Belgian data were mapped on the European PC model (41,083 SNPs). Panels show plots of the first two PCs with indication of the province of inhabitance as indicated. The central map of Belgium shows the geographical location of each province. Panels with Flemish provinces are shown on top while Walloon province panels are shown on the bottom.

Additional file 6: Figure S4. Variant allele frequency correlations between Belgian and other populations. Variant allele frequencies (VAF) for all SNPs were calculated with reference to the human genome build hg19. Plots show the VAF of Belgian versus other continental (a) and European (b) populations. Pearson correlation coefficients are shown on top of each plot. For visualization purposes, only 1000 random points are shown.

Abbreviations

PC: Principal component; SNP: Single nucleotide polymorphism; VAF: Variant allele frequency

Acknowledgements

We wish to thank all colleagues at the WIV-ISP that contributed to the accomplishment of the BelPHG-21 project. We acknowledge the support of the Platform Biotechnology and Bioinformatics, the ICT team, the dispatch center, and the communication team of the WIV-ISP as well as the Federal Service of Internal Affairs for the sample coding. Finally, we acknowledge all study subjects for their participation.

Funding
The BelPHG-21 study is funded by the Belgian Science Policy via the BRAIN.be (Belgian Research Action through Interdisciplinary Network) research program (BR/121/PI/BELPHG-21).

Authors' contributions
MVDB, JT, HVO, and VDW designed the BelPHG-21 study. JVdE analyzed the genomic data and wrote the manuscript. TD and EG did the statistical analysis related to factors determining consent rates. SDK and NR were responsible for the sample preparations and DNA extractions. JRV was responsible for the micro-array analysis. SD and JT performed the NHIS 2013 correlations. ED and VDW prepared the BelPHG-21 applications to ethical and privacy committees. All authors read and approved the final manuscript.

Consent for publication
Not applicable

Competing interests
The authors declare that they have no competing interests.

Author details
[1]Scientific Institute of Public Health, Brussels, Belgium. [2]Department of Medical Biochemistry and Cell Biology, Institute of Biomedicine, The Sahlgrenska Academy, University of Gothenburg, Gothenburg, Sweden. [3]Laboratory of Cytogenetics and Genome Research, Department of Human Genetics, KU Leuven, Leuven, Belgium. [4]Department of Applied Mathematics, Computer Science and Statistics, Ghent University, Ghent, Belgium. [5]Department of Public Health, Ghent University, Ghent, Belgium.

References

1. Abecasis GR, Auton A, Brooks LD, MA DP, Durbin RM, Handsaker RE, Kang HM, Marth GT, McVean GA. An integrated map of genetic variation from 1,092 human genomes. Nature. 2012;491:56–65.
2. Auton A, Abecasis GR, Altshuler DM, Durbin RM, Abecasis GR, Bentley DR, Chakravarti A, Clark AG, Donnelly P, Eichler EE, Flicek P, Gabriel SB, Gibbs RA, Green ED, Hurles ME, Knoppers BM, Korbel JO, Lander ES, Lee C, Lehrach H, Mardis ER, Marth GT, McVean GA, Nickerson DA, Schmidt JP, Sherry ST, Wang J, Wilson RK, Gibbs RA, Boerwinkle E, et al. A global reference for human genetic variation. Nature. 2015;526:68–74.
3. Jiang Y, Epstein MP, Conneely KN. Assessing the impact of population stratification on association studies of rare variation. Hum Hered. 2013;76:28–35.
4. Ameur A, Dahlberg J, Olason P, Vezzi F, Karlsson R, Lundin P, Che H, Thutkawkorapin J, Kusalananda Kahari A, Dahlberg M, Viklund J, Hagberg J, Jareborg N, Jonasson I, Johansson A, Lundin S, Nilsson D, Nystedt B, Magnusson P, Gyllensten U. SweGen: a whole-genome map of genetic variability in a cross-section of the Swedish population. bioRxiv. 2016;
5. Gudbjartsson DF, Helgason H, Gudjonsson SA, Zink F, Oddson A, Gylfason A, Besenbacher S, Magnusson G, Halldorsson BV, Hjartarson E, Sigurdsson GT, Stacey SN, Frigge ML, Holm H, Saemundsdottir J, Helgadottir HT, Johannsdottir H, Sigfusson G, Thorgeirsson G, Sverrisson JT, Gretarsdottir S, Walters GB, Rafnar T, Thjodleifsson B, Bjornsson ES, Olafsson S, Thorarinsdottir H, Steingrimsdottir T, Gudmundsdottir TS, Theodors A, et al. Large-scale whole-genome sequencing of the Icelandic population. Nat Genet. 2015;47:435–44.
6. Genome of the Netherlands Consortium LC, Menelaou A, Pulit SL, van Dijk F, Palamara PF, Elbers CC, PBT N, Ye K, Guryev V, Kloosterman WP, Deelen P, Abdellaoui A, van Leeuwen EM, van Oven M, Vermaat M, Li M, JFJ L, Karssen LC, Kanterakis A, Amin N, Hottenga JJ, Lameijer E-W, Kattenberg M, Dijkstra M, Byelas H, van Setten J, van Schaik BDC, Bot J, Nijman IJ, Renkens I, et al. Whole-genome sequence variation, population structure and demographic history of the Dutch population. Nat Genet. 2014;46:818–25.
7. Jakkula E, Rehnström K, Varilo T, Pietiläinen OPH, Paunio T, Pedersen NL, de Faire U, Järvelin M-R, Saharinen J, Freimer N, Ripatti S, Purcell S, Collins A, Daly MJ, Palotie A, Peltonen L. The genome-wide patterns of variation expose significant substructure in a founder population. Am J Hum Genet. 2008;83:787–94.
8. Novembre J, Johnson T, Bryc K, Kutalik Z, Boyko AR, Auton A, Indap A, King KS, Bergmann S, Nelson MR, Stephens M, Bustamante CD. Genes mirror geography within Europe. Nature. 2008;456:98–101.
9. Nelis M, Esko T, Mägi R, Zimprich F, Zimprich A, Toncheva D, Karachanak S, Piskáčková T, Balaščák I, Peltonen L, Jakkula E, Rehnström K, Lathrop M, Heath S, Galan P, Schreiber S, Meitinger T, Pfeufer A, Wichmann H-E, Melegh B, Polgár N, Toniolo D, Gasparini P, D'Adamo P, Klovins J, Nikitina-Zake L, Kučinskas V, Kasnauskienė J, Lubinski J, Debniak T, et al. Genetic structure of Europeans: a view from the north–east. PLoS One. 2009;4:e5472.
10. Humphreys K, Grankvist A, Leu M, Hall P, Liu J, Ripatti S, Rehnström K, Groop L, Klareskog L, Ding B, Grönberg H, Xu J, Pedersen NL, Lichtenstein P, Mattingsdal M, Andreassen OA, O'Dushlaine C, Purcell SM, Sklar P, Sullivan PF, Hultman CM, Palmgren J, Magnusson PKE. The genetic structure of the Swedish population. PLoS One. 2011;6:e22547.
11. Demarest S, Van der Heyden J, Charafeddine R, Drieskens S, Gisle L, Tafforeau J. Methodological basics and evolution of the Belgian health interview survey 1997–2008. Arch Public Heal. 2013;71:24.
12. Drieskens S, Demarest S, D'Hoker N, Ortiz B, Tafforeau J. Is a Health Interview Survey an appropriate tool to assess domestic violence? Eur J Pub Health. 2017;27:903–9.
13. Martiniello M. Belgium, migration, 1946 to present. In: The encyclopedia of global human migration. Oxford: Blackwell Publishing Ltd; 2013.
14. Durbin RM, Altshuler DL, Durbin RM, Abecasis GR, Bentley DR, Chakravarti A, Clark AG, Collins FS, De La Vega FM, Donnelly P, Egholm M, Flicek P, Gabriel SB, Gibbs RA, Knoppers BM, Lander ES, Lehrach H, Mardis ER, McVean GA, Nickerson DA, Peltonen L, Schafer AJ, Sherry ST, Wang J, Wilson RK, Gibbs RA, Deiros D, Metzker M, Muzny D, Reid J, et al. A map of human genome variation from population-scale sequencing. Nature. 2010;467:1061–73.
15. Bogner HR, Wittink MN, Merz JF, Straton JB, Cronholm PF, Rabins PV, Gallo JJ. Personal characteristics of older primary care patients who provide a buccal swab for apolipoprotein E testing and banking of genetic material: the spectrum study. Community Genet. 2005;7:202–10.
16. Le Marchand L, Lum-jones A, Saltzman B, Marchand L, Visaya V, AMY N, Kolonel LN. Feasibility of collecting buccal cell DNA by mail in a cohort study feasibility of collecting buccal cell DNA by mail in a cohort study 1. Cancer Epidemiol Biomark Prev. 2001;10:701–3.
17. Cozier YC, Palmer JR, Rosenberg L. Comparison of methods for collection of DNA samples by mail in the Black Women's Health Study. Ann Epidemiol. 2004;14:117–22.
18. Moorman PG, Skinner CS, Evans JP, Newman B, Sorenson JR, Calingaert B, Susswein L, Crankshaw TS, Hoyo C, Schildkraut JM. Racial differences in enrolment in a cancer genetics registry. Cancer Epidemiol Biomark Prev. 2004;13:1349–54.
19. Audrain J, Tercyak KP, Goldman P, Bush A. Recruiting adolescents into genetic studies of smoking behavior recruiting adolescents into genetic studies of smoking behavior 1. Cancer Epidemiol Biomark Prev. 2002;11:249–52.
20. Sterling R, Henderson GE, Corbie-Smith G. Public willingness to participate in and public opinions about genetic variation research: a review of the literature. Am J Public Health. 2006;96:1971–8.
21. Royal C, Baffoe-Bonnie A, Kittles R, Powell I, Bennett J, Hoke G, Pettaway C, Weinrich S, Vijayakumar S, Ahaghotu C, Mason T, Johnson E, Obeikwe M, Simpson C, Mejia R, Boykin W, Roberson P, Frost J, Faison-Smith L, Meegan C, Foster N, Furbert-Harris P, Carpten J, Bailey-Wilson J, Trent J, Berg K, Dunston G, Collins F. Recruitment experience in the first phase of the African American Hereditary Prostate Cancer (AAHPC) study. Ann Epidemiol. 2000;10(8 Suppl):S68–77.
22. Oliver JM, Slashinski MJ, Wang T, Kelly PA, Hilsenbeck SG, McGuire AL. Balancing the risks and benefits of genomic data sharing: genome research participants' perspectives. Public Health Genomics. 2012;15:106–14.
23. Audrain J, Tercyak KP, Goldman P, Bush A. Recruiting adolescents into genetic studies of smoking behavior. Cancer Epidemiol Biomark Prev. 2002;11:249–52.

24. Ameur A, Dahlberg J, Olason P, Vezzi F, Karlsson R, Martin M, Viklund J, Kähäri AK, Lundin P, Che H, Thutkawkorapin J, Eisfeldt J, Lampa S, Dahlberg M, Hagberg J, Jareborg N, Liljedahl U, Jonasson I, Johansson Å, Feuk L, Lundeberg J, Syvänen A-C, Lundin S, Nilsson D, Nystedt B, Magnusson PK, Gyllensten U. SweGen: a whole-genome data resource of genetic variability in a cross-section of the Swedish population. Eur J Hum Genet. 2017;25:1253–60.
25. McEvoy BP, Montgomery GW, McRae AF, Ripatti S, Perola M, Spector TD, Cherkas L, Ahmadi KR, Boomsma D, Willemsen G, Hottenga JJ, Pedersen NL, Magnusson PKE, Kyvik KO, Christensen K, Kaprio J, Heikkilä K, Palotie A, Widen E, Muilu J, Syvänen A-C, Liljedahl U, Hardiman O, Cronin S, Peltonen L, Martin NG, Visscher PM. Geographical structure and differential natural selection among North European populations. Genome Res. 2009;19:804–14.
26. Sherry ST, Ward MH, Kholodov M, Baker J, Phan L, Smigielski EM, Sirotkin K. dbSNP: the NCBI database of genetic variation. Nucleic Acids Res. 2001;29:308–11.
27. Zheng X, Levine D, Shen J, Gogarten SM, Laurie C, Weir BS. A high-performance computing toolset for relatedness and principal component analysis of SNP data. Bioinformatics. 2012;28:3326–8.
28. Obenchain V, Lawrence M, Carey V, Gogarten S, Shannon P, Morgan M. VariantAnnotation: a Bioconductor package for exploration and annotation of genetic variants. Bioinformatics. 2014;30:2076–8.

APPLaUD: access for patients and participants to individual level uninterpreted genomic data

Adrian Thorogood[1*], Jason Bobe[2], Barbara Prainsack[3,12], Anna Middleton[4,13], Erick Scott[5], Sarah Nelson[6], Manuel Corpas[7], Natasha Bonhomme[8], Laura Lyman Rodriguez[9], Madeleine Murtagh[10], Erika Kleiderman[1] and on behalf of the Participant Values Task Team of the Global Alliance for Genomics and Health[11]

Abstract

Background: There is a growing support for the stance that patients and research participants should have better and easier access to their raw (uninterpreted) genomic sequence data in both clinical and research contexts.

Main body: We review legal frameworks and literature on the benefits, risks, and practical barriers of providing individuals access to their data. We also survey genomic sequencing initiatives that provide or plan to provide individual access. Many patients and research participants expect to be able to access their health and genomic data. Individuals have a legal right to access their genomic data in some countries and contexts. Moreover, increasing numbers of participatory research projects, direct-to-consumer genetic testing companies, and now major national sequencing initiatives grant individuals access to their genomic sequence data upon request.

Conclusion: Drawing on current practice and regulatory analysis, we outline legal, ethical, and practical guidance for genomic sequencing initiatives seeking to offer interested patients and participants access to their raw genomic data.

Keywords: Ethics, Law, Genomic data, Individual access, Whole genome sequencing, Direct-to-consumer, Privacy, Raw genomic data, Patient engagement, Citizen science

Background

The quantity of genomic data generated about individual patients, research participants, and consumers is rapidly increasing. The Global Alliance for Genomics and Health (GA4GH), an international public-private consortium, develops technical standards and frames policy to facilitate the sharing of health and genomic data between health care, research, and individuals. Analyzing and sharing these data leads to novel health insights and opportunities [1], but it raises ethical questions about the flow of data back to individuals. Debate has centered on what types of individual findings should be reported from testing or research [2] and has tended to focus on the clinical validity and actionability of results, and whether or not individuals want to receive them [3, 4]. A distinct but equally important question is whether or not patients or research participants should be able to access to their "raw" (uninterpreted) genomic sequence data [5, 6].

A task team of the GA4GH on individual access was established to explore how genomic data generated in both clinical and health research contexts can be more readily shared with individual patients and participants. Research participants primarily want data that is clinically relevant to them or their families [7, 8]. They also attach intrinsic value to genomic data and expect to be able to access data that "belongs to them." Of 4140 individuals participating in an ongoing international GA4GH survey, 61% would want to be able to access their raw sequence data (with most having the intention to use the data as the basis of further exploration) [9]. Our task team envisages a standard system that allows interested patients and participants to "pull" their genomic data from clinical laboratories or research projects on

* Correspondence: adrian.thorogood@mcgill.ca; adrian.thorogood@ga4gh.org
[1]Centre of Genomics and Policy, Department of Human Genetics, McGill University Faculty of Medicine, Montreal, Quebec H3A 0G1, Canada

request. Processes allowing individuals to access *uninterpreted data* are different from policies or processes on the return of individual findings. The latter are premised on the information's clinical relevance and/or actionability. The right to access uninterpreted data does not undermine the right not to know where it is provided on request. Even so, there are concerns over the accuracy and utility of uninterpreted data, and fears of misuse by individuals or third party services may result in psychological harms or wasted health care resources [10]. Regardless, various research initiatives are opting to provide individual access, most notably the US "All of Us" [11] and UK 100,000 Genomes [12] initiatives, and participatory research projects such as the Personal Genome Project [13]. Drawing on a review of current practice and analysis of the legal right to access personal health data, this paper supplies practical guidance for clinical laboratories or research projects seeking to provide participants access to uninterpreted genomic data. We recognize that it may not always be feasible or appropriate to provide individual access, especially in some (e.g., legacy) research contexts. We predict, however, that individual access will become expected or required as genomics becomes more clinically oriented and the public begins to insist on participatory data governance.

Current practice

The projects providing or planning to provide individual access to uninterpreted genomic data are listed in Table 1 (adapted from [14]). We were only able to identify one such genomic sequencing project outside of the USA. Data types and formats may differ depending on the context, sequencing platform, analysis pipelines, and evolution of common file formats. The examples of genomic data formats currently provided to participants include reduced BAM, VCF, and FASTQ. The usefulness of the data is enhanced where it is accompanied by rich, standard metadata [15]. Genomic sequencing initiatives may also provide individuals access to their associated health data (phenotypic, clinical, environmental). The choice of file format and the choice of when to provide access should be considered from the perspective of both the project and the individual.

Table 1 Projects providing individual access to genomic data

Project	Dates	Context	# Genomes Sequenced to date	Platform	Lab Report with Signout	Results returned	Report to Health Record	Raw Data to Participants	Accredited lab	Last updated
Harvard PGP	2005-	Research	352	WGS	No	Y Filtered Variants w/ Lit Annot	No	Yes (variants)	No	Nov 2017
BWH/Harvard MedSeq	2011-	Research	110	WGS	Yes	Monogenic, Common, PGx	Yes	FASTQ	Yes	Nov 2017
Mount Sinai HealthSeq	2012–2015	Research	40	WGS	No	Monogenic, Common, PGx	No	BAM, VCF	No	paper
Mayo "10 scientists"	2012–2014	Research	10	WES	No	Monogenic	No	Yes	No	paper
Institute for Systems Biology (ISB) Pioneer 100	2014	Research	108	WGS	No	Monogenic, Common, PGx	No	BAM, VCF	No	paper
BWH/BCH/ Harvard BabySeq Project	2015–	Research	160	WGS	Yes	Monogenic, PGx	Yes	FASTQ	Yes	Nov 2017
Nevada Institute of Personalized Medicine	2015–	Research	0	WES	Yes	Monogenic, PGx	No	BAM, VCF	No	paper
NYGC Seeq.io	2016-	Research	~500	WGS (ultra low coverage)	No	ancestry, microbiome	No	BAM	No	Feb 2017
NIH All of Us	2017-	Research	0	WGS	?	?	?	?	?	Nov 2017
100,000 Genomes (UK)	2015-	Research	44,633	WGS	Yes	Monogenic, PGx	Yes	Yes[a]	Yes	Jan 2018

Compiled as web site: Bobe, Jason. "sharing-genome-studies," online: <http://blog.jasonbobe.net/sharing-genome-studies/>
[a]Does not routinely provide access to BAM of VCF files, but participants are allowed to view the files on-site

A legal right to access?

In many countries, individuals have a legal right to access their personal data held by government bodies and commercial entities [16–18]. A general right to access personal data is included in the EU General Data Protection Regulation (GDPR) (art 15), which comes into force in May 2018 [17]. This internationally recognized right empowers individuals to ascertain what data these entities have about them and how their personal data are used. The right also enables individuals to ensure their data are accurate, up to date, and used in a transparent, fair, and lawful manner. Upon request, individuals must be provided with a copy of their data in a reasonable timeframe, in a useful format, and for a reasonable cost. There is considerable uncertainty and jurisdictional variation over whether or not genetic data is legally considered inherently identifiable. Regardless, genomic data will still fall under broad definitions of personal data used in many jurisdictions (e.g., GDPR art 4(1)), as long as it "relates to" an identifiable individual, which is increasingly the case for linked genomic data in clinical, commercial, and translational research contexts.

Similarly, patients have a legal right to access their health record ([19], art. 19). This ensures transparency in the physician-patient relationship and allows patients to correct inaccurate information (which may be used by third parties such as insurers) or transfer records when changing physicians. Access to health data also empowers patients to take an active role in their health care. Though raw laboratory data are not typically considered part of the health record, this is changing for genomics. In the USA, recent legislative amendments and interpretive guidance extend the right to access under the US federal health privacy law to a broad range of records that may be used to make decisions about individuals, including information generated as part of a laboratory test [20]. For genetic sequencing, this might include "the full gene variant information generated by the test" [21]; for genomic sequencing, the raw sequence data [22]. Genomic sequencing initiatives providing a right to access should indicate this in the consent form, along with the basic information on what is available and how to request access. Consent forms should clearly distinguish between access rights and other communication policies, such as the return of individual findings of clinical relevance [13]. As we discuss below, more detailed guidance can be provided to those individuals requesting access at the point of implementation.

The right of access is generally subject to narrow exceptions: where it would reveal confidential information (about other patients or health professionals), risk serious harm to the individual, or involve disproportionate effort [23]. Providing an individual access to her *own* genomic data would not generally breach professionals' legal duties of confidentiality to third parties or present serious risks to the individual. An important legal distinction for research contexts is that many countries limit individual access to research data, usually to protect commercial interests and scientific validity [24]. It is often unclear, however, if research exceptions in general access to information provisions were meant to restrict participants from accessing their own data [25]. International and national research ethics guidelines are largely silent about individual access to health data. This is surprising, given that many incorporate other data protection principles [26–28]. Some mention that participants have the right to access their clinical data on demand, unless temporary or permanent non-disclosure is approved by a research ethics committee with reasons ([29], Table 2). Regardless, research exceptions are unlikely to apply as sequencing moves to clinical or hybrid clinical-research contexts. Researchers seeking to provide individuals with access to genomic data may also have to contend with clinical services, clinical laboratory, and/or medical product regulations. The US regulations, for example, require any test results used for clinical decision-making to be done in a certified laboratory [30]. While these restrictions may block the return of clinically relevant individual findings from research laboratories, it is not clear why they would also apply to uninterpreted genomic data.

In conclusion, it is likely that clinical laboratories have, or will soon have, a legal obligation to provide individuals their raw genomic data upon request. While it is less likely that a legal right applies in research contexts, we propose that projects should still consider providing a default right of participants to access their own individual-level genomic data upon request. Any exceptions to access should be transparently stated, clearly justified, and approved by a research ethics committee or similar body. If access compromises the primary objective of the study, it could be withheld until the objective is achieved. In both research and clinical contexts, data stewards providing individual access should make efforts to ensure data is of high quality and interoperable. Standard use agreements could accompany access explaining that the data is provided "as is," without implied or express warranties (e.g., that the data is fit for a particular purpose––namely clinical interpretation or decision-making), and disclaiming liability for any harm resulting from the individual's use of the data.

Handling ethical and practical concerns

There are many good reasons for researchers to provide access to individual-level uninterpreted data. Empirical studies show that many people believe that their genomic data belongs to them––that they have a right to access, use, and distribute their data as they see fit

Table 2 Summary of recommendations

Provide access to genomic data in standard formats
1) FASTQ: read-level data
2) BAM: Binary Alignment Map
3) gVCF: Genome Variant Call Format
4) FASTQ: assembled diplotype genome
(Ability to reconstruct genomes: FASTQ ~ = BAM > gVCF > FASTQ)
Provide access upon request unless withholding access is justified (by an Access Office or Research Ethics Committee)
1) Breaching confidentiality of a third party (*could consent from the third party be obtained?*)
2) Imminent and serious harm to the mental or physical health of the individual (*could the harm be mitigated?*)
3) Access compromises a primary objective of a research study (*could access instead be provided at the end of the study?*)
4) Expense compromises the feasibility of a research study (*could participants be asked to cover the costs?*)
Establish appropriate data tracking and security processes
1) Authentication service (e.g., Experian) or in-person account creation
2) Best practices for data security (encryption, user access controls, transfer protocols)
Describe the right to access in the consent form
1) Distinguish from the plan for return of individual findings of clinical relevance
Provide detailed information at the point of access
1) Participant's right to access uninterpreted data
2) Description of access process
3) Description of risks posed by research-grade data
4) Description of benefits provided by uninterpreted genetic data
5) Description of available genetic counseling services
6) Description of how data will be accessed, stored, and transferred
No warranty and disclaimers
1) Clear articulation that data may not meet clinical standards, and should not be used as a basis for clinical interpretation or decision-making without medical advice and confirmatory testing in an accredited laboratory
2) Clear disclaimers that research sponsors do not offer a warranty of the data accuracy and are not liable for harm caused from using the data
3) Research sponsors should still strive to generate the highest quality data
Funders
1) Incentivize projects to provide participants access
2) Support costs of participant access

[31]—–even if this contradicts laws or consent forms [32, 33]. Providing access may also build trust and incentivize participation [34]. Moreover, patients are often experts in their condition and may be more motivated to determine the relevance of their health data than researchers focused on discovery [35]. Access will enable curious citizen scientists to explore the myriad meanings of their DNA. Research may even thrive when individuals themselves share data with patient-led registries [36, 37], research projects, or public repositories like openSNP [38, 39] or Open Humans [39]. The usefulness of raw genomic data for the individual will also increase with improvements in data quality and interoperability, expansion of the knowledge base of genotype-phenotype relationships, and the availability of reliable third party services. The more data that is held by individuals, the more portals to connect users to research initiatives [40, 41]; interpretation services to provide ancestry, genealogy, and health or wellness information; and tools to facilitate citizen science and self-driven interpretation [42].

There are, however, concerns third party interpretation services may provide uncertain, potentially inaccurate information of little benefit and may lead to anxiety or unnecessary medical follow-up [43]. To promote responsible use, data stewards could provide individuals who request access information about the limitations of data quality, the limitations of self-directed or third party interpretations, and the importance of secure storage and responsible sharing. In particular, clarity is needed that the data should not be used as a basis for clinical interpretation or decision-making without seeking medical advice and confirmatory testing in an accredited laboratory. User portals could facilitate download and communication, or even direct transfer/donation to trusted storage platforms or research projects. Data stewards should also ensure access processes are privacy protective and secure. They require basic authentication processes (is this actually the participant?); tracking processes (is this actually the participant's genome?); and a means of re-identifying a genome (how do I break the code?). Researcher confidentiality may be breached if requestors are not properly authenticated, or if data from the wrong genome is returned. Privacy concerns persist after data has been accessed. Individuals may be ill-prepared to keep their own data secure, and third party services may not offer comparable privacy and security protections [44]. Again, research projects could provide individuals with tips on how to safeguard their data. While researchers should do their best to encourage individuals to store and use their data carefully, the ultimate responsibility to do so will rest with the individual.

There are also fears that access may divert resources away from clinical or research activities. Moreover, individuals seeking professional interpretation of their data could be a drain on primary care and genetic services within the health system. This could waste public health system resources and unfairly divert resources to the most proactive, healthy, and educated individuals. Providing access should not, however, necessitate expensive interpretation or counseling, as may be the case for the return of individual results. Costs would be limited to basic tracking, authentication, and communication processes––already common in many laboratory contexts and clinical practices––and download costs.

Currently, many researchers feel they should provide access to individual-level data to patients and participants, but do not have the appropriate resources to do so. To address this problem, research funding bodies could help by providing resources, infrastructure, and incentives. Instead of each project establishing its own system, common data management platforms could be developed to enable individual access (such as those already offered to researchers by direct-to-consumer companies) [45]. Data sharing repositories enabling broad research community access could be modified to enable individual access. Individual access endorsements or badges could recognize laboratory or researcher efforts to share data with interested participants and patients.

Conclusion

We provide a summary of recommendations for sequencing initiatives providing individual access to uninterpreted genomic data in Table 2. More data and experience is needed to definitively refute paternalist concerns about individuals managing their own genomic data. This will only happen if researchers do what they do best: experiment in a responsible manner to understand how to most appropriately support and enable individual access to genomic data. Here, the variable to tweak is not the data analysis, but the participant communication pipeline. The experiment is off to a promising start.

Abbreviations

GA4GH: Global Alliance for Genomics and Health

Acknowledgements

We would like to thank Gratien Dalpé, Academic Associate at the Centre of Genomics and Policy, McGill University, for his assistance preparing the final manuscript.

Funding

Adrian Thorogood and Erika Kleiderman were funded by The CanSHARE project, which is supported by the Genome Quebec, Genome Canada, the government of Canada, the Ministère de l'Économie, Innovation et Exportation du Québec, and the Canadian Institutes of Health Research (fund no. 141210).

Authors' contributions

JB led the work on the survey of projects. AT and SN led the comparative legal review. All authors contributed substantially to the conceptual design of the paper; identification of legal, ethical, and practical challenges and to the recommendations; and drafting of the manuscript. All authors read and approved the final manuscript.

Consent for publication

Not applicable

Competing interests

The authors declare that they have no competing interests.

Author details

[1]Centre of Genomics and Policy, Department of Human Genetics, McGill University Faculty of Medicine, Montreal, Quebec H3A 0G1, Canada. [2]Icahn School of Medicine at Mount Sinai, New York, USA. [3]Department of Political Science, University of Vienna, Vienna, Austria. [4]Society and Ethics Research, Connecting Science, Wellcome Genome Campus, Hinxton, UK. [5]Icahn Institute for Genomics & Multiscale Biology, New York, USA. [6]University of Washington, Seattle, USA. [7]Cambridge Precision Medicine, Cambridge, UK. [8]Genetic Alliance, Washington DC, USA. [9]National Human Genome Research Institute, National Institutes of Health, Bethesda, USA. [10]Newcastle University, Newcastle upon Tyne, UK. [11]https://www.ga4gh.org. [12]Department of Global Health & Social Medicine, King's College London, London, UK. [13]Faculty of Education, University of Cambridge, Cambridge, UK.

References

1. Chatzimichali EA, Brent S, Hutton B, Perrett D, Wright CF, Bevan AP, et al. Facilitating collaboration in rare genetic disorders through effective matchmaking in DECIPHER. Hum Mutat. 2015;36:941–9.
2. Jarvik GP, Amendola LM, Berg JS, Brothers K, Clayton EW, Chung W, et al. Return of genomic results to research participants: the floor, the ceiling, and the choices in between. Am J Hum Genet. 2014;94:818–26.
3. Wolf SM, Crock BN, Van Ness B, Lawrenz F, Kahn JP, Beskow LM, et al. Managing incidental findings and research results in genomic research involving biobanks and archived data sets. Genet Med. 2012;14:361–84.
4. Holm IA, Iles BR, Ziniel SI, Bacon PL, Savage SK, Christensen KD, et al. Participant satisfaction with a preference-setting tool for the return of individual research results in pediatric genomic research. J Empir Res Hum Res Ethics. 2015;10:414–26.
5. Kish LJ, Topol EJ. Unpatients—why patients should own their medical data. Nat Biotechnol. 2015;33:921–4.
6. Nelson S. Geneticists should offer data to participants. Nat News. 2016;539:7.
7. Fernandez CV, Bouffet E, Malkin D, Jabado N, O'Connell C, Avard D, et al. Attitudes of parents toward the return of targeted and incidental genomic research findings in children. Genet. Med. 2014;16:633–40.
8. Facio FM, Eidem H, Fisher T, Brooks S, Linn A, Kaphingst KA, et al. Intentions to receive individual results from whole-genome sequencing among participants in the ClinSeq study. Eur J Hum Genet. 2013;21:261–5.
9. Middleton A, Wright CF, Morley KI, Bragin E, Firth HV, Hurles ME, et al. Potential research participants support the return of raw sequence data. J. Med. Genet. 2015;52:571–74.
10. Janssens ACJ. Raw data: access to inaccuracy. Science. 2014;343:968.
11. All of Us Research Program. Core Protocol V1 (2017). https://allofus.nih.gov/sites/default/files/allofus-initialprotocol-v1_0.pdf
12. Genomics England. 100,000 Genomes Project Protocol (2017). Genomics Engl. https://www.genomicsengland.co.uk/100000-genomes-project-protocol/. Accessed 2 Nov 2017.
13. Lunshof JE, Church GM, Prainsack B. Raw personal data: providing access. Science. 2014;343:373–4.
14. Linderman MD, Nielsen DE, Green RC. Personal genome sequencing in ostensibly healthy individuals and the PeopleSeq Consortium. J Pers Med. 2016;6:14.
15. Corpas M, Whicher C, Kovalevskaya NV, Byers T, McMurray AA, Nielsen FG, et al. 10 simple rules for sharing human genomic data. bioRxiv (2017). https://www.biorxiv.org/content/early/2017/01/12/094110. Accessed 12 Oct 2017.
16. OECD. Recommendation of the Council concerning Guidelines governing the Protection of Privacy and Transborder Flows of Personal Data (2013). Jul. Report No.: C(80)58/FINAL, as amended on 11 July 2013 by C(2013)79. http://sec-law.co.il/images/2013-oecd-privacy-guidelines.pdf

17. European Parliament & Council. Regulation on the protection of natural persons with regard to the processing of personal data and on the free movement of such data, and repealing Directive 95/46/EC (General Data Protection Regulation) (2016). L119 Apr 5, 2016. http://eur-lex.europa.eu/legal-content/EN/TXT/HTML/?uri=CELEX:32016R0679&from=FR.
18. Federal Trade Commission. Privacy & Data Security Update (2016). https://www.ftc.gov/reports/privacy-data-security-update-2016. Accessed 13 Oct 2017.
19. International Covenant on Civil and Political Rights (1966). 999 UNTS 171 Dec 19, 1966. http://www.ohchr.org/EN/ProfessionalInterest/Pages/CCPR.aspx.
20. U.S. DHHS. CLIA program and HIPAA privacy rule: patients' access to test reports. Final rule. Fed. Regist. 2014;79:7289–316.
21. U.S. DHHS. Individuals' right under HIPAA to access their health information (2016). https://www.hhs.gov/hipaa/for-professionals/privacy/guidance/access/index.html. Accessed 12 Oct 2017.
22. Evans BJ. HIPAA's individual right of access to genomic data: reconciling safety and civil rights. Am J Hum Genet. 2018;102:5–10.
23. Organisation for Economic Co-operation and Development (OECD). OECD Guidelines on the Protection of Privacy and Transborder Flows of Personal Data (1980). https://www.oecd.org/sti/ieconomy/oecdguidelinesontheprotectionofprivacyandtransborderflowsofpersonaldata.htm. Accessed 12 Oct 2017.
24. Kaye J. The tension between data sharing and the protection of privacy in genomics research. Annu Rev Genomics Hum Genet. 2012;13:415–31.
25. Ries NM. Research participants' rights to access information about themselves held by public research institutions. Health Law Rev. 2010;18:5.
26. WMA. Declaration of Helsinki––ethical principles for medical research involving human subjects. 2013.
27. United States. Federal policy for the protection of human subjects––final rule (2017). Fed Regist. Vol 82 No 12 Jan 19, 2017 p. 7149–7274. Available from: https://www.gpo.gov/fdsys/pkg/FR-2017-01-19/pdf/2017-01058.pdf.
28. Tri-Council Policy Statement (TCPS). Ethical conduct for research involving humans (2014). http://www.pre.ethics.gc.ca/pdf/eng/tcps2-2014/TCPS_2_FINAL_Web.pdf
29. Council for International Organizations of Medical Sciences (CIOMS) in collaboration with the World Health Organization (CIOMS/WHO). International ethical guidelines for health-related research involving humans (2016). http://cioms.ch/ethical-guidelines-2016/WEB-CIOMS-EthicalGuidelines.pdf
30. CMS. Research Testing and Clinical Laboratory Improvement Amendments of 1988 (CLIA) Regulations [Internet]. 2014 [cited 2017 Oct 12]. Available from: https://www.cms.gov/Regulations-and-Guidance/Legislation/CLIA/index.html?redirect=/clia/
31. Bredenoord AL, Kroes HY, Cuppen E, Parker M, van Delden JJM. Disclosure of individual genetic data to research participants: the debate reconsidered. Trends Genet. 2011;27:41–7.
32. Kelly SE, Spector TD, Cherkas LF, Prainsack B, Harris JM. Evaluating the consent preferences of UK research volunteers for genetic and clinical studies. PLoS One. 2015;10:e0118027.
33. Kaye J, Kanellopoulou N, Hawkins N, Gowans H, Curren L, Melham K. Can I access my personal genome? The current legal position in the UK. Med Law Rev. 2014;22:64–86.
34. Haeusermann T, Greshake B, Blasimme A, Irdam D, Richards M, Vayena E. Open sharing of genomic data: who does it and why?. PLoS ONE (2017). https://www.ncbi.nlm.nih.gov/pmc/articles/PMC5423632/. Accessed 12 Oct 2017).
35. Sanderson SC, Linderman MD, Suckiel SA, Zinberg R, Wasserstein M, Kasarskis A, et al. Psychological and behavioural impact of returning personal results from whole-genome sequencing: the HealthSeq project. Eur J Hum Genet. 2017;25:280–92.
36. Philippakis AA, Azzariti DR, Beltran S, Brookes AJ, Brownstein CA, Brudno M, et al. The matchmaker exchange: a platform for rare disease gene discovery. Hum Mutat. 2015;36:915–21.
37. Home––matchmaker exchange (2017). http://www.matchmakerexchange.org/. Accessed 13 Oct 2017.
38. Greshake B, Bayer PE, Rausch H, Reda J openSNP–a crowdsourced web resource for personal genomics PLOS ONE 2014;9:e89204.
39. Home––Open Humans (2017). Open Hum. https://www.openhumans.org/. Accessed 2 Nov 2017.
40. DNA.LAND (2017). About. https://dna.land/. Accessed 12 Oct 2017.
41. openSNP (2017). Welcome OpenSNP. https://opensnp.org/. Accessed 12 Oct 2017.
42. Promethease (2017). https://promethease.com/ondemandagreed. Accessed 2 Nov 2017.
43. Allen C, Gabriel J, Norkunas Cunningham T, Flynn M, Wang C. The impact of raw DNA availability and corresponding online interpretation services: a mixed-methods study. Transl. Behav. Med. 2017;(in press).
44. Badalato L, Kalokairinou L, Borry P. Third party interpretation of raw genetic data: an ethical exploration. Eur J Hum Genet. 2017;25:1189.
45. 23andMe. 23andMe for researchers (2016). https://researchers.23andme.org/. Accessed 24 Jan 2018.

14

Insights about genome function from spatial organization of the genome

Shuvra Shekhar Roy[1], Ananda Kishore Mukherjee[1,2] and Shantanu Chowdhury[1,2*]

Abstract

Over the last 15 years, development of chromosome conformation capture (3C) and its subsequent high-throughput variants in conjunction with the fast development of sequencing technology has allowed investigators to generate large volumes of data giving insights into the spatial three-dimensional (3D) architecture of the genome. This huge data has been analyzed and validated using various statistical, mathematical, genomics, and biophysical tools in order to examine the chromosomal interaction patterns, understand the organization of the chromosome, and find out functional implications of the interactions. This review summarizes the data generated by several large-scale high-throughput chromosome conformation capture studies and the functional implications obtained from the data analyses. We also discuss emerging results on factors (both CCCTC binding factor (CTCF) related and CTCF independent) that could contribute to looping interactions.

Keywords: Genome architecture, Chromosome conformation capture (3C), Hi-C, Topologically associated domains (TAD), Chromatin looping, Histone modifications, Transcription

Background

Recent advances in the field of *c*hromosome *c*onformation *c*apture (3C) and its subsequent advancements like 4C (chromosome conformation capture-on-chip), 5C (carbon copy chromosome conformation capture), and high-throughput chromosome conformation capture (Hi-C) have greatly expanded the general understanding of the three-dimensional spatial arrangement of the genome. Consequently, the notion of a linear chromosomal distribution of various elements of the coding and the non-coding genome is being revisited. This has brought forth a new perspective on how genomic functions such as replication, transcription, and DNA repair would be investigated. Interestingly, the proximity of various regions of the genome, which were otherwise unintuitive, now opens new possibilities by which various aspects of gene regulation would be studied in future. Along with these, herein, we also discuss recent data suggesting the role of telomeres in looping interactions, both near and with interstitial regions of the genome. In addition, factors other than the chromatin-associated protein CCCTC binding factor (CTCF) that could mechanistically contribute to the formation of chromatin loops are also discussed.

* Correspondence: shantanuc@igib.res.in
[1]Genomics and Molecular Medicine Unit, CSIR-Institute of Genomics and Integrative Biology, Mathura Road, New Delhi 110025, India
[2]Academy of Scientific and Innovative Research, CSIR-Institute of Genomics and Integrative Biology, Mathura Road, New Delhi 110025, India

The spatial organization of the genome from Hi-C data

All to all chromosome interaction matrices reveal compartments within genomes

Hi-C is a high-throughput NGS (next-generation sequencing)-based version of chromosome conformation capture assay commonly known as 3C [1]. It provides a global view of all chromosomal interactions across the genome. In such experiments, tagged nucleotides are used to capture ligated products after restriction digestion and appropriately diluted ligation. This is followed by sequencing of the captured DNA fragments. The resulting reads represent ligation of fragments both adjacent to each other as well as far apart in the linear genome. A genome-wide two-dimensional contact matrix is generated by dividing the genome or each chromosome into bins of equal distance and then aligning these reads into those bins. Such matrices depict the collective average of the interactions present within the population of cells used for Hi-C studies [2]. There are several other 3C-based assays to look at chromosomal interactions, but the advantage of

Hi-C experiments is that they give a global view of all the chromosomal interactions across the genome.

Though we obtain vast knowledge on the biology of chromosomal interactions and genome organization from Hi-C data, there are some drawbacks of the method as well. The Hi-C data, for instance, does not provide much quantitative information on the stability and strength of chromosomal interactions. Most of the reported Hi-C studies so far used asynchronized cells and did not consider the variability that the different cell cycle stages (particularly mitotic phase) would most likely introduce into the chromatin conformation. Several reviews have further described the technical aspects of these 3C-based assays and complimentary microscopy techniques [3–5].

In 2009, Lieberman-Aiden et al. used karyotypically normal GM06990 human lymphoblastoid cells to generate 8.4 million reads that uniquely aligned to the human genome reference sequence in a Hi-C experiment; of these, 6.7 million reads corresponded to long-range contacts between segments of the genome > 20 Kb apart. The long-range contact reads were used to construct genome-wide contact matrices by dividing the genome into 1-Mb-sized bins. This revealed an interesting pattern comprising distinct intra- and inter-chromosomal genomic compartments with contacts that were primarily restricted within compartments [2].

A few years later, using Hi-C contact matrices of bin sizes ranging from 20 to 100 Kb, about 1.7 billion read-pairs of data were analyzed and genome-wide formation of compartments or topologically associated domains (TADs) in mouse and human-differentiated and embryonic stem (ES) cells was reported [6]. This was supported by work from Jin et al. who studied chromatin interactions in response to transient TNF-alpha signaling in IMR90 primary human fibroblasts by performing Hi-C before and after 1 h TNF-alpha treatment [7]. This gave about 3.4 billion uniquely mapped paired-end reads from multiple biological replicates, among which, approximately 1.4 billion were intra-chromosomal reads.

A slightly different approach termed as in situ Hi-C was used by Rao et al. where DNA-DNA proximity ligation was carried out inside intact nuclei—in situ ligation reduced the chances of spurious contacts due to random ligation in dilute solution [8]. This also yielded a higher resolution of up to 1 Kb bin size and 200- to 1000-fold more contacts. Using the high-resolution contact matrices, it was possible to see many smaller contact domains (sub-TADs) with higher intra-domain contact frequency like the topologically associated domains and other similar small domains discussed in several Hi-C studies (Fig. 1) [6, 9, 10].

Contacts prevalent within chromosomes rather than across chromosomes

Chromosomal interactions across the whole genome noted that the average intra-chromosomal contact probability between pairs of loci in a chromosome decreased consistently with increasing genomic distance (Fig. 2) [2]. Interestingly, this suggested polymer-like behavior of the genome where the three-dimensional distance between pairs of loci increases with increasing genomic distance. Other studies on chromosomal organization via 3C and fluorescence in situ hybridization (FISH) also made similar observations [1, 11]. Even at distances greater than 200 Mb, the average intra-chromosomal contact probability was always more than the average contact probability between different chromosomes (inter-chromosomal), implying the existence of chromosomal territories.

Intriguingly, probabilities of inter-chromosomal contacts show that small, gene-rich chromosomes (chromosomes 16, 17, 19, 20, 21, and 22) preferentially interact with each other. Furthermore, FISH studies also observed these chromosomes to frequently localize to the center of the nucleus [12, 13]. On the other hand, another small but gene-poor chromosome, i.e., 18, was

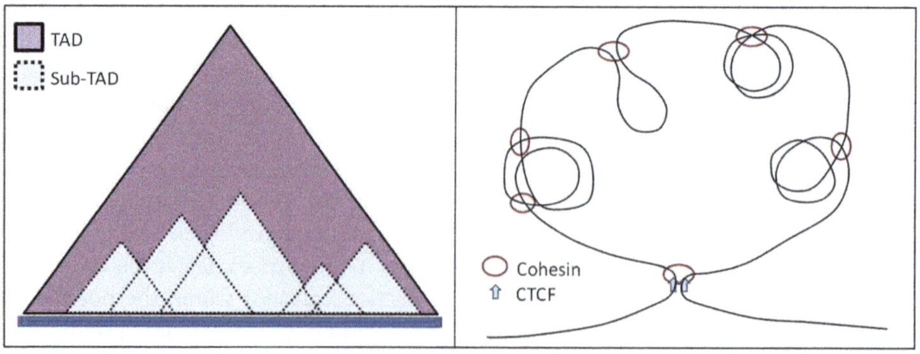

Fig. 1 Schematic representation of how smaller TAD-like structures (sub-TADs) emerge from Hi-C contact matrix and how they might be forming in three dimensions. Dynamic loop extrusion by factors like Cohesin might lead to looping and higher interactions in the extruded loci leading to emergence of sub-TADs

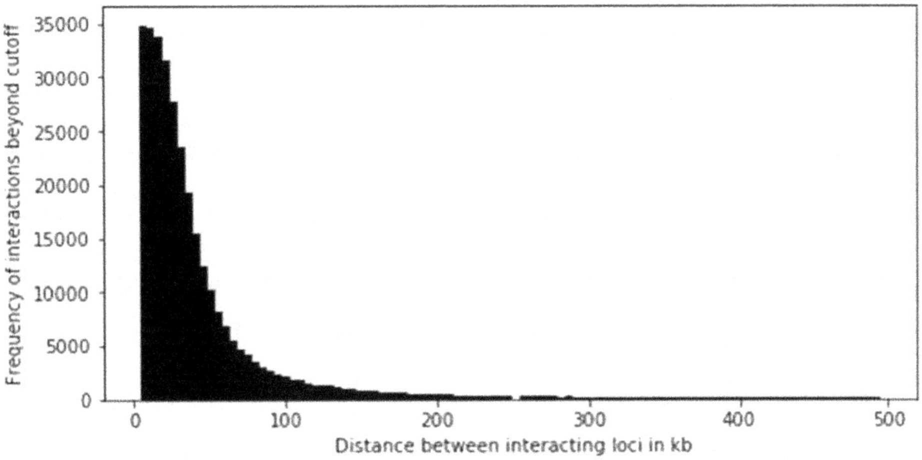

Fig. 2 Frequency distribution of interactions (above cutoff of 10 units of interaction value for a pair of loci) with distance between the interacting loci by analyzing intra-chromosomal Hi-C contact matrix of human chromosome 5 (analysis was done using normalized interaction values from Hi-C data given in Rao et al. [8])

found to interact less with other chromosomes, and FISH studies showed chromosome 18 tends to be located near the nuclear periphery [14].

Genome topologies constructed from interaction matrices

Using principal component analysis, each chromosome could be partitioned into two compartments (termed A and B) such that loci within the same compartment had correlated contact profiles. Through this correlation, even loci belonging to different chromosomes were assigned the same compartment, resulting in the whole genome being divided into two spatial compartments such that greater interactions occurred within than across compartments [15].

It was observed that loci in compartment B had a higher tendency for close spatial localization, suggesting a relatively compact state of chromatin. On the other hand, loci within compartment A showed significant correlation with the presence of genes, higher expression, and chromatin accessibility (DNase I hypersensitivity) and were enriched for activating chromatin marks (H3K36me3). Thus, compartment A could be associated with open, accessible, and actively transcribed chromatin. Together, these suggested open and closed chromatin domains throughout the genome occupy different spatial compartments in the nucleus.

At higher resolution (bin sizes less than 100 Kb), highly self-interacting regions were found to emerge, seen as triangles in the interaction matrix heat map [6]. These regions were termed as topological domains. Topological domains were found to be bound by narrow segments where the chromatin interactions appear to end abruptly. Using a statistic termed directionality index (DI), authors identified 2200 topological domains in mouse ES cells with a median size of 880 Kb that covered ~ 91% of the genome. DI models the difference between a number of upstream and downstream interactions at a given locus along a chromosome, and thereby boundaries of topological domains were detected where there was a significant shift from contact points oriented with upstream vis-à-vis downstream bias. Also, as expected, the frequency of intra-domain interactions was noted to be higher than that of inter-domain interactions.

Boundary demarcations between distinct genome topologies

Consistent with this, fluorescent in situ hybridization (FISH) experiments revealed pairs of loci within a particular topological domain were closer in space than pairs of loci in different topological domains in spite of similar genomic distances between the loci [16]. The genomic regions between the topological domains were defined as either "topological boundary regions" or "unorganized chromatin," depending on their sizes (topological boundary regions: median ~ 0 Kb, 76.3% of the regions < 50 Kb; or unorganized chromatin: median ~ 560 Kb). Moreover, the topological domains correlated with other described components of the genome-like compartments A and B [2], replication time zones [17, 18], and large organized chromatin K9 modification (LOCK) domains [19]. A large subset of the identified domain boundaries also appeared to mark the transition between LAD (lamina-associated domain) and non-LAD regions in the genome [20, 21].

An in situ Hi-C study using high-resolution contact matrices revealed numerous relatively small contact domains with higher intra-domain contact frequency [8] like topologically associated domains or other small domains discussed in other Hi-C studies [6, 9, 10].

Interestingly, the in situ Hi-C data revealed six nuclear sub-compartments based on long-range interaction patterns, both intra-chromosomal and inter-chromosomal, using different approaches to clustering. On comparison with compartment A/B [2], two of the six interaction patterns correlated with loci in compartment A—termed as sub-compartments A1 and A2. These loci were gene dense, harbored highly expressed genes, enriched in activating chromatin marks, and were depleted at the nuclear lamina- and nucleolus-associated domains (NADs). Rest of interaction patterns correlated with loci in compartment B with very different properties than A1 and A2.

The DI data used to identify TADs can vary substantially depending on the sliding window size selected with small window sizes giving smaller TADs and larger ones yielding larger TADs which often nest groups of smaller domains [22]. In fact, reanalysis of the original Hi-C data from which megabase-sized TADs were identified [6] with a different algorithm using smaller window sizes led to the observation that larger conserved TADs tend to consist entirely of smaller domains. These domains were found to be stable across cell lines and persistent across resolutions, with their boundaries having high enrichment in CTCF binding and activating histone marks [22].

Genome compartments: fractal globule versus topologically associated domain architecture

Chromosomal regions have been perceived as an "equilibrium globule"—a compact and densely knotted configuration originally used to describe a polymer in a poor solvent at equilibrium [23, 24]. An alternative model proposes that polymers, including interphase DNA, can self-organize into a long-lived, non-equilibrium conformation described as "fractal globule" [25, 26]. This dense, compact state is adopted by an untangled polymer as it crumbles into a series of small globules in a "beads-on-a-string" configuration. These beads act as monomers in further rounds of spontaneous crumpling until only a single globule of globules of globules remain.

Lieberman-Aiden et al. analyzed the scaling of contact probability of fractal globule and found that it is close to the contact probability observed from the Hi-C data [2]. The predicted scaling of three-dimensional (3D) distance between pairs of loci based on the fractal globule model is also close to the scaling reported by 3D FISH for genomic distances between 500 Kb and 2 Mb [24]. At a scale of several megabases, the data is consistent with a fractal globule model for chromatin organization. Fractal globule is an attractive model to define chromatin organization since they are free of knots [27] and in principle consistent with unfolding and refolding, for instance, during gene regulatory events like activation and repression or processes such as replication and recombination.

Functional implications of domain formation in the genome

Domain boundaries associated with gene promoters and transcription

A strong enrichment of CTCF binding sites was observed at the TAD boundary regions [6], a property also common with many known insulator or barrier elements [20, 28, 29]. Like a classical boundary element is known to stop the spread of heterochromatin, a clear segregation of the heterochromatin marker H3K9me3 modification was observed within the TAD boundary regions (Fig. 3).

Several studies found TADs to be majorly conserved across cell types in a given organism, suggesting TADs to be stable structures of the *3D* genome organization [6, 9, 30]. On the other hand, the smaller scale structures, like sub--TADs, loops, and insulation neighborhoods, all show at least partial variation between different cell lineages, with the variations in their organization appearing to be related with cell type [8, 31–33]. The dynamic chromatin interactions varying

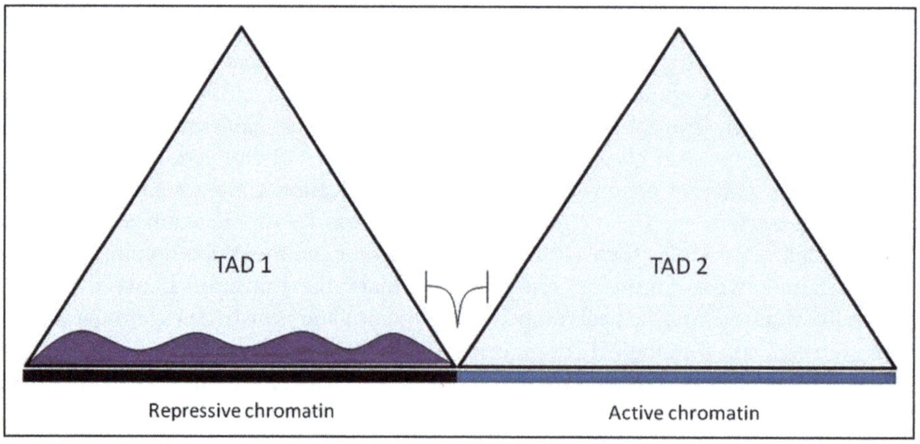

Fig. 3 Schematic representation to show that TAD boundary restricts the spread of heterochromatinization

across cell types were also enriched for differentially expressed genes [6, 34, 35].

These observations suggest that chromatin organization as TADs is mostly stable across cell types, within which specific structures and dynamic interactions can form to play lineage and context-specific regulatory roles contributing to molecular events associated with differentiation [36].

Interestingly, the topological domains do not seem to be the consequence of heterochromatin formation as the detected boundaries were present in both pluripotent cells and their differentiated progeny, i.e., before and after heterochromatinization associated with cellular differentiation. Importantly, this implied that the topological domains along with boundaries delineate the endpoints of heterochromatic spreading [6].

Furthermore, enrichment of chromatin marks associated with promoters and gene bodies and depletion of repressive chromatin marks were detected in the TAD boundaries along with enrichment of housekeeping genes, transcription start sites, and global run on sequencing signal in the boundaries. Together, these suggest a high level of transcriptional activity associated with boundary formation; however, whether boundary formation was a cause or consequence of transcriptional activity was not clear [6].

Point-to-point direct interactions—looping of chromatin affects gene transcription

Interestingly, Hi-C data also revealed pairs of loci that had significantly stronger interaction than any loci lying between them. These were designated as loops, and can appear within topology domains (discussed above), independently and/or across TADs. Interestingly, chromatin loops were not only conserved among human cell lines but also found to be conserved between mouse and human cells [8]; chromatin looping interactions were significantly enriched within cis-regulatory elements like active promoters and enhancers while being depleted at inactive TSS or regions with repressive chromatin marks [6–8]. About 30% of chromatin loops brought promoters and enhancers together (versus 7% expected by chance), and genes with promoters associated with loops were more expressed than genes whose promoters were not associated with any loop (sixfold) [8]. Hi-C data analysis also revealed 55% of distal enhancers interact with at least one active promoter, confirming previous observations that promoters and enhancers often form complex networks to regulate transcription [7, 37]. Interestingly, a particular case study showed many genes without any NF-kappaB (p65) binding site in promoters were induced simultaneously by TNF-alpha (which is known to trigger NF-kappaB signaling), possibly due to sharing of overlapping distal interacting regions containing multiple NF-kappaB binding sites [7].

Somewhat intriguingly, there was little or no change in promoter-enhancer looping interactions at a vast majority of TNF-alpha-responsive enhancers on TNF-alpha treatment. This suggested that in general, promoter-enhancer contacts in untreated cells, which are the existing DNA loops, did not alter upon transient activation or repression of enhancers following treatment. Interestingly, chromatin interactions involving cell type-specific enhancers are variable between cell types indicating context-specific interaction structures. This discrepancy between signal-dependent and cell type-specific enhancers correlates with H3K4me1 chromatin marks, which unlike H3K27me3 remain unchanged upon TNF-alpha treatment [7]. Other studies have also observed pre-existing looping interactions at several loci induced by p53, FOXO3, and glucocorticoid receptors [33, 38, 39].

Chromatin loops marked through CTCF binding

Analysis of ENCODE ChIP-seq data revealed loci with chromatin loops were typically bound by the insulator protein CTCF (86%) and the Cohesin subunits RAD21 (86%) and SMC3 (87%) [8]. This was consistent with many reports which, using a variety of experimental approaches, suggest a role for CTCF and cohesion in mediating DNA loops [28, 40, 41]. As many of these loops demarcate domains, this observation was also concordant with studies showing CTCF delimits structural and regulatory domains [6, 42, 43]. Furthermore, most peak loci possessed a unique DNA site containing a CTCF binding motif to which all the three proteins (CTCF, Rad21, and SMC3) bind. A vast majority of these motif pairs present in the peak loci were oriented in a convergent manner suggesting that a pair of CTCF motifs in the convergent orientation might be required for the formation of a loop [8].

Looping through telomere ends

Studies by Shay and Wright's groups have demonstrated that telomeres loop to specific loci (within 10 Mb)—a phenomenon called TPE-OLD (telomere positioning effect—over long distances). These studies showed that genes close to the telomere were silenced in young primary cells with long telomeres, but were activated when telomeres became short with cellular aging, an effect that was reversed by re-elongation of telomeres upon exogenous expression of the hTERT (human telomerase) gene. Interactions of the telomere and sub-telomere with chromosomal regions up to 10 Mb away from the telomere end were revealed using modified Hi-C and 3C experiments [44–47]. An important function of such looping showed looping of chromosome 5 sub-telomere resulted in heterochromatic silencing of the telomerase gene in young primary cells [44]. Reports also suggest that several telomere binding proteins like TRF2 and

TIN2 can associate with interstitial telomere-like repeats [48]. Such regions of repetitive DNA (often referred to as interstitial telomeric repeats or ITS) may be crucial in forming sub-telomeric loops by recruitment of telomeric factors and structural proteins like Lamins to mediate interaction with telomeres [49, 50].

Single cell Hi-C

Though Hi-C has given insights into the functional chromosomal organization, one can argue that this is an average, probabilistic view of chromosomal interactions with much cell-to-cell variation, such that observed domain organization and chromosomal interactions might represent just a fraction of the cells. Some recent studies attempt to address this with Hi-C at the single cell level. In the first such report, pooled data recapitulated the formation of TAD-like structures indicating these domains are robust and form the basis of chromosome conformation in each cell [51]. However, the observed variability in inter-domain contacts suggested significant differences might be possible in the higher order folding of chromosomes. Another single cell study concluded that, though structures of individual TADs and chromosome loops vary substantially from cell to cell, the higher order organizational signatures like the A/B compartments are mostly retained [15, 52]. In addition, it was also noted that LADs and active enhancers and promoters are consistently organized genome-wide in every cell [52].

Other factors that nucleate chromatin interactions in the 3D genome

The finding that a large fraction of loop boundaries are bound by CTCF and the Cohesin subunits [8] led to popularization of the extrusion model of loop formation. In this model, a pair of factors (Cohesin), possibly with motor function, can dynamically bind to DNA and move along the DNA in opposite directions extruding the chromatin to form loops in between until they dissociate or stall at a boundary element (CTCF motif) [53–55]. This model largely helps to understand the nested nature of TADs and loops as well as the consequences of CTCF motif deletion and inversion [36, 56]. However, many convergent CTCF motifs exist that do not delineate loops and, interestingly, a considerable fraction of identified loops did not appear to possess any CTCF or cohesion binding sites. Together, these suggest other factors that could be involved in loop formation.

The fact that CTCF motifs can act as boundary elements only when they are oriented in convergent fashion indicate dimerization of CTCF is required for stalling extruding cohesion loops. This indicates that interactions between DNA binding proteins can possibly play a role in mediating loop formation and chromatin interactions by bringing distant genomic loci together. Indeed, DNA binding proteins, YY1 and ZNF143, were found to be enriched in loop loci [8]. Both homodimer and heterodimer formation by proteins bound to distant loci can lead to chromosomal looping interactions. In a somewhat similar context, the telomere binding factor TRF2 has already been implicated in mediating telomeric looping into the extra-telomeric interstitial regions including the TERT loci [44]. The fact that TRF2 can bind to many extra-telomeric sites throughout the human genome implicates possibility of it mediating other looping interactions as well [49, 50, 57, 58].

Apart from proteins, nucleic acids like lncRNAs and secondary DNA structures might also contribute to mediating chromosomal looping interactions. lncRNAs called as activating/enhancer RNAs were found to bind with mediator proteins (MED12 and MED1), and their knockdown led to diminished looping between the ncRNA loci and their target genes and a decrease in mediator binding to the genes [59, 60].

Another interesting context that could support loop formation in a CTCF-independent way comes from the possible involvement of DNA secondary structures such as G quadruplexes. G quadruplexes are typically formed by a stretch of DNA with 3-guanines repeated at least four times at close interval and are reported to be involved in various biological functions across life forms [61–64]. The formation of inter-molecular G quadruplexes raises the possibility that such secondary structure DNA motifs can bring together two distant genomic loci. G quadruplexes have been found to be enriched in DNAse hypersensitive (DHS) promoters as well as DHS cis-regulatory elements [65, 66]. Half G quadruplexes (i.e., two runs of guanines, both containing at least three consecutive Gs) were also found to be enriched at the boundaries of these DHS promoters and *cis*-elements but were depleted in the vicinity of these sites. Computational analyses showed such half G quadruplexes, one from DHS promoter and one from DHS enhancer, could come together to form G quadruplexes thereby bringing together promoters and enhancers elements via looping [65]. Several studies have reported enrichment of potential G quadruplex forming sequences in fragile genomic regions associated with pathogenic, cancer-causing breakpoints, and structural variations [67–70]. The possible formation of chromosomal loops mediated by G quadruplexes could be one of the contributing factors in increasing the fragility of such chromosomal regions. Disruption of these DNA structure motifs could give a clear idea regarding their roles in mediating functional chromosomal interactions.

Disruption of TAD boundaries has deleterious effects: clinical implications

Deletion of a TAD boundary, on the other hand, was found to result in spreading of contacts across the

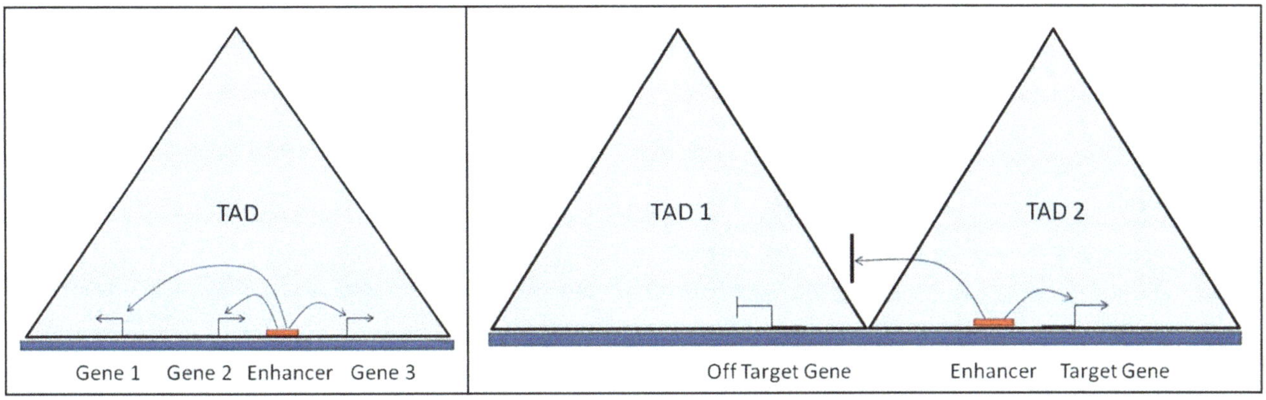

Fig. 4 Schematic representation showing coordinated expression of genes mediated by an enhancer interacting with multiple promoters within the same TAD and a TAD boundary restricts enhancer interaction and activity only to target genes within the same TAD

deleted region and transcriptional misregulation [9]. Several reports of TAD disruptions leading to pathological outcomes have further consolidated TADs as functional units. Structural variants disrupting CTCF-associated TAD boundaries were noted to allow de novo promoter-enhancer interactions and ectopic gene expression causing limb malformation [71] and oncogenic transformation [72]. Oncogenic chromosomal rearrangements were also shown to induce aberrant oncogenic expression in AML and medulloblastoma by causing enhancer trafficking where an enhancer acts on a gene other than its normal target due to genomic rearrangements like TAD disruptions [73, 74]. Moreover, it was also reported from Hi-C in prostate cancer cells that generation of smaller TADs due to establishment of additional boundaries and particular cancer-specific interactions within TADs could be associated with oncogenic transformation [75], and perturbation of CTCF binding by aberrant DNA methylation could cause oncogenic gene expression leading to gliomas by a mechanism similar to "enhancer hijacking" [76]. These pathogenic outcomes based on deregulated enhancer function along with another reporter construct insertion study [77] implicates that cis-regulatory elements can act in a non-specific manner but within a given TAD in the genome (Fig. 4).

Apart from how abrogation of TADs could lead to pathogenic phenotypes, chromatin looping interactions, even when not associated with a particular TAD, might have significant clinical implications. Recently, an interaction has been reported that might be crucial for TERT reactivation across different cancer types. The two cytosine to thymidine single-point mutations in the TERT proximal promoter (- 146 and - 124 bp from the translation start site), recurrent in several cancers, were found to create a de novo consensus binding motif for ETS transcription factor family [78–83]. GABPA (an ETS transcription factor) was shown to bind to the mutant promoter leading to a long-range chromatin interaction with a region 300 Kb upstream of the TERT promoter. As an effect of this binding, the promoter locus was changed into an open, active chromatin region, eventually enhancing TERT expression which has been widely associated with cancer development [84–86]. In another study, it was shown that telomere looping at the 4q35 locus regulates expression of SORBS2, which is disrupted in the age-associated genetic disease facioscapulohumeral muscular dystrophy [45].

Conclusion

Many large-scale studies with vast Hi-C data have given us important insights into possible mechanisms of looping, subsequent higher order chromatin organization, and functional significance of domain formation and chromosomal interactions in different processes including transcriptional regulation. Given that these observations encourage a shift in perspective regarding how *3D* organization of the genome impacts biological processes, it is of much interest to understand the underlying mechanistic rules governing causation of the folded architecture. Though cohesion and CTCF have been implicated in the formation of chromosomal loops, these proteins are absent in a fraction of loop loci (~ 10–14%) suggesting a role of other factors in determining how chromosomal interactions arise and how these contribute to specific cellular and context-specific functions.

Abbreviations

3C: Chromosome conformation capture; 3D: Three-dimensional; 4C: Chromosome conformation capture-o-chip; 5C: Carbon copy chromosome conformation capture; AML: Acute myeloid leukemia; ChIP-seq: Chromatin immunoprecipitation-sequencing; CTCF: CCCTC binding factor; DI: Directionality index; ENCODE: Encyclopedia of DNA elements; ES: Embryonic stem cells; FISH: Fluorescence in situ hybridization; hTERT: Human telomerase reverse transcriptase; LAD: Lamina-associated domains; lncRNA: Long non-coding RNA; LOCK: Large organized chromatin K9 modifications; NAD: Nucleolus-associated domains; NGS: Next-generation sequencing; TAD: Topologically associated domain; TIN2: TERF1-interacting nuclear factor 2; TNF: Tumor necrosis factor; TPE-OLD: Telomere position effect-over long distances; TRF2: Telomere repeat binding factor 2; TSS: Transcription start site

Acknowledgements
Research fellowships from CSIR (to SSR and AKM) are acknowledged. SC is a recipient of Wellcome Trust/DBT India Alliance Fellowship [grant number 500127/Z/09/Z].

Funding
No experimental expenses had been incurred in preparation of this manuscript.

Authors' contributions
SSR contributed to the literature review, Hi-C data analysis (for Fig. 2), making of figures, and manuscript writing; AKM contributed to the making of figures and manuscript writing; and SC contributed to the overall conceptualization of the review article and manuscript writing. All authors read and approved the final manuscript.

Consent for publication
Not applicable

Competing interests
The authors declare that they have no competing interests.

References

1. Dekker J. Capturing chromosome conformation. Science. 2002;295:1306–11.
2. Lieberman-Aiden E, van Berkum NL, Williams L, Imakaev M, Ragoczy T, Telling A, et al. Comprehensive mapping of long-range interactions reveals folding principles of the human genome. Science. 2009;326:289–93.
3. Simonis M, Kooren J, de Laat W. An evaluation of 3C-based methods to capture DNA interactions. Nat Methods. 2007;4:895–901.
4. De Wit E, De Laat W. A decade of 3C technologies-insights into nuclear organization. Genes Dev. 2012;26:11–24.
5. Fraser J, Williamson I, Bickmore WA, Dostie J. An overview of genome organization and how we got there: from FISH to Hi-C. Microbiol Mol Biol Rev. 2015;79:347–72.
6. Dixon JR, Selvaraj S, Yue F, Kim A, Li Y, Shen Y, et al. Topological domains in mammalian genomes identified by analysis of chromatin interactions. Nature. 2012;485:376–80.
7. Jin F, Li Y, Dixon JR, Selvaraj S, Ye Z, Lee AY, et al. A high-resolution map of the three-dimensional chromatin interactome in human cells. Nature. 2013; 503:290–4.
8. Rao SSP, Huntley MH, Durand NC, Stamenova EK, Bochkov ID, Robinson JT, et al. A 3D map of the human genome at kilobase resolution reveals principles of chromatin looping. Cell. 2014;159:1665–80.
9. Nora EP, Lajoie BR, Schulz EG, Giorgetti L, Okamoto I, Servant N, et al. Spatial partitioning of the regulatory landscape of the X-inactivation centre. Nature. 2012;485:381–5.
10. Sexton T. Three-dimensional folding and functional organization principles of the Drosophila genome. Cell. 2012;148:458–72.
11. Yokota H, van den Engh G, Hearst JE, Sachs RK, Trask BJ. Evidence for the organization of chromatin in megabase pair-sized loops arranged along a random walk path in the human G0/G1 interphase nucleus. J Cell Biol. 1995; 130:1239–49.
12. Tanabe H, Habermann FA, Solovei I, Cremer M, Cremer T. Non-random radial arrangements of interphase chromosome territories: evolutionary considerations and functional implications. Mutat Res - Fundam Mol Mech Mutagen. 2002;504:37–45.
13. Boyle S, Gilchrist S, Bridger JM, Mahy NL, Ellis J a, Bickmore W a. The spatial organization of human chromosomes within the nuclei of normal and emerin-mutant cells. Hum Mol Genet. 2001;10:211–9.
14. Croft JA, Bridger JM, Boyle S, Perry P, Teague P, Bickmore WA. Differences in the localization and morphology of chromosomes in the human nucleus. J Cell Biol. 1999;145:1119–31.
15. Lieberman-Aiden E. Comprehensive mapping of long-range interactions reveals folding principles of the human genome. Science. 2009;326: 289–93.
16. Eskeland R. Ring1B compacts chromatin structure and represses gene expression independent of histone ubiquitination. Mol Cell. 2010;38:452–64.
17. Hiratani I. Genome-wide dynamics of replication timing revealed by in vitro models of mouse embryogenesis. Genome Res. 2010;20:155–69.
18. Ryba T. Evolutionarily conserved replication timing profiles predict long-range chromatin interactions and distinguish closely related cell types. Genome Res. 2010;20:761–70.
19. Wen B, Wu H, Shinkai Y, Irizarry RA, Feinberg AP. Large histone H3 lysine 9 dimethylated chromatin blocks distinguish differentiated from embryonic stem cells. Nat Genet. 2009;41:246–50.
20. Guelen L. Domain organization of human chromosomes revealed by mapping of nuclear lamina interactions. Nature. 2008;453:948–51.
21. Peric-Hupkes D. Molecular maps of the reorganization of genome-nuclear lamina interactions during differentiation. Mol Cell. 2010;38:603–13.
22. Filippova D, Patro R, Duggal G, Kingsford C. Identification of alternative topological domains in chromatin. Algorithms Mol Biol. 2014;9:14.
23. Munkel C, Langowski J. Chromosome structure predicted by a polymer model. Phys Rev E. 1998;57:5888–96.
24. Mateos-Langerak J, Bohn M, de Leeuw W, Giromus O, Manders EMM, Verschure PJ, et al. Spatially confined folding of chromatin in the interphase nucleus. Proc Natl Acad Sci U S A. 2009;106:3812–7.
25. Grosberg AY, Nechaev SK, Shakhnovich EI. The role of topological constraints in the kinetics of collapse of macromolecules. J Phys Fr. 1988;49:2095–100.
26. Grosberg a, Rabin Y, Havlin S, Neer a. Crumpled globule model of the three-dimensional structure of DNA. Europhys Lett. 1993;23:373–8.
27. Vasilyev OA, Nechaev SK. Topological correlations in trivial knots: new arguments in favor of the representation of a crumpled polymer globule. Theor Math Phys. 2003;134:142–59.
28. Phillips JE, Corces VG. CTCF: master weaver of the genome. Cell. 2009;137: 1194–211.
29. Handoko L. CTCF-mediated functional chromatin interactome in pluripotent cells. Nat Genet. 2011;43:630–8.
30. Dixon JR, Jung I, Selvaraj S, Shen Y, Antosiewicz-Bourget JE, Lee AY, et al. Chromatin architecture reorganization during stem cell differentiation. Nature. 2015;518:331–6.
31. Dowen JM, Fan ZP, Hnisz D, Ren G, Abraham BJ, Zhang LN, et al. Control of cell identity genes occurs in insulated neighborhoods in mammalian chromosomes. Cell. 2014;159:374–87.
32. Ji X, Dadon DB, Powell BE, Fan ZP, Borges-Rivera D, Shachar S, et al. 3D chromosome regulatory landscape of human pluripotent cells. Cell Stem Cell. 2016;18:262–75.
33. Phillips-Cremins JE, Sauria MEG, Sanyal A, Gerasimova TI, Lajoie BR, Bell JSK, et al. Architectural protein subclasses shape 3D organization of genomes during lineage commitment. Cell. 2013;153:1281–95.
34. Kagey MH. Mediator and cohesin connect gene expression and chromatin architecture. Nature. 2010;467:430–5.
35. Noordermeer D. The dynamic architecture of Hox gene clusters. Science. 2011;334:222–5.
36. Dixon JR, Gorkin DU, Ren B. Chromatin domains: the unit of chromosome organization. Mol Cell. 2016;62:668–80.
37. Smallwood A, Ren B. Genome organization and long-range regulation of gene expression by enhancers. Curr Opin Cell Biol. 2013;25:387–94.
38. Melo CA. eRNAs are required for p53-dependent enhancer activity and gene transcription. Mol Cell. 2013;49:524–35.
39. Tan PY. Integration of regulatory networks by NKX3-1 promotes androgen-dependent prostate cancer survival. Mol Cell Biol. 2012;32:399–414.
40. Splinter E, Heath H, Kooren J, Palstra RJ, Klous P, Grosveld F, et al. CTCF mediates long-range chromatin looping and local histone modification in the beta-globin locus. Genes Dev. 2006;20:2349–54.

41. Hou C, Zhao H, Tanimoto K, Dean A. CTCF-dependent enhancer-blocking by alternative chromatin loop formation. Proc Natl Acad Sci U S A. 2008;105: 20398–403.
42. Xie X, Mikkelsen TS, Gnirke A, Lindblad-Toh K, Kellis M, Lander ES. Systematic discovery of regulatory motifs in conserved regions of the human genome, including thousands of CTCF insulator sites. Proc Natl Acad Sci U S A. 2007; 104:7145–50.
43. Cuddapah S, Jothi R, Schones DE, Roh TY, Cui K, Zhao K. Global analysis of the insulator binding protein CTCF in chromatin barrier regions reveals demarcation of active and repressive domains. Genome Res. 2009;19:24–32.
44. Kim W, Ludlow AT, Min J, Robin JD, Stadler G, Mender I, et al. Regulation of the human telomerase gene TERT by telomere position effect—over long distances (TPE-OLD): implications for aging and cancer. PLoS Biol. 2016;14: e2000016.
45. Robin JD, Ludlow AT, Batten K, Gaillard MC, Stadler G, Magdinier F, et al. SORBS2 transcription is activated by telomere position effect-over long distance upon telomere shortening in muscle cells from patients with facioscapulohumeral dystrophy. Genome Res. 2015;25:1781–90.
46. Robin JD, Ludlow AT, Batten K, Magdinier F, Stadler G, Wagner KR, et al. Telomere position effect: regulation of gene expression with progressive telomere shortening over long distances. Genes Dev. 2014;28:2464–76.
47. Stadler G, Rahimov F, King OD, Chen JC, Robin JD, Wagner KR, et al. Telomere position effect regulates DUX4 in human facioscapulohumeral muscular dystrophy. Nat Struct Mol Biol. 2013;20:671–8.
48. Mignon-Ravix C, Depetris D, Delobel B, Croquette M-F, Mattei M-G. A human interstitial telomere associates in vivo with specific TRF2 and TIN2 proteins. Eur J Hum Genet. 2002;10:107.
49. Wood AM, Danielsen JMR, Lucas CA, Rice EL, Scalzo D, Shimi T, et al. TRF2 and lamin A/C interact to facilitate the functional organization of chromosome ends. Nat Commun. 2014;5:5467.
50. Wood AM, Laster K, Rice EL, Kosak ST. A beginning of the end: new insights into the functional organization of telomeres. Nucleus. Taylor & Francis. 2015;6:172–8.
51. Nagano T, Lubling Y, Stevens TJ, Schoenfelder S, Yaffe E, Dean W, et al. Single-cell Hi-C reveals cell-to-cell variability in chromosome structure. Nature. 2013;502:59–64.
52. Stevens TJ, Lando D, Basu S, Liam P, Cao Y, Lee SF, et al. 3D structures of individual mammalian genomes studied by single-cell Hi-C. Nature. 2017; 544:59–64.
53. Alipour E, Marko JF, Marko JF, Mirny L, Alipour E, Marko J, et al. Self-organization of domain structures by DNA-loop-extruding enzymes. Nucleic Acids Res. 2012;40:11202–12.
54. Fudenberg G, Imakaev M, Lu C, Goloborodko A, Abdennur N, Mirny LA. Formation of chromosomal domains by loop extrusion. Cell Rep. 2016;15: 2038–49.
55. Sanborn AL, Rao SSP, Huang S-C, Durand NC, Huntley MH, Jewett AI, et al. Chromatin extrusion explains key features of loop and domain formation in wild-type and engineered genomes. Proc Natl Acad Sci. 2015;112:201518552.
56. Boyan B, Giacomo C. Organization and function of the 3D genome. Nat Rev Genet. 2016;17:661–78.
57. Yang D, Xiong Y, Kim H, He Q, Li Y, Chen R, et al. Human telomeric proteins occupy selective interstitial sites. Cell Res. 2011;21:1013–27.
58. Simonet T, Zaragosi L-E, Philippe C, Lebrigand K, Schouteden C, Augereau A, et al. The human TTAGGG repeat factors 1 and 2 bind to a subset of interstitial telomeric sequences and satellite repeats. Cell Res. 2011;21:1028–38.
59. Trimarchi T, Bilal E, Ntziachristos P, Fabbri G, Dalla-Favera R, Tsirigos A, et al. Genome-wide mapping and characterization of notch-regulated long noncoding RNAs in acute leukemia. Cell. 2014;158:593–606.
60. Lai F, Orom UA, Cesaroni M, Beringer M, Taatjes DJ, Blobel GA, et al. Activating RNAs associate with Mediator to enhance chromatin architecture and transcription. Nature. 2013;494:497–501.
61. Rawal P, Kummarasetti VBR, Ravindran J, Kumar N, Halder K, Sharma R, et al. Genome-wide prediction of G4 DNA as regulatory motifs: role in Escherichia coli global regulation. Genome Res. 2006;16:644–55.
62. Yadav VK, Abraham JK, Mani P, Kulshrestha R, Chowdhury S. QuadBase: genome-wide database of G4 DNA—occurrence and conservation in human, chimpanzee, mouse and rat promoters and 146 microbes. Nucleic Acids Res. 2008;36:381–5.
63. Lipps HJ, Rhodes D. G-quadruplex structures: in vivo evidence and function. Trends Cell Biol. 2009;19:414–22.
64. Rhodes D, Lipps HJ. Survey and summary G-quadruplexes and their regulatory roles in biology. Nucleic Acids Res. 2015;43:8627–37.
65. Hegyi H. Enhancer-promoter interaction facilitated by transiently forming G-quadruplexes. Sci Rep. 2015;5:1–6.
66. Thurman RE, Rynes E, Humbert R, Vierstra J, Maurano MT, Haugen E, et al. The accessible chromatin landscape of the human genome. Nature. 2012; 489:75–82.
67. Bose P, Hermetz KE, Conneely KN, Rudd MK. Tandem repeats and G-rich sequences are enriched at human CNV breakpoints. PLoS One. 2014;9:1–8.
68. Murat P, Balasubramanian S. Existence and consequences of G-quadruplex structures in DNA. Curr Opin Genet Dev. 2014;25:22–9.
69. De S, Michor F. DNA secondary structures and epigenetic determinants of cancer genome evolution. Nat Struct Mol Biol. 2011;18:950–5.
70. Katapadi VK, Nambiar M, Raghavan SC. Potential G-quadruplex formation at breakpoint regions of chromosomal translocations in cancer may explain their fragility. Genomics. 2012;100:72–80.
71. Lupiáñez DG, Kraft K, Heinrich V, Krawitz P, Brancati F, Klopocki E, et al. Disruptions of topological chromatin domains cause pathogenic rewiring of gene-enhancer interactions. Cell. 2015;161:1012–25.
72. Hnisz D, Weintraub AS, Day DS, Valton A-L, Bak RO, Li CH, et al. Activation of proto-oncogenes by disruption of chromosome neighborhoods. Science. 2016;351:1454–8.
73. Northcott PA, Lee C, Zichner T, Stütz AM, Erkek S, Kawauchi D, et al. Enhancer hijacking activates GFI1 family oncogenes in medulloblastoma. Nature. 2014;511:428–34.
74. Gröschel S, Sanders MA, Hoogenboezem R, De Wit E, Bouwman BAM, Erpelinck C, et al. A single oncogenic enhancer rearrangement causes concomitant EVI1 and GATA2 deregulation in leukemia. Cell. 2014;157:369–81.
75. Taberlay PC, Achinger-Kawecka J, Lun ATL, Fabian A, Bauer DC, Smyth GK, et al. Three-dimensional disorganisation of the cancer genome occurs coincident with long range genetic and epigenetic alterations. Genome Res. 2016;26:719–31.
76. Flavahan WA, Drier Y, Liau BB, Gillespie SM, Venteicher AS, Stemmer-Rachamimov AO, et al. Insulator dysfunction and oncogene activation in IDH mutant gliomas. Nature. 2015;529:110–4.
77. Symmons O, Uslu W, Tsujimura T, Ruf S, Nassari S, Schwarzer W, et al. Functional and topological characteristics of mammalian regulatory domains. Genome Res. 2014;24:390–400.
78. Horn S, Figl A, Rachakonda PS, Fischer C, Sucker A, Gast A, et al. TERT promoter mutations in familial and sporadic melanoma. Science. 2013;339: 959–61.
79. Borah S, Xi L, Zaug AJ, Powell NM, Dancik GM, Cohen SB, et al. TERT promoter mutations and telomerase reactivation in urothelial cancer. Science. 2015;347:1006–10.
80. Huang FW, Hodis E, Xu MJ, Kryukov GV, Chin L, Garraway LA. Highly recurrent TERT promoter mutations in human melanoma. Science. 2013;339:957–9.
81. Vinagre J, Almeida A, Pópulo H, Batista R, Lyra J, Pinto V, et al. Frequency of TERT promoter mutations in human cancers. Nat Commun. 2013;4:2185.
82. Bell RJA, Rube HT, Kreig A, Mancini A, Fouse SD, Nagarajan RP, et al. The transcription factor GABP selectively binds and activates the mutant TERT promoter in cancer. Science. 2015;348:1036–39.
83. Li Y, Zhou QL, Sun W, Chandrasekharan P, Cheng HS, Ying Z, et al. Non-canonical NF-κB signalling and ETS1/2 cooperatively drive C250T mutant TERT promoter activation. Nat Cell Biol. 2015;17:1327–38.
84. Stern JL, Theodorescu D, Vogelstein B, Papadopoulos N, Cech TR. Mutation of the TERTpromoter, switch to active chromatin, and monoallelic TERTexpression in multiple cancers. Genes Dev. 2015;29:2219–24.
85. Xi L, Schmidt JC, Zaug AJ, Ascarrunz DR, Cech TR. A novel two-step genome editing strategy with CRISPR-Cas9 provides new insights into telomerase action and TERT gene expression. Genome Biol. 2015;16:1–17.
86. Akıncılar SC, Khattar E, Boon PLS, Unal B, Fullwood MJ, Tergaonkar V. Long-range chromatin interactions drive mutant TERT promoter activation. Cancer Discov. 2016;6:1276–92.

15

Attitudes of stakeholders in psychiatry towards the inclusion of children in genomic research

Anna Sundby[1,2*], Merete Watt Boolsen[3], Kristoffer Sølvsten Burgdorf[4], Henrik Ullum[4], Thomas Folkmann Hansen[2,5,6] and Ole Mors[1,2]

Abstract

Background: Genomic sequencing of children in research raises complex ethical issues. This study aims to gain more knowledge on the attitudes towards the inclusion of children as research subjects in genomic research and towards the disclosure of pertinent and incidental findings to the parents and the child.

Methods: Qualitative data were collected from interviews with a wide range of informants: experts engaged in genomic research, clinical geneticists, persons with mental disorders, relatives, and blood donors. Quantitative data were collected from a cross-sectional web-based survey among 1227 parents and 1406 non-parents who were potential stakeholders in psychiatric genomic research.

Results: Participants generally expressed positive views on children's participation in genomic research. The informants in the qualitative interviews highlighted the age of the child as a critical aspect when disclosing genetic information. Other important aspects were the child's right to an autonomous choice, the emotional burden of knowing imposed on both the child and the parents, and the possibility of receiving beneficial clinical information regarding the future health of the child. Nevertheless, there was no consensus whether the parent or the child should receive the findings. A majority of survey stakeholders agreed that children should be able to participate in genomic research. The majority agreed that both pertinent and incidental findings should be returned to the parents and to the child when of legal age. Having children does not affect the stakeholder's attitudes towards the inclusion of children as research subjects in genomic research.

Conclusion: Our findings illustrate that both the child's right to autonomy and the parents' interest to be informed are important factors that are found valuable by the participants. In future guidelines governing children as subjects in genomic research, it would thus be essential to incorporate the child's right to an open future, including the right to receive information on adult-onset genetic disorders.

Keywords: Child, Minors, Attitude, Whole genome sequencing, Ethics research, Mental disorders

Background

Genome sequencing, i.e., whole genome and whole exome sequencing, has now become a widely used tool in research, for example, in psychiatric genetic research. However, genome sequencing generates huge amounts of genomic information, and some of the information is clinically useful. There is an ongoing debate regarding the return of individual genetic research findings unrelated to the condition under investigation [1–7]. The community of clinical genetics refers to these types of results as incidental findings [8]. Because incidental findings should be expected as a part of genomic sequencing, other terms as secondary findings, additional findings, and non-pertinent findings have been suggested [8, 9]. However, the increasing number of large-scale genomic research projects and the inclusion of children in genomic research

* Correspondence: anna.sundby@regionh.dk
[1]Department of Clinical Medicine, Psychosis Research Unit, Aarhus University Hospital, Skovagervej 2, 8240 Risskov, Denmark
[2]The Lundbeck Foundation Initiative for Integrative Psychiatric Research, iPSYCH, Copenhagen, Denmark

have made it important to continue the debate regarding the return of individual genetic research findings.

Children are included in genomic research because some disorders are childhood-onset [10]. For example, to study genetic factors for attention deficit hyperactivity disorder (ADHD) or autism spectrum disorder, blood and buccal cell samples from children with ADHD and autism spectrum disorder are often taken in the clinical setting and used for research. Research on children can also be done using already existing biological samples stored in a biobank. For example, the Danish Neonatal Screening Biobank holds archived blood samples from all newborns in Denmark after screening for phenylketonuria (PKU) and several other congenital diseases [11, 12]. The dried blood spot samples from the neonates provide an opportunity to use DNA from children in research on a very large scale [13].

The ethical issues of including children in genomic research differ from those raised when including adults [10, 14, 15]. One of the biggest differences is that a third party (the parent) must mediate the relationship between the researcher and the research subject (the child) [10, 14, 16, 17]. Furthermore, young children do not have the same capacity to understand the research information or the implications of the findings. Common questions related to involvement of children in genomic research include the following: Should children be research subjects at all? Should the findings be returned to the child or to the parents? [5, 10, 14, 17–22].

If samples are to be collected and used for research purposes in Denmark, the research subject must be of legal age (18 years) to voluntarily consent to participate. If the research subject is not able to consent, for example, due to the age, such authority lies with the parents or the guardians [23]. If a research project obtains significant information about the health status of the research subject, the research subject should be informed, unless the subject has specifically expressed that s/he does not want the information [24]. The Danish National Committee on Health Research Ethics has developed a guideline on genomic research [25]. The guideline addresses inclusion of children in genomic research and return of incidental findings. It is emphasized that there must be a direct health benefit for the individual child to justify inclusion in whole genome research and that it is not sufficient to include children simply to study the genetics of a disease. Return of results to research subjects is only the norm for genetic variants with high penetrance that predispose to serious curable or treatable disease, whereas no return of results is usually provided for genetic variants with low or moderate penetrance and uncertain clinical significance [25]. The guideline furthermore addresses that exemptions from obtaining consent from parents can be given if the research project uses biological samples archived in a biobank [25].

The legal regulations in Denmark on the return of findings based on children's DNA are complex and unclear. This makes it difficult for researchers and research subjects to navigate the regulations governing the return of findings from children's DNA. We assume that the severity, treatability, and preventability of the disease; the age of the child; and the decision on whether to return incidental findings are discussed at the time of donation of the sample as all these issues are relevant in the decision-making. It is unclear whether the return of genetic risk variants with high penetrance that predispose to severe, curable, or treatable diseases encompasses all known genetic risk variants, including those for adult-onset disorders. Furthermore, it is unclear if results should be returned to the child or to the parents or whether results should be conveyed only when the child reaches legal age.

Previous studies have found that parents generally think that children should be involved in genomic research and that parents have a right to know about incidental findings concerning their child(ren) [26–28]. However, to our knowledge, no large studies have investigated the attitudes towards the involvement of children in genomic research among potential research subjects. Researching such attitudes is important to create policy strategies and to design future genomic research projects that meet the concerns and expectations of potential research subjects. We chose to study this topic in the context of mental disorders because they are some of the commonest disorders in the population and because large genome sequencing studies have been conducted in this area [13, 29, 30].

This study aims to explore attitudes among stakeholders in psychiatric genetic research towards (i) children as genomic research subjects, (ii) return of pertinent and incidental findings, and (iii) disclosure to participants (i.e., children) or their parents.

Methods

For the purpose of this study, we use pertinent findings for the results that are directly relevant to the condition under investigation and incidental findings for results that are not directly related to the research project but may have health importance for the individual research subject. We use a definition of childhood, starting from birth to 18 years of age, referring to the legal age of majority in Denmark. According to Statistics Denmark, individuals under age 26 living at home are defined as home-living children because the child has parental reference to at least one adult [31]. Therefore, we define a parent as a male or female with children under age 26 years living at home.

Mixed-methods

We used a mixed-methods design as we collected both qualitative and quantitative data [32]. This approach was taken to address the overall aim at different levels and to complement the strengths of a single-method design. The motivation for using a mixed-methods design was to use qualitative interviews to develop and modify the quantitative survey, but the qualitative interviews also helped to approach the field regarding inclusion of children in genomic research from a different angle.

The qualitative interviews

The interviews were conducted to explore the attitudes towards ethical issues regarding the use of genomic sequencing in psychiatric research among experts engaged in genomic research and among potential research participants in psychiatric genetic research. The interview guides were developed on the basis of relevant literature and the research questions of the overall study. The overall interview guides focused on attitudes towards (i) sharing findings, (ii) consenting procedures, (iii) duties of genomic researchers, (iv) including children in genomic research, and (v) misuse of knowledge in genetics. All interviews were conducted from 5 December 2012 to 11 December 2013, and each interview lasted between 30 and 90 min.

The informants

The informants in the interviews were experts (doctors and nurses), working with issues surrounding genomic research in Denmark, including the Faroe Islands, and persons with schizophrenia. The informants in the focus group interviews comprised persons with ADHD, parents to children with ADHD, clinical geneticists, and healthy controls (blood donors from the Danish Blood Donor Study (DBDS)). The persons with schizophrenia and ADHD were all diagnosed by psychiatrists and were stable in daily life. All blood donors who participated in the interview (and in the survey) are part of DBDS and have all given a broad informed consent to participate in a biobank for research and are potential healthy controls in genomic research. To be included in DBDS, the individual is not subjected to any medical treatment [33, 34]. Participants in the interview were all voluntary and informants were recruited through direct email contact, paper flyers at an activity center for mentally ill, invitations posted in an ADHD Facebook group, an ADHD clinic, a psychiatrist working in private practice, Copenhagen Hospital Biobank, and a blood donor Facebook group.

The data analysis

All interviews were audio-recorded, transcribed verbatim, coded, and analyzed using the software package NVivo 11 [35]. NVivo helped to organize and manage the data. We were based on Grounded Theory as a method to analyze the data because this analytical strategy suggests a systematic analysis and enabled us to explore and identify attitudes towards inclusion of children in genomic research [36]. All transcripts were analyzed after a three-step process: open coding, axial coding, and selective coding [36]. We examined the emerging core concepts overall and how they differed across informants. We used some of the core concepts to develop items for the survey. For this paper, we selected relevant and illustrative quotes regarding inclusion of children in genome research and return of results. All quotes were translated from Danish into English by a professional translator.

The cross-sectional survey

After the interviews, we conducted a cross-sectional web-based survey to explore specific attitudes towards the use of genomic research in Denmark. The survey was a translated and modified version of an English web-based survey developed at the Wellcome Trust Sanger Institute in Cambridge, UK [3, 37–40]. Details of the development of both the English and Danish survey versions have been presented elsewhere [39, 41]. On basis of the qualitative interviews, we modified the Danish version of the survey to also include items on informed consent, inclusion of children, and personal and familial experience with mental disorders. Items on children in genomic research were included because it was a key topic in the interviews and because the category of "age" came up in all the interviews. The survey also included self-reported sociodemographic information. Ten explanatory video films with subtitles and voice-over were used to illustrate the ethical issues raised in the survey (https://svaros.dk/holdning). The data collection began in August 2014 and ended April 2015. An average survey took approximately 23 min to complete.

The stakeholders

Potential stakeholders in psychiatric genomic research were recruited: (i) 241 persons with mental disorders (who are potential cases in genomic research), (ii) 671 relatives to individuals with mental disorders (who are potential controls in genomic research), (iii) 1623 blood donors from DBDS (who serve as healthy controls in genomic research) [33, 34], (iv) 28 clinical geneticists (who analyze, return, and explain genomic data to patients and relatives in their clinical work and who must validate sequencing findings obtained in research), and (v) 74 psychiatrists (who diagnose and treat people with mental disorders) ($N = 2637$). We recruited the stakeholders through email, paper flyers at psychiatric hospitals, invitations posted in an ADHD Facebook group, and links at the homepages of the Danish Psychiatric

Association, the Danish Society of Medical Genetics, and user groups of psychiatric patients and their relatives.

Statistical analysis

We expected parents to hold a different attitude from non-parents because of their parenthood and interest in the health of their child. For example, parents might be motivated to participate in genomic research with the hope of receiving crucial information about the health of their child. Therefore, we decided to divide the stakeholders into parents (persons with children under age 26 years living at home) or non-parents (persons without children).

Descriptive statistics were used to characterize the study sample. Unadjusted associations between items and stakeholders were estimated using χ^2 tests. Multivariable logistic regression with 95% confidence interval (CI) and a p value of 0.05 was used to estimate the association between item and stakeholder adjusted for parenthood, gender, age, educational level, marital status, and stakeholder group. "Don't know" answers were omitted from the regression model but are a part of the analysis regarding the questions towards the return of pertinent and incidental finding in children (Figs. 1 and 2). The analysis of the data was carried out using SAS® 9.4 [42].

The two questions regarding return of pertinent and incidental findings from research could be answered in four different ways, and respondents were allowed to choose more than one answer. We analyzed all answers and identified six combinations for pertinent findings and the same six combinations for incidental findings: (i) The child should have the opportunity to get information about pertinent/incidental findings, (ii) The child should not have

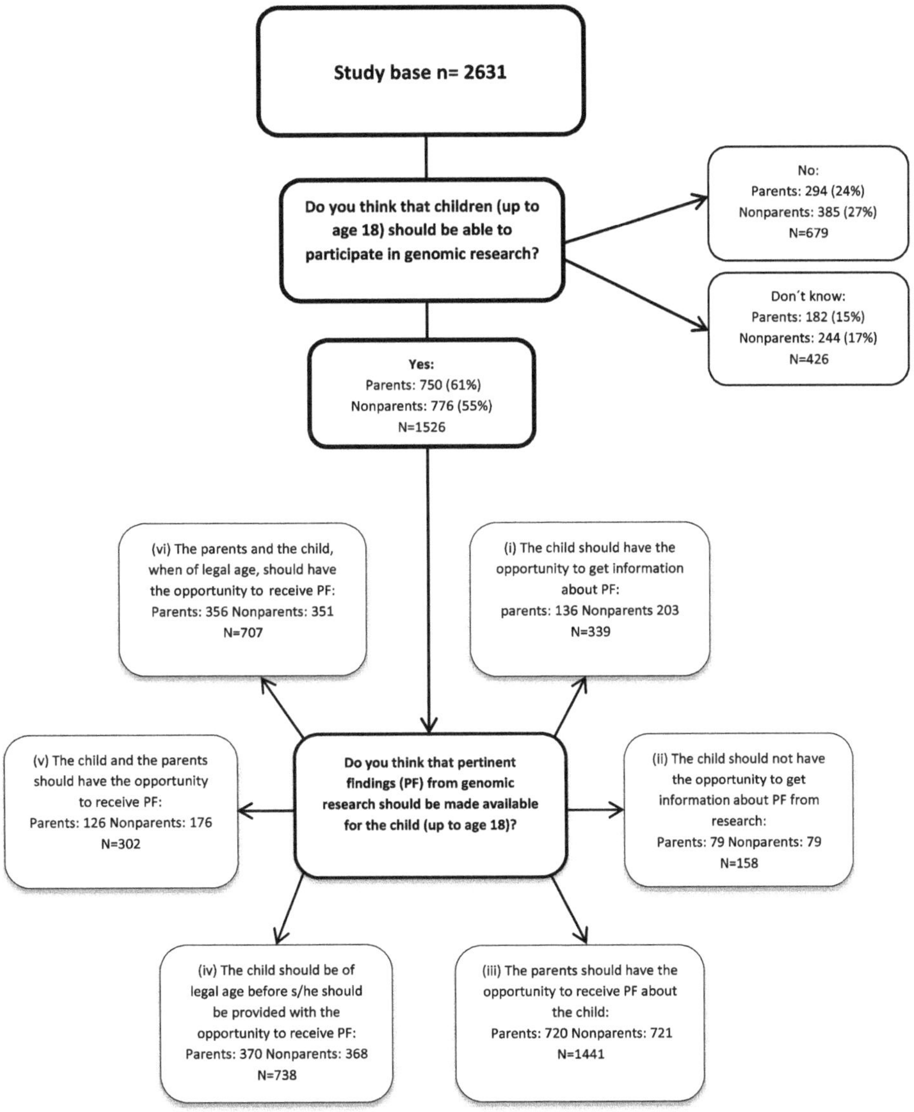

Fig. 1 Flowchart of attitudes towards the return of pertinent findings to children or parents distributed on parents and non-parents, *N* and %

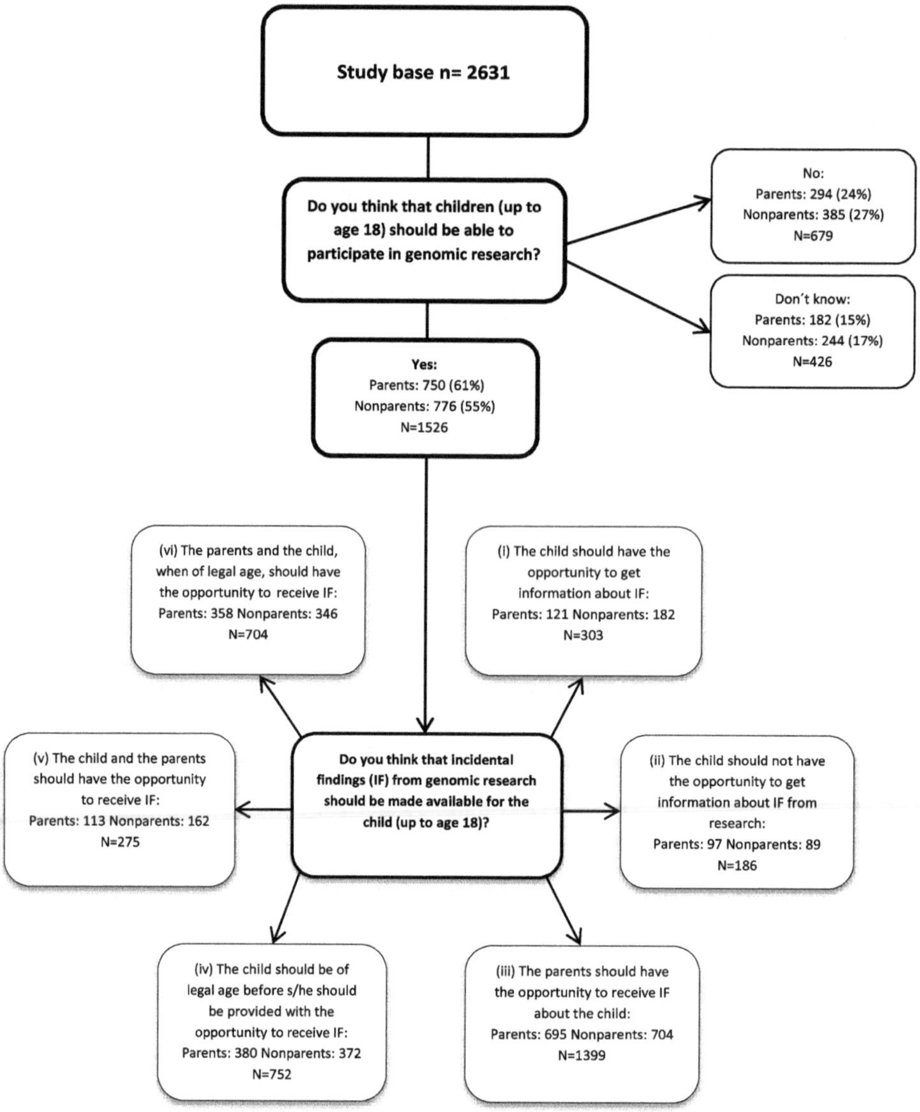

Fig. 2 Flowchart of attitudes towards the return of incidental findings to children or parents distributed on parents and non-parents, *N* and %

the opportunity to get information about pertinent/incidental findings from research, (iii) The parents should have the opportunity to receive pertinent findings about the child if they would like to receive such information, (iv) The child should be of legal age before getting the opportunity to receive pertinent/incidental findings, (v) The child and the parents should have the opportunity to receive pertinent/incidental findings, (vi) The parents and the child, when of legal age (18 years), should have the opportunity to receive pertinent/incidental findings (Figs. 1 and 2).

Results

The qualitative interviews

Sample characteristics

A total of 10 semi-structured qualitative interviews were conducted in this study: four focus-group interviews, one group interview, and five individual interviews. The informants consisted of 12 males and 17 females (Table 1).

Attitudes towards using children as genomic research subjects

The attitudes towards children as genomic research subjects varied. Experts, persons with ADHD, and parents of children with ADHD generally had a positive view, whereas clinical geneticists, persons with schizophrenia, and blood donors tended to have more ambiguous views. As a person with schizophrenia expressed: "Children should not be part of a research project, really. That makes a mess."

The age of the child played a major role in the talks and discussions about children as research subjects. The overall attitudes were that parents should make the decision on whether their child should be included or not, that parents should give informed consent on behalf

Table 1 Overview of sample characteristics in the qualitative interviews

Interview	Informants	Supplementary information
Focus-group interview	Persons with ADHD (n = 5)	Male 4; female 1, Denmark
Focus-group interview	Parents to children with ADHD (n = 6)	Male 1; female 5, Denmark
Focus-group interview	Danish blood donors from DBDS (n = 5)	Male 3; female 2, Denmark
Focus-group interview	Clinical geneticists (n = 6)	Male 1; female 5, Denmark
Group interview	Experts (n = 2)	Females, Faroe Islands
Individual interview	Expert	Female, Faroe Islands
Individual interview	Expert	Male, Denmark
Individual interview	Person with schizophrenia	Female, Denmark
Individual interview	Person with schizophrenia	Male, Denmark
Individual interview	Person with schizophrenia	Male, Denmark

The table presents type of interview and background of informants, their gender, and country of origin

of their child, and that the child should be involved in the decisions depending on the age of the child: the older the child, the more the child should be involved in the decision-making.

The willingness to enroll children in genomic research was often associated with the dilemma of receiving the individual results:

> Is it fair to deny the child the right to receive this knowledge [about genetic findings] from a research project – to say yes or no regarding information about himself or herself? [...] I don't think that we should cancel the research, but I think that the problem occurs when you start to involve the families in the return of individual results. (Clinical geneticist)

During the interview with the clinical geneticists, the discussion addressed the use of PKU samples in research and two informants discussed use of their own children's archived neonatal PKU samples:

Clinical geneticist A, "My children, they should no longer be registered in the Danish Neonatal Screening Biobank. I will make the call next week."

Clinical geneticist B, "What then if they become ill, and you could have known it?"

Clinical geneticist A, "Yes. I know that."

Clinical geneticist B concludes, "But it is exactly post-rationalization that is difficult. I mean, beforehand, we may say that we don't want to know, but what then if we suddenly have a sick child; and we could have done something?"

The clinical geneticists discussed the return of individual research results in relation to a child's right to make an autonomous decision, but the discussion also touched on the possibility of using the research results as clinical information with potential health benefits for their child.

Several informants agreed that there is an emotional burden linked to the recipient's age at the time when information about such findings is received and that returning the findings too early in the child's life "could ruin the person's life." (blood donor)

> There is great difference between a man of 51, such as myself, with certain life experience – both good and bad – and a young person with certain ideas about how hers life should be and who would be terribly disturbed by knowing of an 80 percent risk of getting, for example, breast cancer. So you have to consider when in life you should [return knowledge about findings], I think that is what I believe. (Other blood donor)

Some of the informants also expressed that there is an emotional burden associated with the return of findings to parents:

> Well, I would say that neither I nor the children should know anything before they have turned 18. I really don't think that the children can handle knowledge about it at their age and I... I should obviously not have that knowledge about my children before they are 18. Because I cannot handle such knowledge either. (Mother of a child with ADHD)

Nevertheless, there was no consensus among the informants as to whether the parents or the child should receive the findings:

> There is no doubt that I love him, my son. I love him more than anything, but I would still like to have that knowledge because it is a hard life. (Father of a child with ADHD)

The two parents of children with ADHD have different attitudes to the return of findings from research. The mother states that she should not receive the findings until her child is of legal age, whereas the father focuses on the opportunity to receive information that could help his son get a better life.

In short, the attitudes towards children as genomic research subjects varied among informants. There was no consensus whether children should be genomic research subjects and whether the child or the parents should receive the genomic findings. The attitudes of involving children in genomic research focused on the child's right to take an autonomic choice when of legal age, the emotional burden of knowing of both children and parents, and the possibility of receiving clinical information that could benefit the future health of the child.

The survey

Sample characteristics

A total of 1227 parents and 1406 non-parents completed the survey. Their characteristics are presented in Table 2. There was an almost equal distribution between males and females. Parents were significantly older than non-parents. The mean age was 45 years (standard deviation (SD) 8) for parents and 48 years (SD 15) for non-parents. A higher proportion of parents than non-parents had a long level of education and was married/cohabiting. The majority of persons with mental disorders were diagnosed with depression, anxiety, or obsessive compulsive disorder by a doctor. The majority of relatives was first or second degree relatives (data not shown). No statistically significant differences were found for gender and stakeholder group membership between parents and non-parents (Table 2). Thus, parents were older, better educated (due to their age), and—not surprisingly—married.

Table 2 Socioeconomic characteristics of survey respondents

	Parents (n = 1227) % (n*)	Non-parents (n = 1406) % (n*)	p value**
Sex			0.37
Female	52 (634)	53 (751)	
Male	48 (591)	47 (653)	
Age groups, years			*< 0.0001*
20–40	28 (347)	36 (504)	
41–60	70 (852)	37 (526)	
61–76	2 (25)	27 (373)	
Age, mean (SD)	45 (8)	48 (15)	
Level of high education***			*< 0.0001*
None higher education	2 (17)	3 (34)	
Short higher(< 3 years)	28 (348)	29 (409)	
Medium higher (3–4 years)	29 (356)	32 (449)	
Long higher (> 4 years)	38 (470)	31 (439)	
Other education	3 (35)	5 (74)	
Marital status			*< 0.0001*
Married/cohabiting	84 (1032)	52 (727)	
Partnership	6 (70)	17 (241)	
Single	10 (122)	31 (437)	
Stakeholder groups			0.23
Persons with mental disorders	8 (103)	10 (138)	
Relatives to persons with mental disorders	26 (322)	25 (348)	
Blood donors	61 (749)	62 (871)	
Clinical geneticists	2 (18)	1 (10)	
Psychiatrists	3 (35)	2 (39)	

Socio-economic characteristics of the sample; specified in percentage and number of participants divided by parents, non-parents, and total. Values that were significant at $p < 0.05$ are set in italics
*n varies because of missing data
**χ^2
***Children must receive 10 years of compulsory education in Denmark

Attitudes towards children as research subjects

Table 3 shows the respondents' attitudes towards children as research subjects in genomic research and the association with parenthood, age, educational level, marital status, and stakeholder group.

Significantly more males (76%) than females (63%) and significantly less persons with short higher education (67%) and other education (61%) than persons with long higher education responded that children should be able to participate in genomic research.

A consistent finding was that parents were more positive towards this statement than non-parents. Persons in the young age group (20–40 years), the older age group (61–76 years), single persons, persons with mental disorders, psychiatrists, and clinical geneticists were more negative towards the statement than persons in the mid age group (41–60 years), persons in a partnership, and blood donors (Table 3).

Attitudes towards the return of pertinent and incidental findings in children

As shown in Table 3, a majority of parents and non-parents responded that children should be able to participate in genomic research. However, 24% parents and 27% non-parents responded that children should not be able to participate in genomic research, whereas 15% parents and 17% non-parents did not know (Figs. 1 and 2).

The parents and non-parents who responded in the survey that children should be able to participate in genomic research (N = 1526) were asked about their opinion on the return of pertinent and incidental findings (Figs. 1 and 2). Firstly, both for pertinent (N = 1441) and incidental findings (N = 1399), parents and non-parents agreed that the parents should have the opportunity to receive both types of findings. Secondly, the child should be of legal age before s/he should be provided with the opportunity to receive pertinent (N = 738) and incidental findings (N = 752).

Discussion

Main findings

Two main discussion points emerged from the study. These will be discussed below.

Table 3 Attitudes towards children as research subjects in genomic research

Do you think that children (up to age 18) should be able to participate in genomic research?

	N	Yes % (n)	OR $_{adj}$ (95% CI)*	p value
Parenthood	2205			
Non-parent		67 (776)	1.00	
Parent		72 (750)	1.19 (0.96–1.48)	0.12
Sex	2204			
Female		63 (714)	1.00	
Male		76 (813)	1.77 (1.46–2.14)	*< 0.0001*
Age groups, years	2202			
41–60		70 (806)	1.00	
20–40		69 (501)	0.95 (0.76–1.19)	0.67
61–76		68 (218)	0.92 (0.68–1.24)	0.58
Level of high education	2206			
Long higher (> 4 years)		73 (562)	1.00	
None higher education		60 (25)	0.54 (0.29–1.06)	0.07
Short higher (< 3 years)		67 (424)	0.75 (0.59–0.96)	*0.03*
Medium higher (3–4 years)		69 (456)	0.87 (0.68–1.11)	0.26
Other education		61 (60)	0.59 (0.38–0.93)	*0.02*
Marital status	2204			
Married/cohabiting		70 (1034)	1.00	
Partnership		70 (186)	1.09 (0.79–1.50)	0.62
Single		65 (305)	0.95 (0.75–1.22)	0.69
Stakeholder group	2207			
Blood donors		70 (956)	1.00	
Persons with mental disorder		66 (130)	1.00 (0.73–1.39)	0.99
Relatives		70 (386)	1.05 (0.84–1.31)	0.69
Clinical geneticists		67 (18)	0.88 (0.40–2.12)	0.77
Psychiatrists		69 (38)	0.84 (047–1.59)	0.59

Multivariable logistic regression was used to assess the attitudes towards children as research subjects in genomic research and the association with parenthood, stakeholder group, gender, age, educational level, and marital status, with 95% CI and p value of < 0.05. Values that were significant at $p < 0.05$ are set in italics
*Adjusted for parenthood, stakeholder group, gender, age, educational level, and marital status

Children as genomic research subjects

In a hypothetical scenario, participants are overall likely to hold positive views on the question whether a child should be able to participate in genomic research. Our findings are similar to Fernandez et al. [26], who found that the majority of participants reported that children should be able to take part in genomic research, whether the condition under study began in childhood or not and independent of the existence of effective treatment. We expected that parents in the survey would hold a more positive view and that non-parents would be more indifferent, but we found only slightly higher agreement among parents than among non-parents.

There was a significant higher agreement towards involving children in genomic research among males than females indicating that males have less protective or more chance-taking behaviors than females. Having short higher education and other education was significantly associated with less agreement toward the statement than persons with long higher education. Individuals with low level of education may have greater difficulties in understanding the consequences of genomic data and thus more worries of letting children participate in genomic research.

Return of results: the child's right to know

Participants expressed an attitude and also an expectation towards receiving research findings concerning children. In the interviews, there was no consensus whether the child or the parents should receive the findings. However, the return of findings were discussed in relation to the child's age, the child's right to an autonomous choice, the possibility of the emotional burden of knowing, and the possibility of receiving important clinical information. The majority in the survey agreed that both pertinent and incidental findings should be returned to the parents and to the child when of legal age. As parents might be more interested than non-parents in receiving clinical information, we expected that parents would take a more positive approach than non-parents. Nevertheless, we found that parents and non-parents had very similar attitudes. Thus, having children does not seem to affect stakeholder's attitudes towards return of findings in children.

Furthermore, there were no differences in participant's attitudes towards the return of pertinent or incidental findings, and they are interested in receiving both types of findings regarding the child. Our results illustrate that there are an interest in receiving information and that the child's right to an autonomous choice and the parent's interest to be informed are both important for the participants. The results in the survey study are consistent with those of Kleiderman et al. [28], who found that parents of children affected by a wide range of rare diseases expressed an interest in receiving their child's research results. However, our results further illustrate that the guideline on genomic research from The Danish National Committee on Health Research Ethics [25] is more restrictive than participants' attitudes toward the inclusion of children in genomic research. The participants accept inclusion of children in research without having direct health benefit for the child, including studies of the genetics of a disorder. Additionally, they are more positive towards the return of findings than the guideline.

Although the child does not have full autonomy when joining a genomic research project, the child will have autonomy in the future as an adult. As genomic research may provide clinical findings about both child-onset and adult-onset genetic diseases, it is important to consider the child's future and right to know and not to know. Respecting the right to an open future means that the child can make his or her own autonomous decisions when reaching adulthood [10, 14, 27, 43, 44]. Feinberg [45] holds that children have a right, while they still are children, to remain ignorant of disease predisposition until they reach adulthood, presumably capable of making a well-informed decision. While the child is still a child, these "future interests" include interests that the child will in fact come to have in the future, including future interests that might never happen [45]. The right to an open future protects the child against having important life choices determined by parents (and others) before having the ability to make them for themselves. The parents can disrupt the child's right to an open future if they act in a way that will cut off possibilities for the child in adulthood. For example, if the parents receive findings regarding breast cancer because of mutations in *BRCA1* or *BRCA2* genes, it can conflict with the child's right to an open future if knowledge is disclosed that the child would have wanted to live without as an adult. Nevertheless, withholding discoverable information from the child could also close a child's future if the child were interested in receiving the information as a child. The legal regulations in Denmark are unclear whether the return of genomic results encompasses adult-onset disorders and whether the results should be returned to the child or to the parents. Thus, it is uncertain whether parents have a right to be informed about all their child's genomic research findings, including adult-onset genetic diseases. The Clinical Sequencing Exploratory Research (CSER) Consortium and the Electronic Medical Records and Genomics (eMERGE) Network aim to develop practical strategies for addressing questions concerning the return of results in genomic research [44]. For example, CSER and eMERGE address whether adult-onset findings should be offered for pediatric research participants. They conclude that, during the consent process, the parents should be offered the choice of whether or not to have the adult-onset actionable incidental findings returned along with counseling [44]. In the Danish clinical setting, children are not being offered genetic testing for adult-onset conditions. However, the Danish legislation in this area is unclear concerning the research setting. The results of this study suggest that it is important to incorporate the child's right to an open future and to address the return of information of adult-onset genetic diseases in the (legal) discussion of children as research subjects.

Strengths and limitations

It is a strength of the study design that we used both interview and survey data because the two methods contribute different insights to the topic. The interviews provided an in-depth perspective concerning the attitudes and views of the informants, and the survey data provided a broader and more general attitudinal perspective. It is also a strength that the first part of our data collection was used to inform the second part as the qualitative interviews inspired us to include additional items in the questionnaire survey to focus its scope. In the survey, we were measuring genomic research broadly and not specific at psychiatric genomic research. However, no larger studies have investigated the attitudes of patients, relatives, and health professionals in psychiatry and genetics towards inclusion of children in genomic research. The results illustrate that persons with mental disorder and relatives do not differ in attitudes from blood donors or clinical geneticists. This study thus contributes with new original knowledge in relation to stakeholders in psychiatry in Denmark. Each mode of the data collection provided preferential access to certain parts of the population. It is a strength that we used web-based survey since this method is known to recruit hard-to-reach groups [46, 47]. At the same time, using a web-based survey requires the participants to be familiar and have access to smartphones, tables, laptops, or computers, and this may have excluded potential participants without these skills or without access to such technical equipment or to the internet. Therefore, it is likely that the included stakeholders comprise a more homogeneous group, more positive of genomic research than potential participants who did not participate.

The study had some further limitations. Firstly, the survey and interviews had the inherent weaknesses of measuring hypothetical scenarios as the participants did not have direct experience with genomic research. "Real" requests to use children in genomic research could produce different results. Responses to hypothetical scenarios often anticipate behaviors and future intentions rather than actual behavior [48]. Therefore, studies are needed to test actual behavior in a real-life situation. However, we still think hypothetical scenario methodology is an important tool for predicting interest in research and to understand attitudes. Secondly, participants were included on a voluntary basis, and we cannot assess the effects of nonresponse. Participants may be more favorably inclined towards genomic research than the general population since the participants must be willing to take part in this study and thereby have an influence on the generalizability of the study results. The survey might have been too difficult or too long to maintain the engagement of the participants. This potential selection bias could mean that the results do not represent all possible stakeholders in genomic research in Denmark. Thirdly, persons with schizophrenia are difficult to recruit into trials [49]. In this study, it was difficult to recruit persons with schizophrenia to a focus-group interview. As a result, we decided to conduct individual interviews in this group of informants. A focus-group interview allows for interaction between the informants and may have given some more knowledge about the mechanisms behind certain attitudes expressed by the informants. Furthermore, it is possible that other core concepts would have emerged including persons with different mental disorders and psychiatrists because it is not certain that the focus-group interview went to saturation with the included interviews. Fourthly, we used a broad age category of children in the survey. As the age of the child could influence the participant's attitudes, it is important to divide the children into several age categories in future studies. Fifthly, the participants were not specifically recruited because of their status as parents or non-parents. Instead, in consideration of the overall aim, we focused on recruiting a wide range of stakeholders in psychiatric genomic research: persons with mental disorders, relatives, blood donors, psychiatrists, and clinical geneticists. Finally, it could have been interesting to study the participants' attitudes towards the return of pertinent and incidental findings in children subdivided into childhood-onset and adult-onset diseases.

Conclusion

Participants generally reported positive views on the inclusion of children in genomic research. Additionally, our results illustrate that both the child's right to autonomy and the parent's interest in receiving information are important and valuable factors for the participants. They hold more positive and comprehensive views than the current Danish Guidelines on the inclusion of children and return of findings. However, having children does not affect the participant's attitudes towards the inclusion of children as research subjects in genomic research.

Genomic research on children raises complex ethical issues and it is important to consider the view of potential stakeholders who rarely get a voice. Despite the Danish Guidelines, the issue of when to include children as research subjects and how to deliver research findings is still unclear and subject to legal challenges. We hope that sharing their attitudes will help the mobilization of knowledge on children as genomic research subjects, the return of findings regarding adult-onset genetic diseases, and the children's right to an open future. Similarly, we think it is important to address these issues in the legal discussion of children as research subjects.

Acknowledgements

The authors would like to thank Professor Mette Hartlev for valuable constructive comments on an earlier draft of this article.

Funding

We would like to acknowledge the Deciphering Developmental Disorders (DDD) study for funding the films that were used in this research. The DDD study presents independent research commissioned by the Health Innovation Challenge Fund (grant number: HICF-1009-003), a parallel funding partnership between the Wellcome Trust and the Department of Health, UK. This study was supported financially by the Lundbeck Foundation Initiative for Integrative Psychiatric Research, *i*PSYCH, Denmark (grant number: R155-2014-1724) and Aarhus University. The funding sources were not involved in the study design, data collection and analysis, writing of the manuscript, or decision regarding publication.

Authors' contributions

AS, MWB, TFH, and OM conceived the study and its design. AS, KSB, and HU recruited participants from the Danish Blood Donor Study. AS recruited the participants for the interviews and survey, collected the data, and performed the data analyses and the statistical analyses. AS drafted the manuscript. All authors reviewed and commented on drafts of the manuscript, and all authors approved the final version of the manuscript.

Competing interests

The authors declare that they have no competing interests.

Author details

[1]Department of Clinical Medicine, Psychosis Research Unit, Aarhus University Hospital, Skovagervej 2, 8240 Risskov, Denmark. [2]The Lundbeck Foundation Initiative for Integrative Psychiatric Research, iPSYCH, Copenhagen, Denmark. [3]Department of Political Science, Copenhagen University, Copenhagen, Denmark. [4]Department of Clinical Immunology, Copenhagen University Hospital, Copenhagen, Denmark. [5]Institute for Biological Psychiatry, Mental Health Centre Sct. Hans, Copenhagen University Hospital, Copenhagen, Denmark. [6]Danish Headache Center, Department of Neurology, Rigshospitalet-Glostrup, Copenhagen, Denmark.

References

1. Christenhusz GM, Devriendt K, Dierickx K. To tell or not to tell? A systematic review of ethical reflections on incidental findings arising in genetics contexts. Eur J Hum Genet. 2013;21:248–55.
2. Klitzman R, Appelbaum PS, Fyer A, Martinez J, Buquez B, Wynn J, et al. Researchers' views on return of incidental genomic research results: qualitative and quantitative findings. Genet Med. 2013;15:888–95.
3. Middleton A, Morley KI, Bragin E, Firth HV, Hurles ME, Wright CF, et al. Attitudes of nearly 7000 health professionals, genomic researchers and publics toward the return of incidental results from sequencing research. Eur J Hum Genet. 2016;24:21–9.
4. Parens E, Appelbaum P, Chung W. Incidental findings in the era of whole genome sequencing? Hastings Cent Rep. 2013;43:16–9.
5. Ryan KA, De Vries RG, Uhlmann WR, Roberts JS, Gornick MC. Public's views toward return of secondary results in genomic sequencing: It's (almost) all about the choice. J Genet Couns. 2017;26:1197–212.
6. Solberg B, Steinsbekk KS. Managing incidental findings in population based biobank research. J Epidemiol. 2012;21:195–202.
7. Yu J-H, Harrell TM, Jamal SM, Tabor HK, Bamshad MJ. Attitudes of genetics professionals toward the return of incidental results from exome and whole-genome sequencing. Am J Hum Genet. 2014;95:77–84.
8. Tan N, Amendola LM, O'Daniel JM, Burt A, Horike-Pyne MJ, Boshe L, et al. Is "incidental finding" the best term? A study of patients' preferences. Genet Med. 2017;19:176–81.

9. Mitchell C, Ploem C, Chico V, Ormondroyd E, Hall A, Wallace S, et al. Exploring the potential duty of care in clinical genomics under UK law. Med Law Int. 2017;17:158–82.
10. Holm IA. Pediatric issues in return of results and incidental findings: Weighing autonomy and best Interests. Genet Test Mol Biomark. 2017;21: 155–8.
11. Nørgaard-Pedersen B, Hougaard DM. Storage policies and use of the Danish Newborn Screening Biobank. J Inherit Metab Dis. 2007;30:530–6.
12. Nørgaard-Pedersen B, Simonsen H. Biological specimen banks in neonatal screening. Acta Pædiatrica. 1999;88:106–9.
13. Pedersen CB, Bybjerg-Grauholm J, Pedersen MG, Grove J, Agerbo E, Bækvad-Hansen M, et al. The iPSYCH2012 case–cohort sample: new directions for unravelling genetic and environmental architectures of severe mental disorders. Mol Psychiatry. 2017; https://doi.org/10.1038/mp.2017.196.
14. Hens K. Whole genome sequencing of children's DNA for research: points to consider. J Clin Res Bioeth. 2011; https://doi.org/10.4172/2155-9627.1000106e.
15. Hens K, Van El CE, Borry P, Cambon-Thomsen A, Cornel MC, Forzano F, et al. Developing a policy for paediatric biobanks: principles for good practice. Eur J Hum Genet. 2013;21:2–7.
16. Hens K, Cassiman J-J, Nys H, Dierickx K. Children, biobanks and the scope of parental consent. Eur J Hum Genet. 2011;19:735–9.
17. Sénécal K, Thys K, Vears DF, Van Assche K, Knoppers BM, Borry P. Legal approaches regarding health-care decisions involving minors: implications for next-generation sequencing. Eur J Hum Genet. 2016;24:1559–64.
18. Zawati MH, Rioux A. Biobanks and the return of research results: Out with the old and in with the new? J Law Med Ethics. 2011;39:614–20.
19. Anderson JA, Meyn MS, Shuman C, Zlotnik Shaul R, Mantella LE, Szego MJ, et al. Parents perspectives on whole genome sequencing for their children: qualified enthusiasm? J Med Ethics. 2016; https://doi.org/10.1136/medethics-2016-103564.
20. Dodson DS, Goldenberg AJ, Davis MM, Singer DC, Tarini BA. Parent and public interest in whole-genome sequencing. Public Health Genomics. 2015;18:151–9.
21. Sabatello M, Appelbaum PS. Raising genomic citizens: Adolescents and the return of secondary genomic findings. J Law Med Ethics. 2016;44:292–308.
22. Newson AJ. Whole genome sequencing in children: ethics, choice and deliberation. J Med Ethics. 2017; https://doi.org/10.1136/medethics-2016-103943.
23. The Danish Ministry of Health. Bekendtgørelse af lov om videnskabsetisk behandling af sundhedsvidenskabelige forskningsprojekter (in Danish) [Danish Act on Research Ethics Review of Health Research Projects and later amendments]. (2017). https://www.retsinformation.dk/Forms/R0710.aspx?id= 192671. Accessed 17 Sep 2017.
24. The Danish Ministry of Health. Bekendtgørelse om information og samtykke til deltagelse i sundhedsvidenskabelige forskningsprojekter samt om anmeldelse af og tilsyn med sundhedsvidenskabelige forskningsprojekter (in Danish) [Executive Order no. 1464 of 2 December 2016 on information and consent at inclusion of trial subjects in biomedical research projects and the notification and supervision of health research projects]. (2017).https://www.retsinformation.dk/Forms/R0710.aspx?id=185233. Accessed 12 Jun 2017.
25. The Danish National Committee on Health Research Ethics. Vejledning om genomer af 1. Februar 2017 (in Danish) [Guidelines on genomes of 1 February 2017]. (2017). http://www.nvk.dk/emner/genomer/vejledning-om-genomer. Accessed 16 Sep 2017.
26. Fernandez CV, Bouffet E, Malkin D, Jabado N, O'Connell C, Avard D, et al. Attitudes of parents toward the return of targeted and incidental genomic research findings in children. Genet Med. 2014;16:633–40.
27. Fernandez CV, O'Connell C, Ferguson M, Orr AC, Robitaille JM, Knoppers BM, et al. Stability of attitudes to the ethical issues raised by the return of incidental genomic research findings in children: a follow-up study. Public Health Genomics. 2015;18:299–308.
28. Kleiderman E, Knoppers BM, Fernandez CV, Boycott KM, Ouellette G, Wong-Rieger D, et al. Returning incidental findings from genetic research to children: views of parents of children affected by rare diseases. J Med Ethics. 2014;40:691–6.
29. Yu TW, Chahrour MH, Coulter ME, Jiralerspong S, Okamura-Ikeda K, Ataman B, et al. Using whole-exome sequencing to identify inherited causes of autism. Neuron. 2013;77:259–73.
30. Demontis D, Lescai F, Børglum A, Glerup S, Østergaard SD, Mors O, et al. Whole-exome sequencing reveals increased burden of rare functional and disruptive variants in candidate risk genes in individuals with persistent attention-deficit/hyperactivity disorder. J Am Acad Child Adolesc Psychiatry. 2016;55:521–3.
31. Statistics Denmark. Statistikdokumentation for Boligopgørelsen 2017 (in Danish) [Statistical documentation for the housing inentory]. (2017). www.dst.dk/Site/Dst/SingleFiles/kvaldeklbilag.aspx?filename=794ee8d0-1b80-4eb0-a077-1a0c22aca5ecBoligopg%C3%B8relsen. Accessed 7 Dec 2017
32. Johnson RB, Onwuegbuzie AJ. Mixed methods research: a research paradigm whose time has come. Educ Res. 2004;33:14–26.
33. Pedersen OB, Erikstrup C, Kotzé SR, Sørensen E, Petersen MS, Grau K, et al. The Danish Blood Donor Study: a large, prospective cohort and biobank for medical research. Vox Sang. 2012;102:271.
34. The Danish Blood Donor Study. The Danish Blood Donor Study. http://www.dbds.dk/defaultuk.htm (2016). Accessed 16 Sep 2017.
35. QSR International. Nvivo. Daresbury: Daresbury, Cheshire, WA4 4AB United Kingdom; 2017.
36. Strauss AL, Corbin JM. Basics of qualitative research: techniques and procedures for developing grounded theory. 2nd ed. Thousand Oaks: Sage Publications; 1998.
37. Middleton A, Wright CF, Morley KI, Bragin E, Firth HV, Hurles ME, et al. Potential research participants support the return of raw sequence data. J Med Genet. 2015;52:571–4.
38. Middleton A, Bragin E, Parker M. Finding people who will tell you their thoughts on genomics—recruitment strategies for social sciences research. J Community Genet. 2014:291–302.
39. Middleton A, Bragin E, Morley KI, Parker M. Online questionnaire development: using film to engage participants and then gather attitudes towards the sharing of genomic data. Soc Sci Res. 2013:211–23.
40. Middleton A, Parker M, Wright CF, Bragin E, Hurles ME, On behalf of the DDD Study. Empirical research on the ethics of genomic research. Am J Med Genet A. 2013;161:2099–101.
41. Sundby A, Boolsen MW, Burgdorf KS, Ullum H, Hansen TF, Middleton A, et al. Stakeholders in psychiatry and their attitudes toward receiving pertinent and incident findings in genomic research. Am J Med Genet A. 2017;173: 2649–58.
42. SAS Institute Inc. SAS® 9.4. Cary; 2017.
43. Davis DS. Genetic dilemmas and the child's right to an open future. Hastings Cent Rep. 1997;27:7.
44. Jarvik GP, Amendola LM, Berg JS, Brothers K, Clayton EW, Chung W, et al. Return of genomic results to research participants: The floor, the ceiling, and the choices in between. Am J Hum Genet. 2014;94:818–26.
45. Feinberg J. The child´s right to an open future. In: Whose child? Children´s right, parental authority, and state power. Auth. State Power. 1st ed. New Jersey: Littlefield, Adams & Co; 1980. p. 124–53.
46. Barratt MJ, Ferris JA, Lenton S. Hidden populations, online purposive sampling, and external validity: taking off the blindfold. Field Methods. 2015; 27:3–21.
47. Frippiat D, Marquis N, Wiles-Portier E. Web surveys in the social sciences: an overview. *Population*. 2010;65:285–311.
48. Persky S, Kaphingst KA, Condit CM, McBride CM. Assessing hypothetical scenario methodology in genetic susceptibility testing analog studies: a quantitative review. Genet Med. 2007;9:727–38.
49. Jørgensen R, Munk-Jørgensen P, Lysaker PH, Buck KD, Hansson L, Zoffmann V. Overcoming recruitment barriers revealed high readiness to participate and low dropout rate among people with schizophrenia in a randomized controlled trial testing the effect of a Guided Self-Determination intervention. *BMC Psychiatry.* 2014;14:28.

16

Genome-scale portrait and evolutionary significance of human-specific core promoter tri- and tetranucleotide short tandem repeats

N. Nazaripanah[1], F. Adelirad[1], A. Delbari[1], R. Sahaf[1], T. Abbasi-Asl[2] and M. Ohadi[1*]

Abstract

Background: While there is an ongoing trend to identify single nucleotide substitutions (SNSs) that are linked to inter/intra-species differences and disease phenotypes, short tandem repeats (STRs)/microsatellites may be of equal (if not more) importance in the above processes. Genes that contain STRs in their promoters have higher expression divergence compared to genes with fixed or no STRs in the gene promoters. In line with the above, recent reports indicate a role of repetitive sequences in the rise of young transcription start sites (TSSs) in human evolution.

Results: Following a comparative genomics study of all human protein-coding genes annotated in the GeneCards database, here we provide a genome-scale portrait of human-specific short- and medium-size (≥ 3-repeats) tri- and tetranucleotide STRs and STR motifs in the critical core promoter region between − 120 and + 1 to the TSS and evidence of skewing of this compartment in reference to the STRs that are not human-specific (Levene's test $p < 0.001$). Twenty-five percent and 26% enrichment of human-specific transcripts was detected in the tri and tetra human-specific compartments (mid-$p < 0.00002$ and mid-$p < 0.002$, respectively).

Conclusion: Our findings provide the first evidence of genome-scale skewing of STRs at a specific region of the human genome and a link between a number of these STRs and TSS selection/transcript specificity. The STRs and genes listed here may have a role in the evolution and development of characteristics and phenotypes that are unique to the human species.

Keywords: Short tandem repeat, Core promoter, Human-specific, Trinucleotide, Tetranucleotide

Introduction

Speciation and evolution are, at least in part, due to the plasticity (expansion or contraction) of short tandem repeats (STRs)/microsatellites, which can function as "tuning knobs" in response to the environment or other genes [1–3]. In line with the above, certain STRs are directionally expanded in the human species or co-occur identically in related taxa such as primates [4–8]. Genes that contain STRs in their promoters have higher expression divergence compared to genes with fixed or no STRs in the gene promoters [9]. Recent reports indicate a role of repetitive sequences in the rise of young transcription start sites (TSSs) in human evolution [10–12].

Preliminary data on the sequencing of a number of "exceptionally long" STRs (≥ 6-repeats), which compose 1–2% of all human core promoter STRs [3], support critical evolutionary adaptive roles for a number of these STRs. Human specificity of the predominant allele of the *RIT2* core promoter STR in the human species, the presence of the shortest allele of this STR (5-repeat) in hunter-gatherer humans (BUSHMAN KB1: rs113265205), the lack of this allele in the agricultural modern humans (Genome Aggregation database: gnomad.broadinstitute.org), and its co-occurrence with schizophrenia provide the

* Correspondence: mi.ohadi@uswr.ac.ir; ohadi.mina@yahoo.com
[1]Iranian Research Center on Aging, University of Social Welfare and Rehabilitation Sciences, Tehran, Iran

first indication of STR allele selection in humans [13]. A link between the *CYTH4* core promoter STR (the longest tetranucleotide STR identified in a human gene core promoter) with the Old World monkeys and Apes and evidence of extreme "disease-only" genotypes at this STR with schizophrenia [14] provide the first link between a primate-specific STR and higher-order brain functions in human. The "exceptionally long" CA-repeat in the core promoter of *SCGB2B2* is another example of directional STR expansion in the Old World monkeys and Apes [5]. The *PAXBP1* gene is an extreme example in which expansion of a core promoter CT-repeat occurs in the Old World monkeys and reaches maximum length and complexity in human; OMIM: 617621 [4].

As "exceptionally long" STRs may be subject to natural selection, short- and medium-size alleles ($\geq$ 3-repeats) might have had similar fate. This is indicated by the predominance of specific short- and medium-size penta- and hexanucleotide STRs and their cognate transcription factors (TFs) in the critical core promoter interval [15]. Indeed, shortening of a number of STRs and their identical co-occurrence is linked to the evolution of primates [8]. In line with the above findings, repeats associated with younger human TSSs tend to be shorter than those in older TSSs [10]. In the study reported here, we present genome-scale data on two categories of STRs, i. e., tri- and tetranucleotide STRs, and their implication in human evolution.

Materials and methods

The interval between − 120 and + 1 to the TSS of all human protein-coding genes annotated in the GeneCards database (version 3.0) (www.genecards.org) was screened for tri- and tetranucleotide STRs of $\geq$ 3-repeats, based on the Ensembl database (versions 87-91) (asia.ensembl.org) and using the Microsatellite Repeats Finder at the following link: http://insilico.ehu.es/mini_tools/microsatellites/

The evolutionary status of the identified STRs was analyzed in 25 species (N), including primates ($N = 5$), non-primate mammals ($N = 12$), birds and reptiles ($N = 5$), amphibians ($N = 1$), and fish ($N = 2$), based on the Ensembl database.

Human specificity of transcripts was evaluated based on the multiple and pair-wise %identity scoring of the TSS-flanking 5′ untranslated region (UTR), using the sequence alignment program Clustal Omega (https://www.ebi.ac.uk/Tools/msa/clustalo), and the overall composition of the transcript and encoded protein (i.e., length of the transcript, number of exons and amino acids). The threshold of sequence identity was set at 50%, which was based on the comparison of two randomly selected and unrelated sequences in the human genome.

The p value for the skewing of the human-specific STR compartment was calculated using Levene's equality of variances test.

The p values for transcript enrichment were calculated using the two by two table analysis;

the human-specific tri- and tetranucleotide STR groups were compared against corresponding randomly selected STRs from the non-human-specific STRs. The comparison was set based on the sample size of the human-specific STRs (n) and the sample size of the non-human compartments (1.5n).

Results

Overall prevalence of tri- and tetranucleotide STR motifs across human protein-coding core promoter sequences

In total, 56 and 82 STR motifs were detected for the tri- and tetranucleotide repeats, respectively (Figs. 1 and 2). The most prevalent tri- and tetranucleotide STR motifs across the human protein-coding gene core promoters were GGC and GGGC, respectively (Figs. 1 and 2). In the category of non-GC STRs, GGA and TCCC were the most prevalent tri- and tetranucleotides, respectively.

Skewing of the human-specific core promoter tri- and tetranucleotide STRs

A significant skewing of the tri- and tetranucleotide STR distribution was found in the human-specific tri- (Fig. 1) and tetranucleotide (Fig. 2) compartments (Levene's $p < 0.001$). While the most prevalent tri- and tetranucleotide repeats in the non-human-specific category were the GGC- and GGGC-repeats, respectively, the most prevalent human-specific STRs were of the GCC and CTCC motifs, respectively. Disproportionate distribution of human-specific STRs was also detected in other STRs such as CCT, GAA, CTCC, GTTT, and GAAA.

The human-specific tri- and tetranucleotide STRs were of a wide range of motifs, e.g., the CCA motif in *ADCY6*, the TCCC motif in *ARHGEF35*, GCCC in *DRD2*, and GTTT in *MCTP2* (Tables 1 and 2).

In a number of instances, not only the STR, but also the genes containing those STRs, were human-specific (e.g., *ARHGEF35*, *AMY1C*, and *C1orf204*). Furthermore, a number of the tri- and tetranucleotide STRs were found to be unique to the human species at the specified interval of − 120 to + 1 TSS. For example, in the tetranucleotide compartment, CACC, GACA, CCGG, GATA, TCTG, GGCT, and TTTA STRs were detected in human only.

Enrichment of human-specific transcripts at the human-specific STR compartment

Based on sequence comparison and the overall composition of the transcript and encoded protein, 25

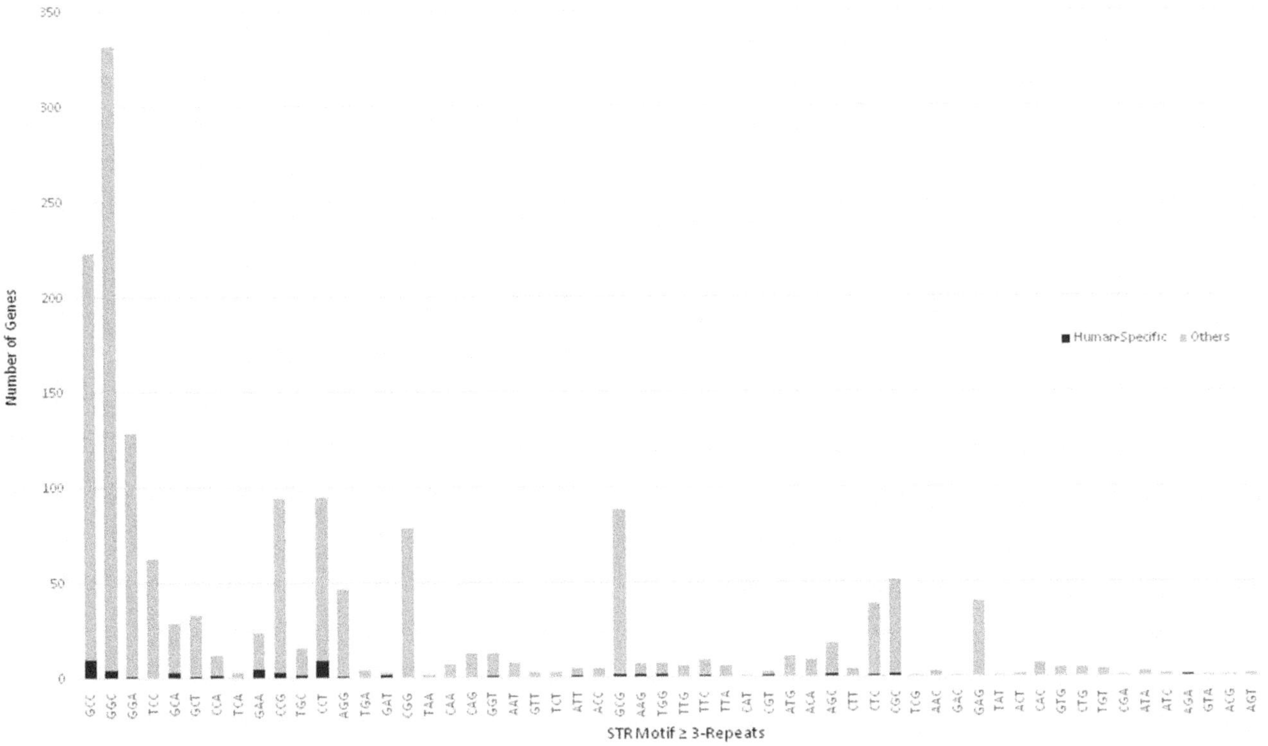

Fig. 1 Genome-scale prevalence of human protein-coding core promoter trinucleotide STRs and significant skewing of the human-specific STR compartment

and 26% of the transcripts in the tri and tetra human-specific compartments were found to be human-specific (mid-$p < 0.00002$ and mid-$p < 0.002$), respectively). The %identity score of multiple sequence alignment for the human-specific transcripts was 0 (exemplified in Fig. 3), and pair-wise analysis (exemplified in Fig. 4) resulted in %identity scores ranging from 37 to 48%. In the trinucleotide category, 14 genes, *MPRIP, NPAS1, PAQR9, PRSS1, R3HDM2, TMEM99, ZSCAN30, C22orf24, ECSCR, AMY1C, DDX58, C1orf204, RGPD6*, and *LCE2B*, contained human-specific transcripts. In the tetranucleotide category, five genes, *DRD2, DUX4, TEAD4, ARL17B*, and *ARHGEF35*, contained human-specific transcripts.

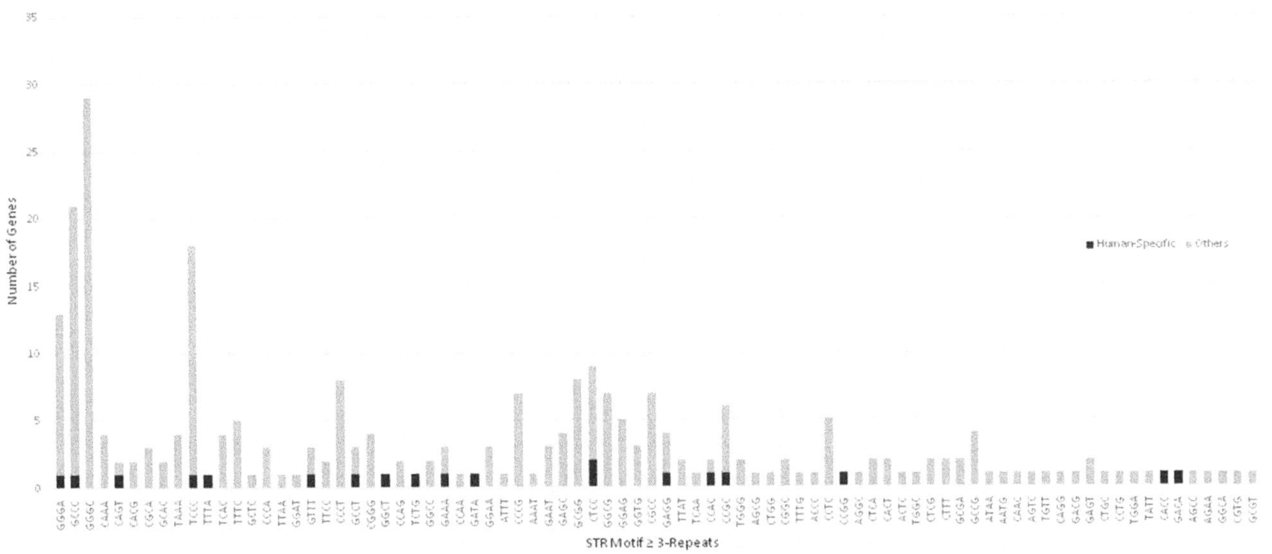

Fig. 2 Genome-scale prevalence of human protein-coding core promoter tetranucleotide STRs and significant skewing of the human-specific STR compartment

Table 1 Genome-scale human-specific core promoter trinucleotide STRs

Human gene symbol	Ensembl transcript ID	Variant no.	STR formula	
ADCY6	ENST00000307885.4	201	− 48 (CCA)3	
AMY1C	ENST00000370079.3	201	− 79 (ATT)3	
APBB1	ENST00000311051.7	202	− 11 (GCC)3	
BRINP2 (FAM5B)	ENST00000361539.4	201	− 26 (GGC)5	
BVES	ENST00000446408.2	203	− 35 (CCT)3	
CCDC178 (C18orf34)	ENST00000300227.12	201	− 79 (AGC)3	− 57 (CGC)5
CDH4	ENST00000543233.2	201	− 31 (CCT)3	
CIAPIN1	ENST00000563341.1	202	− 39 (CCT)3	
CNTNAP2	ENST00000361727.7	201	− 98 (TGC)3	
CST4	ENST00000217423.3	201	− 83 (GGA)3	
CYP4A11	ENST00000310638.8	201	− 86 (CCT)3	
C1orf204	ENST00000368102.5	201	− 11 (AAG)3	
C22orf24	NM_015372.2.1		− 98 (GCA)3	
C3AR1	ENST00000546241.1	202	− 70 (AGA)3	
DDX58	ENST00000379868.5	201	− 33 (CCT)3	
ECSCR	NM_001077693.3.1		− 121 (CCA)3	
GRIN2D	ENST00000263269.3	201	− 67 (GCC)3	
GSDMB	ENST00000394175.6	203	− 55 (GGC)3	
INPP4B	ENST00000503927.5	202	− 88 (CGC)3	
KBTBD12	ENST00000407609.7	204	− 57 (CCT)3	
KIAA1211	ENST00000504228.5	205	− 81 (AAG)3	
KRAS	ENST00000311936.7	202	− 89 (GAA)3	
KTN1	ENST00000395308.5	202	− 70 (GCG)9	
LACTBL1	ENST00000426928.6	201	− 58 (GAA)3	
LCE2B	ENST00000368780.3	201	− 79 (CCT)3	
LCOR (C10orf12)	ENST00000356016.7	202	− 97 (CCT)3	− 108 (GCC)3
MPRIP	ENST00000466186.2	209	− 88 (GCA)11	
MSANTD3 (C9orf30)	ENST00000374885.5	201	− 25 (GCC)7	
NPAS1	ENST00000439365.6	201	− 135 (GAA)9	
OR4X1	ENST00000320048.1	201	− 100 (GAT)3	
PABPC1L2B	ENST00000373521.3	201	− 60 (GCC)3	
PAQR9	ENST00000498470.1	203	− 29 (TGC)3	
PRSS1	ENST00000492062.1	205	− 96 (GAT)3	
RGPD6	ENST00000455695.1	205	− 70 (GGC)5	
RNF215	ENST00000215798.10	201	− 123 (GCT)5	
R3HDM2	ENST00000448732.1	208	− 27 (GCC)3	
SCN3B	ENST00000299333.7	201	− 29 (GGT)3	
SERPINB9	ENST00000380698.4	201	− 12 (GCA)3	
SIGLEC7	ENST00000305628.7	201	− 80 (TTC)3	
SPATC1L (C21orf56)	ENST00000330205.10	202	− 37 (TGG)4	−90 (TGG)4
STUB1	ENST00000219548.8	201	− 91 (GCC)3	
SUMF1	ENST00000272902.9	201	− 102 (AGC)3	
TEX12	ENST00000280358.4	201	− 105 (TGG)3	

Table 1 Genome-scale human-specific core promoter trinucleotide STRs *(Continued)*

Human gene symbol	Ensembl transcript ID	Variant no.	STR formula	
TMEM99	ENST00000301665.7	201	– 32 (CCG)3	– 47 (CCG)3
			– 59 (CCG)3	– 83 (CCG)3
			– 110 (CCG)3	– 125 (CCG)3
			– 48 (GCC)3	– 60 (GCC)3
			– 84 (GCC)3	– 126 (GCC)3
TNNC2	ENST00000372557.1	202	– 53 (GCC)3	
TPTE	ENST00000427445.6	201	– 110 (GCG)3	
TRBJ2-7	ENST00000390419.1	201	– 120 (GGC)3	
TRGV5	NC_000007.14:TRGV5:u_t_1.1		– 48 (CTC)3	
TRIM39	ENST00000376656.8	201	– 58 (CCT)4	
UAP1	ENST00000367926.8	204	– 9 (CGT)3	
VNN2	ENST00000326499.10	201	– 31 (GAA)10	
WRN	NM_000553.4.1		– 67 (GCC)3 – 69 (CCG)4	– 92 (GCC)3
WRNIP1	ENST00000618555.4	205	– 67 (CCG)3	
ZDHHC21	XM_006716760.1.1		– 32 (AGG)3	
ZSCAN30	ENST00000639929.1	212	– 57 (GAA)3	

The numbers before the brackets represent the start site of the STR in respect of the corresponding transcription start site. "Variant no" corresponds to the Ensembl isoform number

Table 2 Genome-scale human-specific core promoter tetranucleotide STRs

Human gene symbol	Ensembl transcript ID	Variant no.	STR formula
ARHGAP5	ENST00000345122.7	202	– 110 (GGGA)4
ARHGEF35	ENST00000378115.2	201	– 22 (TCCC)3
ARL17B	ENST00000622877.4	201	– 99 (CTCC)3
ATP7A	ENST00000343533.9	201	– 27 (GAGG)3
DRD2	ENST00000542616.1	207	– 54 (GCCC)3
DUX4	ENST00000565211.1	203	– 144 (GGCT)6
FAM83G	ENST00000388995.10	202	– 80 (TCTG)3
GTF2IRD2B	ENST00000614064.4	206	– 17 (GAAA)3
JCAD (KIAA1462)	ENST00000375377.1	201	– 13 (CCGG)3
MCTP2	ENST00000451018.7	203	– 102 (GTTT)3
METTL21C (C13orf39)	ENST00000267273.6	201	– 69 (CAGT)3
OR10G6	ENST00000307002.3	201	– 123 (GATA)13
PHYHD1	ENST00000308941.9	201	– 107 (TTTA)3
SAMD1	ENST00000269724.5	201	– 75 (CCGC)3
TEAD4	ENST00000540314.1	206	– 51 (CTCC)3
TRAJ49	ENST00000390488.1	201	– 124 (GCCT)7
TRDJ2	ENST00000390475.1	201	– 86 (CCAC)3
TRAV38-1	ENST00000390464.2	201	– 109 (CACC)3
TRAV7	ENST00000390429.3	201	– 111 (GACA)3

The numbers before the brackets represent the start site of the STR in respect of the corresponding transcription start site. "Variant no" corresponds to the Ensembl isoform number

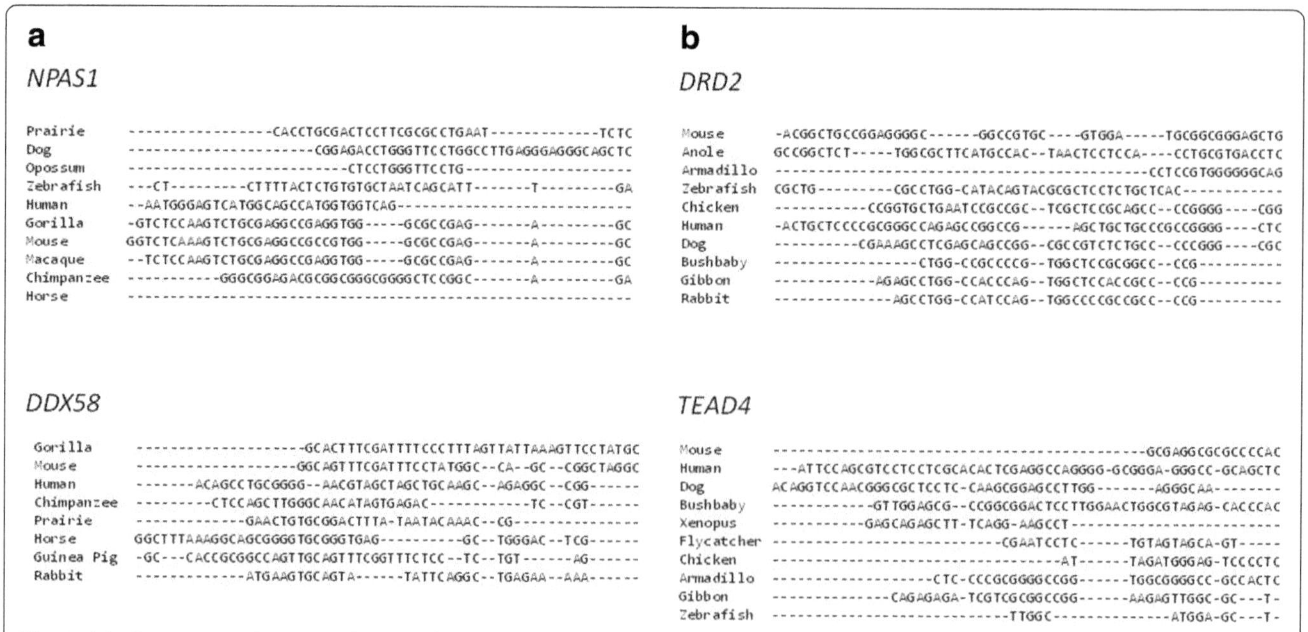

Fig. 3 Multiple sequence alignment of the TSS-flanking 5′UTRs. Examples of ClustAl Omega sequence alignment are represented in the tri- (**a**) and tetranucleotide (**b**) categories. Species inclusion was based on the information available in the Ensembl database

A number of the identified STRs were linked to non-canonical translation in the following genes, *TEAD4, ECSCR, MPRIP, PAQR9, PRSS1*, and *ZSCAN30.*

Discussion

There is an ever-growing literature on the biological and pathological implications of STRs at the inter- and intraspecies levels [16–27]. The STRs listed in the present study are genetic codes that are unique to humans and are likely to be responsible for the human-specific regulation of the relevant genes. The significant enrichment of human-specific transcripts at the human-specific STR compartment indicates a link to a mechanism for TSS selection and transcript specificity.

A number of the identified STRs such as GTTT have established repressor activity [6, 28, 29] and are differentially expanded in certain genes in the Old World

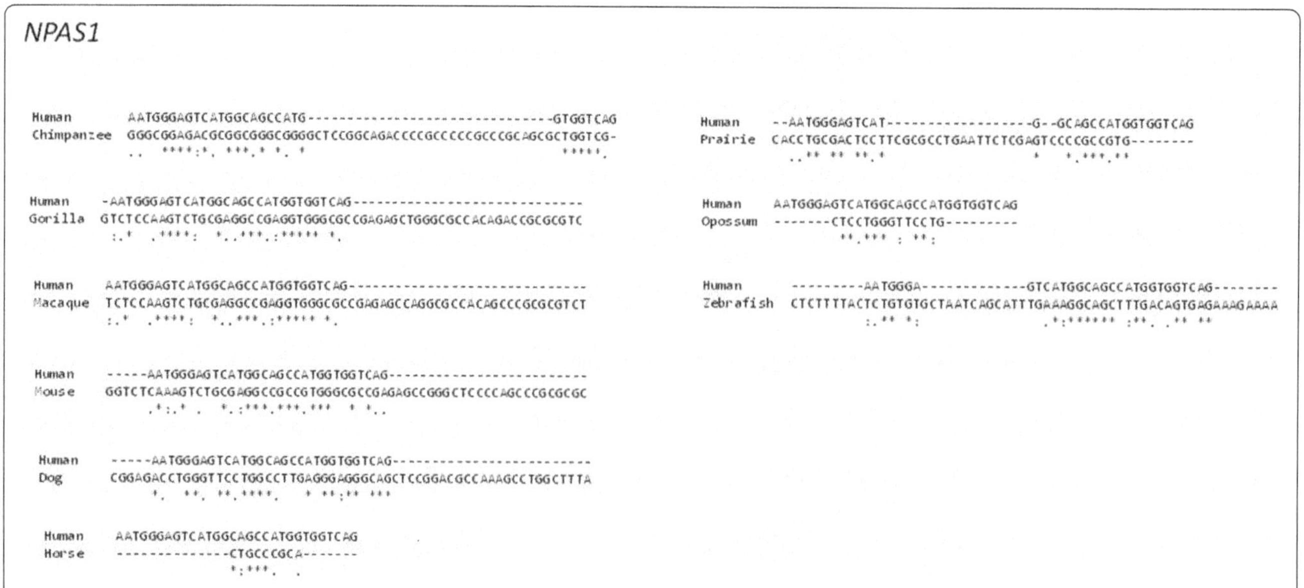

Fig. 4 Pair-wise sequence comparison of the TSS-flanking 5′UTRs. %identity scoring was performed between human and other species. Asterisks represent sequence identity

monkeys and Apes [14]. Purine STRs such as GAAA repeats are also functional in gene expression regulation, and their link to certain diseases unique to humans were previously reported [30, 31]. While the CG-rich STRs (e.g., CCG, GGGC) are subject to DNA methylation and can repress gene expression activity [32], they can also form G4 quadruplex structures, which have significant functions in gene expression regulation [33]. Several other identified STRs can form G4 structures with high overlap fraction (e.g., AGGG/CCCT, GCCC/GGGC).

It is not possible to estimate the number of crucial events that have led to the emergence of the human species. However, only a few genetic changes are needed to spur the evolution of new species in general, exemplified by the highly restricted initial divergence in butterfly hybridization models [34]. Accelerated evolution of a number of the identified genes in the present study (e.g., *DRD2*) has a well-established role in the origin of *Homo sapiens* [35]. Remarkably, a human-specific 7-amino acid transcript of this gene is flanked by a human-specific GCCC-repeat. Human-specific transcripts are increasingly recognized of having a role in the pathogenesis of diseases unique to the human species, such as schizophrenia [25, 36].

In a number of instances, not only the STR and the transcript, but also the gene containing these STRs and transcripts, were unique to humans, e.g., *AMY1C*, which is indicated in the evolution of the human phenotype during the Pleistocene [37].

For a number of the identified genes, sparse literature is available on the relevant function and pathways (e.g., *ARHGEF35*, *CXorf40A*, *C22orf24*, *TMEM99*, and *ARL17B*).

In a number of the identified genes, the STRs were linked to noncanonical (non-AUG) translation. Although the significance of this compartment is unknown for the most part, recent emerging data indicates likely biological functions [38].

The plasticity of STRs confers them unique ability to respond to adaptive evolutionary processes in a more efficient way than the quaternary codes provided by the SNSs. This potential aspect of STRs is vastly unknown at present, and it is expected that identification of STRs that have evolved differentially in humans vs. other species may pave the way for better understanding of the evolutionary implication of these highly mutable motifs.

This study warrants expansion to other vitally important gene regulatory sequences such as the distal promoter, 5′UTR, and 3′UTR. It is also necessary to sequence these STRs in characteristics and diseases that are unique to the human species. The recent reports of mass STR analysis using CRISPR/Cas9 [39] make it particularly more feasible to investigate STRs in the context of human evolution.

Conclusion

Our findings provide the first evidence of genome-scale skewing of STRs at a specific region of the human genome, and support a link between STRs and TSS selection/transcript specificity. The genes and STRs listed here may have a role in the divergence of humans from other species through the development of characteristics and phenotypes that are unique to the human species.

Abbreviations

SNS: Single nucleotide substitution; STR: Short tandem repeat; TF: Transcription factor; TSS: Transcription start site; UTR: Untranslated region

Funding

This research was funded by the University of Social Welfare and Rehabilitation Sciences, Tehran, Iran.

Authors' contributions

NN, FA, and TAA carried out the bioinformatics studies. AD and RS were the advisors of the project and helped in coordination. MO conceived the study, designed the project, supervised the analysis and wrote the manuscript. All authors read and approved the final manuscript.

Consent for publication

Not applicable.

Competing interests

The authors declare that they have no competing interests.

Author details

[1]Iranian Research Center on Aging, University of Social Welfare and Rehabilitation Sciences, Tehran, Iran. [2]Department of Biostatistics, University of Social Welfare and Rehabilitation Sciences, Tehran, Iran.

References

1. Press MO, Carlson KD, Queitsch C. The overdue promise of short tandem repeat variation for heritability. Trends Genet. 2014;30(11):504–12.
2. King DG. Evolution of simple sequence repeats as mutable sites. Adv Exp Med Biol. 2012;769:10-25.
3. Ohadi M, Mohammadparast S, Darvish H. Evolutionary trend of exceptionally long human core promoter short tandem repeats. Gene. 2012; 507(1):61–7.
4. Mohammadparast S, Bayat H, Biglarian A, Ohadi M. Exceptional expansion and conservation of a CT-repeat complex in the core promoter of PAXBP1 in primates. Am J Primatol. 2014;76(8):747–56.
5. Nikkhah M, Rezazadeh M, Khorshid HK, Biglarian A, Ohadi M. An exceptionally long CA-repeat in the core promoter of SCGB2B2 links with the evolution of apes and Old World monkeys. Gene. 2016;576(1):109–14.
6. Rezazadeh M, Gharesouran J, Mirabzadeh A, Khorshid HK, Biglarian A, Ohadi M. A primate-specific functional GTTT-repeat in the core promoter of CYTH4 is linked to bipolar disorder in human. Prog Neuro-Psychopharmacol Biol Psychiatry. 2015;56:161–7.

7. Namdar-Aligoodarzi P, Mohammadparast S, Zaker-Kandjani B, Kakroodi ST, Vesiehsari MJ, Ohadi M. Exceptionally long 5′ UTR short tandem repeats specifically linked to primates. Gene. 2015;569(1):88–94.
8. Ohadi M, Valipour E, Ghadimi-Haddadan S, Namdar-Aligoodarzi P, Bagheri A, Kowsari A, et al. Core promoter short tandem repeats as evolutionary switch codes for primate speciation. Am J Primatol. 2015;77(1):34–43.
9. Sonay TB, Carvalho T, Robinson MD, Greminger MP, Krützen M, Comas D, et al. Tandem repeat variation in human and great ape populations and its impact on gene expression divergence. Genome Res. 2015;25(11):1591–9.
10. Li C, Lenhard B, Luscombe NM. Integrated analysis sheds light on evolutionary trajectories of young transcription start sites in the human genome. bioRxiv. 2017. https://doi.org/10.1101/192757.
11. Kramer M, Sponholz C, Slaba M, Wissuwa B, Claus RA, Menzel U, et al. Alternative 5′untranslated regions are involved in expression regulation of human heme oxygenase-1. PLoS One. 2013;8(10):e77224.
12. Li Y, Seidel K, Marschall P, Klein M, Hope A, Schacherl J, et al. A polymorphic microsatellite repeat within the ECE-1c promoter is involved in transcriptional start site determination, human evolution, and Alzheimer's disease. J Neurosci. 2012;32(47):16807–20.
13. Emamalizadeh B, Movafagh A, Darvish H, Kazeminasab S, Andarva M, Namdar-Aligoodarzi P, et al. The human RIT2 core promoter short tandem repeat predominant allele is species-specific in length: a selective advantage for human evolution? Mol Gen Genomics. 2017;292(3):611–7.
14. Khademi E, Alehabib E, Shandiz EE, Ahmadifard A, Andarva M, Jamshidi J, et al. Support for "disease-only" genotypes and excess of homozygosity at the CYTH4 primate-specific GTTT-repeat in schizophrenia. Genet Test Mol Biomarkers. 2017;21(8):485–90.
15. Bushehri A, Barez MM, Mansouri S, Biglarian A, Ohadi M. Genome-wide identification of human-and primate-specific core promoter short tandem repeats. Gene. 2016;587(1):83–90.
16. Valipour E, Kowsari A, Bayat H, Banan M, Kazeminasab S, Mohammadparast S, et al. Polymorphic core promoter GA-repeats alter gene expression of the early embryonic developmental genes. Gene. 2013;531(2):175–9.
17. Heidari A, Fam ZNS, Esmaeilzadeh-Gharehdaghi E, Banan M, Hosseinkhani S, Mohammadparast S, et al. Core promoter STRs: novel mechanism for inter-individual variation in gene expression in humans. Gene. 2012;492(1):195–8.
18. Bagshaw AT. Functional mechanisms of microsatellite DNA in eukaryotic genomes. Genome Biol Evol. 2017;9(9):2428–43.
19. Bagshaw AT, Horwood LJ, Fergusson DM, Gemmell NJ, Kennedy MA. Microsatellite polymorphisms associated with human behavioural and psychological phenotypes including a gene-environment interaction. BMC Med Genet. 2017;18(1):12.
20. Gymrek M, Willems T, Guilmatre A, Zeng H, Markus B, Georgiev S, et al. Abundant contribution of short tandem repeats to gene expression variation in humans. Nat Genet. 2016;48(1):22–9.
21. Hammock EA, Young LJ. Microsatellite instability generates diversity in brain and sociobehavioral traits. Science. 2005;308(5728):1630–4.
22. Carrat GR, Hu M, Nguyen-Tu M-S, Chabosseau P, Gaulton KJ, van de Bunt M, et al. Decreased STARD10 expression is associated with defective insulin secretion in humans and mice. Am J Hum Genet. 2017;100(2):238–56.
23. Abe H, Gemmell NJ. Evolutionary footprints of short tandem repeats in avian promoters. Sci Rep. 2016;6:19421.
24. Fondon JW, Hammock EA, Hannan AJ, King DG. Simple sequence repeats: genetic modulators of brain function and behavior. Trends Neurosci. 2008; 31(7):328–34.
25. Alizadeh F, Bozorgmehr A, Tavakkoly-Bazzaz J, Ohadi M. Skewing of the genetic architecture at the ZMYM3 human-specific 5′ UTR short tandem repeat in schizophrenia. Mol Gen Genomics. 2018; https://doi.org/10.1007/s00438-018-1415-8.
26. Hannan AJ. Tandem repeats mediating genetic plasticity in health and disease. Nat Rev Genet. 2018; https://doi.org/10.1038/nrg.2017.115.
27. Watts PC, Kallio ER, Koskela E, Lonn E, Mappes T, Mokkonen M. Stabilizing selection on microsatellite allele length at arginine vasopressin 1a receptor and oxytocin receptor loci. Proc Biol Sci. 2017;284(1869):20171896. https://doi.org/10.1098/rspb.2017.1896.
28. Andrioli LPM, Vasisht V, Theodosopoulou E, Oberstein A, Small S. Anterior repression of a Drosophila stripe enhancer requires three position-specific mechanisms. Development. 2002;129(21):4931–40.
29. Andrioli LP, Oberstein AL, Corado MS, Yu D, Small S. Groucho-dependent repression by sloppy-paired 1 differentially positions anterior pair-rule stripes in the Drosophila embryo. Dev Biol. 2004;276(2):541–51.
30. Darvish H, Heidari A, Hosseinkhani S, Movafagh A, Khaligh A, Jamshidi J, et al. Biased homozygous haplotypes across the human caveolin 1 upstream purine complex in Parkinson's disease. J Mol Neurosci. 2013;51(2):389–93.
31. Heidari A, Hosseinkhani S, Talebi S, Meshkani R, Esmaeilzadeh-Gharedaghi E, Banan M, et al. Haplotypes across the human caveolin 1 gene upstream purine complex significantly alter gene expression: implication in neurodegenerative disorders. Gene. 2012;505(1):186–9.
32. Quilez J, Guilmatre A, Garg P, Highnam G, Gymrek M, Erlich Y, et al. Polymorphic tandem repeats within gene promoters act as modifiers of gene expression and DNA methylation in humans. Nucleic Acids Res. 2016; 44(8):3750–62.
33. Sawaya S, Bagshaw A, Buschiazzo E, Kumar P, Chowdhury S, Black MA, et al. Microsatellite tandem repeats are abundant in human promoters and are associated with regulatory elements. PLoS One. 2013;8(2):e54710.
34. Kronforst MR, Hansen ME, Crawford NG, Gallant JR, Zhang W, Kulathinal RJ, et al. Hybridization reveals the evolving genomic architecture of speciation. Cell Rep. 2013;5(3):666–77.
35. Dorus S, Vallender EJ, Evans PD, Anderson JR, Gilbert SL, Mahowald M, et al. Accelerated evolution of nervous system genes in the origin of Homo sapiens. Cell. 2004;119(7):1027–40.
36. Li M, Jaffe AE, Straub RE, Tao R, Shin JH, Wang Y, et al. A human-specific AS3MT isoform and BORCS7 are molecular risk factors in the 10q24. 32 schizophrenia-associated locus. Nat Med. 2016;22(6):649–56.
37. Hardy K, Brand-Miller J, Brown KD, Thomas MG, Copeland L. The importance of dietary carbohydrate in human evolution. Q Rev Biol. 2015;90(3):251–68.
38. Na CH, Barbhuiya MA, Kim MS, Verbruggen S, Eacker SM, Pletnikova O, Troncoso JC, Halushka MK, Menschaert G, Overall CM, Pandey A. Discovery of noncanonical translation initiation sites through mass spectrometric analysis of protein N termini. Genome Res. 2018;28(1):25–36.
39. Shin G, Grimes SM, Lee H, Lau BT, Xia LC, Ji HP. CRISPR–Cas9-targeted fragmentation and selective sequencing enable massively parallel microsatellite analysis. Nat Commun. 2017;8:14291.

17

Better governance, better access: practising responsible data sharing in the METADAC governance infrastructure

Madeleine J. Murtagh[1*], Mwenza T. Blell[2], Olly W. Butters[1], Lorraine Cowley[1], Edward S. Dove[3], Alissa Goodman[4], Rebecca L. Griggs[5], Alison Hall[6], Nina Hallowell[7], Meena Kumari[8], Massimo Mangino[9], Barbara Maughan[9], Melinda C. Mills[7], Joel T. Minion[1], Tom Murphy[4], Gillian Prior[10], Matthew Suderman[5], Susan M. Ring[5], Nina T. Rogers[11], Stephanie J. Roberts[1], Catherine Van der Straeten[12,13], Will Viney[14], Deborah Wiltshire[15], Andrew Wong[11], Neil Walker[16] and Paul R. Burton[1]

Abstract

Background: Genomic and biosocial research data about individuals is rapidly proliferating, bringing the potential for novel opportunities for data integration and use. The scale, pace and novelty of these applications raise a number of urgent sociotechnical, ethical and legal questions, including optimal methods of data storage, management and access. Although the open science movement advocates unfettered access to research data, many of the UK's longitudinal cohort studies operate systems of *managed* data access, in which access is governed by legal and ethical agreements between stewards of research datasets and researchers wishing to make use of them. Amongst other things, these agreements aim to respect the reasonable expectations of the research participants who provided data and samples, as expressed in the consent process. Arguably, responsible data management and governance of data and sample use are foundational to the consent process in longitudinal studies and are an important source of trustworthiness in the eyes of those who contribute data to genomic and biosocial research.

Methods: This paper presents an ethnographic case study exploring the foundational principles of a governance infrastructure for Managing Ethico-social, Technical and Administrative issues in Data ACcess (METADAC), which are operationalised through a committee known as the METADAC Access Committee. METADAC governs access to phenotype, genotype and 'omic' data and samples from five UK longitudinal studies.

Findings: Using the example of METADAC, we argue that three key structural features are foundational for practising responsible data sharing: independence and transparency; interdisciplinarity; and participant-centric decision-making. We observe that the international research community is proactively working towards optimising the use of research data, integrating/linking these data with routine data generated by health and social care services and other administrative data services to improve the analysis, interpretation and utility of these data. The governance of these new complex data assemblages will require a range of expertise from across a number of domains and disciplines, including that of study participants. Human-mediated decision-making bodies will be central to ensuring achievable, reasoned and responsible decisions about the use of these data; the METADAC model described in this paper provides an example of how this could be realised.

Keywords: Data ethics, Data governance, Data access, Data Access Committee (DAC), Governance, Participant involvement, Ethnography, Qualitative research, Interdisciplinarity

* Correspondence: Madeleine.Murtagh@newcastle.ac.uk
[1]Newcastle University, Newcastle upon Tyne, UK
Full list of author information is available at the end of the article

Introduction: managing access to complex and sensitive research resources

Genomic and biosocial research data about humans continue to proliferate, bringing with them questions about who should be able to store, hold and use which data and samples, for which purposes and with what safeguards. The research data landscape is changing with new forms of research data becoming available (e.g. next-generation sequencing, epigenetic and other 'omics' data [1, 2]) and existing data being increasingly accessible for research (e.g. via linkage of administrative, environmental or healthcare data to research data [3]). This complexity will only increase as new informatics technologies that enable citizen-generated data (e.g. social media, direct-to-consumer genetic testing and wearable sensors) become available to combine with research resources in novel ways.

Open science and drivers for data sharing

Expectations surrounding how research data should be shared are based on international commitments to using, and re-using, publicly funded research and its outputs for the public good [4]. Data sharing is also supported by research funding agencies internationally through their policies on open science (https://www.nwo.nl/en/policies/open+science/data+management, http://www.allianzinitiative.de/en/archive/research-data/principles.html, [5–10]). Scientific benefits of data sharing are seen to include verification, replication and the ability to pool analyses, as well as potential cost savings [9–13]. Some of the drive for shared data use comes from patients and publics themselves: already, many rare disease patients advocate for greater sharing of genomic and clinical data; the quantified-self movement promotes and facilitates sharing of data from wearables [14, 15]; citizen scientists demand to access clinical and other randomised controlled trials data and have filed legal claims to realise those demands [16]. International policy positions research data and samples as a public good which can only be fully realised by their wide and appropriate use [17], though some types of research data such as that generated by commercial companies (e.g. pharmaceutical companies) sit outside this definition. Indeed, some have gone even further to argue that not sharing research data is a breach of participant, patient and public rights and is itself a harm [18]. While the open science movement advocates unfettered access to research practices and the data produced by them, when those data are individual-level human data, access and re-use must be managed within relevant legal and ethical requirements, taking account of specific agreements together with the reasonable expectations of the research participants who provided those data and samples. These requirements, agreements and expectations are embodied in the consents that study participants, or their guardians if they are minors, give at the outset of the study and, often, for new collections or sub-studies.

Data sharing, longitudinal studies and the consent process

To date, the majority of genomic and biosocial research data made available for sharing in the UK have been derived originally from publicly funded research studies: longitudinal studies [19–21] (e.g. birth cohorts, household panel studies, other population and disease-based biobanks, clinical trials); survey and other social science datasets; and case-control studies, including case series collected/generated for research purposes and typically compared to a research-generated control group (e.g. Wellcome Trust Case Control Consortium [22]).

In longitudinal studies, the focus of this paper, it is not possible to foresee all possible research uses at the outset of a study. Data and samples may be used for long after their collection, sometimes decades afterwards. For studies whose *raison d'etre* is the provision of a scientific resource for which public good is the intended outcome and impact, renewed consent for each and every new research use would not only be unwieldy but would impede that aim. Instead, broad consent (within specific stipulations reflecting contemporary social values and scientific practice) are sought for future uses of the data and samples collected. In the UK, under the Human Tissue Act 2004 (https://www.legislation.gov.uk/ukpga/2004/30/contents) and Health Research Authority, the principles and contents of information given to study participants in consent and participant information documentation for different forms of data and samples are prescribed [23, 24].

But consent is a process, one which does not finish with the receipt of information and the signing of a form. Ethical approval for studies using broad consent includes mechanisms to ensure that those consents are respected and the expectations inherent in them are maintained, for example, explicitly stating which bodies can approve data and sample access. Appropriate, responsible data management and governance of the uses of those data and samples are foundational to this ongoing consent process in longitudinal studies and an important source of their trustworthiness in the eyes of individuals who have contributed data and continue to permit its storage and use. While trust is crucial for ensuring public support for research generally, for longitudinal studies, this requirement is particularly acute as ongoing research participation is vital to the longevity, productivity and impact of those studies. Research participants quite reasonably expect that their privacy (including confidentiality) will be protected and that uses of the data and samples they contribute will fall within agreed and anticipated parameters.

Data Access Committees

While Research Ethics Committees (RECs, also called Institutional Review Boards or Research Ethics Boards) interpret the ethical, legal and social frameworks within which studies are conducted and determine how data and samples may be used, data access permissions in the UK are governed by a hierarchy of regulatory processes and bodies depending on the data's sensitivity and potential disclosivity (the ease and precision with which individuals may be identified). Light-touch administrative processes manage low-dimensional data with a low risk of disclosure or other participant harms. Data Access Committees (DACs) manage more complex or higher-risk data access. Key assessment criteria for determining the release—and access route—of such data include (1) consistency with original ethical approval, including consents and information materials and the stated aim of the data collection, and the risk of identification of individual participants; (2) issues that may directly concern individual participants, such as the risk of identifying a previously unknown disease or strong disease determinant which may warrant clinical action, or the risk of bringing a long-standing study into disrepute; (3) unreasonably damaging the intellectual property of the project, or otherwise undermining the effort invested by its investigators or data custodians (often more than one generation of researchers).

DACs may consider any risks which may impact the individual participant or otherwise breach their expectations, thereby provoking them to withdraw from a study. In addition to the above criteria, some DACs assess the quality of the science or potential public benefit for all applications, and nearly all committees assess the quality and potential benefit of the science when there is a request for use of a finite resource (e.g. blood samples). Across the research/scientific community in the genomic, health and social sciences, there is a strong consensus that the application of all such criteria (when applied in a proportionate and transparent manner) is good for society *and* 'good for science' [25].

Data access governance in the UK

In the UK, access to data and samples—the 'resources' of longitudinal and other research studies—is operationalised by a networked series of independent data repositories and independent data governance infrastructures, though some longitudinal studies successfully operate in-house or partially in-house governance and data issue mechanisms. Governance of access permissions (ethics and policy oversight) and governance of data issue (technical governance) may be managed jointly or by separate infrastructures. Each of these governance infrastructures is designed to ethically, legally, efficiently and securely manage access (permissions and/or distribution) to research resources of varied levels of complexity and sensitivity. For many UK longitudinal studies, data can be accessed online by bona fide researchers and are governed by end-user licences. Access decisions may also be made based on algorithms and straightforward rules for legitimate access. These are generally rapid mechanisms for access, notwithstanding any additional time required for online training upon registration to a data repository (e.g. UKDA training in data management [24]) or additional administrative processes for managing cost recovery. The data released by these mechanisms are those for which there is deemed to be a low risk of individual disclosure and which are considered not to raise additional ethical issues. Some data will only ever be made available in secure privacy-protected settings—where researchers travel to the data location (e.g. secure air-gapped data centres) or send the analysis to the data (e.g. DataSHIELD) and receive back only the analytical outcomes [26]. Use of finite resources such as blood, urine or other biological samples will always require additional oversight for legal (e.g. Human Tissue Act 2004 requirements), ethical and scientific reasons and because each use of samples necessarily precludes future uses of those samples and therefore must be judged more carefully.

Classes of research data which are potentially sensitive, insofar as they are disclosive and/or raise particular ethical concerns (e.g. incidental/secondary findings), require additional access oversight. Beyond the character of the data themselves, some research uses of data are also potentially controversial and may thereby raise ethical issues and potential harms for research participants themselves, or for the longitudinal studies of which they are a part. Some types of research are particularly likely to raise issues of potential participant and/or study harms, e.g. those involving potentially stigmatising issues, such as mental health, sexuality, criminality and certain diseases, or those in which researchers are perceived to make a commercial gain. Particular forms of or combinations of data also raise risks of disclosure. That individuals can be identified within a genetic dataset has been demonstrated methodologically [27, 28], but only where there is a reference sample/sequence available [29]. For most practical applications identification is unlikely, though this may change with increasing availability of commercial genotyping. Identification of individuals via phenotype data, though less certain than genetic identification, is easier to enact if large numbers or certain classes of variables are combined in analyses (e.g. place of residence if defined too narrowly, minority ethnicity, disability, rare disease status) or, as in longitudinal studies, repeated measures are available. Research questions which further combine genotypic and phenotypic data and/or samples may increase the potential for disclosure of individual identity. Whether the important issue for disclosure is one of identifying the individual per se (i.e. that a particular individual belongs to a longitudinal study) or whether it is about identifying the individual *and* information about them is an empirical question that requires further examination. Where there are higher risks of disclosure or the proposed research raises other potential harms, online registration and end-user licencing are not sufficient safeguards. Decisions about complex societal values are ultimately not amenable to a simple algorithmic approach whether derived heuristically or, for example, via machine learning. We argue that a human-mediated review and decision-making process are then required. Moreover, we argue that study participants must be central to this decision-making, seeing through to its conclusion the ongoing consent dialogue initiated in consent and participant information, procedures and documents.

Methods

In this paper, we present an ethnographic case study [30] exploring the foundational principles of one infrastructure for Managing Ethico-social, Technical and Administrative issues in Data ACcess (METADAC), i.e. the ethics and policy oversight of data access. The METADAC infrastructure (that is, the METADAC Access Committee and associated work of the pilot project described below) aims to put into practice the tenets of responsible data sharing [31, 32] and, in particular, operationalises incorporation of study participant voices and perspectives in access decision-making. The case study approach allows an in-depth consideration of a particular instance of practice, or 'case'. As Stake [33] describes, the case study is 'the study of the particularity and complexity of a single case, coming to understand its activity within important circumstances' (p. 6). Case studies may involve a range of methods, the prime criteria being those most able to elicit in-depth, rather than summary, understandings of the object of study; in this case qualitative methods. The analysis in this case study here was informed by a qualitative ethnographic evaluation study conducted alongside the METADAC pilot. The ethnographic evaluation included observations of METADAC committee meetings ($n = 16$) and policy development workshops ($n = 4$) from June 2016 to October 2017. Meetings were observed by JTM. Twenty-seven interviews were conducted by JTM from March to August 2017 with 18 internal METADAC stakeholders (committee members, funders, study representations, technical review team members) and 9 METADAC applicants (from 31 invitations). Internal stakeholders were asked about METADAC processes and policies. Applicants were asked about their experience of the application process. Meeting observations and interview transcripts were coded thematically using the constant comparative approach [34] and following the steps provided by Braun and Clarke [35] for organising analysis without the imposition of a specific epistemic framework. This approach involves researchers familiarising themselves with the data corpus (e.g. the interview data in this study), generating initial codes and then developing, reviewing and defining themes. Having identified our themes using this approach, we took a constructivist-interpretivist approach to interpret the themes, i.e. 'particular actors, in particular places, at particular times, fashion meaning out of events and phenomena through prolonged, complex processes of social interaction involving history, language and action' and 'that to understand this world of meaning one must interpret it' [36]. We present this analysis with representative extracts from the interviews and description derived from the observational notes of the meetings.

Results: the METADAC infrastructure

The METADAC infrastructure hosts a Data Access Committee (the METADAC Access Committee) governing access to the complex and sensitive genotypic and phenotypic classes of data and biological samples (see Table 1) generated by five longitudinal studies in the UK [37]: the National Child Development Study (a.k.a. 1958 Birth Cohort) [38], 1970 Birth Cohort Study (BCS70) [39], English Longitudinal Study of Ageing (ELSA) [40], Millennium Study [41] and *Understanding Society* (UK Household Longitudinal study) [42, 43]. It also acts as an appeals committee to one other longitudinal study (TwinsUK). METADAC has become the UK's only independent, multi-study data and sample access infrastructure for longitudinal studies. The 3-year METADAC pilot brought together, under one infrastructure, a number of separate data and sample access structures. Meetings to review applications are held at six-weekly intervals, primarily as teleconferences and twice annually through face-to-face meetings, when other planning and policy development workshops also occur. Administratively, the METADAC Secretariat manages logs and performs a quality audit of all incoming applications. The Secretariat serves as the hub for all METADAC communication and interactions with and between stakeholders. Three key structural features provide the foundation for practising responsible data sharing: independence and transparency; interdisciplinarity; and participant-centric decision-making. These features are realised in the access assessment criteria by which the METADAC Access Committee makes its decisions (see Table 2).

Independence and transparency

One way of safeguarding the right to privacy and the right to benefit from science is to ensure a robust and independent data access process, especially for the most complex and sensitive research resources. The METADAC process has promoted fair, consistent and transparent practices of data access, including making these openly available on its website. An independent mechanism for access, particularly for more complex and sensitive data and sample access, supports these processes. The importance of independence is recognised by funders, for example, Wellcome's recent Longitudinal Population Studies Strategy [8], and reflected in its criteria for funding longitudinal studies:

Table 1 Classes of data and their access regulation

	Data type and sensitivity	Data available from	Governance	Data distribution
Class 1	Low risk survey/phenotype-only data, e.g. social, economic, psychological and health data.	UK Data Archive	Requires registered access—online (clickable) end-user licences.	Data is distributed by the UK Data Archive.
Class 2	Low sensitivity genetic-only data, e.g. genome-wide single nucleotide polymorphism [SNP] data generated from a standard chip with no phenotype information except gender and a coarse measure of area of residence.	European Genome-phenome Archive (EGA)	Requires application to the Sanger DAC, administrative decision based on algorithm of criteria for access.	Data are distributed by the EGA.
Class 3	Potentially disclosive survey/phenotype-only data, e.g. small geographic area. Requires application to a study-based Data Access Committee made up of study investigators and study technical staff.	Study DAC	Data may be issued on special licence or may be accessible only in on or off-site data safe havens.	Data is distributed, or the data safe haven hosted, by the study.
Class 4	New forms of genetic-only data, e.g. exome sequence and epigenetic data, are reviewed by METADAC until their ethical implication/sensitivity is established.	METADAC	Application requires review and approval from the METADAC Access Committee. Application includes signed agreement to conditions of use. Data use is monitored annually.	Data are distributed by the EGA.
Class 5	Genetic-only data with known ethical issues, e.g. incidental findings risk in exome data.	METADAC	Application requires review and approval from the METADAC Access Committee. Application includes signed agreement to conditions of use. Data use is monitored annually.	Data are distributed by the EGA.
Class 6	Biological samples. Use of biological samples will always require additional oversight for ethical and scientific reasons.	METADAC	Application requires independent scientific review and approval from the METADAC Access Committee.	Samples are issued by the relevant study under Material Transfer Agreements.
Class 7	Any combination of Classes 1 to 6, e.g. individual-level phenotype data linked to genotype data or samples, is potentially more disclosive than any one class of data alone.	METADAC	Application requires review and approval from the METADAC Access Committee. Application includes signed agreement to conditions of use. Data use is monitored annually.	Combined datasets are issued on unique IDs and distributed by the study and EGA.
Class 8	Non-research data. High risk when multiple variables are combined.	Various	Certain administrative datasets, e.g. education and criminal records, are available for research via the Administrative Data Research Network. Individual-level health data is available for research via NHS Digital).	Various

Table 2: METADAC assessment criteria

1. The application has been submitted by bona fide researchers (using the MRC definition [21]).
2. The application does not risk producing information that may allow individual study participants to be identified.
3. The application does not violate (or potentially violate) any of the ethical permissions granted to the study and/or any of the consent forms signed by the participants or their guardians.
4. The application addresses topics that fall within the acknowledged remit of the project, as understood by participants.
5. There is no significant risk that the application might upset or alienate study members or of reducing their willingness to continue as active participants.
6. There is no significant risk that the application might harm individuals in the study, or the study as a whole.
7. The application does not require access to a depletable finite resource (whole blood extracted DNA, blood, saliva and urine). OR The quality of the science has been reviewed formally by independent external reviewers (agreed by METADAC) and has been judged to merit the use of finite biological samples. NB: Applications for finite bio-samples are seen as being in competition with other potential future uses of those samples, and therefore the quality of the science is reviewed formally.
8. Includes a high quality plain language summary—following METADAC guidance.

> If it is not possible to fully inform participants about how their data could be used, it is essential that robust governance is in place, for example, ensuring there is an independent, accountable decision-making body that can provide assurances that data are being used and linked appropriately and responsibly. (p.6).

Four additional benefits of independent, multi-study access processes are clear and go beyond the values and norms of fair, transparent and ethical practice. First, in the competitive academic environment in which most researchers work, data 'hugging' or hoarding—real or perceived—by researchers internal to a study can result in constraints on sharing [44–46]. Independent access processes, be they via registered access [47] or DACs, support data accessibility. Second, study-independent, multi-study access bodies can facilitate data discovery. Year-on-year increases in applications and approvals through METADAC (a 2.5-fold increase over the first 24 months of operation) are a testament to the increasing international interest in the use of UK longitudinal studies and demonstrate the role of METADAC in enabling access to study resources. Moreover, researchers applying to METADAC are signposted to other similar studies and are able to make data and/or sample access applications to multiple studies simultaneously. Third, an independent committee allows determinations to be made about new uses of data and samples out-with the natural tendency of studies to be risk-averse regarding their own resource. As users of the resources return derived datasets and code for variable creation, rapid assessment of the ethical/governance implications of the uses and governance of new forms of data (e.g. exome sequencing and epigenetic data) is actioned, and the parameters for onward use are established through consensus. But of course, independent does not mean detached. Having representatives from the studies at the table is vital and enables the Access Committee to ensure that the particularities of the study are taken into account and that their future uses are optimised. METADAC has a responsibility to the ongoing 'health and wellbeing' of the studies and their participants and not simply to facilitate the exploitation of their data and samples. Fourth, METADAC enables learning across the studies under its remit, sharing good practice but also providing leverage to change practice. For example, one of the studies under METADAC governance, Understanding Society (the UK Household Longitudinal study), has been able to extend the availability of its genetic data resource internationally. The governance of multiple studies made it evident that Understanding Society's historical restrictions on international access to its data were not a fundamental feature of longitudinal studies in the UK—other studies had no such restrictions. The existence of international access practices in the other studies under METADAC enabled the study directors to reassure their stakeholders that extending their access practices to harmonise with those of others was safe, proportionate and ethically acceptable. In short, there has been a virtuous circle of learning amongst METADAC studies.

Interdisciplinarity

Because no individual or group can possibly hold knowledge of all aspects of research, research ethics and governance [48], the METADAC Access Committee is comprised of committee members with social, biomedical, ethical, legal and clinical expertise and individuals with study-facing expertise. Committee meetings also include, as observers, study PIs or study representatives, funder representatives and members of the Technical Review Team (TRT) who support the decision-making. The METADAC Secretariat and the ethnographic observer (JTM) also attend meetings. Where appropriate, the committee also takes advantage of third-party specialist knowledge, particularly where an applicant seeks to use finite samples. The METADAC Access Committee is chaired by a social scientist with ethics expertise (MJM), and its deputy chair (NW) has bioscience and bioinformatics expertise. The TRT is chaired (PRB) by a biomedical scientist with data science expertise.

The ethnographic study demonstrated that each of the contributors to the METADAC Committee brings particular understandings and perspectives which contribute richness to collaborative decision-making but that these are understandings that are not rigidly bound, with many members contributing across these domains.

Study-facing committee members, currently two members of longitudinal studies not governed by METADAC, bring (i) embedded and embodied subjective experience of longitudinal studies and thereby of study participants' reasonable expectations, and (ii) insights into the project proposals that are not coloured by specialist knowledge of the same field. While researchers may be carried along by the importance and interest of a particular exploration (it answers an interesting research question), study-facing members are more likely to see each application 'anew' and spot where concerns may be raised for participants. We elaborate this contribution further below.

Committee members with legal and ethical expertise bring (i) knowledge of relevant laws and regulation (particularly data protection law, access, consent and issues around data privacy and biomedical research); (ii) understanding of the ethical norms, values and principles underpinning decision-making and their implications for protecting and promoting the rights, interests and welfare of participants, researchers and the broader community; and (iii) awareness of the remit and purpose of DACs as well as ensuring the clarity and transparency of METADAC practices and the documentation which seeks to explain them.

Biomedical scientists bring (i) an awareness of methodology, used not to judge the science but instead to understand whether variables requested are 'necessary and reasonable'; (ii) understanding of the biomedical background to research, to judge the likely potential for incidental or secondary findings in research which at first glance may not raise clinically actionable findings; and (iii) awareness of related published material or similar data resources that may assist applicants in carrying out research.

Social scientists bring (i) methodological expertise to assess the aims, feasibility and study designs being proposed; (ii) understanding of social and environmental contexts when and where data were collected and their potential impact; (iii) awareness of issues that may give rise to ethical or public concern, often reframing scientific descriptions which uncover simplistic or paternalistic assumptions which may trigger participant concerns; and (iv) experience of working with other epidemiological and longitudinal studies.

Clinicians bring understanding of (i) both the seriousness and treatability of genetically influenced illnesses; (ii) the sometimes blurred boundaries between research and clinical practice, highlighting the complexities and impact of communicating uncertainties to patients and their families, especially in the contexts of interpreting genetic variants and reporting unlooked for findings; and (iii) a working appreciation of negotiating meaningful informed consent to genetic testing with patients when results may not be predicted or imminent.

The TRT provides practical insight on the affordances of the studies and a broad evaluation of disclosure risk and incidental finding risk, based on a knowledge of the size of the dataset, prevalence and heritability of the condition and issues that will require further detailed discussion in committee. The TRT includes individuals with expertise about the study data and other resources. They offer the Access Committee insight into what data resources are available, the format and affordances of those resources and, for example, whether there are specific technical or other restrictions on their use.

Study PIs/representatives bring a strong awareness of the field generally and their own studies in particular; the availability and relevance of data, any weaknesses in the data (e.g. a low response rate in a particular questionnaire) and other relevant data not known to the applicant—whether from a new wave or from another longitudinal study with similar data. Similarly, funders contribute understanding of strategic and policy direction nationally and internationally. In practice, any member may bring a contribution from many of these types.

This rich mix of participants brings to bear perspectives and understandings which support independent and transparent decision-making. During the ethnographic evaluation that has run in parallel to the formation of METADAC, internal stakeholders have frequently highlighted the value of having a range of disciplinary expertise in the Access Committee:

> A multi-disciplinary approach is really vital so that we have both the technical expertise, ethical expertise, social sciences expertise, so that all those elements can be brought to one place to determine what is the optimal approach. (from evaluation interview with METADAC Committee member)

> If there are ethical issues, there is someone who is a specialist in that and who's thinking about it and brings issues up if they need to be discussed ...there are lots of different people on [METADAC] from lots of different backgrounds and disciplines ... So there are different people thinking about these things. (from evaluation interview with METADAC Technical Review Team member)

In the ethnographic evaluation, the diversity of the committee was seen to produce higher quality reflection and analysis on matters of concern [49]. Participants described that hearing perspectives from other fields improved their own critical thinking. But, as Fitzgerald and Callard [50] put it: 'There are two current certainties in ... interdisciplinary research: everyone wants to do it, and no one quite knows how' (p.23). While the benefits of multi-, inter- or transdisciplinary working in data governance are still emerging and deserve careful documentation, for METADAC, there appears to be some clear benefits, particularly in supporting a rounded approach that takes account of a plurality of perspectives in coming to consensual decisions. That is not to suggest that consensus is necessarily or always straightforward; it requires work. It is not simply multiple perspectives from distinct disciplines that are brought to bear in decision-making (not least because some members already span disciplines), but also multiple standpoints or subject positions [51, 52]; more colloquially, we say people 'wear many hats'. Individuals taking part in METADAC decision-making inhabit and draw upon multiple subject positions: researcher/researched, data custodian/data producer, etc. We argue that it is these multiple subject positions which are important in decision-making precisely because they call upon the different interests, commitments and expectations which are central to ensuring decisions made about data and sample access are coherent with, and respect the contributions made by, study participants.

Discussion: participant-centred decision-making

Engaging the perspectives of stakeholder communities within research and involving them in decision making are lauded for meeting ethical expectations and norms as well as for its (potential) pragmatic outcomes, improving the alignment of research with societal values and the relevance of research outputs or their translation [48]. Engagement is variously understood as a form of democracy, an act of respect and an acknowledgement of human rights [53–59]. In Braidotti's vision of affirmative ethics [60, 61], the understanding that 'we are all in this together' even if we are 'not one and the same' [62] places upon science an injunction to work *with* the communities, public and research participants who are both the basis and beneficiaries of science. Of course, study participants are already at the centre of longitudinal studies; it is only with their permission and with their contributions of data and samples that the studies endure. Respecting those contributions means that participant-centredness must do more than simply consider them research *subjects*. Engaging study participants in meaningful decision-making is both an ethical and practical solution, enabling alignment with participant expectations, social norms and values and, arguably, improving study outputs as a result. As we argue elsewhere, engagement brings study participants closer to the research but also brings the research and researchers closer to them [48]. In the case of longitudinal studies, there is a long history of study participant involvement in decision-making. The Avon Longitudinal Study of Parents and Children (ALSPAC) has an advisory panel comprised of study participants and has included on its ethics committee original cohort participants (now nearing 26 years of age) for the past decade and parent members for much longer [63]. ALSPAC study participant members now number half of the ethics committee and their involvement in ethical decision-making is normalised and embedded. In METADAC, the need for multiple perspectives, met through the interdisciplinary constitution of its Access Committee, demands the inclusion of study-facing participants. By incorporating study-facing members, METADAC was seen by members and observers of the committee as contributing new and valuable voices to discussions about data access and use and expanding how Data Access Committees have historically functioned:

> They're the ones who have insights, that nobody else has, on the anxieties, the concerns, the hopes about how their data will be used, so they absolutely need to be represented. (from evaluation interview with METADAC Committee member)

Participant-centredness can, of course, take many forms: it can be agonistic or it can be solidaristic; it can be prescribed or it can be negotiated [64]. In scientific contexts, participatory activities can range from unidirectional communication and outreach, through various levels of involvement in decision-making, to participant control, as in citizen science endeavours. Each has advantages and limitations. In METADAC, which requires understanding and expertise from many perspectives in order to make responsible decisions, participant-centredness takes three forms: (1) respecting study participant expectations, (2) involving study participants in decision-making roles and (3) communicating the results of access decisions to participants (and others) in a format that is clear and accessible, for example, in plain language summaries.

Study participant expectations

Participant expectations of longitudinal research are central to METADAC's access decisions. Criterion 5 (Table 2) seeks to ensure that there is no significant risk of *upsetting or alienating* participants. While this is an instrumental value, in that it seeks to reduce loss to a study or studies, it is also primarily a value-based position aimed at respecting the participation and on-going commitment of study participants. Though in a cognate context, Caldicott's injunction that participants (in her report, patients) should not be 'surprised' by the uses of their data and samples also holds true for METADAC. Assessment criterion 5 (Table 2) is called upon in METADAC Access Committee meetings in the form of two questions: (1) Would the research proposed by an applicant be likely to upset or alienate participants? and (2) Would the research breach the reasonable expectations of participants? It is via these questions that issues of potential sensitivity and controversy may be addressed. It might be plausible to construct an algorithm to identify certain classes of research as sensitive, and indeed for one of the studies under METADAC oversight, there is a specific list of research topics and types which are restricted. Specifically, the 1958 Birth Cohort explicitly restricts studies of the genetics of intelligence, sexuality and criminality. This stipulation is based on information provided to nurses during their formal training for taking samples from study participants. While not included in the information material given to study participants during consent, which are always the foundation of legitimate access requests, a previous committee determined that nurses might reasonably have been expected to communicate those exceptions to at least some study participants, and therefore, these continue to be respected. A list of restricted variables has been identified, but decisions about how and whether the research purposes proposed for their use are acceptable and approval granted are often more complicated than simple yes/no determinations. A level of interpretation and judgement is necessary, one which requires human decision-making.

Study participant involvement in decision-making

An ethnographic (observational) study of a predecessor to the METADAC Committee identified, during access decisions, repeated consideration of what the participant or public might 'think' of a particular application for data use [44]. In that DAC, however, understanding of participant's voice was hypothetical only, because no participants or other study-facing members were involved in DAC decision-making and no empirical account of their views was available (e.g. in qualitative studies of their expectations of involvement). The METADAC infrastructure addresses this absence by including two study-facing members who are themselves participants of longitudinal studies. To avoid conflicts of interest, these study-facing members are independent of the studies for which METADAC provides oversight. This decision was taken to avoid the assumption that study-facing members directly represent the study population. While two committee members cannot feasibly represent the individual interests of tens of thousands of study participants, the aim is to ensure that the embodied, material and emotional experience of being a study participant is incorporated into data governance decision-making. In this pilot phase of METADAC, development of study-facing members' involvement is assisted by involving individual study participants who also happen to have active research careers in the social sciences.

METADAC has avoided treating study-facing members' views as 'non-expert'. Involvement of study participants in decision-making roles requires active work to ensure it is meaningful; a learning process that means that the contributions of these study participant members deepen as their knowledge and understanding of governance issues grow. Pre-meetings are held with all new members of the METADAC, whatever their background, and continue for study-facing members as often as they wish, to create a space for longer discussion of applications and other issues that is typically available in a 90-min committee meeting. After each committee meeting, informal debrief sessions are hosted to make sure that any uncertainties are resolved, wider observations can be aired and suggestions made. Active chairing in committee meetings and an inclusive culture facilitates all committee members to express their positions. The Secretariat acts as the first point of call when resolving practical problems, and the Access Committee as a whole can be used to develop insights into particular areas of procedure or policy.

Even within the acknowledged limitations above, decisions made with study-facing members create different outcomes to those made without their insight and experience. For example, the embodied sense of discomfort or unease from the participant perspective (or indeed from other members) warrants greater scrutiny of applications by the Access Committee and can and does encourage different decisions to be made. Some areas of sensitivity are not easily exposed through schematic, algorithmic or technical evaluation. We call upon study-facing members to scrutinise areas of uncertainty, along with other members, to articulate a range of potential study participants' responses in order to highlight potential risks and harms of any given application and to anticipate how even well-intentioned research outcomes can produce harms, including those caused by sensationalised media reactions. Some of this reasoning can be messy and speculative; it stays with the troubling or unforeseen parts of an application and tries to consider how the expectations of science may align with or compromise those of study participants. In this regard, study-facing members bring a unique perspective and position to the committee. But they do more than this. In the METADAC, the study-facing members have been central in developing policy, particularly as new forms of data become available. Deep and often lively discussions during METADAC development workshops become the basis of policy and processes (e.g. recent access policy developed for exome sequence data and epigenetic data). In this way, decision-making in METADAC occurs *with* study participants as part of an ongoing, collaborative and negotiated process.

Plain language summaries

Although plain language summaries (PLSs), or 'lay summaries', are an expectation of communication of much research—and most funders now require these for their applications—researchers often do not communicate their work in ways which are clear or accessible to non-expert audiences. METADAC owes a duty to communicate the outputs of their contributions to the participants of those studies for which it provides oversight and to their funders. PLSs are published on the METADAC website [1]. Development of PLS policy, processes and guidance, therefore, has been taken seriously, and METADAC's study-facing committee members assess each PLS according to guidance provided to applicants [65]. As a reflection of the care by which PLSs are reviewed, and to ensure data use is transparent to study participants and the wider community, many applicants are asked to improve their PLS. While guidance is readily available, time-poor researchers do not always fully absorb the instructions or do not recognise the importance attached to PLSs by METADAC. The chief reason applicants are asked to revise their applications is on the basis of inadequate descriptions of their research.

Limitations

This ethnographic case study was informed by the evaluation of the METADAC pilot. As such, it represents findings at an interim stage of METADAC's development and should be read with this limitation in mind. The ethnographic methods used here provide a deep understanding of a phenomenon. Though they are not statistically generalizable, the findings, as expected of qualitative studies employing quality criteria for rigour and trustworthiness, can be considered theoretically generalizable to cognate settings.

Conclusions: future considerations for data and samples access

At present, there is a strong strategic investment in ensuring not only the optimisation of research data use, but also of routine data generated by health and social care services and other administrative data services. Moving forward, the international research community is proactively working towards creating a world in which all such sources of data can, where beneficial, be integrated, linked and jointly interpreted; this potential has been demonstrated in the UK in projects such as The Farr Institute, the Administrative Data Research Network, Connected Health Cities, the pan-London Health Information Exchange and the Medical Research

Council's Health Data Research UK initiative [3, 66–69]. The aims of such initiatives and the enhanced data utility they provide are three-fold: (1) to better use data to directly inform front-line clinical care and public health, (2) to enhance the evidence base for health and social care planning and strategic development and formally evaluate key initiatives in these domains and (3) to enhance the quality and scope of research in health science, bioscience and the social sciences. Given the desirability of such aims and current investment by governments and funders, it is inevitable that over the next decade, DACs are going to have to deal with increasingly large volumes of data generated from the health, social care and other administrative data sources.

At least some DACs in the future are going to have to be able to deal simultaneously, and effectively, with the data access criteria they already apply to research data *and* the laws, rules and regulations applicable, in the UK, to National Health Service (NHS), social care and administrative data (e.g. education, criminal records, national statistics). Whether research DACs evolve to deal with NHS data (or as gatekeepers for NHS data), taking account of the real risk of identification of at least some individuals, even after pseudonymisation and similar privacy-protecting mechanisms, will demand a profound change in perspective. In addition, governing these data will create a substantive additional workload and a need to rapidly explore and develop appropriate systems shortcuts and technological solutions to some aspects of the assessment. Empirical evidence demonstrates that DACs face serious challenges when a new class of data suddenly becomes available: part of the argument for creating METADAC was that a predecessor DAC was not set up and therefore not able to deal effectively with newly available genomic data. METADAC and several studies under it have had to work hard (and rapidly) to modify the systems and structures for releasing conventional SNP-based genome-wide association data to deal with full DNA sequence data and with methylation data—even though many of the principles are unaltered. Other studies, not part of METADAC, have invested extensive time in considering how—if at all—to manage and govern new data classes such as data generated via social media or image data.

The evolution of DACs that can deal jointly and effectively with all of the criteria and rules applying to routinely collected health, social care or administrative data and all of those already applied in dealing with research data will be an equivalent challenge but of a much higher order. This will particularly prove to be the case when DACs have to start considering data generated and interpreted in real time; indeed, many new applications of health data in the future will require that research data from individuals can be combined with real-time health service data to enable rapid clinical responses for diagnosis or management. If both the research community and health and social care services are to be ready when the new data start flowing, there is a need for experts across all the domains and disciplines relating to data access and sharing, including study participants and users of health, social care and other services, to have worked through all of these issues. Human-mediated decision-making bodies will undoubtedly be central to ensuring the trustworthiness of the most sensitive of those data, and achievable, reasoned and responsible decisions about their use. The METADAC model we describe here provides a template for such future governance.

Abbreviations

ALSPAC: Avon Longitudinal Study of Parents and Children; DAC: Data Access Committee; METADAC: Managing Ethico-social, Technical and Administrative issues in Data ACcess; NHS: National Health Service; PLS: Plain language summary; REC: Research Ethics Committee; TRT: Technical Review Team; UKDA: UK Data Archive

Acknowledgements

The authors would like to thank all members of METADAC, past and present, for their contributions to the review of data access applications and to the development of policy and procedures.

Funding

The METADAC is funded jointly by the Medical Research Council (MRC), the Wellcome Trust and the Economic and Social Research Council (ESRC) -MR/N01104X/1 & MR/N01104X/2 - and is supported by the Wellcome Trust and MRC funded Fostering new Opportunities for Researchers via Wider Access to Research Data and Samples (58FORWARDS) study - 108439/Z/15/Z. The ethnographic study identified is similarly funded, though no funder played a role in the design of the study or in the collection, analysis and interpretation of the data.

Authors' contributions

MJM conceived the METADAC, and all authors (except MTB, OWB and JTM) contributed to the development of the METADAC infrastructure, application review process and/or development of policy and procedure. MJM also conceived the ethnographic study. The study protocol was developed by MJM, JTM and SJR. Recruitment and data collection were conducted by JTM, with data analysis completed by MJM, JTM and SJR. MJM wrote an initial draft of the manuscript, to which all authors provided feedback and approved the final version.

Competing interests

The authors declare that they have no competing interests.

Author details

[1]Newcastle University, Newcastle upon Tyne, UK. [2]University of Cambridge, Cambridge, UK. [3]University of Edinburgh, Edinburgh, UK. [4]University College London Institute of Education, London, UK. [5]University of Bristol, Bristol, UK. [6]PHG Foundation, Cambridge, UK. [7]University of Oxford, Oxford, UK. [8]University of Essex, Colchester, UK. [9]King's College London, London, UK. [10]NatCen, Brentwood, London, UK. [11]University College London, London, UK. [12]Imperial College London, London, UK. [13]Ghent University Hospital, Ghent, Belgium. [14]Goldsmiths, University of London, London, UK. [15]UK Data Archive, University of Essex, Colchester, UK. [16]NIHR Bioresource, Cambridge, UK.

References

1. METADAC [www.metadac.ac.uk - Accessed 5 Jan 2018].
2. European_Genome-Phenome_Archive. www.ebi.ac.uk/ega/home. Accessed 22 Apr 2018.
3. Administrative Data Research Network [https://adrn.ac.uk/ - Accessed 5 Jan 2018].
4. Pilat D, Fukasaku Y. OECD principles and guidelines for access to research data from public funding. Data Science Journal. 2007;6:OD4–OD11.
5. Guidelines to the rules on open access to scientific publication and open access to research data in Horizon 2020 [http://ec.europa.eu/research/participants/data/ref/h2020/grants_manual/hi/oa_pilot/h2020-hi-oa-pilot-guide_en.pdf].
6. Understanding Patient Data [https://understandingpatientdata.org.uk/ - Accessed 5 Jan 2018].
7. National Institutes of Health Data Sharing Policies [https://grants.nih.gov/policy/sharing.htm - Accessed 5 Jan 2018].
8. Longitudinal Population Studies Strategy [https://wellcome.ac.uk/sites/default/files/longitudinal-population-studies-strategy_0.pdf - Accessed 5 Jan 2018].
9. Munafò MR, Nosek BA, Bishop DV, Button KS, Chambers CD, du Sert NP, Simonsohn U, Wagenmakers E-J, Ware JJ, Ioannidis JP. A manifesto for reproducible science. Nature Human Behaviour. 2017;1:0021.
10. Ioannidis JP. How to make more published research true. PLoS Med. 2014; 11(10):e1001747.
11. Ross JS, Krumholz HM. Ushering in a new era of open science through data sharing: the wall must come down. JAMA. 2013;309(13):1355–6.
12. Chan A-W, Song F, Vickers A, Jefferson T, Dickersin K, Gøtzsche PC, Krumholz HM, Ghersi D, Van Der Worp HB. Increasing value and reducing waste: addressing inaccessible research. Lancet. 2014;383(9913):257–66.
13. Gøtzsche PC. Why we need easy access to all data from all clinical trials and how to accomplish it. Trials. 2011;12(1):249.
14. Nafus D, Sherman J. Big data, big questions| this one does not go up to 11: the quantified self movement as an alternative big data practice. International journal of communication. 2014;8:11.
15. Prainsack B. Personalized medicine: empowered patients in the 21st century? Vancouver: NYU Press; 2017.
16. Geraghty KJ. 'PACE-Gate': when clinical trial evidence meets open data access. London, England: SAGE Publications Sage UK; 2016.
17. Organisation for Economic C-o, Development: OECD principles and guidelines for access to research data from public funding. Vancouver: OECD Pub; 2007.
18. Jones KH, Laurie G, Stevens L, Dobbs C, Ford DV, Lea N. The other side of the coin: harm due to the non-use of health-related data. Int J Med Inform. 2017;97:43–51.
19. UK_Data_Service. www.ukdataservice.ac.uk. Accessed 22 Apr 2018.
20. Department_of_Education. Summary of Longitudinal Surveys. https://assets.publishing.service.gov.uk/government/uploads/system/uploads/attachment_data/file/444794/DFE-RR458_Summary_of_longitudinal_surveys.pdf. Accessed 22 Apr 2018.
21. Medical_Research_Council. Cohort Directory. https://mrc.ukri.org/research/facilities-and-resources-for-researchers/cohort-directory/. Accessed 22 Apr 2018.
22. Burton PR, Clayton DG, Cardon LR, Craddock N, Deloukas P, Duncanson A, Kwiatkowski DP, McCarthy MI, Ouwehand WH, Samani NJ. Genome-wide association study of 14,000 cases of seven common diseases and 3,000 shared controls. Nature. 2007;447(7145):661–78.
23. 2018 HTAAJ
24. HRA Consent decision tools [http://www.hra-decisiontools.org.uk/consent/index.html - Accessed 5 Jan 2018].
25. Burton PR, Banner N, Elliot MJ, Knoppers BM, Banks J. Policies and strategies to facilitate secondary use of research data in the health sciences. Vancouver: Oxford University Press; 2017.
26. Gaye A, Marcon Y, Isaeva J, LaFlamme P, Turner A, Jones EM, Minion J, Boyd AW, Newby CJ, Nuotio M-L. DataSHIELD: taking the analysis to the data, not the data to the analysis. Int J Epidemiol. 2014;43(6):1929–44.
27. Homer N, Szelinger S, Redman M, Duggan D, Tembe W, Muehling J, Pearson JV, Stephan DA, Nelson SF, Craig DW. Resolving individuals contributing trace amounts of DNA to highly complex mixtures using high-density SNP genotyping microarrays. PLoS Genet. 2008;4(8): e1000167.
28. Masca N, Burton PR, Sheehan NA. Participant identification in genetic association studies: improved methods and practical implications. Int J Epidemiol. 2011;40(6):1629–42.
29. Gymrek M, McGuire AL, Golan D, Halperin E, Erlich Y. Identifying personal genomes by surname inference. Science. 2013;339(6117):321–4.
30. Harrison H, Birks M, Franklin R, Mills J. Case study research: foundations and methodological orientations. In: *Forum Qualitative Sozialforschung/Forum: Qualitative Social Research*; 2017.
31. Knoppers BM. Framework for responsible sharing of genomic and health-related data. HUGO J. 2014;8(1):3.
32. Framework for Responsible Sharing of Genomic and Health-Related Data. https://www.ga4gh.org/docs/ga4ghtoolkit/rsgh/Framework-Version-10September2014.pdf. Accessed 5 Jan 2018.
33. Stake RE. The art of case study research. Vancouver: Sage; 1995.
34. Glaser BG. The constant comparative method of qualitative analysis. Soc Probl. 1965;12(4):436–45.
35. Braun V, Clarke V. Using thematic analysis in psychology. Qual Res Psychol. 2006;3(2):77–101.
36. Schwandt TA. Constructivist, interpretivist approaches to human inquiry. Handbook of Qual Res. 1994;1:118–37.
37. METADAC Studies [https://www.metadac.ac.uk/studies-2/ - Accessed 5 Jan 2018].
38. Power C, Elliott J. Cohort profile: 1958 British Birth Cohort (national child development study). Int J Epidemiol. 2005;35(1):34–41.
39. Elliott J, Shepherd P. Cohort profile: 1970 British birth cohort (BCS70). Int J Epidemiol. 2006;35(4):836–43.
40. Steptoe A, Breeze E, Banks J, Nazroo J. Cohort profile: the English longitudinal study of ageing. Int J Epidemiol. 2012;42(6):1640–8.
41. Connelly R, Platt L. Cohort profile: UK millennium Cohort study (MCS). Int J Epidemiol. 2014;43(6):1719–25.
42. Buck N, McFall S. Understanding society: design overview. Longitudinal and Life Course Studies. 2011;3(1):5–17.
43. Benzeval M, Davillas A, Kumari M, Lynn P. Understanding society: the UK household longitudinal study biomarker user guide and glossary. Vancouver: Institute for Social and Economic Research, University of Essex; 2014.
44. Murtagh MJ, Turner A, Minion JT, Fay M, Burton PR. International data sharing in practice: new technologies meet old governance. Biopreservation and biobanking. 2016;14(3):231–40.
45. Murtagh MJ, Thorisson GA, Wallace SE, Kaye J, Demir I, Fortier I, Harris JR, Cox D, Deschênes M, LaFlamme P. Navigating the perfect [data] storm. Norsk epidemiologi. 2012;21(2):203–9.
46. Lowrance WW: Access to collections of data and materials for health research. A report to the Medical Research Council and the Wellcome Trust London: MRC and Wellcome Trust 2006.
47. Dyke SOM, Kirby E, Shabani M, Thorogood A, Kato K, Knoppers BM. Registered access: a 'triple-A'approach. Eur J Hum Genet. 2016;24(12):1676–80.
48. Murtagh MJ, Minion JT, Turner A, Wilson RC, Blell M, Ochieng C, Murtagh B, Roberts S, Butters OW, Burton PR. The ECOUTER methodology for stakeholder engagement in translational research. BMC medical ethics. 2017;18(1):24.
49. Murtagh M, Roberts SJ, Minion J, JT. METADAC (Managing Ethico-social, Technical and Administrative issues in Data ACcess). In: European Biobank Week. Vienna: European, Middle Eastern & African Society for Biopreservation and Biobanking; Biobanking and Biomolecular Resources Research Infrastructure - European Research Infrastructure Consortium; 2016.
50. Fitzgerald D, Callard F. Social science and neuroscience beyond interdisciplinarity: experimental entanglements. Theory, Culture & Society. 2015;32(1):3–32.
51. Davies B, Harré R. Positioning: the discursive production of selves. J Theory Soc Behav. 1990;20(1):43–63.
52. Harré R, Moghaddam FM. The self and others: positioning individuals and groups in personal, political, and cultural contexts. Vancouver: Greenwood publishing group; 2003.
53. Kim SY, Wall IF, Stanczyk A, De Vries R. Assessing the public's views in research ethics controversies: deliberative democracy and bioethics as natural allies. Journal of Empirical Research on Human Research Ethics. 2009;4(4):3–16.
54. Gottweis H, Gaskell G, Starkbaum J. Connecting the public with biobank research: reciprocity matters. Nat Rev Genet. 2011;12(11):738–9.
55. Winickoff DE. From benefit sharing to power sharing: partnership governance in population genomics research. UC Berkeley: Center for the Study of Law and Society Jurisprudence and Social Policy Program; 2008. Retrieved from https://escholarship.org/uc/item/845393hh

56. Stirling A. Power, truth and progress: towards knowledge democracies in europe. Future Directions For Scientific Advice In. 2015:133.
57. Solberg B: Biobank consent models—are we moving toward increased participant engagement in biobanking? Journal of Biorepository Science for Applied Medicine 2015, 3:23–33.
58. Rial-Sebbag E, Cambon-Thomsen A: Governing biobanks through a European infrastructure. In: Ethics, Law and Governance of Biobanking. Springer; 2015: 139–151.
59. O'Doherty K, Einsiedel E. Public engagement and emerging technologies. Vancouver: UBC Press; 2012.
60. Braidotti R: Transpositions: on nomadic ethics: polity; 2006.
61. Braidotti R: On putting the active back into activism. New Formations 2010, 68(68):42–57.
62. Aspirations of a posthumanist [https://www.youtube.com/watch?v=LNIYOKfRQks].
63. Minion JT, Butcher F, Timpson NJ, Murtagh MJ. The ethics conundrum in Recall by Genotype (RbG) research: perspectives from birth cohort participants. bioRxiv. 2017:124636.
64. Dean RJ. Beyond radicalism and resignation: the competing logics for public participation in policy decisions. Policy & Politics. 2017;45(2):213–30.
65. METADAC Plain Language Summary Guidance. https://www.metadac.ac.uk/files/2017/06/v1.0-Plain-language-guidance-for-METADAC-applications.pdf. Accessed 5 Jan 2018.
66. The Farr Institute. http://www.farrinstitute.org/. Accessed 5 Jan 2018.
67. Connected Health Cities. https://www.connectedhealthcities.org/. Accessed 5 Jan 2018.
68. London Information Exchange. https://www.digitalhealth.net/2015/03/london-information-exchange-on-way/. Accessed 5 Jan 2018.
69. MRC Health Data Research UK. https://www.mrc.ac.uk/about/institutes-units-centres/uk-institute-for-health-and-biomedical-informatics-research/. Accessed 5 Jan 2018.

Evaluation of a bone morphogenetic protein 6 variant as a cause of iron loading

Cameron J. McDonald[1†], Gautam Rishi[2†], Eriza S. Secondes[2], Lesa Ostini[1], Daniel F. Wallace[2], Darrell H. G. Crawford[3], Hanlon Sia[4], Paul Clark[5] and V. Nathan Subramaniam[2*]

Abstract

Background: Atypical iron overload without variation in the five clinically associated hereditary hemochromatosis genes is now recognized; however, their etiology remains unknown. Since the identification of iron overload in the bone morphogenetic protein 6 (*Bmp6*) knockout mouse, the search has been on for clinically pathogenic variants in the *BMP6* gene. A recent report proposes that variants in the pro-peptide region of *BMP6* are the underlying cause of several cases of iron overload. We performed targeted next-generation sequencing on three cases of atypical iron overload with Asian ethnicity and identified a p.Q118dup (aka p.E112indelEQ, p.Q115dup, p.Q118_L119insQ) variant in *BMP6*. The purpose of this study was to characterize the molecular function of the identified BMP6 variant. Molecular characterization by immunofluorescence microscopy and Western blotting of transfected cells, bioinformatics, and population analyses was performed.

Results: In contrast to reports for other BMP6 pro-peptide variants in this region, our data indicates that this variant does not affect the function of the mature BMP6 protein.

Conclusions: Our data suggest that assignment of disease causation in clinical cases of iron overload to pro-peptide variants in BMP6 should thus be treated with caution and requires biological characterization.

Keywords: Next-generation sequencing, Hemochromatosis, Causal variant, Iron overload, BMP6

Background

Hereditary hemochromatosis (HH) is relatively common in individuals of northern European descent, with the majority of cases caused by mutations in the *HFE* (hemochromatosis) gene [1, 2]. Conversely, mutations in *HFE* are rare in non-European populations, and mutations in other genes involved in the regulation of iron homeostasis are more common, such as hemojuvelin (*HFE2*), hepcidin (*HAMP*), transferrin receptor 2 (*TFR2*), and ferroportin (*SLC40A1*), collectively termed non-*HFE* HH [3]. Potentially, all regulators of iron homeostasis are functional gene candidates for non-*HFE* HH [4]. BMP6 (bone morphogenetic protein 6) is a key regulator of iron homeostasis [5]. *Bmp6* null mice have significant iron overload due to deficiency of the iron-regulatory hormone hepcidin [6, 7].

Under normal conditions, *Bmp6* expression is upregulated by increased iron and increases the expression of hepcidin which in turn limits iron absorption and recycling [8]. A recent report associated heterozygosity for three variants p.P95S, p.L96P, and p.Q113E in the BMP6 pro-peptide region with nine clinical cases of unexplained atypical iron overload [9]. The functional link between BMP6 and iron overload remains to be explained, however, and further validation is necessary to better assess causality of these variants [10]. In addition, p.E112Q and p.R257H have also been associated with more severe iron overload in patients with *HFE* mutations [11]. Here, we report on three clinical cases of unexplained atypical iron overload carrying an amino acid insertion within the same region of the BMP6 pro-peptide.

Results

Three cases are reported in this study; in each case, *HFE* genotyping was negative, and liver specialist assessment was undertaken. Case 1 was a 62-year-old male who over the past 7 years had raised serum ferritin (SF) ranging

* Correspondence: nathan.subramaniam@qut.edu.au
†Equal contributors
[2]Institute of Health and Biomedical Innovation and School of Biomedical Sciences, Queensland University of Technology (QUT), 60 Musk Avenue, Kelvin Grove, Brisbane, Queensland 4059, Australia
Full list of author information is available at the end of the article

between 1261 and 1675 μg/L (normal range 30–300 μg/L), with transferrin saturation (TS) 40 to 53% (normal range 20–55%). His alanine aminotransferase (ALT), aspartate aminotransferase (AST), and gamma-glutamyl transferase (GGT) levels all remained normal during this time. Liver biopsy revealed grade 4 stainable iron in hepatocytes, with accentuation of iron in the pericanalicular region (Fig. 1). No other pathology was noted. In case 2, a male aged 51 who did not consume any alcohol, presented with a SF of 1714 μg/L and TS of 34%. ALT, AST, and GGT were all normal although ultrasound revealed a diffusely echogenic liver suggesting fatty liver. Ferriscan identified a hepatic iron concentration of 44 μmol/g (normal range 3–33 μmol/g). In case 3, also male aged 62 presented with a SF of 1133 μg/L and TS of 90%. His ALT, AST, and GGT levels were mildly elevated. Liver biopsy demonstrated grade 1 iron with a hepatic iron content of 39 μmol/g and F1 fibrosis. Case 3's pathophysiology is complex with positivity for α-thalassemia trait and hepatitis C virus (genotype 2).

Targeted next-generation sequencing revealed that all three patients were negative for mutations in the five currently clinically associated HH genes. Further analysis identified a tri-nucleotide duplication in *BMP6*, c.353-355dupAGC (NM_001718.4), resulting in the in-frame duplication of glutamine p.Q118dup (NP_001709.1), which was confirmed by Sanger sequencing. The nomenclature for this variant is noted in three different forms across various datasets: dbSNP rs771616962 refers to this variant as p.Q118_L119insQ, wANNOVAR, and the NHLBI Exome Sequencing Project (http://evs.gs.washington.edu/EVS/) (ESP) dataset as p.E112delinsEQ and Exome Aggregation Consortium database (ExAC) [12] as p.Gln115dup. While all of these references result in the same outcome, extension of the poly-Q stretch, the various designations may cause underestimation of its population frequency. This variant, or its pseudonyms, is not present in the 1000 Genomes database; however, it is present in the ESP (MAF = 0.016) and ExAC (MAF = 0.041).

To investigate possible functional effects of the p.Q118dup variant on BMP6, immunofluorescence microscopy studies were conducted using BMP6 wildtype and p.Q118dup proteins with a pro-peptide N-terminus HA tag and a mature protein C-terminus V5 tag. We used both CHO cells and liver endothelial (SK-HEP-1) cells as a model for examining the molecular and functional effects of the p.Q118dup variant. CHO cells (Fig. 2a) and SK-HEP-1 cells (Fig. 3a) stably expressing these constructs showed some peri-nuclear Golgi localization of p.Q118dup pro-peptide; however, this was very mild and not consistent within all cells. No difference in localization was observed for the mature V5-tagged protein in either the CHO or endothelial cell models.

Discussion

BMP6 is an important signaling molecule required for iron homeostasis [6, 8]. In response to increase in body iron levels *BMP6* levels increase, this results in an increase in BMP-SMAD signaling. The increased BMP_SMAD signaling then leads to an increase in hepcidin production. Recently, Daher and colleagues reported on three BMP6 pro-domain variants associated with iron overload and attributed the cause to reduced export of BMP6 [9]. We compared levels of the full-length and mature wildtype and mutant BMP6 protein in cell lysate and conditioned media (Fig. 2b, c). This revealed no difference in the capacity of cells to export p.Q118dup BMP6 protein compared to wildtype. Recent studies have suggested that the endothelial cells may be responsible for BMP6 production in response to iron [13]. Using an endothelial cell model, we show that there is

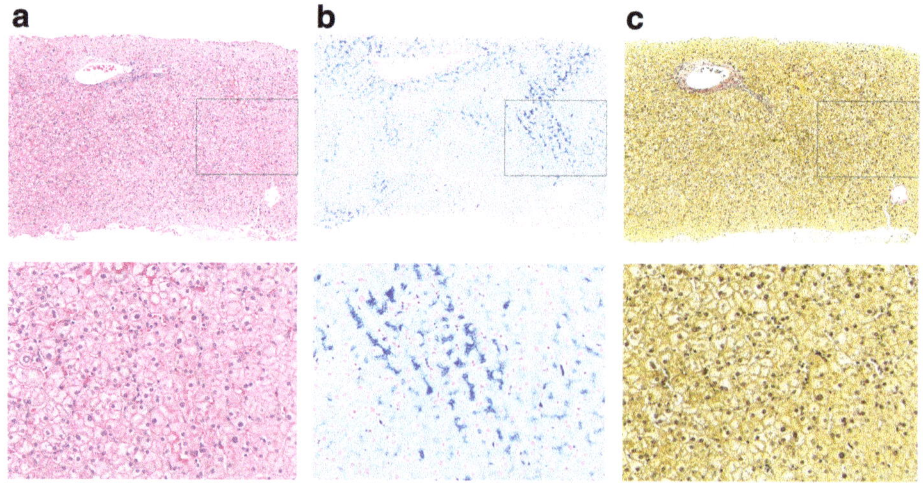

Fig. 1 Liver biopsy histology of case 1. Histological sections stained with **a** H&E, **b** Perls, and **c** Van Gieson's stain showing significant stainable iron, without pathology

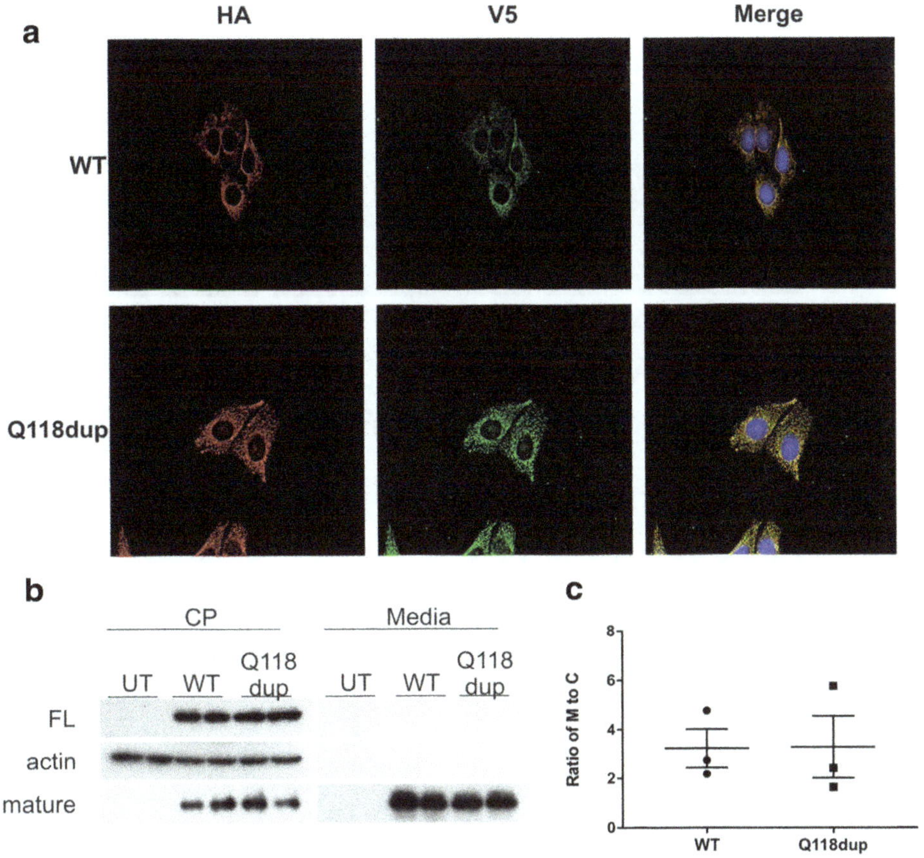

Fig. 2 Cellular characterization of p.Q118dup BMP6 protein in CHO cells. **a** Representative immunofluorescence image of stably transfected CHO cells. The HA tag on the N-terminus of the pro-peptide, and the V5 tag on the C-terminus of the mature protein co-localize, and no difference in localization of the wildtype (WT) and p.Q118dup variant of BMP6 was observed. **b** Western blot identifying the full-length (FL) and mature BMP6 proteins in cell lysate (CP) and conditioned media (M) from untransfected (UT), wildtype (WT), and p.Q118dup (Q118dup) BMP6 transfected CHO cells. **c** Quantification of Western blot shown in **b** ($n = 3$). The ratio of mature proteins in conditioned media (M) to cell lysate (CP) shows no difference in protein secretion between WT and p.Q118dup BMP6 transfected cells

no difference in the expression of the mature BMP6 protein for either the wildtype or p.Q118dup variant of BMP6 (Fig. 3b).

As population frequency can provide insight into the likelihood of a variant being benign, we investigated the frequency of p.Q118dup in the ExAC dataset (http://exac.broadinstitute.org/) which contains exome sequences from 60,706 unrelated individuals, and carrier rates were then calculated using the Hardy-Weinberg equation. This analysis reveals that in East Asians, p.Q118dup has a carrier rate of 1 in 3 (MAF = 0.191), with other ethnicities also showing lower though still significant carrier rates, confirming the experimental result that this variant is unlikely to affect gene function. Finally, we examined the classification of the p.Q118dup variant according to the "standards and guidelines for the interpretation of sequence variants" as set out by the American College of Medical Genetics and Genomics [14]. Based on the above data plus bioinformatic functional impact prediction using several tools available through ANNOVAR, it confirmed a categorization of the BMP6 pro-peptide p.Q118dup variant as benign.

Conclusions

The data from this study suggest no functional relationship between the p.Q118dup BMP6 pro-peptide variant and iron overload. Recognition of this variant as non-deleterious and not causative is notably important for this specific variant, as cursory literature and database analysis may easily lead to the alternate and incorrect assumption that this variant is likely causative of the disease.

Firstly, the alternative nomenclature of p.E112indelEQ (which while technically incorrect is the annotation returned by several common sequence alignment programs) places this variant at the same location as the p.Q113E variant recently reported by Daher et al. [9] to be deleterious and the cause of atypical iron overload in two members of a single family. Secondly, the absence of p.Q118dup from the two most commonly known variant data sets, dbSNP and 1000 Genomes, erroneously suggests

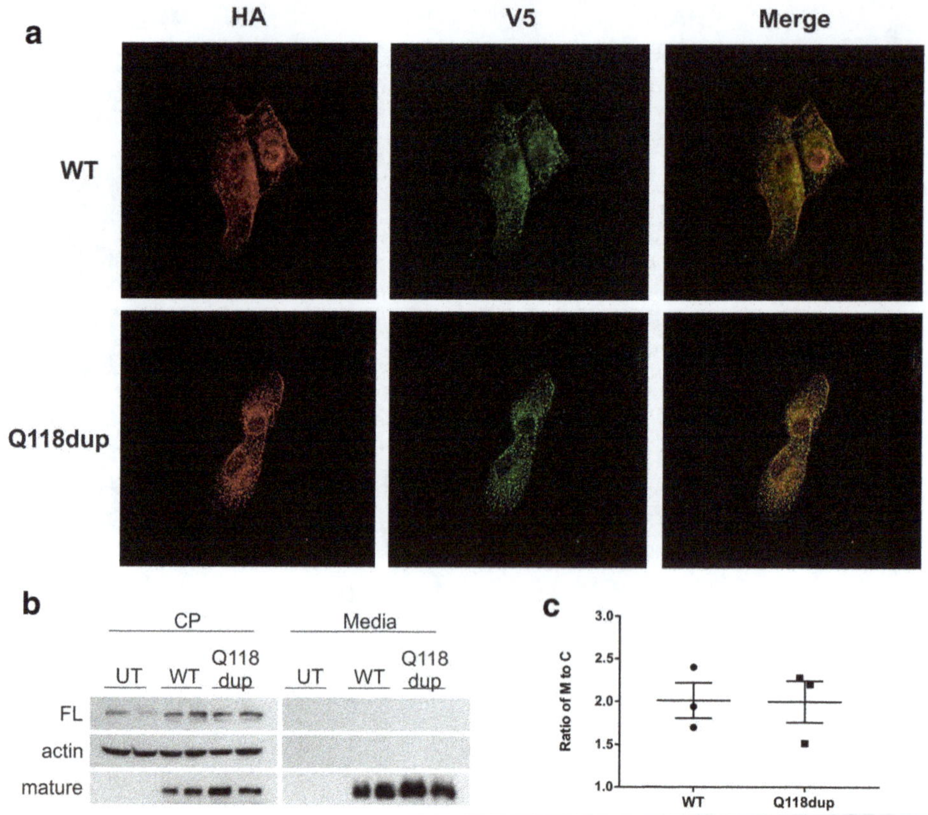

Fig. 3 Cellular characterization of p.Q118dup BMP6 protein in a hepatic endothelial cell model, SK-HEP-1. **a** Immunofluorescence of stably transfected SK-HEP-1 cells. No difference in localization of the wildtype and p.Q118dup variant of BMP6 was observed. HA tag, N-terminus of pro-peptide; V5 tag, C-terminus of mature protein. **b** Western blot identifying the full-length (FL) and mature BMP6 proteins in cell lysate (CP) and conditioned media from untransfected (UT), wildtype (WT), and p.Q118dup BMP6 transfected SK-HEP-1 cells. **c** Quantification of Western blot shown in **b** ($n = 3$). The ratio of mature proteins in conditioned media to cell lysate shows no difference in protein secretion between WT and p.Q118dup BMP6 transfected cells

that this variant is novel and thus is not carried in the healthy population. Finally, using CHO and hepatocyte endothelial cell models, we showed no difference in the localization or secretion between the wildtype and p. Q118dup BMP6 variant.

These data demonstrated no relationship between *BMP6* variants and atypical iron overload and provided no functional evidence of causality for the identified variant. Furthermore, the analysis raises uncertainty about the population prevalence of *BMP6* variants and their potential role in iron overload conditions. This has important clinical implications, for monitoring and treatment, and should lead to pause before suggesting genetic and family counseling based on *BMP6* variation. Atypical, non-HFE-related iron overload remains an important question; however, despite prior literature and the appeal of its underlying biological plausibility, the genetic causes of some forms of HH appear elusive. Further research will be required to uncover the potential genetic causes of these atypical iron overload disorders.

Methods

Patients and consent

This study was approved by and performed in accordance with the ethical standards of the QIMR Berghofer Human Research Ethics Committee and with the Helsinki Declaration of 1975, as revised in 1983. Informed and written consent was obtained for the studies described in this report.

Cases of clinically identified atypical iron overload of unknown cause were referred to our laboratory for sequencing by primary clinicians. Cases were required to demonstrate persistently (two or more occasions) abnormal iron indices (including significantly elevated serum ferritin > 1000 μg/L for males and > 700 for females) and to be negative for p. C282Y homozygosity. Referring clinicians were also requested to undertake a detailed iron metabolism-specific patient history for other possible attributing causes, such as high alcohol consumption. In total, 31 referred patients have been sequenced to date at our center [15].

Next-generation sequencing

Sequencing of genomic DNA using an iron-associated gene panel, covering 39 genes and 11 promoters associated with iron overload or anemia, was performed as previously described [15]. Samples were sequenced to an average of 400-fold coverage using an Ion Torrent PGM, with post run QC/QA filtering performed using Torrent Suite (v3.6; Life Technologies). Sequences were mapped to the Human Genome (HG19) and variants identified using the Torrent Variant Caller (version 3.6.59049 with Germ Line - Low Stringency configuration). Identified variants were annotated using the wANNOVAR tool [16], and candidate mutations confirmed by Sanger sequencing.

BMP6 expression constructs

Constructs containing double-tagged human *BMP6* coding sequence were synthesized and validated by BioMatik (Delaware, USA). Constructs were generated for both the wildtype and p.Q118dup sequence and contained a hemagglutinin (HA)—tag inserted between amino acids 23 and 24 (N-terminus of the pro-protein) and a V5-tag at the carboxyl-terminus of the mature protein.

Transfections

Chinese hamster ovary (CHO) cells and the hepatic endothelial cell line, SK-HEP-1 [17] (HTB-52, ATCC, Manassas, Virginia), were transiently transfected with the *BMP6* expression constructs by reverse transfection using Viafect (Promega, Alexandria, NSW, Australia) according to the manufacturer's instructions. Stably expressing cells were obtained by placing transfected cells under selection with puromycin at 10 μg/ml (CHO cells) or 1 μg/ml (SK-HEP-1 cell).

Immunofluorescence

Transfected cells were plated on glass coverslips to a confluency of 50–70% for 24 h. Cells were washed twice in PBSCM (phosphate buffered saline with 1 mM $CaCl_2$, 1 mM $MgCl_2$) and then fixed (in methanol at – 20 °C for 5 min for CHO cells then at room temperature (RT) for 10 min and in 3% paraformaldehyde for SK-HEP Hep-1 cells which were then permeabilized using 0.1% saponin in PBSCM for 15 min). Primary antibodies were then applied: anti-HA at 1:10 (12CA5, tissue culture supernatant) and anti-V5 at 1:3000 (Merck-Millipore). After three PBSCM washes, cells were incubated in Alexa Fluor donkey secondary antibodies: anti-mouse 488/594 and anti-rabbit 594/488 (Invitrogen). Following three final PBSCM washes, coverslips were mounted in Prolong Gold anti-fade mounting media (Invitrogen) and imaged by microscopy at × 63 magnification using a Zeiss Axioscope NLO-780 or a Leica TCS SP5. Z stacks were imaged for each focal point, and the final images were processed using the in-built deconvolution algorithms in the Zen software. The immunofluorescence studies were performed at least three times for each cell line.

Western blotting

Conditioned media from transfected cells were precipitated with 10% trichloroacetic acid overnight at 4 °C, and the precipitate was washed with cold acetone and resuspended in SDS sample buffer (SB) (50 mm Tris pH 6.8, 2% SDS, 10% glycerol, 0.1% bromophenol blue, 5% β-mercaptoethanol). Cell lysates were also resuspended in SDS SB, and all samples were loaded onto a 12% SDS-PAG and electrophoresed. Following transfer onto 0.2 μm nitrocellulose (Bio-Rad), blots were blocked for 3 h at RT in Western blocking buffer WBB (10% skim milk powder in 0.5% Tween-20 in TBS (TBST)). Primary anti-V5 (2.5 ng/ml) or anti-HA (1:100) diluted in WBB was applied overnight at 4 °C. Following three washes in TBST, blots were incubated with secondary anti-mouse horseradish peroxidase (1:10,000 in WBB) for 1 h at RT. Blots were washed extensively with TBST, Luminata Forte Western Chemiluminescent HRP Substrate (Merck-Millipore) was applied for 5 min, and blots were then exposed to film. Western blotting was repeated three times for each cell line and quantified. Band volume and density were quantitated using ImageJ software (National Institutes of Health, Maryland, USA).

Abbreviations

ALT: Alanine aminotransferase; AST: Aspartate aminotransferase; BMP: Bone morphogenetic protein; ESP: NHLBI Exome Sequencing Project 6500 database; ExAC: Exome Aggregation Consortium database; GGT: Gamma-glutamyl transferase; SF: Serum ferritin; TS: Transferrin saturation

Acknowledgements

The authors gratefully acknowledge the support of the patients in this study.

Funding

This work was supported in part by the project grants (APP1031325 and APP1048000) from the National Health and Medical Research Council (NHMRC) of Australia to VNS. VNS is the recipient of an NHMRC Senior Research Fellowship (APP1024672 and APP1118888).

Authors' contributions

CJM, DFW, GR, and VNS designed the studies. CJM, GR, ES, and LO performed the experiments. CJM, GR, LO, DFW, and VNS analyzed the data. DHGC, HS, and PC contributed to clinical aspects of the study. CJM, GR, and VNS wrote the manuscript. CJM, GR, DFW, DHGC, HS, and PC critically reviewed the manuscript. VNS provided the funding. All authors read and approved the final manuscript.

Consent for publication

Informed and written consent was obtained for publication.

Competing interests

The authors declare that they have no competing interests.

Author details
[1]QIMR Berghofer Medical Research Institute, Brisbane, Australia. [2]Institute of Health and Biomedical Innovation and School of Biomedical Sciences, Queensland University of Technology (QUT), 60 Musk Avenue, Kelvin Grove, Brisbane, Queensland 4059, Australia. [3]Faculty of Medicine and Biomedical Sciences, The University of Queensland, Brisbane, Australia. [4]ICON Cancer Care, Southport, Australia. [5]Princess Alexandra and Mater Hospitals, Brisbane, Australia.

References

1. Allen KJ, Gurrin LC, Constantine CC, Osborne NJ, Delatycki MB, Nicoll AJ, McLaren CE, Bahlo M, Nisselle AE, Vulpe CD, et al. Iron-overload-related disease in HFE hereditary hemochromatosis. N Engl J Med. 2008;358(3):221–30.
2. Feder JN, Gnirke A, Thomas W, Tsuchihashi Z, Ruddy DA, Basava A, Dormishian F, Domingo R Jr, Ellis MC, Fullan A, et al. A novel MHC class I-like gene is mutated in patients with hereditary haemochromatosis. Nat Genet. 1996;13(4):399–408.
3. McDonald CJ, Wallace DF, Crawford DH, Subramaniam VN. Iron storage disease in Asia-Pacific populations: the importance of non-HFE mutations. J Gastroenterol Hepatol. 2013;28(7):1087–94.
4. Wallace DF, Subramaniam VN. Non-HFE haemochromatosis. World J Gastroenterol. 2007;13(35):4690–8.
5. Rishi G, Subramaniam VN. The liver in regulation of iron homeostasis. Am J Physiol Gastrointest Liver Physiol. 2017;313(3):G157–65.
6. Meynard D, Kautz L, Darnaud V, Canonne-Hergaux F, Coppin H, Roth MP. Lack of the bone morphogenetic protein BMP6 induces massive iron overload. Nat Genet. 2009;41(4):478–81.
7. Andriopoulos B Jr, Corradini E, Xia Y, Faasse SA, Chen S, Grgurevic L, Knutson MD, Pietrangelo A, Vukicevic S, Lin HY, et al. BMP6 is a key endogenous regulator of hepcidin expression and iron metabolism. Nat Genet. 2009;41(4):482–7.
8. Kautz L, Besson-Fournier C, Meynard D, Latour C, Roth MP, Coppin H. Iron overload induces BMP6 expression in the liver but not in the duodenum. Haematologica. 2011;96(2):199–203.
9. Daher R, Kannengiesser C, Houamel D, Lefebvre T, Bardou-Jacquet E, Ducrot N, de Kerguenec C, Jouanolle AM, Robreau AM, Oudin C, et al. Heterozygous mutations in BMP6 pro-peptide lead to inappropriate hepcidin synthesis and moderate iron overload in humans. Gastroenterology. 2016;150(3):672–683 e674.
10. McDonald CJ, Rishi G, Wallace DF, Subramaniam VN. Genetic variants in the BMP6 pro-peptide may not cause iron loading and should be interpreted with caution. Gastroenterology. 2016;151(4):770–1.
11. Piubelli C, Castagna A, Marchi G, Rizzi M, Busti F, Badar S, Marchetti M, De Gobbi M, Roetto A, Xumerle L, et al. Identification of new BMP6 pro-peptide mutations in patients with iron overload. Am J Hematol. 2017;92(6):562–8.
12. Lek M, Karczewski KJ, Minikel EV, Samocha KE, Banks E, Fennell T, O'Donnell-Luria AH, Ware JS, Hill AJ, Cummings BB, et al. Analysis of protein-coding genetic variation in 60,706 humans. Nature. 2016;536(7616):285–91.
13. Canali S, Zumbrennen-Bullough KB, Core AB, Wang CY, Nairz M, Bouley R, Swirski FK, Babitt JL. Endothelial cells produce bone morphogenetic protein 6 required for iron homeostasis in mice. Blood. 2017;129(4):405–14.
14. Richards S, Aziz N, Bale S, Bick D, Das S, Gastier-Foster J, Grody WW, Hegde M, Lyon E, Spector E, et al. Standards and guidelines for the interpretation of sequence variants: a joint consensus recommendation of the American College of Medical Genetics and Genomics and the Association for Molecular Pathology. Genet Med. 2015;17(5):405–24.
15. McDonald CJ, Ostini L, Wallace DF, Lyons A, Crawford DH, Subramaniam VN. Next-generation sequencing: application of a novel platform to the analysis of atypical iron disorders. J Hepatol. 2015;63(5):1288-93.
16. Kircher M, Witten DM, Jain P, O'Roak BJ, Cooper GM, Shendure J. A general framework for estimating the relative pathogenicity of human genetic variants. Nat Genet. 2014;46(3):310–5.
17. Heffelfinger SC, Hawkins HH, Barrish J, Taylor L, Darlington GJ. SK HEP-1: a human cell line of endothelial origin. In Vitro Cell Dev Biol. 1992; 28A(2):136–42.

19

Mutational analysis of epidermal and hyperproliferative type I keratins in mild and moderate psoriasis vulgaris patients: a possible role in the pathogenesis of psoriasis along with disease severity

Tamilselvi Elango[1,5*†], Jingying Sun[1], Caihong Zhu[1], Fusheng Zhou[1], Yaohua Zhang[4], Liangdan Sun[1,2,3], Sen Yang[1,2,3] and Xuejun Zhang[1,2,3,4,5*†]

Abstract

Background: Mutations in keratin proteins have been vastly associated with a wide array of genodermatoses; however, mutations of keratins in psoriasis have not been fully investigated. The main aim of the current research was to identify the mutation in K14, K10, K16, and K17 genes in two stages of psoriasis patients.

Methods: Ninety-six psoriatic skin biopsies were collected. mRNA transcript of K14, K10, K16, and K17 was prepared, amplified, and sequenced. Sanger sequences of all keratins were further validated for mutational analysis using Mutation Surveyor and Alamut Visual. Then, in silico analysis of protein stability and protein and gene expression of all keratins was performed and validated.

Results: Out of 44 mutations, about 75% of keratins are highly pathogenic and deleterious. Remaining 25% mutations are less pathogenic and tolerated in nature. In these 33 deleterious mutations were immensely found to decrease keratin protein stability. We also found a correlation between keratin and Psoriasis Area and Severity Index score which added that alteration in keratin gene in skin causes severity of psoriasis.

Conclusions: We strongly concluded that acanthosis and abnormal terminal differentiation was mainly due to the mutation in epidermal keratins. In turn, disease severity and relapsing of psoriasis are mainly due to the mutation of hyperproliferative keratins. These novel keratin mutations in psoriatic epidermis might be one of the causative factors for psoriasis.

Keywords: Epidermal keratins, Hyperproliferative keratins, Psoriasis, Mutation, Sanger sequence, Immunofluorescence

Background

Keratinocytes play a special role in sensing epidermal barrier and regulating immune homeostasis [1]. Its functions mainly depend on structural proteins including keratins [2, 3]. Keratins are the main intermediate filament cytoskeleton in all epithelia [4]. In normal skin, basal cells express two keratins, K5 and K14, whereas suprabasal cells express differentiation-related keratins K1 and K10. These pairs act as a flexible scaffold enabling cells to resist physical stress and also play different cellular functions including protection from apoptosis and regulation of epithelial organization [5, 6]. Consequently, defects in intermediate filaments (Ifs) can lead to cell fragility and are linked to a wide array of genodermatoses and cancers [7, 8].

Psoriasis is a genetically determined chronic inflammatory skin disease characterized by acanthosis, abnormal keratinization, and inflammatory cell infiltrates [9]. In psoriatic lesions, terminal differentiation of keratinocytes is incomplete, which leads to preferential activation and

* Correspondence: tami24684@gmail.com; ayzxj@vip.sina.com
†Equal contributors
[1]Institute and Department of Dermatology, The First Affiliated Hospital, Anhui Medical University, Hefei, China

proliferation of cells that do not mature properly. These changes are mainly associated with altered expression of keratin in psoriasis [10]. In psoriasis, the changes in keratin expression include a reduction in K1 and K10 and the induction of hyperproliferation-associated keratins K6, K16, and K17 [11–13]. Furthermore, the expression levels of K5 and K14 in the basal cell layer are also altered in the psoriatic epidermis. The hyperproliferation in psoriasis seemed to result from an increase in the number of transit amplifying cells, following depletion of the stem cell compartment [14, 15]. As a whole, these changes suggest that each keratin pair provides specific functional requirements to epidermal keratinocytes.

Over 90% of pathogenic mutations in keratinopathies are missense mutations with a small number of small in-frame insertions vs. deletion mutations and a few intronic splice site defects leading to larger in-frame deletions. Various cutaneous disorders are identified with mutated keratin proteins namely epidermolysis bullosa simplex (EBS) (K5, K14), ichthyosis (Kl, K2, and K10), palmoplantar keratoderma (K9), type I pachyonychia congenita (PC) (K6a and K16), seven type II pachyonychia congenita (K6b and K17), and monilethrix (K81, K83, and K86). [16–18]. In most of these conditions, the associated pathology results from fragile keratinocytes expressing the mutated keratin protein.

Since genome-wide association studies (GWAS), connecting the psoriasis to the late cornified envelope gene cluster has specified that epidermal abnormalities along with hyperproliferative keratin pattern play a major role in the pathogenesis of psoriasis [19, 20]. The main aim of this study is to identify the mutation of keratin in two stages of psoriasis, which mainly causes hyperproliferation along with defects in the cornification process. We speculated that mutation in K14 and K10 might serve as an important mechanism that affects keratinocyte proliferation in all stages of psoriasis, and also, mutation in hyperproliferative keratin K16 and K17 might be the foremost cause for the incurability and exacerbation of this disease. To verify our hypotheses, we examined Sanger sequence of the CDS region of all these keratins in both stages. For mutational analysis, we performed a computational analysis of all these keratins in both stages of psoriasis by using Mutation Surveyor and Alamut Visual software. In addition, effects of these mutations on protein stability were predicted by using in silico prediction tools. According to our result, the predicted deleterious mutations were mainly clustered in the rod domain on the keratin protein, which is crucial for keratin function. Most of the deleterious mutations were predicted to decrease protein stability which might cause the changes in protein expression of these keratins. All these changes might trigger or exacerbate psoriasis.

Materials and methods

Patients' details

Patients with psoriasis vulgaris ($n = 96$) who visited the Institute of Dermatology, Anhui Medical University (AHMU), between 2015 and 2017 were recruited in this study. The age of all patients ranged from 16 to 71 years (mean ± SD, 37.97 ± 14.03 years). In this study, based on body surface area involvement, we have categorized patients into two, (i) mild psoriasis patients ($n = 48$) and (ii) moderate psoriasis patients ($n = 48$). Severity of plaque psoriasis was graded into mild and moderate to severe disease. Mild disease was defined as body surface area (BSA) ≤ 10, Psoriasis Area and Severity Index (PASI) ≤ 10, and dermatology life quality index (DLQI) ≤ 10 and moderate to severe psoriasis as BSA > 10 or PASI > 10 and DLQI > 10 [21]. Patients were included in the study based on the following criteria: (i) at least one well-demarcated, erythematous, scaly lesion verified by at least two dermatologists; (ii) each lesion tissue was confirmed by clinical histopathology; (iii) no systematic anti-psoriatic treatment 2 weeks before skin biopsy; and (iv) no topical anti-psoriatic treatments for 1 week prior to biopsy. Informed consent was obtained from all individuals, under an AHMU-approved protocol. The study was approved by the institutional ethical committee and conducted according to the Declaration of Helsinki principles.

Collection of tissue samples

In all patients, 10 mm of lesional and nonlesional skin biopsies were taken after local anesthesia, lidocaine hydrochloride, and adrenaline bitartrate IP were given intradermally. Nonlesional skin biopsy served as control. Biopsies of psoriatic lesional skin were taken within a lesion, 1 cm from the edge of the plaque border. Biopsies of nonlesional skin were taken 2 cm beyond the plaque border. Skin biopsies were immediately frozen in liquid nitrogen.

Histology analysis

Formalin-fixed skin biopsy was embedded in paraffin and processed routinely. Hematoxylin-eosin staining was used to examine the histological changes in mild and moderate psoriatic skin.

RNA isolation and qRT-PCR

RNA from both skin tissues were isolated by TRIZOL Method, and RT-PCR was conducted using the High-Capacity cDNA Reverse Transcription Kits (Applied Biosystems, USA) according to the manufacturer's protocol. Taqman Master Mix (Applied Biosystems, Bedford, MA) was used with Taqman probes. Real-time quantitative PCR was performed with the ABI PRISM7700 Sequence Detection System. All expression values were normalized against

GAPDH. Relative mRNA expression levels of all examined genes were measured using the comparative $2^{-\Delta\Delta CT}$ [22]. All amplifications were done three times in triplicate.

PCR amplification, sequencing, and mutation screening

cDNA prepared from mild and moderate psoriatic RNA samples were amplified using a polymerase chain reaction (PCR). The primers were designed from the CDS region of all these keratins (Table 1). PCR reaction was performed using Biorad Thermal Cycler. The PCR conditions were as given in Table 1. Amplified PCR products were electrophoresed through a 1.5% agarose gel, to control the quality of fragments. Sanger sequencing reactions were performed on the purified PCR fragments using a BigDye® Terminator v3.1 Cycle Sequencing Kit (Applied Biosystems, USA) with the same primers as for the PCRs, forward and reverse primers in separate experiments. All genes were sequenced from both directions. Reactions were processed on the ABI3730xl instrument. Sequence reads were analyzed using Applied Biosystems. Sequence reads were aligned to human genome by using Bioedit software.

Mutation Surveyor is a useful in silico tool developed by SoftGenetics that assists the detection of sequence variations within Sanger sequencing traces. This tool can process up to 400 lanes of data at a time with high accuracy and sensitivity. It can effectively detect SNPs and mutation in their homozygous or heterozygous states as well as mosaicism [23]. In this study, we did sequence analysis and pathogenic variant identification by using Mutation Surveyor DNA variant analysis software (SoftGenetics, USA). For prediction of the mutation's pathogenicity, Alamut Visual software (Interactive Biosoftware, France) was used [24]. Studies have shown that Alamut Visual is the original decision-support software application, used by leading genetic clinicians and researchers around the world. This software was used for alignment, conservation, SIFT/PolyPhen dbSNP, and Exome Sequencing Project data collection to evaluate the variant in its surrounding genomic context [25, 26]. All variants were further annotated with the Exome Aggregation Consortium (ExAC) [27].

Prediction of deleterious mutations

Six best performing tools were combined into a consensus classifier PredictSNP1.0, which gives significantly improved prediction performance and at the same time returned results for all mutations, confirming that consensus prediction represents an accurate and robust alternative to the predictions delivered by individual tools. Also, this is a user-friendly web interface for all researchers and clinician which enables an easy access to all eight prediction tools, the consensus classifier PredictSNP, and annotations from the Protein Mutant Database and the UniProt database. Based on the above details, we used PredictSNP1.0 (http://loschmidt.chemi.muni.cz/predictsnp1), classifier webserver, to predict the effects of mutations on protein function [28]. PredictSNP1.0 runs and integrates MAPP, nsSNPAnalyzer, PANTHER, PhD-SNP, Polyphen-1, Polyphen-2, SIFT, and SNAP to predict disease-related amino acid mutations. The four tools use machine-learning methods; nsSNPAnalyzer uses random forest, PhD-SNP uses support vector machines, PolyPhen-2 uses Naive Bayes, and SNAP uses neural network. SIFT, MAPP, and PANTHER consider alignment score information, and PolyPhen-1 uses an expert set of empirical rules to predict possible impact of amino acid substitutions. PredictSNP1.0 displays the confidence scores generated by each tool and a consensus prediction as percentages by using their observed accuracy values to ease comparisons [29].

Table 1 Primer sequences and reaction conditions used for PCR

S. no.	Primer	Sequences (5′-3′)	Length (bp)	Reaction conditions
1	K14 F K14 R	GTGGGCAGTGAGAAGGTGAC AGAGGAGAACTGGGAGGAGG	966	Denaturation, 30 s at 94 °C Annealing, 45 s at 59 °C Extension, 60 s at 72 °C No. of cycles, 35
2	K10 F K10 R	GGCTCATCAGGTGGCTAT CAGGCTTCAGCATCTTTG	812	Denaturation, 30 s at 94 °C Annealing, 45 s at 58 °C Extension, 60 s at 72 °C No. of cycles, 35
3	K16 F K16 R	GTGAAGATCCGTGACTGG TGCTGGGAGGAAAGGTGG	856	Denaturation, 30 s at 94 °C Annealing, 45 s at 56 °C Extension,60 s at 72 °C No. of cycles, 35
4	K17 F K17 R	CTTCCGCACCAAGTTTGA TTGCCATCCTGGACCTCT	753	Denaturation, 30 s at 94 °C Annealing, 45 s at 55 °C Extension, 60 s at 72 °C No. of cycles, 35

The effects of predicted deleterious mutations on protein stability

The effect of deleterious missense amino acid substitutions on K14, K10, K16, and K17 protein stability was analyzed with MUpro (http://mupro.proteomics.ics.uci.edu) [30] and I-Mutant2.0 (http://folding.biofold.org/i-mutant/i-mutant2.0.html) [31]. MUpro uses support vector machine (SVM) and neural network to predict the effect of amino acid changes on protein stability and calculates a score between − 1 and 1 as the confidence of prediction. A confidence score < 0 indicates the mutation decreases the protein stability, while a confidence score > 0 means the mutation increases the protein stability. I-Mutant 2.0 uses SVM to predict protein stability alterations upon mutations and provides the predicted free energy change value (DDG) and the sign of the prediction as increase or decrease. DDG is calculated as unfolding Gibbs free energy value (mutated protein) – unfolding Gibbs free energy value (wild-type protein) in Kcal/mole. DDG > 0 means increased protein stability, and DDG < 0 means decreased protein stability.

Visualization of predicted deleterious mutations on 3D keratin structure

The location of predicted deleterious mutations was detected by STRUM (http://zhanglab.ccmb.med.umich.edu/STRUM/), which constructs 3D models according to the iterative threading assembly refinement (I-TASSER) simulation [32].

Western blot analysis

Western blot was performed as previously described [33] using primary antibodies as follows: K14, K10, K16, and K17 and β-actin (Thermo-scientific, USA). Band intensity was analyzed using ImageJ software (http://rsb.info.nih.gov/ij/).

Immunofluorescence and confocal microscopy

Immunofluorescence assay was performed as described previously [34]. Skin biopsies were obtained from the lesional and nonlesional skin of each patient (control samples). Tissues were sectioned (4 μm) and stained with rabbit monoclonal anti-cytokeratin 14 and mouse monoclonal anti-cytokeratin10, 16, and 17 (Thermo-scientific, USA). An appropriate Alexa Fluor 488-conjugated anti-rabbit IgG and anti-mouse IgG (Thermo-scientific, USA) and DAPI staining of nuclei were used. The images were examined by the pathologist using an Olympus–Ix Microscope and merged using Image J Software.

Statistical analysis

Results were expressed as mean ± SD. Data were analyzed using SPSS 16.0 software. Nonlesional and lesional skin biopsy samples were compared by Student's *T* test. Spearman's coefficient was used to identify correlations between keratins and PASI score. $P < 0.05$ was considered to be significant.

Results

Baseline analysis of psoriatic lesion

The histological examination of nonlesional skin tissue shows normal rete ridges with stratum corneum whereas that of mild and moderate lesional skin revealed elongated rete ridges with deregulated stratum corneum (Fig. 1a). Histological analysis was further supported by PASI score. Forty-eight patients with mild psoriasis have a PASI score of 4.06 ± 2.46, and the remaining 48 patients with moderate psoriasis have a PASI score of 17.34 ± 5.64 (Fig. 1b).

Mutated keratin sequences identified through Sanger sequencing

Almost in all the keratin mutation and sequence studies, NCBI database is widely used, NM_000526.4 (K14), NM_000421.3 (K10), NM_005557.3 (K16), and NM_000422.2 (K17), as publicly available sequence data for healthy population (control). According to the above keratins, nucleotide accession number of these sequences is also retrieved from the Chinese population [35–39]. So, in order to predict the mutation in two stages of psoriasis, we have taken the K14, K10, K16, and K17 coding sequences of healthy population available in NCBI database with accession numbers NM_000526.4 (K14), NM_000421.3 (K10), NM_005557.3 (K16), and NM_000422.2 (K17), which were further aligned respectively with the Sanger sequences of K14, K10, K16, and K17 obtained in our patients by using BioEdit software (Additional file 1: Figs. S1 and S2). In mild psoriasis patients, we have identified nine and seven missense mutations in K14 and K10, respectively, as well as no missense mutation was found in K16 and K17. However, in moderate psoriasis, we found nine, six, five, and eight mutations of K14, K10, K16, and K17, respectively.

Some of our mutated K14, K10, and K17 sequences were matched to previously reported sequences in other skin disease (Table 2). In order to compare the mutation identified in psoriasis patients with other skin disease patients, we have used the following Clinvar data in the NCBI website (https://www.ncbi.nlm.nih.gov/clinvar/) (Tables 3 and 4). According to the Clinvar data of NM_000526.4 (K14), 11 missense mutations of K14 were found in EBS disease; comparing these mutations with our K14 results, three mutations were found to be identical in both diseases. Likewise, NM_000421.3 (K10) compared with K10 of our patients showed only one identical mutation. Compared with our results, no identical mutation was

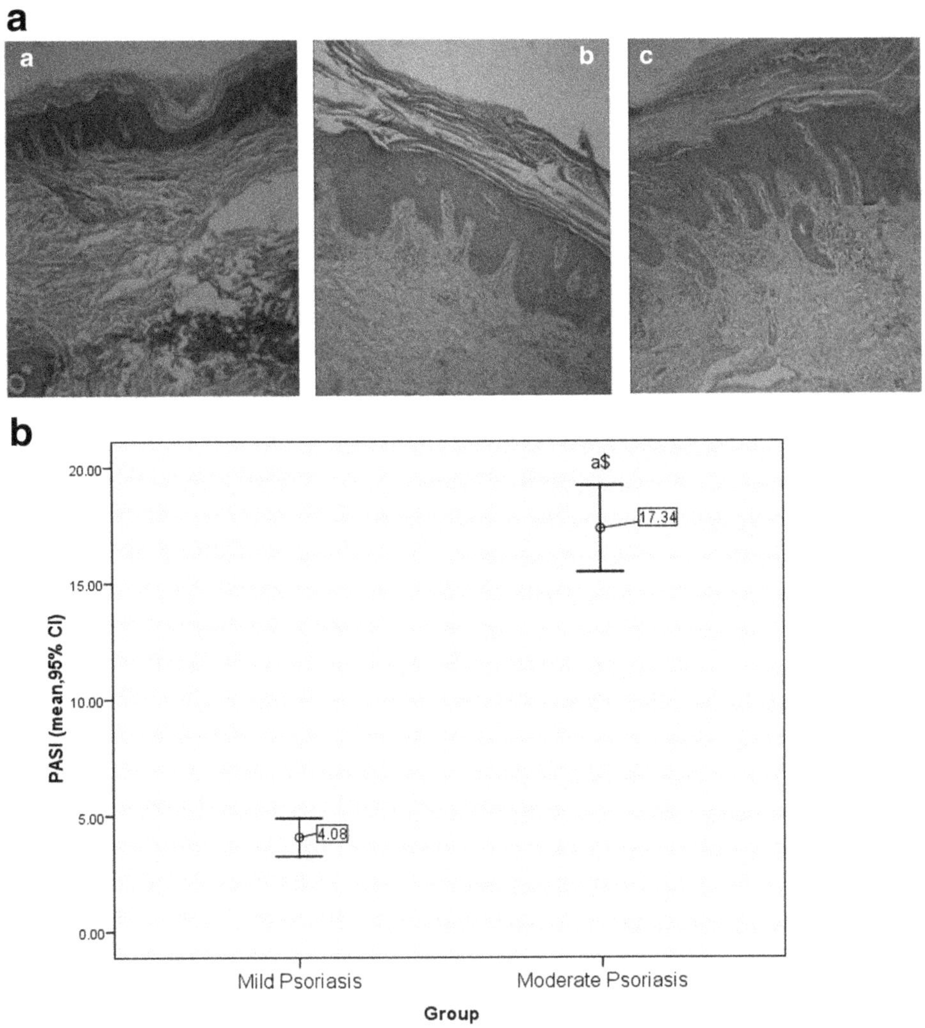

Fig. 1 Histological and clinical examinations of mild and moderate psoriatic skin. **a** Histological examination of skin biopsy. (A) Photomicrograph of nonlesional skin biopsy. (B), (C) Photomicrograph of mild and moderate psoriasis vulgaris showing Munro microabscess, acanthosis, rete ridge elongation, and acute infiltration of dermis. **b** Clinical evidence of mild and moderate psoriatic lesion by PASI score. All values were expressed as mean ± SD. "a" denotes comparison of PASI score of mild and moderate psoriasis. The dollar sign denotes $P < 0.001$

Table 2 Mutated epidermal and hyperproliferative keratin sequences identified in this study consistent with previous reports (clinvar and dbSNP in the NCBI website)

S. no.	Gene	Mutation region	Accession number	
			Clinvar	dbSNP
1	Keratin 14	C442 > T	RCV000056740.1	rs58378809
2		A1234 > T	RCV000056680.1	rs267607403
3		G1237 > A	RCV000056682.1	rs59780231
4		C1246 > T		rs777067461
5	Keratin 10	C520 > A	RCV000056502.1	
6	Keratin 17	C986 > T		rs780535087

found in both keratin 16 and 17. Based on per nucleotide sequence variation, percentage of background mutation rate was calculated. The high background mutation rates of K14 mutated nucleotide sequences in our patients were 33.3% G > C, followed by 22.2% A > T, 16.6% C > T, 11.1% G > T, and 5.5% of C > A, G > A, and C > G, whereas EBS disease condition obtained from the NCBI website showed high mutation rate of 36.3% G > A. Background mutation rates of psoriasis K10 were as follows: G > T, T > A, G > C, A > C, C > G, and C > A with percentages of 30.7, 15.3, and 7.6%. In EI disease, A > C nucleotide K10 mutation is found to be 23% followed with 15.3% of G > A and T > C. The most highest K16 mutated nucleotide found in our study is 40% of C > G whereas 25% of A > T, C > T, and G > T was found in K17.

Table 3 Comparison of publicly available mutated keratin 14 and keratin 10 sequence data of various skin disease patients with mutated sequence obtained in our study (psoriasis) from Clinvar NCBI database

S. no.	Name	Gene(s)	Condition(s)	Clinical significance (Last reviewed)	GRCh37		GRCh38		Variation ID	Allele ID(s)
					Chromosome	Location	Chromosome	Location		
1	NM_000526.4(KRT14): c.1264G > A (p.Glu422Lys)	KRT14	EBH-DM	Pathogenic (Apr 1, 2000)	17	39739497	17	41583245	14623	29662
2	NM_000526.4(KRT14): c.1256T > A (p.Leu419Gln)	KRT14	EBH-DM	Pathogenic (Apr 1, 2000)	17	39739505	17	41583253	14622	29661
3	NM_000526.4(KRT14): c.1243T > C (p.Tyr415His)	KRT14	EBH-DM	Pathogenic (Mar 27, 2015)	17	39739518	17	41583266	14621	29660
4	*NM_000526.4(KRT14): c.1237G > A (p.Ala413Thr)*	*KRT 14*	*Not provided*	*Pathogenic (Jul 31, 2012)*	*17*	*39739524*	*17*	*41583272*	*66319*	*77216*
5	*NM_000526.4(KRT14): c.1234A > T (p.Ile412Phe)*	*KRT14*	*Not provided*	*Pathogenic (Apr 2, 2018)*	*17*	*39739527*	*17*	*41583275*	*66317*	*77214*
6	NM_000526.4(KRT14): c.1228C > T (p.Gln410Ter)	KRT14	Not provided	Pathogenic (Sep 15, 2016)	17	39739533	17	41583281	66313	77210
7	NM_000526.4(KRT14): c.1162C > T (p.Arg388Cys)	KRT14	Not provided	Pathogenic (Oct 13, 2017)	17	39739599	17	41583347	66306	77203
8	NM_000526.4(KRT14): c.1151T > C (p.Leu384Pro)	KRT14	EBS, Koebner type	Pathogenic (Nov 22, 1991)	17	39739610	17	41583358	14611	29650
9	*NM_000526.4(KRT14): c.442C > T (p.Arg148Cys)*	*KRT14*	*EBS, Koebner type*	*Pathogenic (Apr 2, 2018)*	*17*	*39742645*	*17*	*41586393*	*66368*	*77265*
10	NM_000526.4(KRT14): c.374G > A (p.Arg125His)	KRT14	EBH-DM	Pathogenic (Oct 30, 2017)	17	39742713	17	41586461	14613	29652
11	NM_000526.4(KRT14): c.357G > A (p.Met119Ile)	KRT14	EBS, autosomal recessive	Pathogenic (Sep 1, 1997)	17	39742730	17	41586478	14620	29659
12	NM_000421.3(KRT10): c.1374-2A > C	KRT10	Not provided	Pathogenic (Nov 25, 2015)	17	38975415	17	40819163	449615	445767
13	NM_000421.3(KRT10): c.1374-2A > G	KRT10	EI, congenital reticular	Pathogenic (Oct 1, 2010)	17	[illegible]	17	40819163	14581	29620
14	NM_000421.3(KRT10): c.1373 + 2T > C	KRT10	Not provided	Pathogenic (May 9, 2017)	17	38975767	17	40819515	432261	426219
15	NM_000421.3(KRT10): c.1373 + 1G > A	KRT10	EI, congenital reticular	Pathogenic (Oct 1, 2010)	17	38975768	17	40819516	14582	29621
16	NM_000421.3(KRT10): c.1325T > A (p.Leu442Gln)	KRT10	BIE	Pathogenic (Feb 1, 1994)	17	38975817	17	40819565	14575	29614
17	NM_000421.3(KRT10): c.1300C > T (p.Gln434Ter)	KRT10	BIE	Pathogenic (Apr 1, 2006)	17	38975842	17	40819590	29764	38719
18	NM_000421.3(KRT10): c.1281C > A (p.Cys427Ter)	KRT10	BIE	Pathogenic (Jul 1, 2008)	17	38975861	17	40819609	29765	38720
19	NM_000421.3(KRT10): c.494G > C (p.Arg165Pro)	KRT10	Not provided	Likely pathogenic (Sep 1, 2016)	17	38978344	17	40822092	432167	426220
20	NM_000421.3(KRT10): c.482T > C (p.Leu161Ser)	KRT10	BIE	Pathogenic (Aug 21, 1992)	17	38978356	17	40822104	14569	29608
21	*NM_000421.3(KRT10): c.479A > C (p.Tyr160Ser)*	*KRT10*	*Not provided*	*Pathogenic (Apr 2, 2018)*	*17*	*38978359*	*17*	*40822107*	*66178*	*77075*
22	NM_000421.3(KRT10): c.478T > G (p.Tyr160Asp)	KRT10	BIE	Pathogenic (Feb 1, 1994)	17	38978360	17	40822108	14572	29611
23	NM_000421.3(KRT10): c.467G > A (p.Arg156His)	KRT10	BIE	Pathogenic (Sep 20, 2017)	17	38978371	17	40822119	14573	29612
24	NM_000421.3(KRT10): c.460A > C (p.Asn154His)	KRT10	BIE	Pathogenic (Feb 1, 1994)	17	38978378	17	40822126	14571	29610

EBH-DM epidermolysis bullosa herpetiformis-Dowling Meara, *EBS* epidermolysis bullosa simplex, *EI* erythroderma ichthyosiform, *BIE* Bullous ichthyosiform erythroderma

Italicized words indicate comparison of identical mutation obtained in our study (psoriasis) with other skin disease

Validation of mutation and their pathogenicity using Mutation Surveyor and Alamut Visual software

To validate and confirm the mutation, we did in silico analysis of all keratin sequences obtained in these patients by using Mutation Surveyor software. Similar to Sanger sequence results, in silico analysis also showed nine and seven missense mutations in K14 and K10, respectively, as well as no missense mutation was found in

Table 4 Comparison of publicly available mutated keratin 16 and keratin 17 sequence data of various skin disease patients with mutated sequence obtained in our study (psoriasis) from Clinvar NCBI database

S. no.	Name	Gene(s)	Condition(s)	Clinical significance (last reviewed)	GRCh37		GRCh38		Variation ID	Allele ID(s)
					Chromosome	Location	Chromosome	Location		
1	NM_005557.3(KRT16):c.395T > C (p.Leu132Pro)	KRT16	PC 1	Pathogenic (Jan 29, 2016)	17	39768546	17	41612294	14600	29639
2	NM_005557.3(KRT16):c.379C > G (p.Arg127Gly)	KRT16	Not provided	Pathogenic (May 11, 2015)	17	39768562	17	41612310	265217	260165
3	NM_000422.2(KRT17):c.1163T > C (p.Leu388Pro)	KRT17	Not provided	Not provided	17	39776929	17	41620677	66181	77078
4	NM_000422.2(KRT17):c.1112T > C (p.Leu371Pro)	KRT17	Not provided	Not provided	17	39776980	17	41620728	66180	77077
5	NM_000422.2(KRT17):c.325A > G (p.Asn109Asp)	KRT17	Not provided	Not provided	17	39780437	17	41624185	66188	77085
6	NM_000422.2(KRT17):c.309T > C (p.Arg103=)	KRT17	Not provided	Not provided	17	39780453	17	41624201	66187	77084
7	NM_000422.2(KRT17):c.304G > A (p.Val102Met)	KRT17	PC 2	Pathogenic (Mar 1, 2002)	17	39780458	17	41624206	14599	29638
8	NM_000422.2(KRT17):c.296T > C (p.Leu99Pro)	KRT17	PC 2	Pathogenic (Jun 9, 2015)	17	39780466	17	41624214	14598	29637
9	NM_000422.2(KRT17):c.292T > G (p.Tyr98Asp)	KRT17	PC 2	Pathogenic (Feb 1, 1997)	17	39780470	17	41624218	14588	29627
10	NM_000422.2(KRT17):c.284T > C (p.Leu95Pro)	KRT17	PC 2	Pathogenic (May 14, 2015)	17	39780478	17	41624226	14596	29635
11	NM_000422.2(KRT17):c.284T > A (p.Leu95Gln)	KRT17	PC 2	Pathogenic (May 1, 2001)	17	39780478	17	41624226	14595	29634
12	NM_000422.2(KRT17):c.281G > C (p.Arg94Pro)	KRT17	PC 2	Pathogenic (May 1, 2001)	17	39780481	17	41624229	14594	29633
13	NM_000422.2(KRT17):c.281G > A (p.Arg94His)	KRT17	PC 2| SM	Pathogenic (Dec 1, 2001)	17	39780481	17	41624229	14590	29629
14	NM_000422.2(KRT17):c.280C > T (p.Arg94Cys)	KRT17	PC 2|	Pathogenic(Aug 15, 2016)	17	39780482	17	41624230	14591	29630
15	NM_000422.2(KRT17):c.275A > G (p.Asn92Ser)	KRT17	PC 2	Pathogenic (Jul 26, 2016)	17	39780487	17	41624235	14587	29626
16	NM_000422.2(KRT17):c.274A > C (p.Asn92His)	KRT17	SM	Pathogenic (Feb 1, 1997)	17	39780488	17	41624236	14589	29628
17	NM_000422.2(KRT17):c.274A > G (p.Asn92Asp)	KRT17	PC 2	Pathogenic (Nov 3, 2016)	17	39780488	17	41624236	14586	29625
18	NM_000422.2(KRT17):c.263T > A (p.Met88Lys)	KRT17	Not provided	Not provided	17	39780499	17	41624247	66182	77079
19	NM_000422.2(KRT17):c.263T > C (p.Met88Thr)	KRT17	PC 2	Pathogenic (Nov 1, 1999)	17	39780499	17	41624247	14592	29631
20	NM_000422.2(KRT17):c.-6G > C	KRT17	Not specified	Benign (Apr 28, 2016)	17	39780767	17	41624515	380281	375886

PC pachyonychia congenita, *SM* steatocystoma multiplex

K16 and K17 in mild psoriasis (Fig. 2). Also, in moderate psoriasis, a similar mutation pattern was shown (Figs. 3 and 4) using Mutation Surveyor software. The mutation's pathogenicity was predicted by using Alamut Visual software; we found about 34 and 15% of K14 and K10 mutated sequences in mild psoriasis that were predicted to be tolerated, and the remaining sequences were predicted to have damaging effect. Based on Mutation Assessor, all mutated sequences in mild psoriasis have medium effect on the alteration of these protein functions (Table 5; Additional file 1: Fig. S1) where, as in moderate psoriasis, all mutated sequences are predicted to have a damaging role with high effect on the alteration of protein functions (Table 6; Additional file 1: Fig. S2). Totally, we identified 44 mutations among which 16 mutations were found in both mild and moderate psoriasis, whereas 28 mutations were found only in moderate psoriasis.

Comparison of mutated keratins with ExAC databases

When we compared K14, K10, K16, and K17 missense mutated variants in ExAC database with their respective K14, K10, K16, and K17 mutated variants observed in our psoriasis patients, we perceived that one K14 c1237G > A variant with MAF 0.01 and one K17 c906C > T variant with MAF < 0.001 existed in the ExAC database; additionally, none of the K10 and K16 variants were found to occur in the ExAC database. Based on the analysis, our result showed that 99% of K14 and K17 and 100% of K10 and K16 missense mutations in our psoriasis patients were considered to be as novel missense mutation which did not exist in the ExAC database. These novel mutations might be one of the reasons causing severe damage in psoriasis patients (Additional file 2: Table S1).

Identification of deleterious mutations

Keratin 14, 10, 16, and 17 CDS region of amino acid sequence in Fasta format and 44 missense mutations, which were obtained from Mutation Surveyor and Visual Alamut, were loaded to PredictSNP1.0, and all available integrated tools were selected for prediction. PredictSNP1.0 provided predictions for each integrated tool and a consensus prediction as percentages (expected accuracies) and the effect of mutation on protein function as "neutral" and "deleterious" by PredictSNP and all the integrated tools, except for nsSNPAnalyzer, which did not give any prediction for any mutations. In mild psoriasis, about 70% of mutated sequences in both K14 and K10 were found to be deleterious, whereas in moderate psoriasis, all K14 and K10; one K16 and six K17 mutated sequences were found to be deleterious. Out of 44 mutations, 33 mutations were predicted to be as deleterious by PredictSNP and all the integrated tools, except for nsSNPAnalyzer, which did not give any prediction for any mutations (Tables 7 and 8).

Location of the deleterious mutations on the secondary structure and 3D structure of keratin protein

Position of these deleterious mutations in the keratin secondary and 3D structure is important because amino acid sequences of keratin are mainly involved in the assembly of keratin filaments and also binding of keratins and keratin filaments to cell adhesion complexes or signaling molecules. The secondary structure of these keratin proteins has three sub-domains, namely, head domain, central α-helical rod domain, and tail domain. The rod domain is mainly composed of sub-domain coils 1A, 1B, 2A, and 2B connected by linkers L1, L12, and L2 (Fig. 5a). We detected that these 33 mutations were located in the rod domain of these keratin proteins. The deleterious mutations in the rod domain were

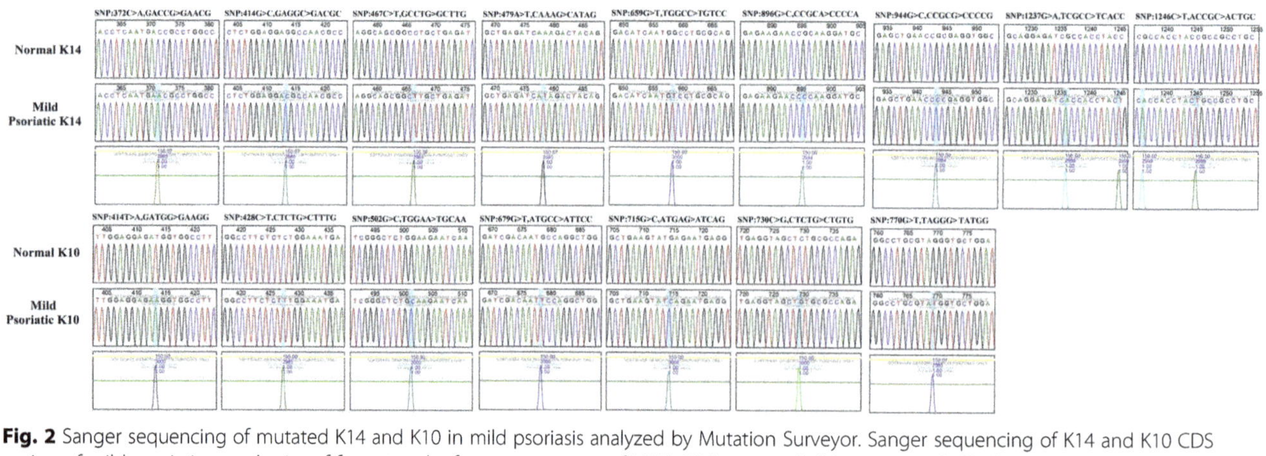

Fig. 2 Sanger sequencing of mutated K14 and K10 in mild psoriasis analyzed by Mutation Surveyor. Sanger sequencing of K14 and K10 CDS region of mild psoriatic samples in scf format and reference sequence of NCBI CDS region of all keratins in Gb file format were analyzed by using Mutation Surveyor; the positions are annotated according to the genetic code

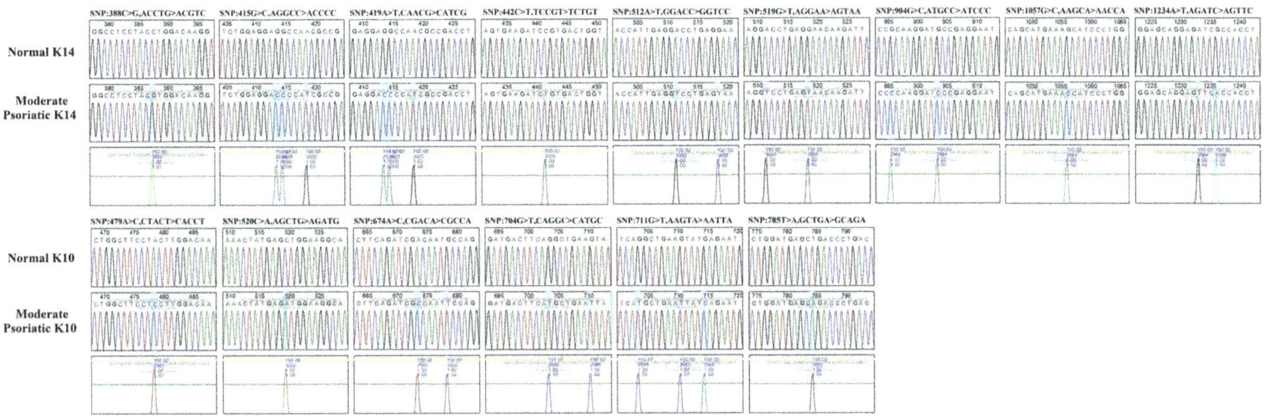

Fig. 3 Sanger sequencing of mutated K14 and K10 in moderate psoriasis analyzed by Mutation Surveyor. Sanger sequencing of K14 and K10 CDS region of moderate psoriatic samples in scf format and reference sequence of NCBI CDS region of all keratins in Gb file format were analyzed by using Mutation Surveyor; the positions are annotated according to the genetic code

mainly noticed in coil 1 and 2 regions, which are critical for the protein activity and structure of these keratins.

Pathogenic K14 mutated sequences are mainly located in exon 1 and exon 6. In keratin 14, exon 1 (1–586 nucleotide) mainly forms the head and rod (coil 1A, L1, and 1B) region and exon 6 (1115–1335 nucleotide) forms the coil 2 of the rod domain. In K14, coil 1A region has five mutated amino acids, L1 region has one mutated amino acid, coil 1B has three mutated amino acids, and coil 2 region has six mutated amino acids. All appeared in close proximity in the folded protein (Fig. 5b).

Exons 1 and 2 are the major regions affected by the mutation in K10; they mainly form the head and rod domain region of 1A and 1B. We found three mutations in coil 1A region and eight mutations in coil 1B region of the rod domain. In K16, one mutation in coil 1B region was found to be deleterious. However, in K17, one mutation in coil 1B region of peptide epitope S2, two mutations in 1B region of peptide epitope S3, and three mutations in coil 2 region of peptide epitope S4 of the rod domain were found (Fig. 5b) The deleterious mutations in the rod domain of all these keratins were mainly noticed in coil 1 and 2 regions, which are critical for the protein activity and structure of these keratins.

Our results showed that the predicted deleterious mutations highly located in the α-helical rod domain, which forms a coiled structure to these keratins, are important to maintain the structural integrity of the skin.

Effect of deleterious mutation on keratin protein stability

After predicting the deleterious effect and position of these mutated sequences in protein, it is imperative to analyze the stability of this protein. Therefore, we have

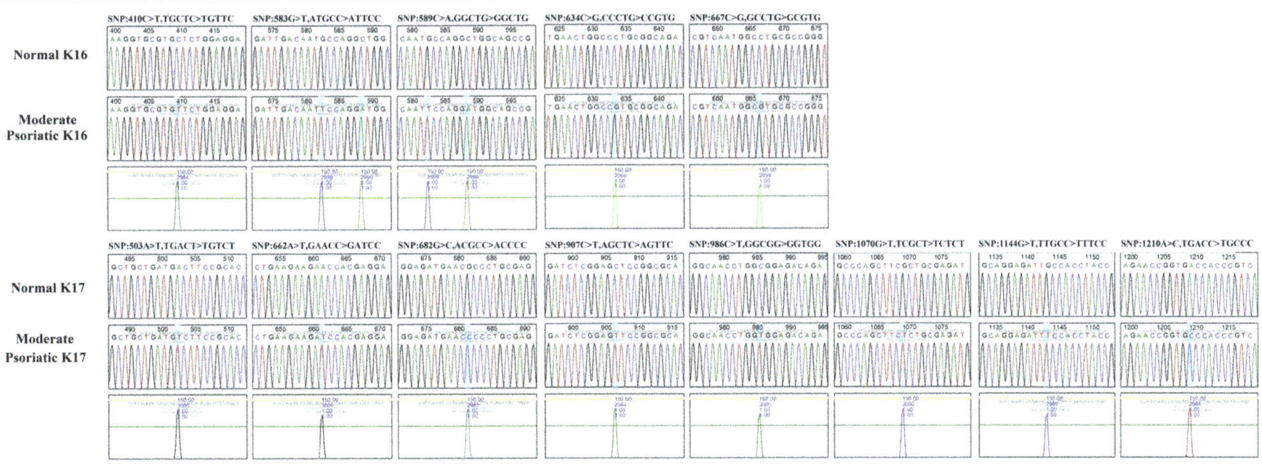

Fig. 4 Sanger sequencing of mutated K16 and K17 in moderate psoriasis analyzed by Mutation Surveyor. Sanger sequencing of K16 and K17 CDS region of moderate psoriatic samples in scf format and reference sequence of NCBI CDS region of all keratins in Gb file format were analyzed by using Mutation Surveyor; the positions are annotated according to the genetic code

Table 5 Prediction of epidermal keratin mutation's pathogenicity in mild psoriasis using Aamut Visual software

S. no.	MT position	Exon	seq.MT	Type of MT	AA change	Variant in protein domain	SIFT			Mutation taster			Mutation assessor		
							Prediction	Score	Range score	Prediction	Range score	*P* value	Prediction	Score	Range score
	K14														
1	372	1	C > A	Missense	p.(Asp124Glu)	IF	Damagin	0.017	0.5119	DC	0.4153	0.992	Medium	2.875	0.8364
2	414	1	G > C	Missense	p.(Glu138Asp)	IF	Tolerated	0.057	0.3792	DC	0.4441	0.998	Medium	2.0099	0.5492
3	467	1	C > T	Missense	p.(Pro156Leu)	IF	Damagin	0.006	0.6138	DC	0.8103	1	Medium	3.335	0.9109
4	479	1	A > T	Missense	p.(Lys160Ile)	IF	Damagin	0.019	0.5008	Polymorphism	0.08979	1	Medium	3.125	0.8823
5	659	2	G > T	Missense	p.(Gly220Val)	IF protein, keratin type I	Damagin	0	0.9122	DC	0.8103	1	Medium	2.565	0.7524
6	896	4	G > C	Missense	p.(Arg299Pro)	IF protein, prefoldin	Damagin	0.008	0.5857	DC	0.4637	0.999	Medium	3.1099	0.8799
7	944	4	G > C	Missense	p.(Arg315Pro)	IF protein, prefoldin	Damagin	0.006	0.6138	DC	0.3752	0.946	Medium	2.3399	0.6737
8	1237	6	G > A	Missense	p.(Ala413Thr)	IF protein (conserved site), IF	Tolerated	0.099	0.3061	Polymorphism	0.4614	0.001	Medium	2.2999	0.6591
9	1246	6	C > T	Missense	p.(Arg416Cys)	IF protein (conserved site), IF	Tolerated	0.132	0.2648	DC	0.8103	1	Medium	2.9349	0.8489
	K10														
10	414	1	T > A	Missense	p.(Asp138Glu)		Tolerated	0.828	0.02894	DC, polymorphism	0.8103	0.992	Medium	2.1549	0.6042
11	428	1	C > T	Missense	p.(Ser143Phe)		Damagin	0.016	0.518	DC	0.8103	0.55	Medium	2.14	0.6007
12	502	1	G > C	Missense	p.(Glu168Gln)	IF protein	Damagin	0.005	0.6317	DC	0.8103	1	Medium	3.4849	0.9277
13	679	2	G > T	Missense	p.(Ala227Ser)	IF protein, keratin type I	Damagin	0.023	0.4813	DC	0.8103	1	Medium	2.6349	0.7736
14	715	2	G > C	Missense	p.(Glu239Gln)	IF protein, keratin type I	Damagin	0.001	0.7842	DC	0.8103	1	Medium	2.7799	0.8138
15	730	2	C > G	Missense	p.(Leu244Val)	IF protein	Damagin	0.005	0.6317	DC	0.8103	0.97	Medium	2.8199	0.8237
16	770	3	G > T	Missense	p.(Arg257Met)	IF protein, keratin type I	Damagin	0	0.9122	DC	0.8103	0.916	Medium	3.25	0.9002

DC disease causing, *IF* intermediate filament, *MT* mutation, *AA* amino acid

Table 6 Prediction of epidermal and hyperproliferative keratin mutation's pathogenicity in moderate psoriasis using Alamut Visual software

S. no.	MT position	Exon	seq.MT	Type of MT	AA change	Variant in protein domain	SIFT			Mutation taster			Mutation assessor		
							Prediction	Score	Range score	Prediction	Range score	*P* value	Prediction	Score	Range score
	K14														
1	388	1	C > G	Missense	p.(Leu130Val)	IF protein	Damaging	0.001	0.7842	DC	0.537	1	High	3.8199	0.9566
2	415	1	G > C	Missense	p.(Ala139Pro)	IF protein	Damaging	0.006	0.6138	DC	0.8103	1	High	4.3049	0.9832
3	419	1	A > T	Missense	p.(Asn140Ile)	IF protein	Damaging	0	0.9122	DC	0.5881	1	High	4.425	0.9876
4	442	1	C > T	Missense	p.(Arg148Cys)	IF protein	Damaging	0	0.9122	DC	0.5881	1	High	3.75	0.9513
5	512	1	A > T	Missense	p.(Asp171Val)	IF protein	Damaging	0	0.9122	DC	0.8103	1	High	3.835	0.9576
6	519	1	G > T	Missense	p.(Arg173Ser)	IF protein	Damaging	0.001	0.7842	DC	0.4122	0.991	High	4.025	0.9697
7	904	4	G > C	Missense	p.(Ala302Pro)	IF protein, prefoldin	Damaging	0.001	0.7842	Polymorphism	0.2979	0.735	High	4.085	0.9729
8	1057	6	G > C	Missense	p.(Ala353Pro)	IF protein, keratin type I	Tolerated	0.189	0.2119	DC	0.4816	0.999	High	3.875	0.9604
9	1234	6	A > T	Missense	p.(Ile412Phe)	IF protein (conserved site), IF	Damaging	0	0.9122	DC	0.8103	1	High	4.65	0.9939
	K10														
10	479	1	A > C	Missense	p.(Tyr160Ser)	IF protein	Damaging	0	0.9122	DC	0.8103	1	High	4.5549	0.9916
11	520	1	C > A	Missense	p.(Leu174Met)	IF protein	Damaging	0.001	0.7842	DC	0.8103	0.991	High	3.5099	0.9302
12	674	2	A > C	Missense	p.(Asp225Ala)	IF protein	Damaging	0	0.9122	DC	0.8103	1	High	3.9549	0.9656
13	704	2	G > T	Missense	p.(Arg235Met)	IF protein, keratin type I	Damaging	0	0.9122	DC	0.8103	0.84	High	4.2849	0.9824
14	711	2	G > T	Missense	p.(Lys237Asn)	IF protein, keratin type I	Damaging	0	0.9122	DC	0.8103	0.983	High	3.66	0.944
15	785	3	T > A	Missense	p.(Leu262Gln)	IF protein, prefoldin, keratin I	Damaging	0	0.9122	DC	0.8103	0.834	High	4.2399	0.9805
	K16														
16	410	1	C > T	Missense	p.(Ala137Val)	IF	Damaging	0.002	0.7209	Polymorphism	0.3052	0.65	Medium	3.39	0.9174
17	583	1	G > T	Missense	p.(Ala195Ser)	IF protein, keratin type I	Damaging	0.029	0.457	DC	0.4734	1	Medium	2.805	0.8202
18	589	1	C > A	Missense	p.(Leu197Met)	IF protein, keratin type I	Damaging	0.028	0.4608	DC	0.4467	0.998	Medium	2.88	0.8371
19	634	1	C > G	Missense	p.(Leu212Val)	IF	Damaging	0.009	0.5743	DC	0.4067	0.988	Medium	2.9849	0.8585
20	667	1	C > G	Missense	p.(Leu223Val)	IF protein, keratin type I	Damaging	0.001	0.7842	DC	0.4434	1	Medium	3.3	0.9406
	K17														
21	503	1	A > T	Missense	p.(Asp168Val)	IF protein, keratin type I	Damaging	0	0.9122	DC	0.8103	1	Medium	3.3	0.9473
22	662	2	A > T	Missense	p.(Asn221Ile)	IF protein, prefoldin	Damaging	0.01	0	Polymorphism	0.4721	0.999	Medium	2.605	0.7645
23	682	3	G > C	Missense	p.(Ala228Pro)	IF protein, prefoldin	Damaging	0.016	0.518	Polymorphism	0.1948	1	Medium	3.25	0.9409
24	907	4	C > T	Missense	p.(Leu303Phe)	IF protein, prefoldin	Damaging	0.03	0, 0.001	Polymorphism	0.5164	1	Medium	3.4749	0.9267
26	986	5	C > T	Missense	p.(Ala329Val)	IF protein	Damaging	0.51	0.017	Polymorphism	0.4593	0.999	Medium	3.17	0.8891
26	1070	5	G > T	Missense	p.(Arg357Leu)	IF protein, keratin type I	Damaging	0	0.9122	DC	0.8103	1	Medium	3.17	0.9796
27	1144	6	G > T	Missense	p.(Ala382Ser)	IF protein (conserved site), IF	Tolerated	0	0.122	Polymorphism	0.4437	0.998	Medium	2.265	0.6465
28	1210	6	A > C	Missense	p.(Thr404Pro)	IF protein (conserved site), IF	Tolerated	0	0.075	Polymorphism	0.5199	1	Medium	2.6099	0.7658

DC disease causing, *IF* intermediate filament, *MT* mutation, *AA* amino acid

Table 7 The expected accuracy results of the missense mutations of epidermal keratins in mild psoriasis predicted as deleterious in PredictSNP and integrated tool

Gene	CDS mutation	AA mutation	Predict SNP		MAPP		PhD-SNP		PolyPhen-1		PolyPhen-2		SIFT		SNAP		PANTHER	
			Prediction	Accuracy	Prediction	Accuracy	Prediction	Accuracy	Prediction	Accuracy	Prediction	Accuracy	Prediction	Accuracy	Prediction	Accuracy	Prediction	Accuracy
K14	c.372C > A	p.D124E	DEL	0.55	DEL	0.59	DEL	0.68	NEU	0.67	DEL	0.59	DEL	0.53	NEU	0.67	DEL	0.69
	c.414G > C	p.E138D	NEU	0.65	NEU	0.64	DEL	0.59	NEU	0.67	NEU	0.72	DEL	0.43	NEU	0.71	DEL	0.66
	c.467C > T	p.P156L	NEU	0.60	DEL	0.48	NEU	0.66	NEU	0.67	DEL	0.40	DEL	0.53	NEU	0.58	DEL	0.67
	c.479A > T	p.K160I	DEL	0.55	DEL	0.41	NEU	0.66	DEL	0.74	NEU	0.61	DEL	0.45	DEL	0.62	DEL	0.57
	c.659G > T	p.G220V	DEL	0.51	NEU	0.73	NEU	0.45	DEL	0.74	DEL	0.65	DEL	0.79	NEU	0.55	DEL	0.74
	c.896G > C	p.R299P	DEL	0.87	DEL	0.75	DEL	0.88	DEL	0.59	DEL	0.50	DEL	0.53	DEL	0.72	DEL	0.74
	c.944G > C	p.R315P	DEL	0.76	DEL	0.62	DEL	0.82	DEL	0.59	DEL	0.50	DEL	0.79	NEU	0.50	DEL	0.70
	c.1237G > A	p.A413T	NEU	0.65	DEL	0.62	NEU	0.51	NEU	0.67	DEL	0.40	NEU	0.67	NEU	0.71	NEU	0.47
	c.1246C > T	p.R416C	DEL	0.61	DEL	0.57	DEL	0.88	DEL	0.59	DEL	0.50	NEU	0.71	NEU	0.55	DEL	0.78
K10	c.414T > A	p.D138E	NEU	0.83	NEU	0.77	NEU	0.66	0	0	0	0	NEU	0.81	NEU	0.83	NEU	0.67
	c.428C > T	p.S143F	NEU	0.60	NEU	0.75	DEL	0.61	0	0	0	0	DEL	0.53	NEU	0.55	DEL	0.74
	c.502G > C	p.E168Q	DEL	0.76	DEL	0.59	DEL	0.86	DEL	0.59	DEL	0.63	DEL	0.53	NEU	0.50	DEL	0.74
	c.679G > T	p.A227S	DEL	0.52	DEL	0.41	DEL	0.77	NEU	0.67	DEL	0.50	DEL	0.46	NEU	0.71	DEL	0.57
	c.715G > C	p.E239Q	DEL	0.76	DEL	0.48	DEL	0.68	DEL	0.59	DEL	0.63	DEL	0.79	NEU	0.55	DEL	0.74
	c.730C > G	p.L244V	DEL	0.55	DEL	0.59	DEL	0.68	NEU	0.67	DEL	0.50	DEL	0.53	NEU	0.71	DEL	0.66
	c.770G > T	p.R257M	DEL	0.76	DEL	0.51	DEL	0.73	DEL	0.74	DEL	0.54	DEL	0.79	NEU	0.58	DEL	0.74

DEL deleterious, *NEU* neutral

Table 8 The expected accuracy results of the missense mutations of epidermal and hyperproliferative keratins in moderate psoriasis predicted as deleterious in PredictSNP and integrated tool

Gene	CDS mutation	AA mutation	Predict SNP		MAPP		PhD-SNP		PolyPhen-1		PolyPhen-2		SIFT		SNAP		PANTHER	
			Prediction	Accuracy	Prediction	Accuracy	Prediction	Accuracy	Prediction	Accuracy	Prediction	Accuracy	Prediction	Accuracy	Prediction	Accuracy	Prediction	Accuracy
K14	c.388C > G	p.L130V	DEL	0.76	DEL	0.72	DEL	0.86	DEL	0.74	DEL	0.65	DEL	0.79	NEU	0.58	DEL	0.77
	c.415G > C	p.A139P	DEL	0.87	DEL	0.59	DEL	0.82	DEL	0.74	DEL	0.65	DEL	0.53	DEL	0.56	DEL	0.74
	c.419A > T	p.N140I	DEL	0.87	DEL	0.84	DEL	0.86	DEL	0.74	DEL	0.81	DEL	0.79	DEL	0.62	DEL	0.87
	c.442C > T	p.R148C	DEL	0.76	DEL	0.59	DEL	0.88	DEL	0.74	DEL	0.68	DEL	0.79	NEU	0.55	DEL	0.87
	c.512A > T	p.D171V	DEL	0.76	DEL	0.59	DEL	0.82	DEL	0.74	DEL	0.56	DEL	0.79	NEU	0.50	DEL	0.74
	c.519G > T	p.R173S	DEL	0.76	DEL	0.59	DEL	0.61	DEL	0.59	DEL	0.43	DEL	0.79	NEU	0.50	DEL	0.74
	c.904G > C	p.A302P	DEL	0.61	DEL	0.75	DEL	0.82	DEL	0.59	NEU	0.64	DEL	0.79	NEU	0.50	DEL	0.84
	c.1057G > C	p.A353P	DEL	0.61	DEL	0.57	DEL	0.86	DEL	0.74	DEL	0.47	NEU	0.74	NEU	0.61	DEL	0.68
	c.1234A > T	P.I412F	DEL	0.87	DEL	0.63	DEL	0.77	DEL	0.74	DEL	0.81	DEL	0.79	DEL	0.56	–	0.00
K10	c.479A > C	p.Y160S	DEL	0.87	DEL	0.86	DEL	0.86	DEL	0.74	DEL	0.81	DEL	0.79	DEL	0.72	–	0.00
	c.520C > A	p.L174M	DEL	0.76	DEL	0.72	DEL	0.59	DEL	0.74	DEL	0.65	DEL	0.79	NEU	0.55	DEL	0.72
	c.674A > C	p.D225A	DEL	0.87	DEL	0.66	DEL	0.68	DEL	0.74	DEL	0.68	DEL	0.79	DEL	0.56	DEL	0.77
	c.704G > T	p.R235M	DEL	0.76	DEL	0.59	DEL	0.82	DEL	0.74	DEL	0.81	DEL	0.79	NEU	0.55	DEL	0.78
	c.711G > T	p.K237N	DEL	0.76	DEL	0.75	DEL	0.88	DEL	0.59	DEL	0.63	DEL	0.79	NEU	0.50	DEL	0.74
	c.785T > A	p.L262Q	DEL	0.76	DEL	0.41	DEL	0.61	DEL	0.74	DEL	0.81	DEL	0.79	NEU	0.50	DEL	0.78
K16	c.410C > T	p.A137V	NEU	0.63	NEU	0.70	NEU	0.45	DEL	0.59	NEU	0.61	DEL	0.79	NEU	0.71	DEL	0.68
	c.583G > T	p.A195S	NEU	0.65	NEU	0.66	NEU	0.51	NEU	0.67	DEL	0.47	DEL	0.45	NEU	0.71	DEL	0.72
	c.589C > A	p.L197M	NEU	0.60	NEU	0.79	DEL	0.59	NEU	0.67	DEL	0.59	DEL	0.45	NEU	0.83	DEL	0.67
	c.634C > G	p.L212V	NEU	0.60	DEL	0.51	NEU	0.51	NEU	0.67	DEL	0.45	DEL	0.53	NEU	0.77	DEL	0.71
	c.667C > G	p.L223V	DEL	0.76	DEL	0.75	DEL	0.77	DEL	0.74	DEL	0.65	DEL	0.79	NEU	0.55	DEL	0.71
K17	c.503A > T	p.D168V	DEL	0.87	DEL	0.63	DEL	0.82	DEL	0.59	DEL	0.68	DEL	0.79	DEL	0.56	DEL	0.87
	c.662A > T	p.N221I	DEL	0.72	NEU	0.77	DEL	0.59	DEL	0.74	DEL	0.81	DEL	0.79	DEL	0.56	DEL	0.74
	c.682G > C	p.A228P	DEL	0.72	DEL	0.48	DEL	0.88	DEL	0.74	DEL	0.50	DEL	0.46	NEU	0.61	DEL	0.73
	c.907C > T	p.L303F	DEL	0.51	NEU	0.75	NEU	0.51	DEL	0.74	DEL	0.65	DEL	0.79	NEU	0.50	DEL	0.78
	c.986C > T	p.A329V	DEL	0.55	DEL	0.48	DEL	0.77	NEU	0.67	DEL	0.40	DEL	0.46	NEU	0.61	DEL	0.72
	c.1070G > T	p.R357L	DEL	0.72	DEL	0.77	DEL	0.68	DEL	0.59	NEU	0.68	DEL	0.79	DEL	0.56	DEL	0.76
	c.1144G > T	p.A382S	NEU	0.83	NEU	0.64	NEU	0.51	NEU	0.67	NEU	0.70	NEU	0.71	NEU	0.71	DEL	0.57
	c.1210A > C	p.T404P	NEU	0.63	DEL	0.51	DEL	0.68	NEU	0.67	NEU	0.71	NEU	0.65	NEU	0.50	NEU	0.48

DEL deleterious, *NEU* neutral

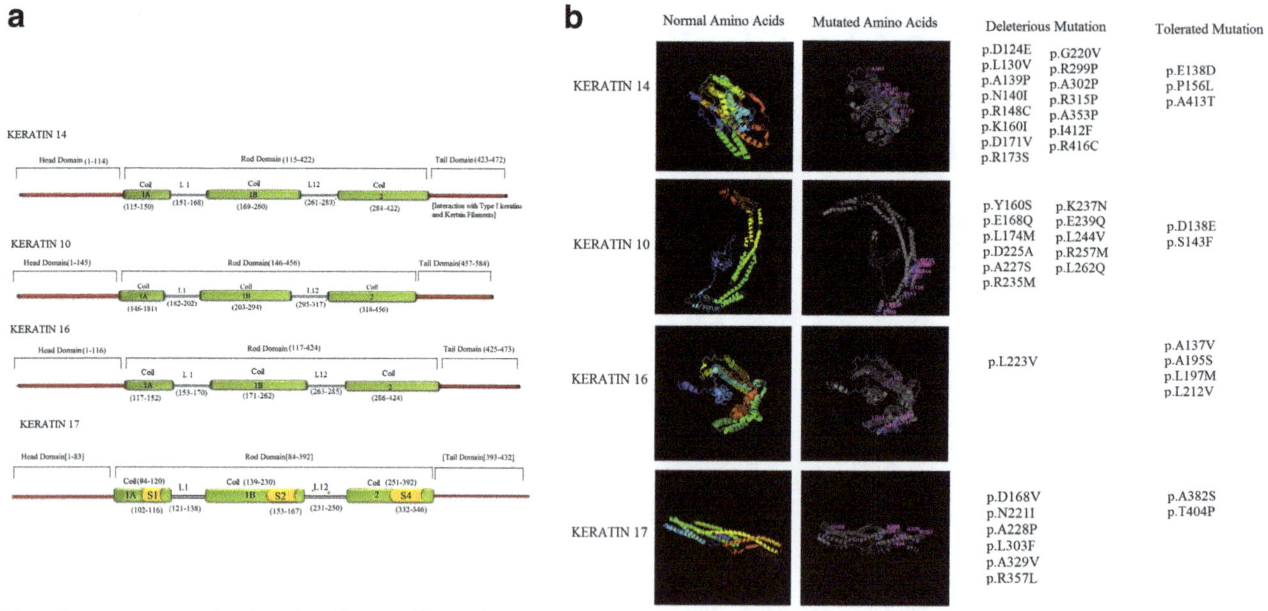

Fig. 5 Protein structure of epidermal and hyperproliferative keratins. **a** Schematic representation of secondary structure of all four keratins with their domain and sub-domain, whereas, S = peptide epitope and L = linker. **b** Specific regions of 33 deleterious mutation localizations. The figures in the left side are the 3D structure of all four keratin CDS regions. The figures in the right side are the 3D structures of domains with mutated residues. The mutated residues are numbered according to their position on the regions. The position of mutated amino acids (aa) on all four keratins and on CDS regions are provided

analyzed the protein stability of these 33 deleterious mutated sequences on MUpro server and I-Mutant 2.0. Our results showed that in mild psoriasis, about two third of K14 and K10 mutation decrease the protein stability. However, in moderate stage, 90% of all mutated K14, K10, K16, and K17 sequences showed decreased protein stability (Table 9).

Immunofluorescence analysis of K14, K10, K16, and K17 antibodies

Our immunofluorescence analysis of normal epidermis shows nuclei stained blue with DAPI throughout all the epidermal layers, whereas K14-positive cells are found only in the basal epidermal cell. Merging discloses co-expression of K14 and DAPI nuclear stain in most of the basal epidermal cells in control samples. In lesional samples, no staining of K14 was seen in the basal layer, whereas DAPI staining was seen throughout the elongated rete ridges, and merged results show there is no staining of K14 in the basal layer (Fig. 6).

DAPI stain was seen throughout all the epidermal layers, and K10 antibodies were stained in the spinosum of the epidermal layer in normal epidermis. K10-positive cells were seen marginally in the elongated rete ridges of the suprabasal epidermal layer. Merging of two stains shows marginal expression of K10 and strong DAPI staining in the nuclei of cells of the suprabasal epidermal layer (Fig. 6).

To confirm the expression of abnormal proteins K16 and K17 on the protein level, we performed immunofluorescence analyses in the corresponding control and lesional psoriatic skin samples. These analyses clearly reflected that there is no K16 expression, and blue DAPI stain stained the nuclei present in the normal epidermal layer. Psoriasis samples showed a strong expression of K16 throughout the suprabasal with nuclei stained with DAPI (Fig. 6). In addition, we analyzed there is no K17 protein in the normal epidermal layer, but nuclei were stained with DAPI and merging showed the blue nuclei staining throughout the epidermis, whereas in psoriasis, the K17 protein was seen in the stratum corneum along with DAPI staining (Fig. 6).

Expression of keratins and their correlation with PASI in both mild and moderate psoriases

In mild psoriatic samples, the mRNA expression of K14 and K10 was significantly elevated to 2.5- and 4.4-folds compared to nonlesional, respectively, whereas in moderate case, these expressions were decreased significantly. In contrast to K14 and K10 expression, decrease in K16 and K17 expression was found in mild psoriasis; as the disease progresses in moderate psoriasis, the expression of K16 and K17 significantly elevated at 4.9- and 4.4-folds compared to nonlesional, respectively (Fig. 7a). Further, these results were confirmed by the Western blot analysis (Fig. 7b). We strongly believed that the changes in keratin

Table 9 Protein stability change prediction results of MUpro and I-Mutant2.0 for the 33 deleterious missense mutations of epidermal and hyperproliferative keratins in mild and moderate psoriasis patients

Stages of psoriasis	Gene	CDS mutation	AA mutation	MUpro				I-Mutant 2.0	
				SVM	SVMCS	NN	NNCS	Stability	DDG
Mild	K14	c.372C > A	p.D124E	Increase	0.05	Increase	0.71	Decrease	– 0.52
		c.479A > T	p.K160I	Decrease	– 0.54	Decrease	– 0.89	Decrease	0.61
		c.659G > T	p.G220V	Increase	0.41	Increase	0.65	Decrease	– 0.46
		c.896G > C	p.R299P	Decrease	– 0.57	Decrease	– 0.75	Decrease	– 2.22
		c.944G > C	p.R315P	Decrease	– 0.53	Decrease	– 0.86	Decrease	– 1.26
		c.1246C > T	p.R416C	Decrease	– 0.16	Decrease	– 0.58	Decrease	– 0.3
	K10	c.502G > C	p.E168Q	Increase	0.39	Increase	0.70	Decrease	– 0.01
		c.679G > T	p.A227S	Decrease	– 1	Decrease	– 0.71	Decrease	– 0.19
		c.715G > C	p.E239Q	Decrease	– 0.12	Decrease	– 0.64	Decrease	– 0.28
		c.730C > G	p.L244V	Decrease	– 1	Decrease	– 0.94	Decrease	– 2.03
		c.770G > T	p.R257M	Increase	0.53	Increase	0.81	Decrease	– 0.47
Moderate	K14	c.388C > G	p.L130V	Decrease	– 1	Decrease	– 1	Decrease	– 1.67
		c.415G > C	p.A139P	Decrease	– 0.63	Decrease	– 0.88	Decrease	– 1.89
		c.419A > T	p.N140I	Decrease	– 0.23	Decrease	– 0.67	Increase	1.57
		c.442C > T	p.R148C	Decrease	– 1	Decrease	– 1	Decrease	– 0.28
		c.512A > T	p.D171V	Decrease	– 0.37	Decrease	– 1	Decrease	– 0.36
		c.519G > T	p.R173S	Decrease	– 1	Decrease	– 1	Decrease	– 2.37
		c.904G > C	p.A302P	Increase	0.63	Increase	0.73	Decrease	– 3.34
		c.1057G > C	p.A353P	Decrease	– 0.58	Decrease	– 0.65	Increase	– 0.65
		c.1234A > T	P.I412F	Decrease	– 1	Decrease	– 0.69	Decrease	– 0.95
	K10	c.479A > C	p.Y160S	Decrease	– 0.47	Decrease	– 0.64	Decrease	– 1.16
		c.520C > A	p.L174M	Decrease	– 1	Decrease	– 1	Decrease	0.12
		c.674A > C	p.D225A	Decrease	– 0.39	Decrease	– 0.78	Decrease	– 0.76
		c.704G > T	p.R235M	Decrease	– 0.45	Decrease	– 0.89	Decrease	– 0.59
		c.711G > T	p.K237N	Decrease	– 0.60	Decrease	– 1	Increase	0.26
		c.785T > A	p.L262Q	Decrease	– 1	Decrease	– 1	Decrease	– 1.82
	K16	c.667C > G	p.L223V	Decrease	– 1	Decrease	– 1	Decrease	– 1.39
	K17	c.503A > T	p.D168V	Decrease	– 0.30	Decrease	– 0.60	Decrease	– 0.73
		c.662A > T	p.N221I	Decrease	– 0.70	Increase	0.62	Increase	1.64
		c.682G > C	p.A228P	Decrease	– 0.04	Increase	0.61	Decrease	– 2.48
		c.907C > T	p.L303F	Decrease	– 1	Decrease	– 1	Decrease	0.95
		c.986C > T	p.A329V	Decrease	– 0.34	Decrease	– 0.88	Decrease	0.12
		c.1070G > T	p.R357L	Increase	0.15	Increase	0.68	Increase	0.66

CDS coding sequences, *AA* amino acid, *SVM* support vector machine, *SVMCS* SVM confidence score, *NN* neural network, *NNCS* neural network confidence score, *DDG* free energy change value

expression in the psoriatic epidermis might be due to some pathogenic variants in these keratin genes.

Correlation analysis of keratins and PASI in two stages of psoriasis is shown in Fig. 8. There was a significant inverse correlation between normal epidermal keratins K14 and K10 and PASI score ($r = -0.998$; $r = -0.996$; $P < 0.001$), respectively, in mild psoriasis, whereas the abnormal hyperproliferative keratins K16 and K17 showed a positive correlation with PASI score ($r = 0.999$; $r = 0.997$; $P < 0.001$) in mild psoriasis. In moderate case, Spearman's coefficient correlation analysis showed a negative correlation between K14 and K10 and PASI score ($r = -0.993$; $r = -0.997$; $P < 0.001$), respectively, and showed a strong positive correlation between K16 and K17 with PASI score, which was statistically significant ($r = 0.998$; $r = 0.998$; $P < 0.001$), respectively. We report that the

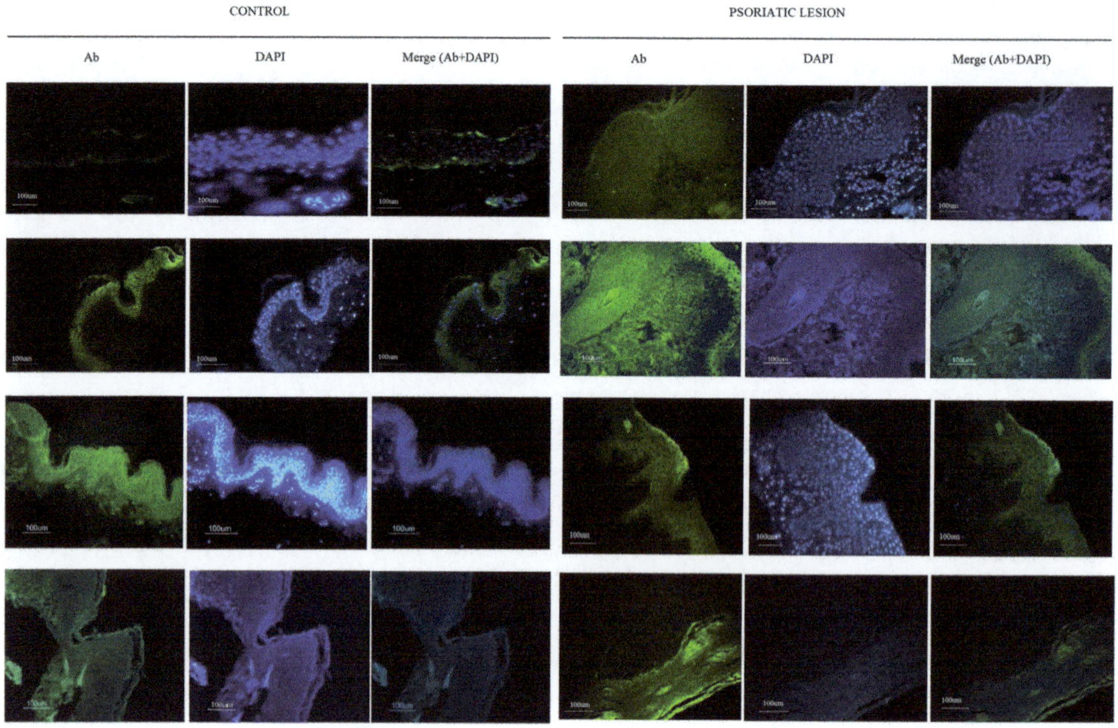

Fig. 6 Immunofluorescence analysis of keratins in nonlesional and lesional psoriatic skin. Immunofluorescence analysis of epidermal and hyperproliferative type I keratins in frozen skin sections from patients with psoriasis in nonlesional and lesional skin (× 20 magnification, respectively), and nuclei were visualized with 4′-6-diamidino-2-phenylindole (DAPI). Bar = 100 μm; Ab antibody

degree of severity of the clinical phenotype has been directly linked to the position of the pathogenic mutation along the keratin polypeptide backbone.

Discussion

The epidermal barrier is formed by keratinocytes contributing tight junctions and the cornified envelope [1]. Barrier dysfunction and cutaneous sensitization can give rise to chronic inflammatory disorders like atopic eczema and psoriasis. Keratin plays an important role in maintaining epidermal barrier as well as in the formation of tight junction and cornified envelope in skin [40]. However, mutations of keratins in psoriasis have not been fully investigated. In this present study, we have identified 18 mutations in the CDS region of K14 out of which nine mutations are slightly pathogenic variants in the psoriatic samples. Numerous mutations in K5 and K14 identified in EBS are dominant pathogenic mutations. There are a few cases of recessive mutations which nearly all lead to loss of function or a natural knockout in K14 [41–45]. Related to our result, studies showed that in EBS, the pathogenic mutations of K14 usually occur within two segments (1A and 2B) of the rod domain [46]. Cells expressing mutated K14 are more prone to stress damage and resistance to apoptosis [47, 48]. Thus, we propose that the mutation of K14 in psoriatic skin might lead to "unnatural" keratin pair; this will drastically alter the assembly of network filament as well as mechanical support in epidermal skin and also increase the degradation of keratin filaments by various stressors, all these leading to acanthosis in psoriasis.

Moreover, we have also identified 13 mutated sequences in keratin 10 protein. Of these, six sequences are considered as highly pathogenic, and the remaining seven sequences are considered as slightly pathogenic. Similar to our study, in epidermal hyperkeratosis, most of the pathogenic missense mutations of K10 usually occur within highly conserved regions of the α-helical rod domains and the non-helical HI domain [49, 50]. Reichelt et al. [51] reported that defects in keratin 10 gene may lead to hyperproliferation of basal cells, induction of c-Myc, cyclin D1, 14-3-3σ, K6, and K16. K10 is essential to control cell proliferation and also for the expression of filaggrin and cornified envelope proteins [52]. In epidermolytic ichthyosis (EI) patients, mutation of K10 proteins showed increased proliferation of suprabasal keratinocytes which leads to ichthyosiform lesions, suprabasal cytolysis, blister formation, and hyperkeratosis [53, 54]. Thus, the mutation in K10 might cause irregularly shaped pathognomonic keratin intermediate filament (KIF) clumps in suprabasal keratinocytes which disrupt the structural stability in the suprabasal

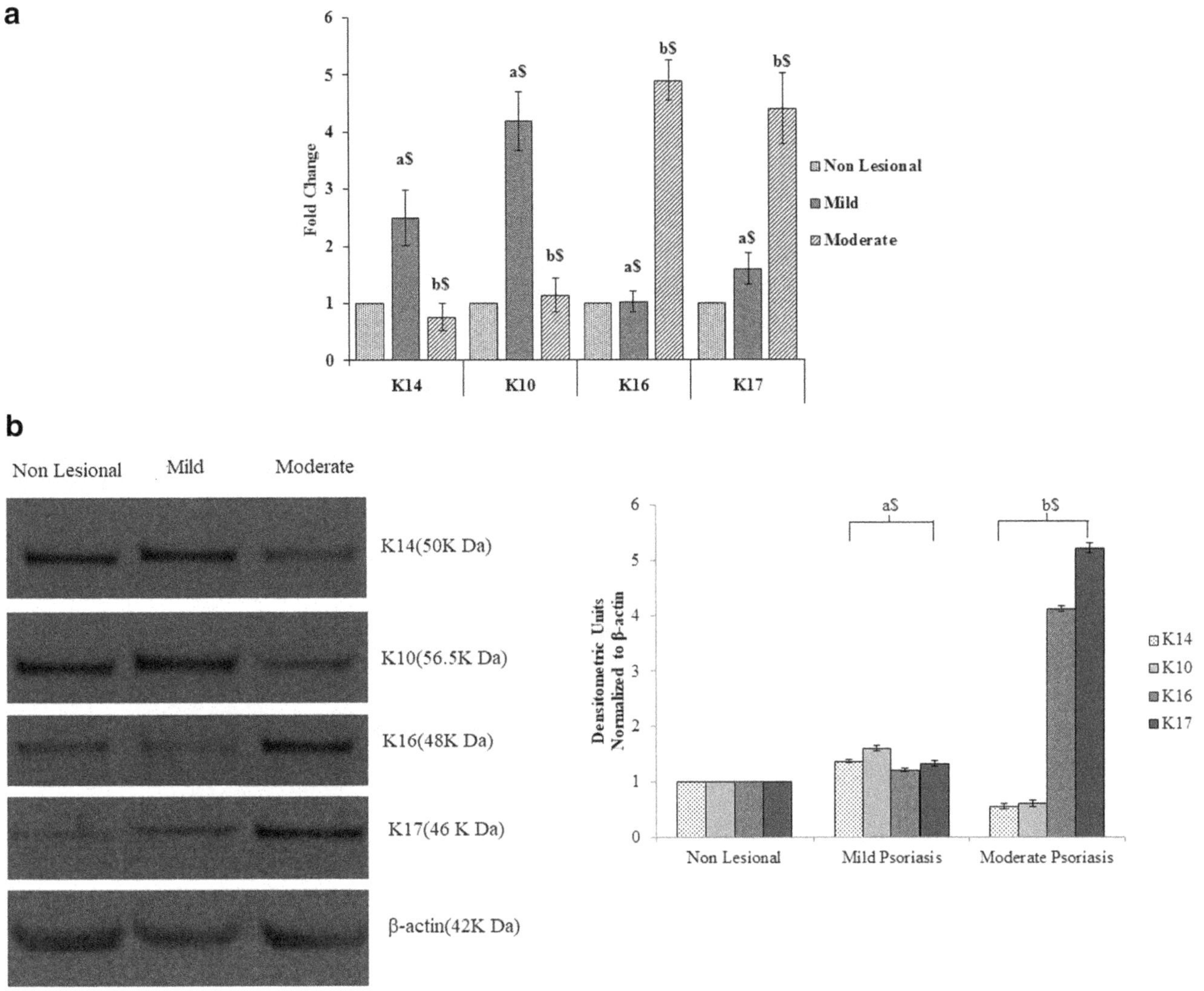

Fig. 7 mRNA and protein expression of epidermal and hyperproliferative type I keratins in mild and moderate psoriasis. **a** Quantification of keratin 14, 10, 16, and 17 mRNA by real-time PCR using the comparative C_T method. The C_T values were normalized on the basis of GADPH and analyzed relative to that of the normal control. Gene expressions were given as fold change. **b** Western blot and densitometric analysis of keratins in mild and moderate psoriasis. All values were expressed as mean ± SD, where the dollar sign denotes $P < 0.001$, "a" denotes comparison of non-lesional with mild psoriatic lesional skin biopsy, and "b": denotes comparison of mild and moderate psoriatic lesional skin biopsy

keratinocytes and also increase cyclin D, cyclin E, and phosho-Akt levels, all of which drive to hyperproliferation of keratinocytes in psoriasis.

Rapid psoriasis recurrence or new morphologies have been reported after discontinuation of traditional therapy or biological [55–58]. Interestingly, psoriatic lesions preferentially appeared at sites of wound healing (Koebner phenomenon). These relapse and Koebner phenomenon in psoriasis might be due to the variants in hyperproliferative keratin genes. In order to reveal this, we studied K16 and K17 psoriatic sequences for mutational analysis. In this, we did not find any mutation and expression of these keratins in mild psoriasis. As disease progresses, we found a pathogenic variant of K16 and K17 in moderate psoriasis. Early studies showed that an inherited dominant mutation in K16 and K17 is causative for pachyonychia congenita (PC) [59–63]. Lessard et al. [64] showed that in PC, loss of K16 function in mice causes the development of prominent calluses on the plantar side of the front and hind paws, which significantly compromise mobility and eventually lead to overt loss of barrier properties and also deletion of K16-produced spontaneously arising palmoplantar keratoderma-like lesions in mice. However, some studies have reported that in the context of epidermal wound healing, K16 mainly promotes a reorganization of the cytoplasmic array of keratin filaments, an event that precedes the onset of keratinocyte migration into the wound site [65], and also, K17-associated

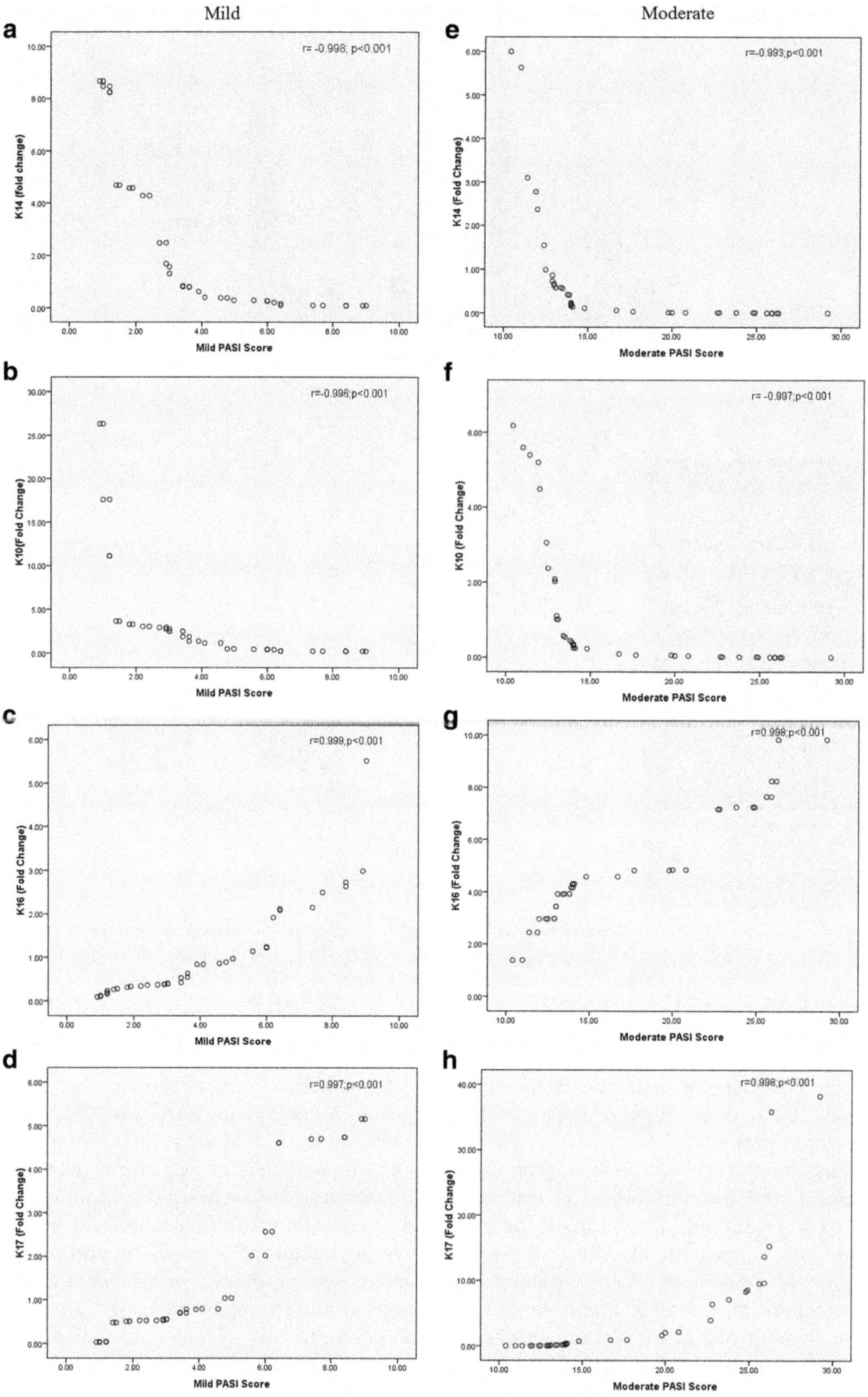

Fig. 8 Correlation between keratins with PASI score. **a** Keratin 14 vs. mild PASI score. **b** Keratin 10 vs. mild PASI score. **c** Keratin 16 vs. mild PASI score. **d** Keratin 17 vs. mild PASI score. **e** Keratin 14 vs. moderate PASI score. **f** Keratin 10 vs. moderate PASI score. **g** Keratin 16 vs. moderate PASI score. **h** Keratin 17 vs. moderate PASI score. The correlation coefficient (*r*) and significant value (*P*) are given in this figure

activation process is believed to be essential in reepithelization of the injured area [66]. In contrast to wound healing, K17 and K16 expression in psoriasis failed to resolve the deregulated inflammatory response which leads to the persistent activation of keratinocytes; these responses might be due to the mutation in these genes. Thus, mutation in K16 and K17 in psoriasis might be one of the factors for disease severity and Koebner phenomenon, accompanied with early relapse even with therapy.

In this study, we showed increased K14 and K10 expression at protein and mRNA levels in mild psoriatic skin biopsies compared to nonlesional skin biopsies; in contrast to this, moderate lesional biopsy showed decreased expression of K14 and K10. Similar to our results, other studies also showed that in lesional biopsy, the level of K14 was considerably higher with downregulation of K10 compared with normal epidermis [11, 13, 67]. Also, we found that the hyperproliferative keratin K16 and K17 expressions were low in mild psoriasis; as diseases progress, these expressions were increased in moderate psoriasis compared to nonlesional. Consistent with our results, many authors also showed an upregulation of K6 and K16 [68, 69] in psoriatic epidermis. Similarly, a study by de Jong et al. [69] also showed the overexpression of K17 in psoriatic epidermis. Earlier reports said that the K17 is considered as a therapeutic target and marker of antipsoriatic therapies used for the treatment of psoriasis [70, 71]. These finding suggests that the changes in keratin expressions might trigger and exacerbate psoriasis.

Finally, we also found a negative correlation between epidermal keratins and PASI score and positive correlation between hyperproliferative keratins and PASI score; these further added that alteration in these keratins in skin might cause severity of psoriasis. Previous studies have reported that the degree of severity of the clinical phenotype has been directly linked to the position of the pathogenic mutation along the keratin polypeptide backbone, although more recent reports provide some exceptions to this, whereby also milder disease phenotypes are caused by pathogenic mutations in the conserved hot spot region of the keratin genes [72, 73]. We demonstrate that the novel mutation of keratins found in psoriasis patients after comparison with ExAC database will be considered as efficient filtering of candidate psoriasis-causing variants and are useful for the discovery of human "knockout" psoriasis variants in protein-coding genes. Overall, we strongly believed that these novel keratin mutations in the psoriatic epidermis might be one of the main causative factors for psoriasis.

Conclusion

Even though there are many highly psoriasis-risk genes identified, in 2009, GWAS have found the importance of keratin pattern in psoriasis. Also, studies have found that keratin plays a critical role in the pathogenesis of psoriasis. Based on all these aspects, we focused on this first and foremost study in which we have identified mutation in K14, K10, K16, and K17 genes in the two stages of psoriasis. Our findings suggest that this mutation leads to decreased expression of highly regulated keratins which in turn causes defective barrier accompanied with acanthosis, hyperplasia, and infiltration of inflammatory mediators in psoriatic epidermis. These findings will offer a clue to better understanding the pathogenesis of psoriasis.

Additional files

Additional file 1: Figure S1. Comparison of the identified K14 and K10 nucleotide sequences of mild psoriasis obtained in this study with those of respective NCBI CDS sequences using BioEdit software (version 7.2). The mutated sequences are indicated by red arrows. **Figure S2.** Comparison of the identified K14, K10, K16, and K17 nucleotide sequences of moderate psoriasis obtained in this study with those of respective NCBI CDS sequences using BioEdit software (version 7.2). The mutated sequences are indicated by red arrows.

Additional file 2: Table S1. K14, K10, K16, and K17 missense mutated variants retrieved from ExAC database. The mutated sequences identical both in our study and ExAC database are indicated by boldfaced words.

Abbreviations

CDS: Coding sequences; EBS: Epidermolysis bullosa simplex; EI: Epidermolytic ichthyosis; GWAS: Genome-wide association studies; IF: Intermediate filament; K10: Keratin 10; K14: Keratin 14; K16: Keratin 16; K17: Keratin 17; KIF: Keratin intermediate filament; PC: Pachyonychia congenita; SNP: Single nucleotide polymorphism

Acknowledgements

We thank all patients who have assisted with this study.

Funding

The study was supported by the postdoctoral funding grant given by Anhui Medical University, National Natural Science Foundation of China (81602397), and the Natural Science Foundation of Shanghai (15ZR1405700).

Authors' contributions

TE performed the experiments and data analysis, interpreted the results, and drafted the manuscript. JS and CZ participated in the sample collection and performed the experiments and data analysis. FZ and YZ contributed to the sequence analysis. LS and SY contributed to the patients' recruitment. XJZ provided guidance on the experimental design and received funding. All authors read and approved the final manuscript.

Competing interests

The authors declare that they have no competing interests.

Author details

[1]Institute and Department of Dermatology, The First Affiliated Hospital, Anhui Medical University, Hefei, China. [2]Key Laboratory of Dermatology, Ministry of Education, Anhui Medical University, Hefei, China. [3]Collaborative

Innovation Center for Complex and Severe Dermatosis, Anhui Medical University, Hefei, China. [4]Institute of Dermatology, Huashan Hospital, Fudan University, Shanghai, China. [5]Anhui Medical University, 81 Meishan Road, Hefei, Anhui Province, China.

References

1. Simpson CL, Patel DM, Green KJ. Deconstructing the skin: cytoarchitectural determinants of epidermal morphogenesis. Nat Rev Mol Cell Biol. 2011;12: 565–80.
2. Kim S, Coulombe PA. Intermediate filament scaffolds fulfill mechanical, organizational, and signaling functions in the cytoplasm. Genes Dev. 2007; 21:1581–97.
3. Magin TM, Vijayaraj P, Leube RE. Structural and regulatory functions of keratins. Exp Cell Res. 2007;313:2021–32.
4. Lane EB, McLean WH. Keratins and skin disorders. J Pathol. 2004;204(4): 355–66.
5. Omary MB, Ku NO, Strnad P, Hanada S. Toward unraveling the complexity of simple epithelial keratins in human disease. J Clin Invest. 2009;119:1794–805.
6. Oriolo AS, Wald FA, Ramsauer VP, Salas PJ. Intermediate filaments: a role in epithelial polarity. Exp Cell Res. 2007;313:2255–64.
7. Fuchs E, Cleveland DW. A structural scaffolding of intermediate filaments in health and disease. Science. 1998;279:514–9.
8. Osborn M. Intermediate filaments as histologic markers: an overview. J Invest Dermatol. 1983;81:104s–9s.
9. Nestle FO, Kaplan DH, Barker J. Psoriasis. N Engl J Med. 2009;61:496–509.
10. Fuchs E, Green H. Changes in keratin gene expression during terminal differentiation of the keratinocyte. Cell. 1980;19:1033–42.
11. Thewes M, Stadler R, Korge B, Mischke D. Normal psoriatic epidermis expression of hyperproliferation-associated keratins. Arch Dermatol Res. 1991;283(7):465–71.
12. Korver JE, van Duijnhoven MW, Pasch MC, et al. Assessment of epidermal subpopulations and proliferation in healthy skin, symptomless and lesional skin of spreading psoriasis. Br J Dermatol. 2006;155(4):688–94.
13. Elango T, Thirupathi A, Subramanian S, et al. Methotrexate normalized keratinocyte activation cycle by overturning abnormal keratins as well as deregulated inflammatory mediators in psoriatic patients. Clin Chim Act. 2015;451(Pt B):329–37.
14. Arnold I, Watt FM. c-Myc activation in transgenic mouse epidermis results in mobilization of stem cells and differentiation of their progeny. Curr Biol. 2001;11:558–68.
15. Waikel RL, Kawachi Y, Waikel PA, Wang XJ, Roop DR. Deregulated expression of c-Myc depletes epidermal stem cells. Nat Genet. 2001;28:165–8.
16. Schweizer J, Bowden PE, Coulombe PA, et al. New consensus nomenclature for mammalian keratins. J Cell Biol. 2006;174:169–74.
17. Schweizer J, Langbein L, Rogers MA, Winter H. Hair follicle-specific keratins and their diseases. Exp Cell Res. 2007;313:2010–20.
18. Uitto J, Richard G, McGrath JA. Diseases of epidermal keratins and their linker proteins. Exp Cell Res. 2007;313:1995–2009.
19. de Cid R, Riveira-Munoz E, Zeeuwen PL, et al. Deletion of the late cornified envelope LCE3B and LCE3C genes as a susceptibility factor for psoriasis. Nat Genet. 2009;41(2):211–5.
20. Zhang XJ, Huang W, Yang S, et al. Psoriasis genome-wide association study identifies susceptibility variants within LCE gene cluster at 1q21. Nat Genet. 2009;41(2):205–10.
21. Mrowietz U, Kragballe K, Reich K, et al. Definition of treatment goals for moderate to severe psoriasis: a European consensus. Arch Dermatol Res. 2011 Jan;303(1):1–10.
22. Livak KJ, Schmittgen TD. Analysis of relative gene expression data using real-time quantitative PCR and the 2(-delta deltaC(T)) method. Methods. 2001;25:402–8.
23. Dong C, Yu B. Mutation surveyor: an in silico tool for sequencing analysis. Methods Mol Biol. 2011;760:223–37.
24. Kumar P, Henikoff S, Ng PC. Predicting the effects of coding non-synonymous variants on protein function using the SIFT algorithm. Nat Protoc. 2009;4:1073–81.
25. Alamut®: gene mutation interpretation software: http://www.interactive-biosoftware.com. Accessed 8 Dec 2014.
26. Karbassi I, Maston GA, Love A, et al. A standardized DNA variant scoring system for pathogenicity assessments in Mendelian disorders. Hum Mutat. 2016;37(1):127–34.
27. Lek M, Karczewski KJ, Minikel EV, et al. Analysis of protein-coding genetic variation in 60,706 humans. Nature. 2016;536(7616):285–91.
28. Bendl J, Stourac J, Salanda O, et al. Predict SNP: robust and accurate consensus classifier for prediction of disease related mutations. PLoS Comput Biol. 2014;10(1):e1003440.
29. Sucularli C, Arslantas M. Computational prediction and analysis of deleterious cancer associated missense mutations in DYNC1H1. Mol Cell Probes. 2017;34:21–9.
30. Cheng J, Randall A, Baldi P. Prediction of protein stability changes for single-site mutations using support vector machines. Proteins. 2006;62(4):1125–32.
31. Capriotti E, Fariselli P, Casadio R. I-Mutant2.0: predicting stability changes upon mutation from the protein sequence or structure. Nucleic Acids Res. 2005;33(Web Server issue):W306–10.
32. Quan L, Lv Q, Zhang Y. STRUM: structure-based prediction of protein stability changes upon single-point mutation. Bioinformatics. 2016;32(19): 2936–46.
33. Johansen C, Funding AT, Otkjaer K, et al. Protein expression of TNF-alpha in psoriatic skin is regulated at a post transcriptional level by MAPK-activated protein kinase 2. J Immunol. 2006;176:1431–8.
34. Raul U, Sawant S, Dange P, et al. Implications of cytokeratin 8/18 filament formation in stratified epithelial cells: induction of transformed phenotype. Int J Cancer. 2004;111:662–8.
35. Meng L, Du J, Li W, et al. Study of a family with epidermolysis bullosa simplex resulting from a novel mutation of KRT14 gene. Zhonghua Yi Xue Yi Chuan Xue Za Zhi. 2017;34(4):504–8.
36. Wang WH, Zhang L, Li LF, et al. Ichthyosis hystrix Lambert type and Curth-Macklin type are a single entity with affected (KRT1 mutation) or unaffected (KRT10 mutation) palms and soles? Eur J Dermatol. 2016;26(5):493–5.
37. Chen PJ, Li CX, Wen J, et al. S159P mutation of keratin 10 gene causes severe form of epidermolytic hyperkeratosis. J Eur Acad Dermatol Venereol. 2016;30(10):e102–4.
38. Dai L, Wu J, Guo H, et al. Mutation p.Leu128Pro in the 1A domain of K16 causes pachyonychia congenita with focal palmoplantar keratoderma in a Chinese family. Eur J Pediatr. 2014;173(6):737–41.
39. Jiang M, Sun Z, Dang E, et al. TGFß/SMAD/microRNA-486-3p signaling axis mediates keratin 17 expression and keratinocyte hyperproliferation in psoriasis. J Invest Dermatol. 2017;137(10):2177–86.
40. Segre JA. Epidermal barrier formation and recovery in skin disorders. J Clin Invest. 2006;116(5):1150–8.
41. Rugg EL, McLean WH, Lane EB, Pitera R, McMillan JR, Dopping-Hepenstal PJ, et al. A functional "knockout" of human keratin 14. Genes Dev. 1994;8(21): 2563–73.
42. Chan Y, Anton-Lamprecht I, Yu QC, Jäckel A, Zabel B, Ernst JP, et al. A human keratin 14 "knockout": the absence of K14 leads to severe epidermolysis bullosa simplex and a function for an intermediate filament protein. Genes Dev. 1994;8(21):2574–87.
43. Jonkman MF, Heeres K, Pas HH, van Luyn MJ, Elema JD, Corden LD, et al. Effects of keratin 14 ablation on the clinical and cellular phenotype in a kindred with recessive epidermolysis bullosa simplex. J Invest Dermatol. 1996;107(5):764–9.
44. Schuilenga-Hut PH, Vlies PV, Jonkman MF, Waanders E, Buys CH, Scheffer H. Mutation analysis of the entire keratin 5 and 14 genes in patients with epidermolysis bullosa simplex and identification of novel mutations. Hum Mutat. 2003;21(4):447.
45. Homberg M, Ramms L, Schwarz N, Dreissen G, Leube RE, Merkel R, et al. Distinct impact of two keratin mutations causing epidermolysisbullosa simplex on keratinocyte adhesion and stiffness. J Invest Dermatol. 2015; 135(10):2437–45.
46. Szeverenyi I, Cassidy AJ, Chung CW, et al. The Human Intermediate Filament Database: comprehensive information on a gene family involved in many human diseases. Hum Mutat. 2008;29:351–60.
47. D'Alessandro M, Russell D, Morley SM, et al. Keratin mutations of epidermolysis bullosa simplex alter the kinetics of stress response to osmotic shock. J Cell Sci. 2002;115:4341–51.
48. Russell D, Ross H, Lane EB. ERK involvement in resistance to apoptosis in keratinocytes with mutant keratin. J Invest Dermatol. 2010;130:671–81.
49. Syder AJ, Yu QC, Paller AS, et al. Genetic mutations in the K1 and K10 genes of patients with epidermolytic hyperkeratosis. Correlation between location and disease severity. J Clin Invest. 1994;93:1533–42.

50. McLean WH, Morley SM, Higgins C, et al. Novel and recurrent mutations in keratin 10 causing bullous congenital ichthyosiformerthroderma. Exp Dermatol. 1999;8:120–3.
51. Reichelt J, Magin TM. Hyperproliferation, induction of c-Myc and 14-3-3σ, but no cell fragility in keratin-10-null mice. J Cell Sci. 2002;115:2639–50.
52. Reichelt J, Furstenberger G, Magin TM. Loss of keratin 10 leads to mitogen-activated protein kinase (MAPK) activation, increased keratinocyte turnover, and decreased tumor formation in mice. J Invest Dermatol. 2004;123(5):973–81.
53. Ishidayamamoto A, Iizuka H, Manabe M, et al. Altered distribution of keratinization markers in epidermolytic hyperkeratosis. Arch Dermatol Res. 1995;287:705–11.
54. Porter RM, Lane EB. Phenotypes, genotypes and their contribution to understanding keratin function. Trends Genet. 2003;19:278–85.
55. Cacoub P, Artru L, Canesi M, Koeger AC, Camus JP. Lifethreatening psoriasis relapse on withdrawal of cyclosporin. Lancet. 1988;2:219–20.
56. Georgala S, Koumantaki E, Rallis E, Papadavid E. Generalized pustular psoriasis developing during withdrawal of short-term cyclosporin therapy. Br J Dermatol. 2000;142:1057–8.
57. Fabiano A, De Simone C, Gisondi P, Piaserico S, Lasagni C, Pellacani G, et al. Management of patients with psoriasis treated with biological drugs needing a surgical treatment. Drug Dev Res. 2014;75(Suppl 1):S24–6.s.
58. Masson Regnault M, Konstantinou MP, Khemis A, Poulin Y, Bourcier M, Amelot F, et al. Early relapse of psoriasis after stopping brodalumab: a retrospective cohort study in 77 patients. J Eur Acad Dermatol Venereol. 2017;6 [Epub ahead of print]
59. McLean WH, Rugg EL, Lunny DP, et al. Keratin 16 and keratin 17 mutations cause pachyonychia congenita. Nat Genet. 1995;9:273–8.
60. Connors JB, Rahil AK, Smith FJ, et al. Delayed-onset pachyonychia congenita associated with a novel mutation in the central 2B domain of keratin 16. Br J Dermatol. 2001;144:1058–62.
61. Smith FJ, Corden LD, Rugg EL, et al. Missense mutations in keratin 17 cause either pachyonychia congenita type 2 or a phenotype resembling steatocystoma multiplex. J Invest Dermatol. 1997;108:220–3.
62. Smith FJ, Fisher MP, Healy E, et al. Novel keratin 16 mutations and protein expression studies in pachyonychia congenita type and focal palmoplantar keratoderma. Exp Dermatol. 2000;9:170–7.
63. Smith FJ. Nail that mutation-keratin 17 defect in late-onset pachyonychia. J Invest Dermatol. 2004;122:x–xi.
64. Lessard JC, Coulombe PA. Keratin 16-null mice develop palmoplantar keratoderma, a hallmark feature of pachyonychia congenita and related disorders. J Invest Dermatol. 2012;132(5):1384–91.
65. Paladini RD, Takahashi K, Bravo NS, Coulombe PA. Onset of re-epithelialization after skin injury correlates with a reorganization of keratin filaments in wound edge keratinocytes: defining a potential role for keratin 16. J Cell Biol. 1996;132(3):381–97.
66. Kim S, Wong P, Coulombe PA. A keratin cytoskeletal protein regulates protein synthesis and epithelial cell growth. Nature. 2006;441:362–5.
67. Castelijns FA, Gerritsen MJ, van Vlijmen-Willems IM, et al. Proliferation is the main epidermal target in the treatment of psoriatic plaques with once daily application of tacalcitol ointment. Acta Derm Venereol. 1999; 79(2):111–4.
68. van de Kerkhof PCM, JornBovenschen H, et al. In vivo effects of topical anti-psoriatic treatments on cutaneous inflammation, epidermal proliferation and keratinisation. Current Drug Therapy. 2007;2:21–6.
69. de Jong EMG, Ferrier CM, de Zwart A, et al. Effects of topical treatment with budesonide on parameters for epidermal proliferation, keratinization and inflammation in psoriasis. J Dermatol Sci. 1995;9:185–94.
70. de Jong EM, van Vlijmen IM, van Erp PE, Ramaekers FC, Troyanovski SM, van de Kerkhof PC. Keratin 17: a useful marker in anti-psoriatic therapies. Arch Dermatol Res. 1991;283(7):480–2.
71. Fu M, Wang G. Keratin 17 as a therapeutic target for the treatment of psoriasis. J DermatolSci. 2012;67(3):161–5.
72. Bowden PE, Knight AG, Liovic M. A novel mutation (p.Thrl98Ser) in the 1A helix of keratin 5 causes the localized variant of epidermolysis bullosa simplex. ExpDermatol. 2009;18(7):650–2.
73. Glasz-Bona A, Medvecz M, Sajo R, Lepesi-Benko R, Tulassay Z, Katona M, et al. Easy method for keratin 14 gene amplification to exclude pseudogene sequences: new keratin 5 and 14 mutations in epidermolysis bullosa simplex. J Invest Dermatol. 2009;129:229–31.

Genetic studies of Polish migraine patients: screening for causative mutations in four migraine-associated genes

Izabela Domitrz[1*†], Michalina Kosiorek[1,2*†], Cezary Żekanowski[2] and Anna Kamińska[1]

Abstract

Background and aim: Migraine is the most common neurological disorder, affecting approximately 12 % of the adult population worldwide, caused by both environmental and genetic factors. Three causative genes have been identified in familial hemiplegic migraine (FHM) families: *CACNA1A*, *ATP1A2*, and *SCNA1A*. Recently, several mutations in *KCNK18* have also been found as causative factors in migraine development. The aim of our study was to identify the genetic background of migraine in the Polish population.

Material and methods: Sixty patients with migraine without aura (MO) or with different types of migraine with aura (MA), including sporadic hemiplegic, familial hemiplegic, and probable familial hemiplegic, were screened for mutations in the four genes previously linked with different types of migraine (*ATP1A2*, *CACNA1A*, *SCN1A*, and *KCNK18*).

Results: Two missense mutations were found. One novel mutation in *SCN1A*, encoding α subunit of sodium channel, causing amino acid change M1500V localized to a region encoding inactivation loop between transmembrane domains III and IV of the channel, was detected in a female FHM patient. The M1500V mutation was absent in a group of 62 controls, as well as in the ExAC database. The second, already known missense mutation S231P in *KCNK18* was found in a female MA patient. Additionally, a novel intronic polymorphism possibly affecting alternative splicing of *SCN1A*, at chr2:16685249, g.77659T>C, and c.4581+32A>G, located between exons 24 and 25, in a region encoding the inactivation loop of the sodium channel was found in a female MO patient. No mutations in *ATP1A2* or *CACNA1A* were found in the study group.

Conclusions: The presence of *SCN1A* mutations and absence of mutations in *ATP1A2 or CACNA1A* suggest that the Polish patients represent FHM type 3. On the other hand, the presence of *KCNK18* mutation indicated another FHM subtype. It could be speculated that contrary to other European populations, the genetic basis of migraine in the Polish population involves mutations in genes not included in the study. Next-generation sequencing methods should be implemented to identify other migraine-associated variants.

Keywords: Familial hemiplegic migraine, Aura migraine, *SCN1A*, *KCNK18*, Missense mutations, Splice variants, SNP, Genetic background

Introduction

Migraine is the most common neurological disorder, affecting approximately 12 % of the adult population worldwide, caused by both environmental and genetic factors. Its heritability is estimated at ca. 40–70 %. Numerous linkage studies and candidate-gene studies have identified causative genes in migraine without aura (MO) and migraine with aura (MA). Three such genes have been identified in familial hemiplegic migraine (FHM). The first FHM type (FHM1; MIM141500) is related to *locus* 19p13 and caused by mutations in *CACNA1A*, encoding alpha 1A subunit of voltage-dependent calcium channel type P/Q (MIM601011) [1]. Mutations in *CACNA1A* contribute also to episodic ataxia (EA2; MIM108500) [2] and spinocerebellar ataxia type 6 (SCA6; MIM183086) [3]. The second FHM type (FHM2; MIM602481) is caused by mutations in *ATP1A2* encoding alpha 2 subunit of Na+/K+–

* Correspondence: idomitrz@wum.edu.pl; mkosiorek@imdik.pan.pl
†Equal contributors
[1]Department of Neurology, Warsaw Medical University, 61 Żwirki i Wigury Street, 02-091 Warsaw, Poland
Full list of author information is available at the end of the article

ATPase (MIM182340) [4]. Finally, the third FHM *locus* (FHM3; MIM609634) is at 2q24, and the implicated gene *SCNA1A* (MIM182389) encodes α subunit of the neuronal voltage-gated Na+ channel [5]. Mutations in *SCN1A* were also recognized as a cause of epilepsy [6]. Recently, *KCNK18* at *locus* 10q25.3, encoding K+ channel subfamily K member 18, has been found to be connected with FHM classified as other than types 1, 2, and 3 (MIM613655) [7]. Other studies have provided evidence on the influence of the genetic variants of *SLC6A4* encoding a serotonin transporter on a migraine occurrence [8]. Moreover, recent genome-wide association studies (GWAS) have identified four novel genetic variants associated with migraine [9]: rs1835740 modulating glutamate homeostasis and specific for migraine with aura; rs11172113 implicating lipoprotein receptor LRP1, which may interact with neuronal glutamate receptors; rs10166942 in close proximity to *TRPM8*, encoding a cold and pain sensor; and rs2651899 in *PRDM16*, with an unclear role [9]. All these variants only confer a small to moderate risk for migraine, which concurs with migraine being a heterogeneous disorder [9]. A list of single nucleotide polymorphisms in 10 genes was reported in a large German case-control cohort study of migraine with aura [10], and recently, several genetic variants in *APOA1BP, TBC1D7, FUT9, STAT6, ATP5B, AJAP1, TSPAN2, FHL5, C7orf10*, and *MMP16* have been found in GWAS performed in Western Europe [11].

To identify the genetic background of migraine in Polish patients, we screened *ATP1A2, CACNA1A, SCN1A*, and *KCNK18* for mutations in a group of 60 patients with migraine without aura or with different types of aura migraine: typical, sporadic hemiplegic and familial hemiplegic or probable familial hemiplegic. All patients were Caucasians of Polish origin.

Materials and methods

Ethics statement

Written consent was obtained from probands, diagnosed family members, and probands' relatives according to the Declaration of Helsinki [12]. The genetic study was approved by the Ethics Committee of the Medical University of Warsaw (Warsaw, Poland) in compliance with the national legislation and the Code of Ethical Principles for Medical Research Involving Human Subjects of the World Medical Association.

Patients' clinical description

Sixty patients (47 women and 13 men), mean age 48 ± 13 years, of the Headache Outpatient Clinic and Neurological Department, with different types of migraine diagnosed according to the International Headache Society (IHS) criteria, 3rd edition [13], were analyzed. The study group consisted of several subgroups: 4 families (Table 1—4a, b, c; 7a, b, c; 43a, b, c, d; 47a, b, c) with familial hemiplegic migraine (FHM3 families) and sporadic hemiplegic migraine (SHM1 family), 6 patients with probable familial hemiplegic migraine (F?HM—families of patients with hemiplegic migraine not seen), 7 patients with sporadic hemiplegic migraine (SHM), 11 patients with MA, and 23 patients with MO. Two members from FHM family no. 7 were also diagnosed with epilepsy with grand mal attacks (father and older son). One patient diagnosed SHM also had mental retardation. One patient with MO also had episodic ataxia (EA). Family no. 43 was examined for the presence of mutations in Leiden factor V. This analysis revealed the presence of the variant G1691A, c.1746G>A p.Arg534Gln, rs6025 in this family and in one family member (mother) was possibly associated with ischemic stroke. The study was performed at the Institute of Hematology and Blood Transfusion in Warsaw, according to standard protocols based on Sanger sequencing. Only the family 43 was examined for Leiden factor V genetic variants, and these tests were part of routine diagnostic practice at the Institute of Hematology and Blood Transfusion in Warsaw and were made available under standard medical care. The migraine patients were interviewed and examined in the Headache Outpatient Clinic or at the Department of Neurology of Warsaw Medical University. The diagnosis was performed by an experienced neurologist. Duration of the disease ranged between 1 and 40 years (mean duration 21 + 11 years), and the frequency of migraine attacks ranged between 1 attack/1 week to 1 attack/12 months. Patient profile is presented in Table 1. As a control group, we used our existing database of 62 samples of whole-exome sequencing, which were screened for the presence of identified variants.

Sanger sequencing

Genomic DNA was extracted from peripheral blood leukocytes using standard salting-out method. Exons 4, 5, 13, 16, 17, 24, 26, 29, 32, and 36 of *CACNA1A*, exons 6, 14, 15, 16, 18, 20, 23, 24, and 26 of *SCNA1A*, exons 9, 16, 17, 18, 19, and 22 of *ATP1A2*, and exons 1, 2, and 3 of *KCN1K18* (all with flanking intronic sequences) were sequenced using automated, fluorescent sequencing method. Genes and the exons were chosen on the basis of the frequencies of known migraine-causing mutations (Leiden Familial Hemiplegic Migraine Variation Database). In brief, the amplicons were amplified with Master Mix Kit and Hot Start Master Mix Kit (Qiagen) using genomic DNA from each patient. Primers were designed using Primer Premier software (Additional file 1). The annealing temperature of 58 °C was applied for all primers. Exon 26 of *SCN1A* and exon 3 of *KCNK18* were amplified in overlapping fragments. Resulting amplicons were purified with Exonuclease I/FastAP (Fermentas) and sequenced using ABI PRISM 3130 Genetic Analyzer (Applied Biosystems)

Table 1 Clinical phenotypes and polymorphisms in migraine-related genes (*SCN1A, CACNA1A, ATP1A2, KCNK18)* in the study group

ID	Year of birth	Sex	Clinical phenotype	*SCN1A, CACNA1A, ATP1A2, KCNK18*
1	1980	M	MA	
2	1976	F	F?HM	c.4498A>G, p.M1500V, rs376885324 (ex. 24 SCN1A)
3	1978	F	MA	rs2298771 (ex. 16 SCN1A)
4a	1952	F	FHM	rs17846715 (ex. 16 ATP1A2)
4b	1978	F	FHM	
4c	1980	F	FHM	
5	1967	F	MO	
6	1967	F	MO	
7a	1954	M	FHM + EPI	
7b	1982	M	FHM + EPI	rs2298771 (ex. 16 SCN1A), rs16012 (ex. 13 CACNA1A)
7c	1984	M	MO	rs2298771 (ex. 16 SCN1A)
8	1980	M	SHM	
9	1958	F	F?HM	
10	1944	F	MO	Novel intronic polymorphism c.4581 + 32A>G, between ex. 24 and 25
11	1946	F	MO	
12	1954	F	MO	
13	1982	F	MA	
14	1966	F	MO	rs2298771 (ex. 16 SCN1A)
15	1962	M	MO	
16	1972	F	MA	c.691T>C, p.S231P, rs363315 (ex. 3 KCNK18), rs16016 (ex. 16 CACNA1A)
17	1962	M	MO	
18	1968	F	MO	
19	1943	F	MO	rs2298771 (ex. 16 SCN1A)
20	1963	F	F?HM	rs41288127, rs55884181 (ex. 22 ATP1A2)
21	1957	F	F?HM	
22	1978	F	MO	rs2298771 (ex. 16 SCN1A)
23	1968	F	MA	rs2298771 (ex. 16 SCN1A)
24	1934	F	MO	rs2298771 (ex. 16 SCN1A)
25	1960	F	MO	
26	1953	F	MA	rs16012 (ex. 13 CACNA1A)
27	1973	F	SHM	rs41276894 (ex. 4 CACNA1A)
28	1977	F	F?HM	
29	1988	M	SHM + MR	
30	1969	F	F?HM	rs61734524 (ex. 9 ATP1A2)
31	1973	F	SHM	rs17846715 (ex. 16 ATP1A2)
32	1976	M	MO	rs2298771 (ex. 16 SCN1A)
33	1987	F	SHM	rs17846715 (ex. 16 ATP1A2), rs41288127 (ex. 22 ATP1A2)

Table 1 Clinical phenotypes and polymorphisms in migraine-related genes (*SCN1A, CACNA1A, ATP1A2, KCNK18)* in the study group *(Continued)*

ID	Year of birth	Sex	Clinical phenotype	*SCN1A, CACNA1A, ATP1A2, KCNK18*
34	1970	F	MO	rs2298771 (ex. 16 SCN1A)
35	1972	F	MO	
36	1968	F	MO	
37	1966	F	MA	rs2298771 (ex. 16 SCN1A)
38	1969	F	MA	
39a	1988	F	MO	
39b	1960	F	MO	
40	1970	F	MO	
41	1974	F	MO	
42	1958	F	MO	
43a	1989	M	MO (mutation in Leiden factor V)	rs61734524 (ex.9 ATP1A2), rs16016 (ex. 16 CACNA1A)
43b	1994	F	FHM (mutation in factor V Leiden)	rs61734524 (ex.9 ATP1A2)
43c	1970	F	FHM (mutation in factor V Leiden)	rs61734524 (ex.9 ATP1A2), rs16016 (ex. 16 CACNA1A)
43d	1950	M	Ischemic stroke (mutation in factor V Leiden), mother with FHM + MA	
44	1958	F	MO + EA	
45	1960	F	MA	
46a	1943	F	SHM mother	rs61734524 (ex. 9 ATP1A2)
46b	1984	M	SHM	Not tested
46c	1943	M	SHM father	
47	1962	F	MA	rs17846715 (ex. 16 ATP1A2)
48	1967	M	SHM	
49	1963	F	MA	
50	1972	F	SHM	rs61734524 (ex. 9 ATP1A2), rs17846715 (ex. 16 ATP1A2)
51	1959	F	MA	rs2298771 (ex. 16 SCN1A)

Description of clinical phenotype: *EA*, episodic ataxia, *Epi* epileptic seizures, *F?HM* probable familial hemiplegic migraine, *FHM* familial hemiplegic migraine, *MA* migraine with typical aura, *MO* migraine without aura, *MR* mental retardation, *SHM* sporadic hemiplegic migraine

with the BigDye Terminator v1.1 Cycle Sequencing Kit (Applied Biosystems).

Results

SCN1A

Among the 60 Polish patients studied, we found a single nucleotide polymorphism c.4498A>G, rs376885324, in exon 24 of *SCN1A* leading to amino acid substitution M1500V (Fig. 1a). This variant was found in a female FHM patient and had not been linked to an FHM clinical phenotype before (see Table 2). The family history of the patient suggested autosomal dominant inheritance pattern, and according to clinical interview, seven members of

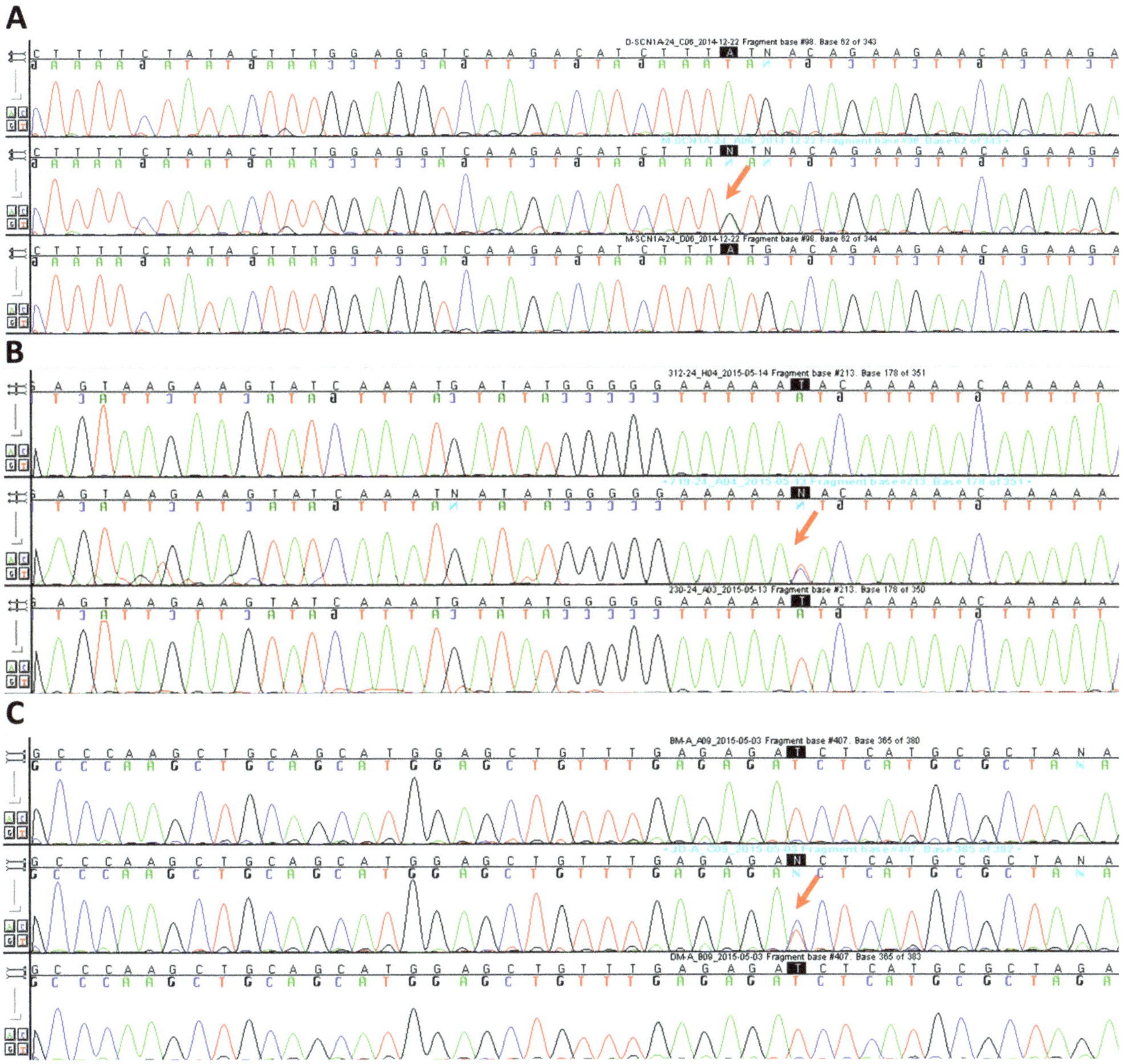

Fig. 1 Mutations in migraine-causing genes. **a** Single nucleotide substitution at c.4498 A>G, resulting in novel amino acid change M1500V, was found in *SCN1A* in a patient with FHM (rs121918632, NG_011906.1, Gene ID: 6323). **b** A novel intronic polymorphism at chr2:16685249, g.77659T>C, c.4581 + 32A>G, located between exons 24 and 25, predicted to affect alternative splicing of *SCN1A*, in a gene region encoding inactivation loop of sodium channel was found in a patient with MO. **c** Mutation at c.691T>C, resulting in amino acid change S231P, in *KCNK18* (rs363315, NG_028085.1, Gene ID 338567) was found in a patient with MA

the proband's family were also affected by migraine, including a sister, the father, two father's brothers, a cousin, and the grandmother (Fig. 2). Unfortunately, the material for genetic testing from other family members was unavailable, but the patient's sister and father were examined clinically and diagnosed with FHM in our clinic.

Bioinformatic assessment using PolyPhen-2 HumDiv [14] showed that the M1500V variant is probably damaging with a score of 0.982 (sensitivity, 0.75; specificity, 0.96). According to the Protein Variation Effect Analyzer (PROVEAN) and Sorting Intolerant From Tolerant (SIFT) tools [15, 16], the variant is damaging (score −3.68) and deleterious (score 0.001). The MutationTaster prediction tool [17] also describes the variant as disease-causing. M1500V was not detected in our 62 controls or in the ExAC database, which contains data for the genomic region of interest from 60,000 DNA samples, including more than 30,000 samples from European populations. However, the mutation is present in the Ensembl database with a frequency of 1:8593.

Table 2 Clinical phenotype of FHM patient with mutation M1500V in exon 24 of *SCN1A*

Hemiplegic migraine (HM) onset (years)	22
Triggering factors of HM	Emotional stress
Hemiplegia during aura phase	+
Sensory disturbances during aura phase	+
Visual disturbances during aura phase	+
Aphasia during aura phase	+
Aura duration (min)	60
Headache duration (h)	4–24
Side of headache and hemiplegia	Bilateral—variable sides
Character	Pulsating
Nausea/vomiting/photophobia/phonophobia during migraine attack	+/–/–/–
Other types of headache	MO, MA

In addition to the M1500V mutation, we identified several synonymous single nucleotide polymorphisms in *SCN1A* in the study group and one missense variant at chr2:166892788, g.37362G>A, c.3199G>A, rs2298771, which the frequency of occurrence, according to data from the dbSNP database, is 24 %, and in our study group was 20 % (in 12 patients with different phenotypes), resulting in amino acid change A1067T (Table 1, Fig. 3). According to PROVEAN and SIFT, this amino acid substitution is neutral with a score of −0.068, and according to PolyPhen-2 mutation, is benign with a score of 0.000. The MutationTaster scored the variant as a neutral polymorphism that might change protein feature. This variant has been assigned to myoclonic astatic epilepsy phenotype [18], which is in accordance with our study showing its presence in one family (no. 7; three patients, 5 % of the study group) with epilepsy and FHM.

In a female patient with migraine without aura and with onset at puberty, we found a novel intronic polymorphism possibly affecting alternative splicing of *SCN1A*, at 2:166852491, g.77659T>C, c.4581+32A>G, located between exons 24 and 25, in a region encoding inactivation loop of the channel (Fig. 1b). The patient had migraine with a typical course: high-intensity attacks lasting for a few hours with frequency 1 per month. The patient's familial history is surprising. Her two sisters were diagnosed with a probable CADASIL (cerebral autosomal dominant arteriopathy with subcortical infarcts and leukoencephalopathy), but with no mutation in *NOTCH3*. One of the sisters suffered from dementia and was diagnosed with disseminated and diffuse changes in brain MRI interpreted as the beginning of demyelination (she died at the age of 60 years, and no medical records were available). The patient's mother died at the age of 69, likely from a stroke. The patient's father and brother and the patient's siblings' children are healthy. The variant could not be evaluated by the PROVEAN, SIFT, or Poly-Phen2 tools due to a lack of data. This variant is absent in the ExAC database, but insertion A/AT has been found at this nucleotide position (http://exac.broadinstitute.org/region/2-166852491-166852491), with a frequency of ca. 0.000008. The substitution reported here could be a novel rare intronic variant. According to the MutationTaster, the variant could alter various protein features, including domains and motifs, downstream of the altered splice site, as presented in Table 3.

CACNA1A and *ATP1A2*

We found several synonymous single nucleotide polymorphisms in the *CACNA1A* and *ATP1A2* genes, as summarized in a Table 1.

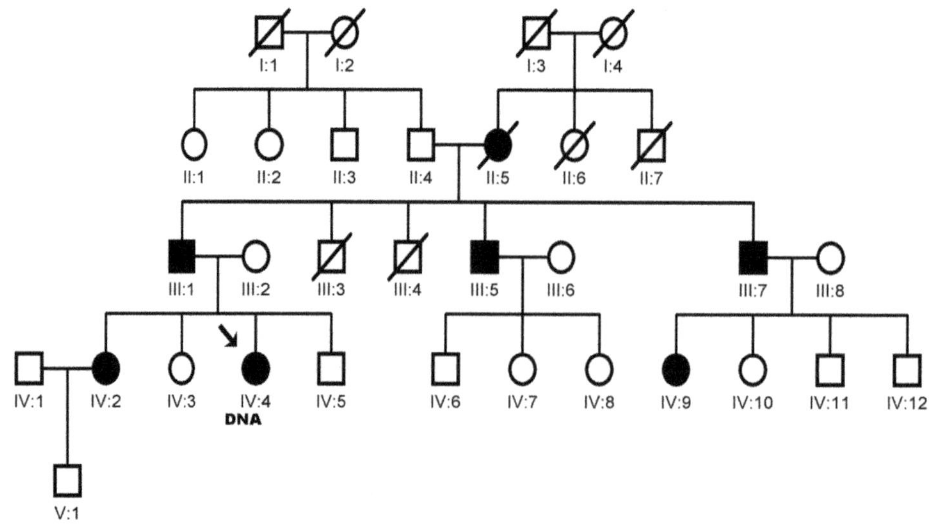

Fig. 2 Pedigree of FHM patient with M1500V mutation in exon 24 of *SCN1A*

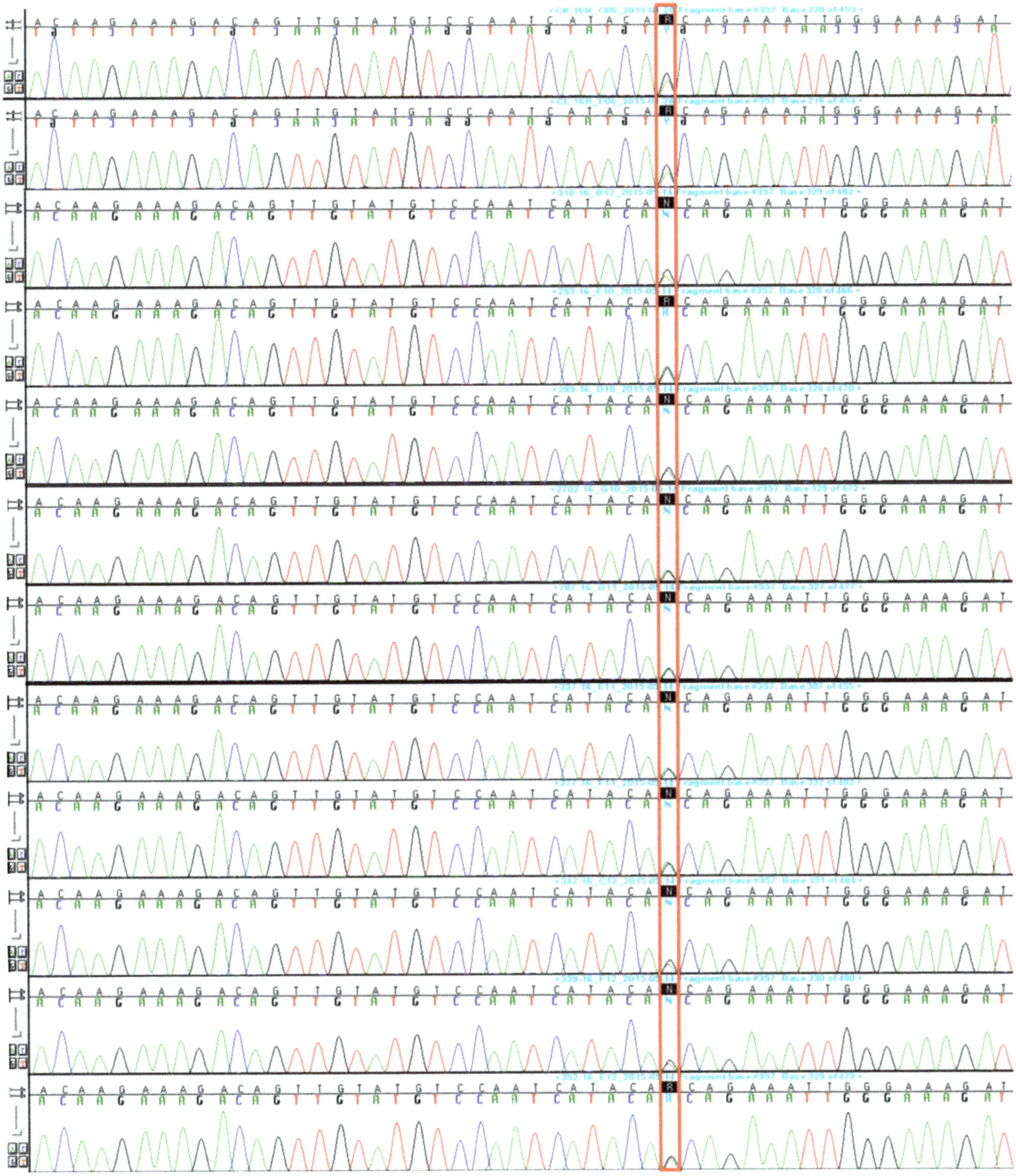

Fig. 3 A1067T mutation in *SCN1A*. Exon 16 with flanking intronic regions of *SCN1A* gene was analyzed as described in the "Materials and methods" section. The missense variant at chr2:166892788, g.37362G>A, c.3199G>A, rs2298771 resulting in amino acid change A1067T found in 12 patients with different phenotypes (20 % of the study group) and present at the high frequency of 24 % in dbSNP database is indicated by frame

Table 3 Predicted loss of protein features downstream of the splice site altered by c.4581+32A>G variant in *SCN1A*

Start (aa)	End (aa)	Feature	Details
1523	1821	Repeat	IV
1537	1537	Conflict	F ->L (in [7]; CAA46439/M91803).
1537	1560	Transmem	Helical; name = S1 of repeat IV; (by similarity).
1572	1595	Transmem	Helical; name = S2 of repeat IV; (by similarity).
1602	1625	Transmem	Helical; name = S3 of repeat IV; (by similarity).
1636	1657	Transmem	Helical; voltage sensor; name = S4 of repeat IV; (by similarity).
1673	1695	Transmem	Helical; name = S5 of repeat IV; (by similarity).
1762	1786	Transmem	Helical; name = S6 of repeat IV; (by similarity).
1788	1788	Carbohyd	N-linked (GlcNAc...) (potential).
1915	1944	Domain	IQ

KCNK18

We identified a known mutation at c.691T>C, rs363315, in exon 3 of *KCNK18* gene leading to amino acid substitution S231P in a female patient diagnosed with migraine with aura (Fig. 1c).

Additionally, in our study group, one family (no. 43; four patients, 6.6 % of the study group) had mutation in Leiden factor V, associated with blood clots. In this family, we did not find any mutations in *SCNA1A*, *CACN1A*, *KCNK18*, and *ATP1A2* reported before as migraine-causing.

Discussion

Migraine is the most common neurological disorder with a well-defined genetic background, but until now, no migraine-causing mutations have been studied in Polish patients [19, 20]. For genetic screenings in this study, we have selected exons 4, 5, 13, 16, 17, 24, 26, 29, 32, and 36 of CACNA1A; exons 6, 14, 15, 16, 17, 23, 24, and 26 of SCNA1A; exons 9, 16, 17, 18, 19, and 22 of ATP1A2; and exons 1, 2, and 3 of KCN1K18 (all with flanking intronic sequences). We have selected these genetic regions according to the frequencies of found disease-causing genetic variants stated in Leiden Familial Hemiplegic Migraine Variation Database and to the highest probability of genetic variant occurrence.

Overall, our study sheds the first light on the genetic background of migraine in the group of Polish patients, and we believe that this study will help us to stratify our pool of patients in terms of their genetic status of the most common variants of the four genes associated with migraine for future large-scale genetic studies.

We have identified a novel FHM-linked M1500V mutation in *SCN1A*, putatively affecting the inactivation loop of the sodium channel protein and confirming previous reports found the S231P mutation in *KCNK18* in a patient with migraine with aura. Surprisingly, no disease-causing variants in *ATP1A2* and *CACNA1A* were found in the study group.

The novel mutation M1500V substitutes evolutionarily highly conserved methionine residue in sodium channel protein type 1 subunit alpha (P35498). The functionally important amphipathic methionine involved in playing binding/recognition of hydrophobic ligand methionine is substituted with hydrophobic valine with a largely non-reactive aliphatic side chains and packed in the protein interior. There is no direct data indicating that M1500V is causally linked to the disease, but it localizes next to the known F1499L mutation (dbSNP:rs121918632), involved in episodes of elicited repetitive daily blindness (ERDB) co-segregating with familial hemiplegic migraines [21] and with I1498M [22]. The Met > Val substitution is in the IFM motif responsible for fast channel inactivation, located in the inactivation intracellular gate between domains III and IV, each composed of six transmembrane segments called S1–S6 (Fig. 4). Since normal fast inactivation of the channel results from occlusion of the intracellular part of the channel by a short loop between domains III and IV, mutation in this region could disturb the inactivation. For example, in oocyte expression studies, R1648H accelerated recovery from inactivation, W1204R shifted the voltage-dependence of activation and inactivation in the negative direction, R859C shifted the voltage-dependence of activation in the positive direction, and T875M enhanced slow inactivation [23]. The alterations due to R1648H and W1204R are predicted to increase sodium channel activity and neuronal excitability, whereas those due to R859C and T875M should decrease the channel activity and neuronal excitability [23]. Further, in vitro studies are necessary to assess the impact of the M1500V mutation on sodium channel inactivation.

The second mutation we found is S231P in *KCNK18*. This mutation has already been well characterized and shown to be migraine-related but not to influence the activity of the TRESK K2P potassium channel encoded by *KSNK18* [24]. Other studies on this variant also showed relationship to migraine with typical aura [25].

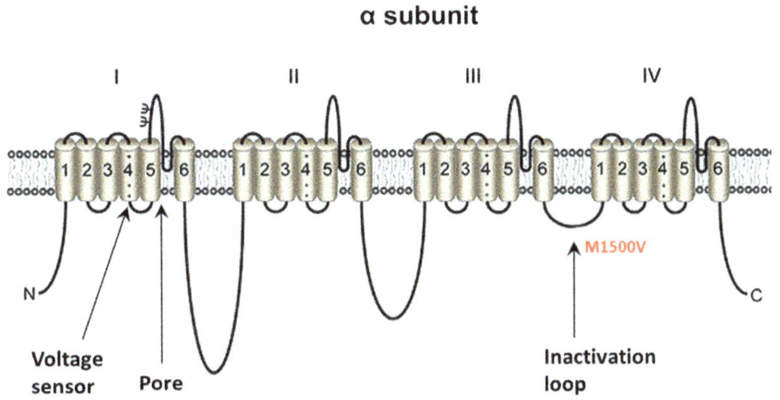

Fig. 4 Transmembrane organization of sodium channel α subunit. The alpha subunit of the channel comprises four repeat domains (I–IV), each containing six membrane-spanning segments (1–6). The highly conserved segment 4 acts as the voltage sensor of the channel close to the pore for ion entry. After Na+ has entered the cell and the membrane's potential reached its maximum, the channel inactivate by closing the inactivation loop. Mutation M1500V could abolish channel inactivation and lead to its constitutive activity and firing of the synapses; however, further functional studies are needed to verify this assumption. Adapted from Brackenbury and Isom [29]

Accordingly, we found the *KCNK18* mutation in a patient with migraine without aura.

The novel intronic polymorphism identified in a family with a CADASIL-like phenotype and migraine could affect the splicing of *SCN1A* transcript. According to the MutationTaster, this variant may influence splicing (e.g., increase or gain of splice donor site at position g.77658) and affect protein features (for details, see Table 3). Voltage-dependent inactivation of Na+ channels is a consequence of voltage-dependent activation, and an inactivated Na+ channel enters a non-conducting and non-transmitting signals state, as the inactivation gate, the cytoplasmic loop linking domains III and IV of the α subunit, obstructs the open pore [26]. Due to its the location in the exon 24, encoding the inactivation loop, the variant seems likely to produce a constitutively inactive form of sodium channel alpha subunit leading to the permanent shut down of synaptic transmission. Detailed functional studies are needed to elucidate the link between generation of various alternative splicing forms and biological pathomechanisms and clinical phenotype of migraine.

Additionally, based on our studies, it could be speculated that the factor Leiden V mutation found in one family carrying no causative mutations in *SCNA1A*, *CACN1A*, *KCNK18*, or *ATP1A2* is a sole cause of migraine with aura and vision disability and strokes in that family, in line with several reports showing a linkage between migraine and Leiden factor variants [27, 28].

Future perspectives

We suggest that mutations located in other parts of the analyzed genes or in other genes, not yet shown to be connected with migraine, cause various clinical forms of the disease in the Polish population. We postulate that next-generation sequencing (NGS) methods, e.g., whole-exome sequencing or whole-genome sequencing, should be used to identify comprehensively migraine-causative mutations. This is reinforced by the fact that recent large-scale studies from European populations have provided an extensive list of susceptibility genes linked to migraine. The list of candidate genes linked with FHM occurrence includes: *APOA1BP*, *TBC1D7*, *FUT9*, *STAT6*, *ATP5B*, *AJAP1*, *TSPAN2*, *FHL5*, *C7orf10*, and *MMP16* [11].

Competing interests

The authors declare that they have no competing interests.

Authors' contributions

ID, MK, CZ, and AK conceived and designed the study. ID, MK, CZ, and AK drafted and revised the manuscript. ID and MK performed the data analysis. ID and AK collected the samples. MK performed RT-PCR and sequencing experiments. All authors read and approved the final manuscript.

Acknowledgements

This study was supported by the statutory grant from the Ministry of Science and Higher Education to the Department of Neurology, Medical University of Warsaw. The funders had no role in the study design, data collection, and analysis; decision to publish; or preparation of the manuscript. The authors would like to acknowledge help with the DNA preparation and sharing WES data for the control group comparison, made by Małgorzata Kobryś and Jakub Fichna, respectively.

Author details

[1]Department of Neurology, Warsaw Medical University, 61 Żwirki i Wigury Street, 02-091 Warsaw, Poland. [2]Department of Neurodegenerative Disorders, Mossakowski Medical Research Centre PAS, 5 Pawinskiego Street, 02-106 Warsaw, Poland.

References

1. Ophoff RA, Terwindt GM, Vergouwe MN, van Eijk R, Oefner PJ, Hoffman SM, et al. Familial hemiplegic migraine and episodic ataxia type-2 are caused by mutations in the Ca2+ channel gene CACNL1A4. Cell. 1996;87:543–52.
2. Denier C, Ducros A, Durr A, Eymard B, Chassande B, Tournier-Lasserve E, et al. Missense CACNA1A mutation causing episodic ataxia type 2. Arch Neurol. 2001;58:292–5.

3. Pietrobon D. CaV2.1 channelopathies. Pflugers Arch. 2010;460:375–93.
4. Gallanti A, Cardin V, Tonelli A, Bussone G, Bresolin N, Mariani C et al. The genetic features of 24 patients affected by familial and sporadic hemiplegic migraine. Neurol Sci. 2011;32 Suppl 1:S141–2.
5. Dichgans M, Freilinger T, Eckstein BE, Babini E, Lorenz-Depiereux B, Biskup S et al. Mutation in the neuronal voltage-gated sodium channel SCN1A in familial hemiplegic migraine. Lancet. 2005;366:371–7.
6. Meisler MH, Kearney JA. Sodium channel mutations in epilepsy and other neurological disorders. J Clin Investig. 2005;115:2010–7.
7. Lafrenière RG, Cader MZ, Poulin JF, Andres-Enguix I, Simoneau M, Gupta N et al. A dominant-negative mutation in the TRESK potassium channel is linked to familial migraine with aura. Nat Med. 2010;16:1157–60.
8. Hamel E. Serotonin and migraine: biology and clinical implications. Cephalalgia. 2007;27:1293–300.
9. Schürks M. Genetics of migraine in the age of genome-wide association studies. J Headache Pain. 2012;13:1–9.
10. Todt U, Netzer C, Toliat M. New genetic evidence for involvement of the dopamine system in migraine with aura. Hum Genet. 2009;125:265–79.
11. Anttila V, Winsvold BS, Gormley P, Kurth T, Bettella F, McMahon G, North American Brain Expression Consortium, UK Brain Expression Consortium, International Headache Genetics Consortium, et al. Genome-wide meta-analysis identifies new susceptibility loci for migraine. Nat Genet. 2013;45: 912–7.
12. Style Matters: Statements from the Vancouver Group. http://dx.doi.org/10.1136/bmj.302.6786.1194. BMJ 1991; 302:1194.
13. The International Classification of Headache Disorders, 3rd edition (beta version) Part one: the primary headaches. Cephalalgia 2013;33(9):644–658.
14. Adzhubei I, Jordan DM, Sunyaev SR. Predicting functional effect of human missense mutations using PolyPhen-2. In: Curr Protoc Hum Genet, Chapter 7:Unit7.20. 2013.
15. Choi Y, Chan AP. PROVEAN web server: a tool to predict the functional effect of amino acid substitutions and indels. Bioinformatics. 2015; 31(16):2745–7.
16. Kumar P, Henikoff S, Ng PC. Predicting the effects of coding non-synonymous variants on protein function using the SIFT algorithm. Nat Protoc. 2009; 4:1073–81.
17. Schwarz JM, Cooper DN, Schuelke M, Seelow D. MutationTaster2: mutation prediction for the deep-sequencing age. Nat Methods. 2014;11:361–2.
18. Escayg A, MacDonald BT, Meisler MH, Baulac S, Huberfeld G, An-Gourfinkel I et al. Mutations of SCN1A, encoding a neuronal sodium channel, in two families with GEFS + 2. Nat Genet. 2000;24(4):343–5.
19. Witkowska-Olearska K. Epidemiology of vascular headache and migraine in the population of Warsaw. Neurol Neurochir Pol. 1975;9:495–502.
20. Szczudlik A. Migraine awareness, treatment, and education in Poland. Cephalalgia. 1998;18:65–6.
21. Vahedi K, Depienne C, Le Fort D, Riant F, Chaine P, Trouillard O et al. Elicited repetitive daily blindness: a new phenotype associated with hemiplegic migraine and SCN1A mutations. Neurology. 2009;72:1178–83.
22. Weller CM, Pelzer N, de Vries B, López MA, De Fàbregues O, Pascual J, et al. Two novel SCN1A mutations identified in families with familial hemiplegic migraine. Cephalalgia. 2014;34:1062–9.
23. Kahlig KM, Misra SN, George Jr AL. Impaired inactivation gate stabilization predicts increased persistent current for an epilepsy-associated SCN1A mutation. J Neurosci. 2006;26:10958–66.
24. Andres-Enguix I, Shang L, Stansfeld PJ, Morahan JM, Sansom MS, Lafrenière RG et al. Functional analysis of missense variants in the TRESK (KCNK18) K channel. Sci Rep. 2012;2:237.
25. Rainero I, Rubino E, Gallone S, Zavarise P, Carli D, Boschi S et al. KCNK18 (TRESK) genetic variants in Italian patients with migraine. Headache. 2014;54: 1515–22.
26. Yu FH, Catterall WA. Overview of the voltage-gated sodium channel family. Genome Biol. 2003;4:207.
27. Intiso D, Crociani P, Fogli D, Grandone E, Cappucci G, Di Rienzo F et al. Occurrence of factor V Leiden mutation (Arg506Gln) and anticardiolipin antibodies in migraine patients. Neurol Sci. 2002;22:455–8.
28. Soriani S, Borgna-Pignatti C, Trabetti E, Casartelli A, Montagna P, Pignatti PF et al. Frequency of factor V Leiden in juvenile migraine with aura. Headache. 1998;38:779–81.
29. Brackenbury WJ, Isom LL. Na channel β subunits: overachievers of the ion channel family. Front Pharmacol. 2011;28(2):53.

Fine-scale population structure of Malays in Peninsular Malaysia and Singapore and implications for association studies

Boon-Peng Hoh[1,2,3*], Lian Deng[1], Mat Jusoh Julia-Ashazila[2], Zakaria Zuraihan[2], Ma'amor Nur-Hasnah[2], Ab Rajab Nur-Shafawati[4], Wan Isa Hatin[4], Ismail Endom[5], Bin Alwi Zilfalil[6], Yusoff Khalid[2,7] and Shuhua Xu[1,8,9*]

Abstract

Fine scale population structure of Malays - the major population in Malaysia, has not been well studied. This may have important implications for both evolutionary and medical studies. Here, we investigated the population sub-structure of Malay involving 431 samples collected from all states from peninsular Malaysia and Singapore. We identified two major clusters of individuals corresponding to the north and south peninsular Malaysia. On an even finer scale, the genetic coordinates of the geographical Malay populations are in correlation with the latitudes (R^2 = 0.3925; P = 0.029). This finding is further supported by the pairwise F_{ST} of Malay sub-populations, of which the north and south regions showed the highest differentiation ($F_{ST\ [North-south]}$ = 0.0011). The collective findings therefore suggest that population sub-structure of Malays are more heterogenous than previously expected even within a small geographical region, possibly due to factors like different genetic origins, geographical isolation, could result in spurious association as demonstrated in our analysis. We suggest that cautions should be taken during the stage of study design or interpreting the association signals in disease mapping studies which are expected to be conducted in Malay population in the near future.

Keyword: Malay, Population sub-structure, F_{ST}, Latitude-PC correlation, GWAS simulation

Background

Malaysia, a multi-ethnic, multi-lingual, multi-cultural and multi-religious country, is located at the crossroads of Southeast Asia. It is separated by the South China Sea into two land masses namely, the Peninsular Malaysia and East Malaysia (also known as the Borneo island). Malaysia has a total population of about 30 million people, of which approximately 26 million populate the Peninsular Malaysia. Among the major populations in Peninsular Malaysia, the Malays are the largest ethnic group and make up to 63% of the total population follow by Chinese, Indians and other minority ethnic groups.

Many Malays are of Malayo-Polynesian (Austronesian) origin that are culturally and historically heterogeneous [1]. The Malays from the west coast of Peninsular Malaysia are historically linked to Sumatera across the Straits of Malacca; while those from the south are thought to have migrated from Jawa, Sulawesi and other parts of Indonesia [2]. The Malays from the north Peninsular have a closer affinity to the Malay Muslims from the Southern Thai due to geographical location. The history of Singapore has never been separated from Peninsular Malaysia since the first century until the year 1965 when Singapore became an independent republic. Therefore, it is very likely that the Malays of Singapore have a similar history of origin as those from the southern part of Peninsular Malaysia [3].

Earlier studies had indicated potential genetic sub-structure among the different groups of Malays from Peninsular Malaysia [2, 4, 5], which could be possibly attributed to the migration history of these respective sub-groups. However, fine-scale sub-structure of the

* Correspondence: hoh.boonpeng@gmail.com; xushua@picb.ac.cn
[1]Chinese Academy of Sciences (CAS) Key Laboratory of Computational Biology, Max Planck Independent Research Group on Population Genomics, CAS-MPG Partner Institute for Computational Biology (PICB), Shanghai Institutes for Biological Sciences, Chinese Academy of Sciences, Shanghai 200031, China

Malay population remained poorly described, especially, previous studies were based on very limited sample size. Indeed, this potentially poses confounding factors to the genetic association studies, in particular genome-wide association studies (GWAS), leading to spurious association signals [6]. Being one of the major populations in the Southeast Asia, characterizing population substructure is crucial in designing, analyzing and interpreting any genetic association study in this region.

In this study, we showed that the genetic diversity and population sub-structure of the Malays from Peninsular Malaysia are correlated to the geographical latitude. Notably, we observed the main differentiations between populations corresponding to the north and south Peninsular Malaysia. In addition, simulation analyses carried out also revealed that the genetic association is greatly affected by population sub-structure, suggesting that consideration of population stratification of samples at the stage of study design and careful interpretation of the association signals are necessary when mapping complex diseases in Malay populations.

Results

Population substructure

We first compared the genetic diversity of the Peninsular Malays from a global scale with 6 populations from HapMap3 including YRI, CHB, JPT, CEU, MEX and GIH. PC plot indicated that the Malays clustered closely to the East Asian populations as expected, and showed a rather small genetic diversity. Several Malay individuals from northern Peninsular Malaysia (PMM) showed closer affinity to the South Asia populations (GIH) (Fig. 1a). We then performed PCA for the Peninsular and Singapore Malays, and revealed a seemingly homogenous cluster (Additional file 1: Figure S1). However, some level of differentiations were observed corresponding to three geographical regions (north, center and south), despite samples from center region that was scattered around (Fig. 1b). We subsequently excluded the samples form the center regions (Pahang and Selangor), and re-ran the *smartPCA*. Two clusters were observed representing the north and south regions, respectively (Fig. 1c).

In ADMIXTURE analysis, a significant difference was observed between the Malays from the north and south in the major component, with 57% and 65% in the north and south, respectively ($P < 0.0001$; Fig. 2). At K=3, the newly appeared component (denoted in green) was seen slightly higher in the central Malays than in the south Malays (6.8% vs 3%; $P = 0.0415$).

Correlation of genetic and geographic coordinates

Given the fact that the PC1 as well as the ADMIXTURE analysis showed significant differences between northern and southern Malay samples, we then investigated if the genetic diversity between these sub-structure of Malays in Peninsular Malaysia were attributed to geographical coordinates. Average PC1 values of southern Malay samples (corresponding to Fig. 1b) were generally less than 0 (except for Johor), whilst all geographically defined northern regions with PC1 >0 (Fig. 3). When we compared the PC1 with geographical latitude of these sample locations, a significant correlation was observed ($R^2 = 0.3925$; $P = 0.029$; Fig. 4). Due to the geographical nature, Peninsular Malaysia is divided into west coast and east coast by the Titiwangsa Ranges. We therefore asked if the genetic diversity could be attributed to the geographical longitude as well. Analysis between PC1 and geographical longitude, however showed no significant correlation (R^2=0.0066; $P = 0.7924$; Addional file 1: Figure S2). We also evaluated if the genetic diversity was related to geographical distance between two populations, but found no significant correlation of F_{ST} between populations and the geographical distances between them ($R^2 = 0.01918$; $P = 0.1385$; Additional file 1: Figure S3).

Genetic differentiation between northern and southern Malays

The regional F_{ST} values indicated the highest regional diversity between the north and the south after 1,000 times bootstrapping repeats ($F_{ST} = 0.001$; CI = 2.07E - 04) (Table 1; Additional file 2: Table S1). To further identify the genomic regions that are highly differentiated between northern and southern Peninsular Malay, we computed the F_{ST} values of the 41,400 SNPs between northern and southern samples, and identified 428 SNPs listed in the top 1% of the F_{ST} (Additional file 2: Table S2); of which 80 (0.1%) had an F_{ST} value >0.05 (Table 2). SNP with the highest F_{ST} value was rs4149264, residing in the candidate gene *ABCA1* - a major gene responsible for high-density lipolipoprotein cholesterol (HDL-c) synthesis. Another highly differentiated SNP, rs4148475, is located at the candidate gene *ABCC4*. This gene is a member of the superfamily of ATP-binding Cassette (ABC) transporters, which may play a role in cellular detoxification [7]. A missense variant rs1056836 appears to be one of the four highly differentiated SNPs, leading to a change of valine to leucine in candidate gene *CYP1B1*, which had a minor allele frequency of 0.48 and 0.19 in northern and southern peninsular Malays, respectively ($F_{ST} = 0.2037$). This candidate gene is responsible in drug metabolism and synthesis of cholesterols, steroids and lipids. It was found to play a role in the susceptibility of glaucoma [8, 9]. We performed an enrichment analysis with DAVID (http://david.abcc.ncifcrf.gov/) by including the top 1% highly differentiated variants, but identified no significant enrichment after Benjamini correction (Additional file 2: Table S3).

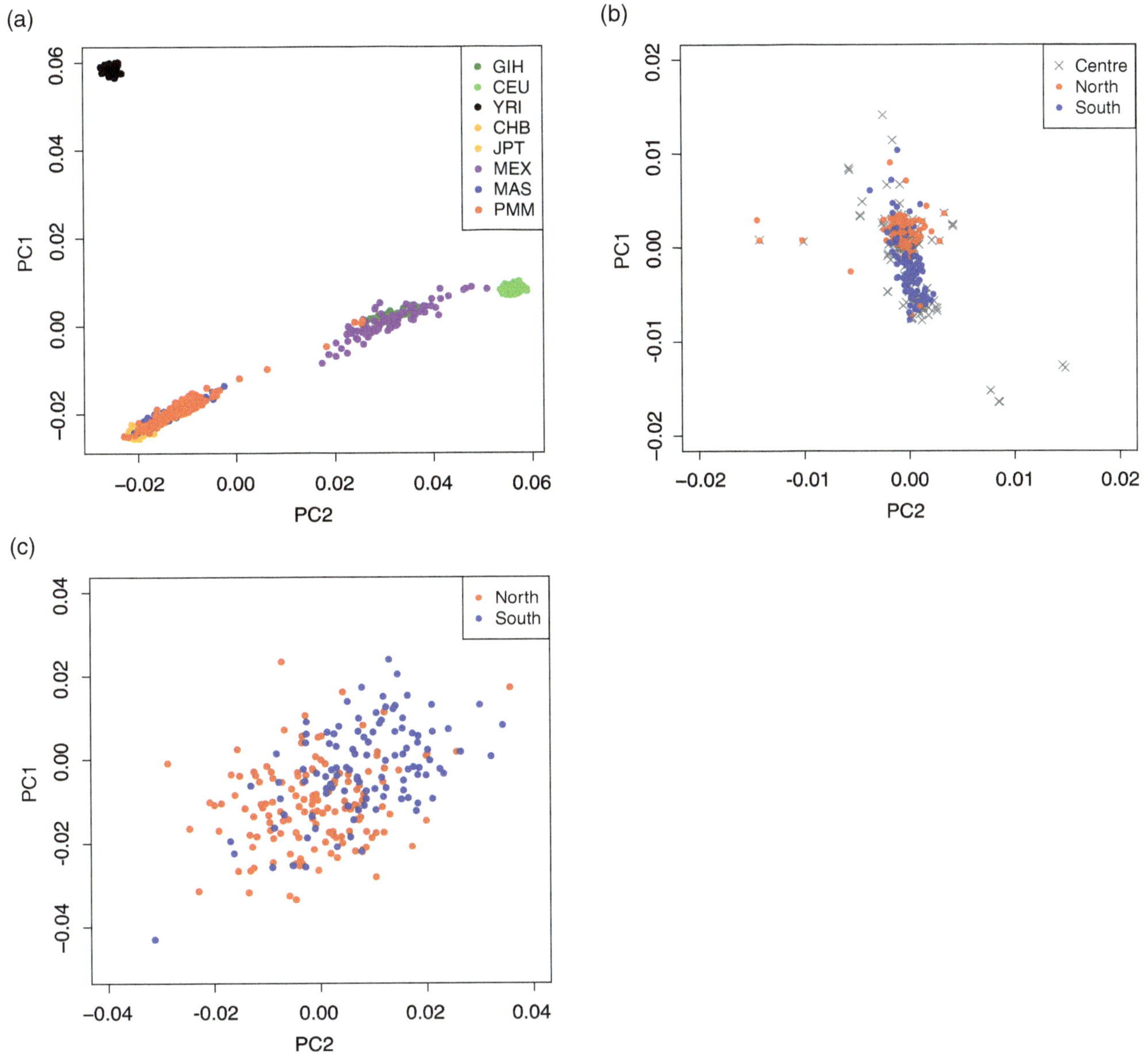

Fig. 1 Principle Component Analysis (PCA) (**a**) Global PCA including populations from HapMap3. GIH, Gujarati India Houston; CEU, Northern and Western European from CEPH collection; YRI, Yoruba Ibadan from Nigeria; CHB, Chinese Beijing; JPT, Japanese Tokyo; MEX, Mexican ancestry from Los Angeles; MAS, Metropolitan Malays from Singapore; PMM, Malays from Peninsular Malay. The Malay populations are of East Asian descendant. (**b**) PCA plot including samples categorized into North vs Centre vs South; (**c**) PCA plot which included only North vs South. Symbols in red represent the northern region; symbols in blue represent southern region. Several outliers were excluded from the PCA plot

We observed that 1,666 SNPs were presented in different minor alleles between the north and south Malays, and their allele frequencies in Malays were compared with that in South Asian (GIH) and East Asian (CHB) (Additional file 2). Although not substantial, differences in allele frequencies were observed between the South- and East- Asians, as well as the between the Malays and both South- and East- Asians. Notably, rs1126965 located at the candidate gene *SLC39A11* revealed an alternative allele frequency of 0.8178 in the northern Malays and 0.4965 in the southern Malays. This gene has been reported to play a role in liver enzyme and smoking initiation [10, 11]. Whether or not this gene is under positive selection in the Malays, however, remain further investigation. We subsequently assessed if these SNPs play a role in phenotypic association, and found that 19 of these SNPs were reported in GWAS catalogue (Additional file 4).

To evaluate the potential effect of population substructure on a disease association study, a series of

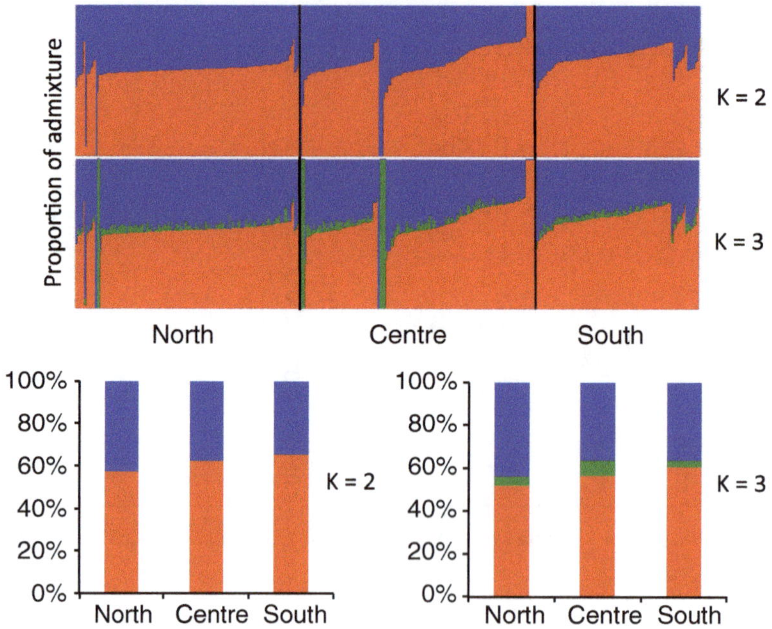

Fig. 2 ADMXITURE analysis of the Malay populations classified according to regions. The bottom plots represented by percentages (Y-axis) indicates the average ADMIXTURE values for each region

computer simulation studies were carried out with PLINK following a case–control GWAS design (Additional file 2: Table S4). The GWAS simulations revealed that the effect on false positive rate and statistical power were greater than expected [12].

Discussion

We demonstrated in this study, that the Malays from Peninsular Malaysia and Singapore are essentially sub-structured. Although genetic correlation with geographical latitude had been previously reported in the Chinese populations [12, 13], it is indeed surprising to reveal such differentiation among the Malay populations even within a small region in Peninsular Malaysia and Singapore (~800 KM from north to south). In addition to that, the F_{ST} between the north and south Malays were similar to those of the earlier report between the northern and southern Han Chinese (F_{ST} = 0.0011) [12] but lower than those within Europeans (F_{ST} = 0.0033) [14] However, we observed higher diversity within the substructures of the Malays. For instance, the F_{ST} between two northern Peninsular Malays from Kedah and Kelantan was 0.017 (Table S4), which is in line with the finding in a recent study [4]. This suggests higher heterogeneity among Malays than previously expected, possibly be due to the recent migration and gene flow from the surrounding populations in this region.

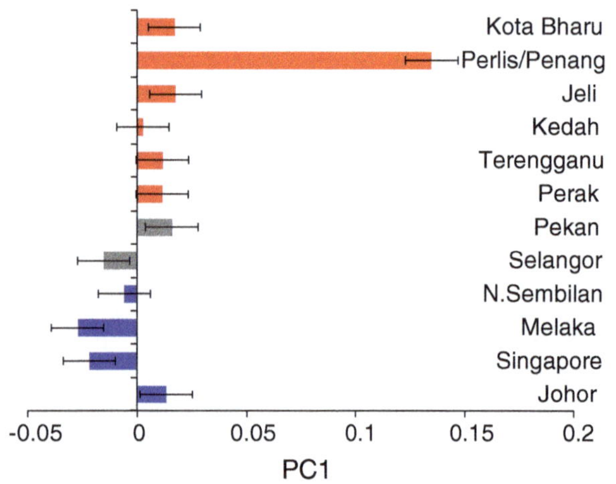

Fig. 3 Average PC1 values of the Malay sub-populations from Peninsular Malaysia and Singapore. Standard error of each population is indicated. The PC1 values correlated well to the geographical locations of each population except for Johor

The Pahang Malays were found to have a closer affinity to the north, although they were classified as the central region in this study. This is likely due to the reason that samples were collected from the Federal Land Development Authority (FELDA) settlers in the Pahang state, of which the majority of them were originated from Kelantan. On a separate note, Selangor, being as the most advanced and most populated state of Malaysia, is where the metropolitan city Kuala Lumpur located. PCA revealed that samples from this population was scattered across both the north and south regions (Fig. 1b & c). We believe that urbanization had likely blurred the boundaries. Similar findings were observed

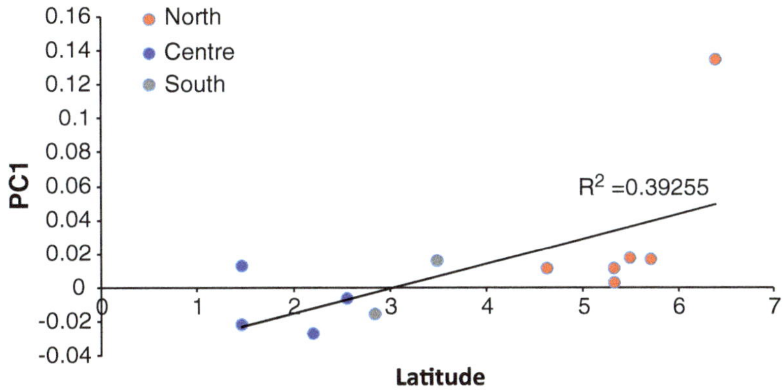

Fig. 4 Correlation between PC1 and latitude coordinate (P = 0.029)

in Xu et al. (2009), where the populations from metropolitan areas showed more complicated composition with multiple ancestral origins compared with those from the rest of the area.

Essentially, identification a panel of ancestry informative markers (AIMs) would be an ideal strategy to correct the population stratification in future genetic association studies [15]. However, the SNP coverage and the sample size in the current study are insufficient for such purpose. Those highly differentiated SNPs between the north and south Malays could be possibly due to genetic drift or, to a lesser extent, natural selection. These SNPs, however may be considered as the putative set of variants as the AIMs for the Malay populations. The candidate gene *ABCA1* is a major gene that plays an important role in high-density lipoprotein cholesterol (HDL-c) synthesis and cholesterol transport [16]. However, whilst we suspect the genetic drift is likely to be the cause, the reason of this gene being highly differentiated between northern and southern Malays remains further investigated. Cautions should be taken though when positive signals of HDL-c and *ABCA1* are identified in the genetic association study of Malays.

We acknowledge several limitations in this study. Sample collection from several locations were small, hence might have resulted into outliers which confounded the outcome of the correlation between genetic differentiation and geographical coordinates. In addition, self-reported ancestry might have also confounded the analysis when assigning to their respective state of origin. However, the number of samples covering all states in Peninsular Malaysia (and Singapore), and the marker utilized in our study are larger than the previous reports, thus provides further insights into the genetic structure of the Malays in Peninsular Malaysia. Notably, we revealed close relationship between genetic and geographical coordinates in the Malay populations. In addition, our results and to which extent the admixtures in Southeast Asia could impact the population stratification thus affect the genetic association studies. Therefore we call for attention to look into alternative strategies for disease mapping in genetically complex populations particularly from Southeast Asia.

Table 1 Pairwise F_{ST} bootstrap values of the Malay between the 3 regions of Peninsular Malaysia

	North	Centre	South
North	-	0.00083315 (CI =2.0684E-04)	0.00111661 (CI = 2.68108E-06)
Centre		-	0.00058556 (CI = 4.0972E-06)
South			-

Pairwise F_{ST} values calculated by bootstrap resampling 1,000 replications

Conclusion

In summary, we revealed that the population substructure of the Malays was correlated to the latitude coordinate. The genetic diversity of the Malays is more heterogeneous than previously expected, and that we proved that such population sub-structure occurred even though within a small geographical region may potentially lead to spurious signals in disease based genetic association studies. Therefore cautions should be taken when carrying out such study design.

Methods

Population and samples

A total of 431 Malay samples were included in this study. These samples were self-identified Malays from Peninsular Malaysia, 116 of which were genotyped with Affymetrix Genome-Wide Human SNP Array 6.0, whilst the remaining samples were genotyped with Illumina 660W (Sample size, N = 90) and Illumina Omni Express (N = 119). The additional 17 Malays samples from Kelantan genotyped with Affymetrix Genome-Wide Human SNP Array 6.0 [17], and 89 samples of metropolitan Malays from Singapore (SGVP) were also included in this analysis [3]. The studies were approved by the research and ethics committees of Universiti Teknologi

Table 2 Top 0.1 % SNPs that are highly differentiated between the Malays from northern and southern region of Peninsular (total SNP = 42633)

rsID	Chr	Position	Minor allele	F_{ST}	MAF_North	MAF_South	Gene	Category
rs4149264	9	107,677,211	C	0.2256	0.4856	0.1682	*ABCA1*	intronic
rs10102377	8	83,762,822	T	0.2251	0.4097	0.2336		
rs4148475	13	95,853,574	A	0.2242	0.4676	0.184	*ABCC4*	intronic
rs1056836	2	38,298,203	G	0.2037	0.4757	0.1934	*CYP1B1*	coding
rs1126965	17	70,642,790	G	0.1931	0.5	0.1822	*SLC39A11*	3utr
rs17769090	15	70,630,120	A	0.1636	0.4648	0.2381		
rs6974363	7	47,633,187	G	0.1421	0.493	0.2333		
rs837395	1	47,269,338	A	0.1400	0.4897	0.2383	*CYP4B1*	intronic
rs4646430	2	38,306,415	G	0.1384	0.4621	0.1981		
rs215101	16	16,052,973	G	0.1206	0.4752	0.271	*ABCC1*	intronic
rs12920607	16	73,728,620	C	0.1183	0.475	0.2736		
rs837398	1	47,266,422	A	0.1124	0.4897	0.2664	*CYP4B1*	intronic
rs809367	10	89,741,806	A	0.1088	0.4307	0.2009		
rs316133	6	52,847,551	C	0.0957	0.4823	0.2594	*GSTA4*	intronic
rs6130511	20	42,681,088	A	0.0916	0.2801	0.1	*TOX2*	intronic
rs2132845	4	140,587,125	T	0.0910	0.4255	0.215	*MGST2*	5utr
rs5761313	22	26,313,745	T	0.0887	0.4964	0.2804	*MYO18B*	intronic
rs10485805	20	54,945,783	G	0.0853	0.4397	0.2336	*AURKA*	intronic
rs10489142	1	7,363,310	G	0.0835	0.4507	0.3364	*CAMTA1*	intronic
rs2274928	13	24,044,546	A	0.0779	0.3601	0.4346	*LINC00327*	intronic
rs11935505	4	145,226,122	A	0.0758	0.04643	0.1682		
rs1566869	12	52,266,348	A	0.0695	0.3406	0.1682		
rs1884897	20	6,612,832	G	0.0695	0.2812	0.1215		
rs4530975	7	104,415,415	T	0.0679	0.1862	0.05607	*LHFPL3*	intronic
rs6024831	20	54,938,464	G	0.0675	0.4161	0.3915	*FAM210B*	intronic
rs1160798	6	112,438,446	C	0.0661	0.06897	0.1934	*LAMA4*	intronic
rs2158196	4	114,416,596	C	0.0658	0.2391	0.09434	*CAMK2D*	intronic
rs16961766	13	103,899,499	A	0.0645	0.3143	0.1524		
rs10962015	9	15,387,949	A	0.0638	0.2482	0.1028		
rs6884962	5	172,682,382	A	0.0623	0.4896	0.3271		
rs17126776	12	39,311,625	A	0.0603	0.1759	0.3318		
rs2755209	13	41,137,804	C	0.0602	0.25	0.1075	*FOXO1*	intronic
rs9783586	13	108,361,559	T	0.0601	0.2671	0.4393	*FAM155A*	intronic
rs10968093	9	27,753,227	A	0.0594	0.06338	0.1776		
rs2458286	8	103,978,699	T	0.0591	0.4424	0.3774		
rs4875364	8	4,444,592	C	0.0583	0.3094	0.1557	*CSMD1*	intronic
rs11145506	9	80,264,584	T	0.0578	0.3821	0.217	*GNA14*	intronic
rs11604366	11	28,887,766	C	0.0577	0.2695	0.4387		
rs9881633	3	112,881,539	T	0.0574	0.3	0.472	*RP11_572M11.3*	intronic
rs5762448	22	28,408,444	C	0.0573	0.3514	0.1916	*TTC28*	intronic
rs6467991	7	83,954,737	C	0.0570	0.344	0.4811		
rs10089677	8	122,660,248	A	0.0569	0.2329	0.09813		
rs1923254	13	41,084,241	G	0.0567	0.3776	0.2143		

Table 2 Top 0.1 % SNPs that are highly differentiated between the Malays from northern and southern region of Peninsular (total SNP = 42633) *(Continued)*

rs7813806	8	5,142,665	C	0.0567	0.2817	0.1355		
rs7625411	3	112,811,428	A	0.0564	0.3403	0.486		
rs2922249	6	127,954,614	C	0.0564	0.09441	0.2196		
rs2294088	8	124,526,607	A	0.0564	0.4514	0.2804	*FBXO32*	intronic
rs12289262	11	12,894,758	T	0.0560	0.3986	0.2336	*TEAD1*	intronic
rs4937523	11	130,347,190	T	0.0558	0.2937	0.4626	*ADAMTS15*	intronic
rs10807768	7	13,662,014	A	0.0554	0.3169	0.1651		
rs976272	14	61,449,328	A	0.0551	0.4897	0.3178	*SLC38A6*	coding
rs13027801	2	143,602,503	C	0.0551	0.2862	0.4533		
rs17701834	19	22,121,458	G	0.0549	0.2172	0.08879		
rs7193843	16	54,677,292	G	0.0548	0.1448	0.285		
rs7097885	10	16,506,501	C	0.0542	0.2832	0.4486	*PTER*	intronic
rs2791398	1	245,965,551	G	0.0540	0.05944	0.1651	*SMYD3*	intronic
rs10486802	7	39,723,768	A	0.0531	0.1884	0.07009	*RALA*	intronic
rs8031676	15	96,910,440	C	0.0529	0.4306	0.2664		
rs7186479	16	82,602,736	C	0.0527	0.2517	0.1168		
rs6054383	20	6,584,604	T	0.0526	0.3986	0.2383		
rs4460308	7	104,420,060	C	0.0521	0.1866	0.07009	*LHFPL3*	intronic
rs3775779	4	70,709,207	A	0.0520	0.476	0.3551	*SULT1E1*	intronic
rs9375877	6	132,690,239	G	0.0517	0.4862	0.3458	*MOXD1*	intronic
rs2180691	20	54,964,361	A	0.0517	0.45	0.2857	*AURKA*	intronic
rs7778955	7	39,740,487	G	0.0517	0.1438	0.04206	*RALA*	intronic
rs4608114	12	92,384,658	A	0.0517	0.4366	0.2736	*C12orf79*	intronic
rs6946733	7	106,670,288	A	0.0514	0.4281	0.2664		
rs816650	10	601,089	T	0.0514	0.114	0.2406	*DIP2C*	intronic
rs6490805	13	24,084,809	C	0.0507	0.08394	0.1981		
rs17171480	7	35,585,669	A	0.0506	0.09286	0.2103		
rs17015112	3	77,319,487	G	0.0505	0.4545	0.3785	*ROBO2*	intronic
rs573186	3	124,178,276	C	0.0503	0.2483	0.1168	*KALRN*	intronic
rs10752609	1	154,791,128	A	0.0501	0.3147	0.1698	*KCNN3*	intronic
rs1862737	16	75,281,964	C	0.0500	0.4155	0.257	*BCAR1*	5utr

MARA and Universiti Sains Malaysia, and the design of this study followed the Helsinki Declaration 1975, as revised in year 2000. The collected samples covered all 11 states of Peninsular Malaysia (Fig. 5), of which were divided into 3 geographical regions for the purpose of this study namely, North, South and the Centre regions, according to their respective latitude coordinate (Table 3). The number of samples and their geographical locations are listed in Table 3. Six selected populations involving 805 samples from the International HapMap Project 3 (HapMap 3) [18] were included in the analysis to characterize the genetic variation of the Malays on a global scale: YRI, GIH, CEU, CHB, JPT and MEX.

Data assemblage

Data QC and assemblage were carried out with PLINK 1.07. Datasets from each platform were first filtered for individuals with >10% missing rate, > 10% SNP missing rate, minor allele frequencies (MAF) < 0.05, and Hardy-Weinberg Equilibrium (HWE) $P < 0.002$. Then the filtered datasets were subsequently merged, consisting 42,633 SNPs shared among all the 402 Malay samples. The dataset was further pruned down by removing any SNP with $r^2>0.8$, leaving a total SNP of 41,400 for further analyses.

Analysis of population structure

Principal Component Analysis (PCA) was first carried out using the *smartPCA* in EIGENSOFT (ver 4.0) package.

Fig. 5 The geographical map of Peninsular Malaysia. The sampling locations are shown in red dots

Table 3 Regional categorization of the Peninsular Malaysia states according to geographical locations and final number of sample included after QC

Region	States	Latitude coordinate*	No. subjects
North	Perlis	6°23′40.06″N (6.394462)	7
	Kedah	5°19′45.2″N (5.329221)	
	Pulau Pinang	5°19′45.2″N (5.329221	
	Perak	4°37′8.46″N (4.619018)	5
	Kelantan (Kota Bharu)	5°42′55.13″N (5.715314)	56
	Kelantan (Jeli)	5°29′49.4″N (5.497056)	74
	Terengganu	5°19′30.33″N (5.325092)	4
Centre	Pahang (Pekan)	3°29′32″N (3.492092)	51
	Selangor	2°50′34.7″N (2.842971)	98
South	Negeri Sembilan	2°33′12.75″N (2.553541)	9
	Melaka	2°12′11.81″N (2.203281)	5
	Johor	1°27′41.98″N (1.461662)	4
	Singapore (Malay)	1°17′13.35″N (1.287043)	89

*Latitude coordinate from Yandex (http://map.yandex.com)

The genetic component of the Malay populations was inferred with ADMIXTURE ver 1.22 (Alexander et al., 2009) [19], with the 41,400 SNPs overlapped across all samples.

Latitude-PC correlation

Pearson's correlation coefficient was calculated to evaluate the relationship between the genetic coordinates (PC values) and the geographic latitudes.

Pairwise F_{ST}

Unbiased estimation of F_{ST} was calculated according to Weir and Hill (2002) [20], with confidence intervals estimated by bootstrapping with 1,000 replications.

GWAS simulation

Simulations on genome-wide association study (GWAS) were performed using PLINK 1.07, following the procedure of Xu et al. (2009) [12].

Additional files

Additional file 1: Figure S1. PCA plot including samples from the 11 states of Peninsular Malaysia. **Figure S2.** Correlation between PC1 and longitude. Figure S3. Correlation between geographical distance and F_{ST}.

Additional file 2: Table S1. Regional pair-wise F_{ST} of the populations collected from different states in Malaysia. **Table S2.** Top 1% SNPs that are highly differentiated between the Malays from northern and southern region of Peninsular. **Table S3.** Gene ontology and enrichment analysis of the candidate genes underlying the top 0.1% highly differentiated SNPs between the Malays from the north and south Peninsular Malaysia. **Table S4.** Simulation analysis for GWAS before and after removing 1,666 SNPs with different minor allele present between the north and south Malays.

Additional file 3: SNPs with different minor allele between northern and southern Malays.

Additional file 4: GWAS annotation.

Abbreviations

GWAS: Genome wide association; FELDA: Federal land development authority; SNP: Single nucleotide polymorphism; AIMs: Ancestry informative markers; HDL-c: High density lipoprotein cholesterol; HDL-c: High density lipoprotein cholesterol; HapMap 3: Hapmap project 3; Punadj: Unadjusted P value; GC: Genomic control; Bonf: Bonfferoni correction; FDR: False discovery rate.

Competing interests

The authors declare that they have no competing interests.

Author's contribution

BPH and SX conceived and designed the study; BPH and LD performed the data analysis; MJJA, ZZ, MNH, NSAR, WIH, IE, ABZ, YK involved in sample collection, BPH, SX and YK funded the experimental and computational costs. BPH, SX and LD prepared and wrote the manuscript. All authors have read and approved the manuscript.

Acknowledgent

The authors would like to acknowledge the medical staffs from Hospital Kota Bharu, Hospital Sungai Buloh, and Hospital Selayang, as well as the staffs and scientists from the PURE-REDISCOVER Study whom assisted in the sample collection. The authors also thank all subjects who had participated this study. B.P. Hoh is supported by the Chinese Academy of Sciences Fellowship for Visiting Scientists from the Developing Countries (2914FFSA0009). The computational analyses of this study were supported by the Strategic Priority Research Program of the Chinese Academy of Sciences (CAS) (XDB13040100), and by the National Science Foundation of China (NSFC) grants (91331204 and 31171218). The experimental funding of this study was supported by the Long-term Research Grant Scheme (LRGS) for Infectious Disease 2011 [600-RMI/LRGS 5/3(3/2011)] funded by Ministry of Education Malaysia, and the Long-term Research Grant Scheme (LRGS) for the REsponDing to IncreaSing CardiOVascular disease pRevalence (REDISCOVER) [600-RMI/LRGS 5/3(2/2011)] funded by Ministry of Education Malaysia. We also would like to acknowledge the contributions the staffs of Human Genome Center, Universiti Sains Malaysia (USM). This study was also supported by USM Apex grant: 1002 / PPSP / 910343 and USM short term grant: 304 / PPSP / 61311034. S.X. is Max-Planck Independent Research Group Leader and member of CAS Youth Innovation Promotion Association. S.X. also gratefully acknowledges the support of the National Program for Top-notch Young Innovative Talents of the "*Wanren Jihua*" Project.

Author details

[1]Chinese Academy of Sciences (CAS) Key Laboratory of Computational Biology, Max Planck Independent Research Group on Population Genomics, CAS-MPG Partner Institute for Computational Biology (PICB), Shanghai Institutes for Biological Sciences, Chinese Academy of Sciences, Shanghai 200031, China. [2]Institute of Medical Molecular Biotechnology, Faculty of Medicine, Universiti Teknologi MARA, Sungai Buloh Campus, Selangor, Malaysia. [3]UCSI University, Kuala Lumpur Campus, Cheras, Kuala Lumpur, Malaysia. [4]Human Genome Centre, School of Medical Sciences, Universiti Sains Malaysia, Kelantan 16150, Malaysia. [5]Faculty of Science and Technology, School of Biosciences and Biotechnology, Universiti Kebangsaan Malaysia, Bangi 43600, Malaysia. [6]Department of Pediatrics, School of Medical Sciences, Universiti Sains Malaysia, Kelantan 16150, Malaysia. [7]UCSI University, Jalan Menara Gading, Taman Connaught, 56000 Kuala Lumpur, Wilayah Persekutuan, Malaysia. [8]School of Life Science and Technology, ShanghaiTec University, Shanghai 200031, China. [9]Collaborative Innovation Center of Genetics and Development, Shanghai 200438, China.

References

1. Bellwood P, Gamble C, Le Blanc SA, Pluciennik M, Richards M, Terrell JE. First Farmers: the Origins of Agricultural Societies, by Peter Bellwood. Malden (MA): Blackwell, 2005; ISBN 0-631-20565-9 hardback £60; ISBN 0-631-20566-7 paperback £17.99, xix+360 pp., 59 figs., 3 tables. Cambridge Archaeological Journal 2007:87
2. Hatin WI, Nur-Shafawati AR, Zahri M-K, Xu S, Jin L, Tan S-G, et al. Population genetic structure of peninsular Malaysia Malay sub-ethnic groups. PLoS One. 2011;6, e18312.
3. Teo Y-Y, Sim X, Ong RTH, Tan AKS, Chen J, Tantoso E, et al. Singapore Genome Variation Project: a haplotype map of three Southeast Asian populations. Genome Res. 2009;19:2154–62.
4. Hatin W, Nur-Shafawati A, Etemad A, Jin W, Qin P, Xu S, et al. A genome wide pattern of population structure and admixture in peninsular Malaysia Malays. Hugo J. 2014;8:5.
5. Consortium HP-AS. Mapping human genetic diversity in Asia. Science. 2009;326:1541–5.
6. Helgason A, Yngvadóttir B, Hrafnkelsson B, Gulcher J, Stefánsson K. An Icelandic example of the impact of population structure on association studies. Nat Genet. 2005;37:90–5.
7. Nandi S, Ma L, Denis M, Karwatsky J, Li Z, Jiang X-C, et al. ABCA1-mediated cholesterol efflux generates microparticles in addition to HDL through processes governed by membrane rigidity. J Lipid Res. 2009;50:456–66.
8. Micheal S, Ms HA, Frcs SNZ, Bs BB, Fcps MA, Fcps FA, et al. Original Article Identification of novel CYP1B1 gene mutations in patients with primary congenital and primary open-angle glaucoma. Clin Experiment Ophthalmol. 2015;43(1):31–9.
9. Bouyacoub Y, Ben Yahia S, Abroug N, Kahloun R, Kefi R, Khairallah M, et al. CYP1B1 gene mutations causing primary congenital glaucoma in Tunisia. Ann Hum Genet. 2014;78:255–63.
10. Park T-J, Hwang J-Y, Go MJ, Lee H-J, Jang HB, Choi Y, et al. Genome-wide association study of liver enzymes in korean children. Genomics Inform. 2013;11:149–54.
11. Argos M, Tong L, Pierce BL, Rakibuz-zaman M, Islam T, Rahman M, et al. Genome-wide association study of smoking behaviours among Bangladeshi adults. J Med Genet. 2014;51:327–33.
12. Xu S, Yin X, Li S, Jin W, Lou H, Yang L, et al. Genomic dissection of population substructure of Han Chinese and its implication in association studies. Am J Hum Genet. 2009;85:762–74.
13. Suo C, Xu H, Khor C-C, Ong RT, Sim X, Chen J, et al. Natural positive selection and north–south genetic diversity in East Asia. Eur J Hum Genet. 2012;20:102–10.
14. Lao O, Lu TT, Nothnagel M, Junge O, Freitag-Wolf S, Caliebe A, et al. Correlation between Genetic and Geographic Structure in Europe. Curr Biol. 2008;18:1241–8.
15. Qin P, Li Z, Jin W, Lu D, Lou H, Shen J, et al. A panel of ancestry informative markers to estimate and correct potential effects of population stratification in Han Chinese. Eur J Hum Genet. 2014;22(2013):248–53.
16. Wu Y, Waite LL, Jackson AU, Sheu WH-H, Buyske S, Absher D, et al. Trans-ethnic fine-mapping of lipid loci identifies population-specific signals and allelic heterogeneity that increases the trait variance explained. PLoS Genet. 2013;9:e1003379.
17. Deng L, Hoh BP, Lu D, Fu R, Phipps ME, Li S, et al. The population genomic landscape of human genetic structure, admixture history and local adaptation in Peninsular Malaysia. Hum Genet. 2014;133:1169–85.
18. Altshuler DM, Gibbs RA, Peltonen L, Altshuler DM, Gibbs RA, Peltonen L, et al. Integrating common and rare genetic variation in diverse human populations. Nature. 2010;467(7311):52–8.
19. Alexander DH, Novembre J, Lange K. Fast model-based estimation of ancestry in unrelated individuals. Genome Res. 2009;19:1655–64.
20. Weir BS, Hill WG. Estimating F-statistics. Annu Rev Genet. 2002;36:721–50.

Rank-based genome-wide analysis reveals the association of Ryanodine receptor-2 gene variants with childhood asthma among human populations

Lili Ding[1], Tilahun Abebe[2], Joseph Beyene[3], Russell A Wilke[4], Arnon Goldberg[5], Jessica G Woo[1], Lisa J Martin[1], Marc E Rothenberg[1], Marepalli Rao[6], Gurjit K Khurana Hershey[1], Ranajit Chakraborty[7] and Tesfaye B Mersha[1*]

Abstract

Background: The standard approach to determine unique or shared genetic factors across populations is to identify risk alleles in one population and investigate replication in others. However, since populations differ in DNA sequence information, allele frequencies, effect sizes, and linkage disequilibrium patterns, SNP association using a uniform stringent threshold on *p* values may not be reproducible across populations. Here, we developed rank-based methods to investigate shared or population-specific loci and pathways for childhood asthma across individuals of diverse ancestry. We performed genome-wide association studies on 859,790 SNPs genotyped in 527 affected offspring trios of European, African, and Hispanic ancestry using publically available asthma database in the Genotypes and Phenotypes database.

Results: Rank-based analyses showed that there are shared genetic factors for asthma across populations, more at the gene and pathway levels than at the SNP level. Although the top 1,000 SNPs were not shared, 11 genes (*RYR2, PDE4D, CSMD1, CDH13, ROBO2, RBFOX1, PTPRD, NPAS3, PDE1C, SEMA5A*, and *CTNNA2*) mapped by these SNPs were shared across populations. Ryanodine receptor 2 (*RYR2*, a statin response-related gene) showed the strongest association in European (p value = 2.55×10^{-7}) and was replicated in African (2.57×10^{-4}) and Hispanic (1.18×10^{-3}) Americans. Imputation analyses based on the 1000 Genomes Project uncovered additional *RYR2* variants associated with asthma. Network and functional ontology analyses revealed that *RYR2* is an integral part of dermatological or allergic disorder biological networks, specifically in the functional classes involving inflammatory, eosinophilic, and respiratory diseases.

Conclusion: Our rank-based genome-wide analysis revealed for the first time an association of *RYR2* variants with asthma and replicated previously discovered *PDE4D* asthma gene across human populations. The replication of top-ranked asthma genes across populations suggests that such loci are less likely to be false positives and could indicate true associations. Variants that are associated with asthma across populations could be used to identify individuals who are at high risk for asthma regardless of genetic ancestry.

Keywords: Asthma, GWAS, Ancestry, Trans-ancestral analysis, Rank analysis, Imputation, dbGaP, 1000 Genomes project, Networks/pathways, *RYR2*

* Correspondence: tesfaye.mersha@cchmc.org
[1]Department of Pediatrics, Cincinnati Children's Hospital Medical Center, University of Cincinnati, Cincinnati, OH 45229, USA
Full list of author information is available at the end of the article

Background

Asthma [MIM 600807] is a disease of chronic airway inflammation that affects over 300 million individuals worldwide, including 24.6 million in the USA [1]. It is estimated that asthma-related health care costs the US economy US$56 billion a year [2]. Asthma has important racial disparities in prevalence, morbidity, mortality, and drug response. In the USA, the prevalence of asthma varies between racial groups, ranging from 7.8% in European-Americans to 11.1% in African-Americans and up to 16.6% in Hispanic-Americans [3]. While differences in lifestyle and socioeconomic status between racial groups may contribute to differences in asthma prevalence, population genetic variation may be partly responsible for the current disparities in asthma susceptibility.

As of December 4, 2012, there are 28 genome-wide association studies (GWAS) that identified 78 SNP-asthma associations. The main strength of GWAS is its ability to systematically explore truly novel candidate SNPs/genes associated with chronic diseases. However, many SNPs identified by GWAS explain only a small fraction of the genetic risk [4-6]. Furthermore, there is selection bias toward 'top hits' in GWAS. As reported by Baye et al. [7], the problem of focusing on few top-hit SNPs is that if the p value threshold is set too low, genes that have little effect individually but are relevant to complex traits when they interact with other genes are not detectable. Recently, Torgerson et al. [8] conducted genetic association studies across asthmatic populations with a cutoff p value of 10^{-6} and discovered 34 SNPs in European-Americans, 4 SNPs in African-Americans and African-Caribbeans, 32 in the Hispanic-Americans, and 75 in the combined meta-analysis. Although such study can discover markers with large effect sizes, stringent cutoff values may not be realistic for across-population comparison given that each population has a unique genetic and demographic history and that populations vary in DNA sequence information, allele frequencies, effect sizes as well as exhibit heterogeneity in linkage disequilibrium (LD) patterns between the identified variants and the causative functional variants that underlie disease risk [9-12].

Studies based on gene sets (a) have a larger effect size on complex trait than individual SNPs, (b) have a greater power to detect functionally relevant genes, and (c) improve the interpretability and reproducibility of genetic studies on complex diseases [13]. Approaches that include genetic signals at all levels, for example, loci/gene and pathways, without an arbitrary threshold of statistical significance are needed. Such methods are capable of extracting more information from GWAS data by identifying loci that have functional similarities. We hypothesized that such an approach could generate sound biological bases for subsequent studies compared with studies that rely on single markers with low p values.

Currently, across-population studies on asthma genetics are limited and several questions are not properly addressed, including the following: How often are the same sets of SNPs, genes, or pathways associated with asthma across populations? To what degree are asthmatic subjects of different populations enriched for common sets of susceptible loci? Answering these questions systematically will allow us to understand risk variants for asthma that are population-specific or shared across populations and implement better interventions for asthma. If a common set of loci are associated with asthma across populations, then it is reasonable to hypothesize that those loci are more likely to share one or more pathways compared to loci that are not shared or associated with asthma. With the availability of data generated on the same commercial SNP chips (i.e., high level overlap in the SNP sets), we have the opportunity to compare genome-wide associations with asthma across populations directly and sift the wheat from the chaff [14]. Therefore, the objective of the current study was to systematically analyze the presence of shared or population-specific genetic risk factors for asthma among European, African-American, and Hispanic asthmatic children at the locus and pathway levels. To accomplish this, we performed genome-wide association analysis of childhood asthma using 859,790 SNP markers genotyped in a sample of 527 affected offspring trios of different racial groups. Affected offspring trio or family design is robust against population substructure, which is of particular concern when studying African-Americans or Hispanic-Americans with diverse ancestry contributions. Comparing with case–control studies, where cases and controls are often unrelated, affected offspring trio design avoids population and DNA quality differences between the cases and controls and the possibility that some controls borrowed from other studies might be affected with the phenotype of interest.

Results

Description of study subjects and association

Table 1 shows racial distribution and the number of SNPs genotyped for each population, shared by all three populations, and specific to each population in the Childhood Asthma Research and Education (CARE) and Childhood Asthma Management Program (CAMP) databases. There are 859,790 autosomal SNPs genotyped in the three populations that passed inclusion criteria; among them 786,195 SNPs (91.4%) are shared by all three populations. The number of population-specific SNPs that are polymorphic in only one of the three populations is 688 for European-American, 3,705 for African-American, and 180 for Hispanic-American population.

Table 1 The number of affected offspring trios and number of SNPs by population

	Number of affected offspring trios			Number of SNPs	
	CAMP	CARE	CAMP and CARE	Genotyped	Population-specific[a]
Population					
EA	334	95	429	842,915	688
AA	42	10	52	855,949	3,705
HA	30	16	46	846,188	180
Total	406	121	527	859,790[b]	4,573
Shared				786,195[c]	

EA, European-American; *AA*, African-American; *HA*, Hispanic-American populations. [a]Number of SNPs that passed inclusion criteria and polymorphic in only one of the three populations. [b]Number of SNPs that passed inclusion criteria and polymorphic in at least one of the three populations. [c]Number of SNPs that passed inclusion criteria and polymorphic in all three populations.

Population-specific associations

Figure 1 shows the Manhattan plots of transmission disequilibrium test (TDT) results for each population separately and all three populations combined (mega-analysis). More significant results were obtained for European-American than African or Hispanic ancestry populations. Although we see little genomic inflation factors (λ) (ranging from 1.03 to 1.2) in this study, our ranked-based approach was not influenced by inflation, and global genomic correction factor is not relevant. In other words, the order of the markers based on rank analysis did not change with correction for the genomic inflation factor (data not shown). Additional file 1: Tables S1, S2, S3, and S4 show the top 100 SNPs of each population and the mega-analysis of the combined data. The *p* values of the top 100 SNPs ranged from 1.81×10^{-7} to 7.74×10^{-6} for European-American, 9.76×10^{-6} to 2.72×10^{-4} for African-American, 4.20×10^{-7} to 2.57×10^{-4} for Hispanic-American population, and 8.33×10^{-9} to 9.58×10^{-6} for the mega-analysis. The top 100 SNPs were not shared between any two or among all three populations. Among the top 100 SNPs of the combined data, 18, 1, and none are shared in the top 100 SNPs of European-American, African-American, and Hispanic-American population, respectively. From mega-analysis, we showed that simply combining diverse data may not result in the identification of variants important in all populations.

Among the genes that mapped to the top 100 loci of European-Americans, *CNTN1* (Contactin 1, [MIM 600016]) was proposed to have an important function in the invasion and metastasis of lung adenocarcinoma cells [15]. The gene *STAT5A* (Signal transducer and activator of transcription 5A, [MIM 601511]) was indicated to be critical in STAT6-independent Th2 cell differentiation and allergic airway inflammation [16]. Among the genes mapped from the top 100 loci of African-Americans, Ryanodine receptor-2 (*RYR2* [MIM 180902]) has been implicated in the calcium response that leads to increased airway contraction [17,18] and extensive airway narrowing, which characterizes a key event underlying asthma. Two other top genes in African-Americans (*CDH13* [MIM 601364] and *PTPRD* [MIM 601598]) are related to lung cancer and childhood asthma [19-21]. Among the top genes in Hispanic-American population, *RBFOX1* (RNA binding protein, fox-1 homolog, *C. elegans*, 1 [MIM 605104]) was reported to be related to survival in lung cancer patients [22]. *DAPK1* (Death-associated protein kinase 1, [MIM 600831]) was shown to be associated with cell death and inflammatory and immunological diseases. *DOCK1* (Dedicator of cytokinesis 1 [MIM 601403]) was reported to be moderately associated with asthma [23].

Additional file 1: Table S5 shows the number of SNPs with *p* values below the cutoffs (0.05, 0.01, 10^{-4}, 10^{-5}, and 10^{-6}) in each population, in the combined sample, and shared between populations. The three populations shared 180 and 2 SNPs at a *p* value cutoff of 0.05 and 0.01, respectively. No SNPs were shared by any two or all three of the populations with more stringent *p* value cutoffs ($<10^{-4}$), which may be partly due to the genetic heterogeneity across the study populations and the small sample size of the study populations. Thus, instead of *p* value cutoff, we focused on top-ranked SNPs/genes and pathways/gene ontologies (GOs) in the rest of our analyses.

Little overlapping among top-ranked SNPs

The left side of Table 2 shows the number of SNP markers shared by any two populations or all three populations among the top-ranked SNPs. Among the top 1,000 SNPs, 2 were shared by European-Americans and African-Americans, 2 by European-Americans and Hispanic-Americans, and none by African-Americans and Hispanic-Americans. For the sets of top 2,000 SNPs, European-Americans and African-Americans shared 4 SNPs, European-Americans and Hispanic-Americans shared 7 SNPs, and African-Americans and Hispanic-Americans shared none. When the top 10,000 SNPs were considered, three SNPs were shared by all three populations, 2 more than what would be expected by chance alone. Table 3 shows these SNPs and their *p* values in each population. Among these shared sites, *ARSB* (Arylsulfatase B, [MIM 611542]) was shown to regulate colonic epithelial cell migration [24].

Although the same loci were not shared among the top-ranked SNPs across populations, many top-ranked SNPs of one population were replicated in the other two populations at a nominal *p* value of 0.05. Table 4 shows among the top 100 SNPs in each population how many had *p* value less than 0.05 in the other two populations or the mega-analysis, where the expectation by chance

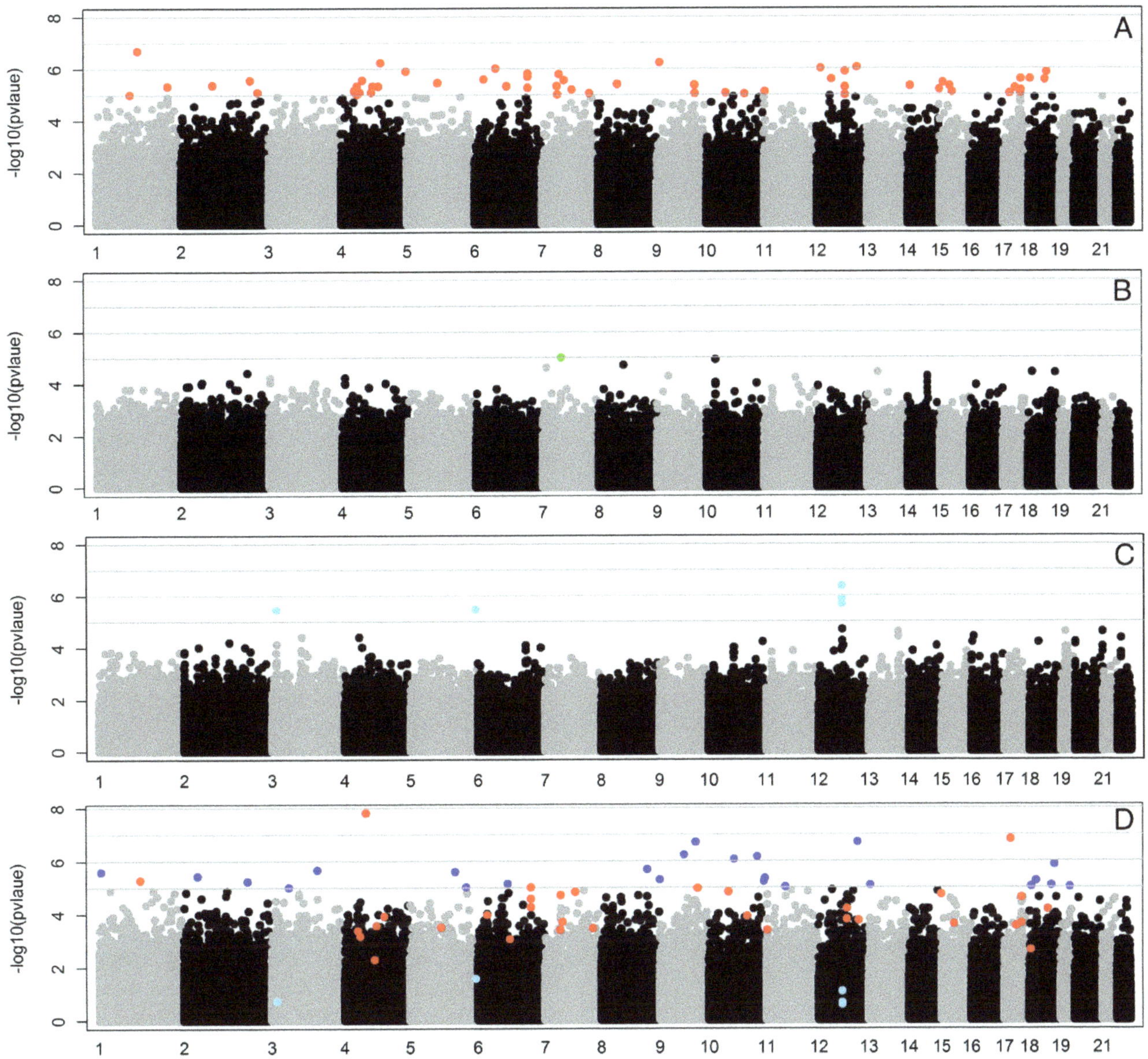

Figure 1 Manhattan plots of the trans-ancestral analysis. (A) European-American, **(B)** African-American, **(C)** Hispanic-American population, and **(D)** all three populations combined. The *y*-axis displays the negative logarithm of the *p* value for each SNP marker; the *x*-axis displays the markers' genomic coordinates by chromosome. In **(A)**, **(B)**, and **(C)**, colored dots (red for European-Americans, green for African-Americans, and light blue for Hispanic-Americans) indicate markers with *p* value $<1 \times 10^{-5}$. Among these markers, those that passed inclusion criteria in the mega-sample are also indicated with their respective colors in **(D)**, where the additional markers with *p* value $<1 \times 10^{-5}$ in the mega-analysis are indicated in blue.

alone is 5. Among the top 100 SNPs of European-American population, 8 in African-American and 18 in Hispanic-American population had *p* value < 0.05. Among the top 100 SNPs of African-American, 7 in European-American and 2 in Hispanic-American population had *p* value < 0.05. Among the top 100 SNPs of Hispanic-American population, 10 in European-Americans and 5 in African-Americans had *p* value <0.05. Additional file 1: Tables S1, S2, S3, and S4 show these SNPs and their *p* values in all three populations and in the mega-analysis. For example, the most significant SNP in European-Americans (rs16929097, *p* value = 1.81×10^{-7}) was replicated in African-Americans with a *p* value of 0.0114, and the second most significant SNP in European-Americans (rs17036023, *IGSF3* [MIM 603491], *p* value = 2.04×10^{-7}) was replicated in Hispanic-American population with a *p* value of 0.0143. One of the top 100 SNPs of European-Americans (rs16863100, *p* value = 7.10×10^{-6}) was replicated in both African-Americans (*p* value = 0.0412) and Hispanic-Americans (*p* value = 0.0143). One of the top 100 SNPs in Hispanic-Americans (rs13486, *p* value = 8.77×10^{-5}) was replicated in both European-Americans

Table 2 SNP level and gene level overlap

Number of top SNPs	SNP level				Gene level						
	Overlap				Mapped genes			Overlap[a]			
	EA and AA	EA and HA	AA and HA	EA, AA, and HA	EA	AA	HA	EA and AA	EA and HA	AA and HA	EA, AA, and HA
1,000	2	2	0	0	328	299	252	41/5	30/4	29/4	11/0
2,000	4	7	0	0	616	531	450	110/16	77/12	72/14	34/0
5,000	39	40	33	0	1,272	1,121	961	322/71	266/54	242/61	126/3
10,000	170	235	145	3	2,120	1,882	1,677	686/199	624/158	575/178	353/18
50,000	3,549	4,228	3,463	284	5,600	5,515	5,193	3,163/1,544	3,041/1,431	2,962/1,454	2,151/401

EA, European-American; *AA*, African-American; *HA*, Hispanic-American populations. [a]Observed/expected by chance.

(p value = 0.0365) and African-Americans (p value = 0.0039).

Replication of top-ranked asthma genes

Among the top 1,000 SNPs, 403, 417, and 405 SNPs were mapped to 328, 299, and 252 genes, and interquartile ranges (IQR) of the numbers of SNPs mapped to each gene were (1, 1), (1, 1), and (1, 2) for European-American, African-American, and Hispanic-American population, respectively. Among the mapped genes, European-American and African-American populations shared 41 genes, European-American and Hispanic populations shared 30 genes, and African-American and Hispanic populations shared 29 genes (Table 2). In addition, there are 11 genes shared by all the three populations. These 11 genes are *RYR2, CSMD1* [MIM 608397], *CDH13, ROBO2* [MIM 602431], *RBFOX1* [MIM 605104], *PTPRD, NPAS3* [MIM 609430], *PDE1C* [MIM 602987], *SEMA5A* [MIM 609297], *CTNNA2* [MIM 114025], and *PDE4D* [MIM 600129]. Table 5 lists the p values of these genes in the three populations, which ranged from 2.55×10^{-7} to 1.62×10^{-3}. *RYR2* is a statin response-related gene that showed the strongest association in European-Americans (p value = 2.55×10^{-7}) and was replicated in African-Americans (2.57×10^{-4}) and Hispanic-Americans (1.18×10^{-3}). *PDE4D* was identified as an asthma susceptibility gene, and PDE4 inhibitors have been developed as medications for asthma [25]. Variants in *PTPRD* were reported to be associated with childhood asthma in Taiwanese population [20].

To detect additional variants in the top-ranked *RYR2* asthma gene across populations, we imputed untyped SNPs in *RYR2* using haplotypes from the 1000 Genomes Project as reference panels. The number of SNPs in this gene that passed the filtering criteria in European-American, African-American, and Hispanic-American populations was 262, 382, and 371, respectively, before imputation, and 2,533, 2,884, and 2,304 after imputation. Post-imputation SNPs were then ranked from most significant to least significant according to their association with asthma. Figure 2 shows the p value and LD plot of the top 1% of the post-imputation SNPs for each population. Additional file 1: Table S6 lists these SNPs, their p values in each population, and possible functional effects. In European-Americans, no imputed SNPs exceeded the strongest association inferred from the genotyped SNP

Table 3 Shared SNPs among top-ranked SNPs

rs ID	Gene	Chr	p value		
			EA	AA	HA
Shared SNPs among top 2,000 SNPs of EA and AA					
rs7045156		9	3.74×10^{-5}	2.50×10^{-3}	3.17×10^{-1}
rs1048329	*LRP2BP*	4	8.89×10^{-4}	1.86×10^{-3}	1
rs12359404	*SORCS1*	10	4.07×10^{-4}	5.32×10^{-4}	3.17×10^{-1}
rs16875946	*ARSB*	5	4.65×10^{-4}	1.70×10^{-3}	4.55×10^{-2}
Shared SNPs among top 2,000 SNPs of EA and HA					
rs2928442		10	1.39×10^{-4}	1.17×10^{-1}	1.62×10^{-3}
rs2272266	*PLA1A*	3	1.23×10^{-4}	6.55×10^{-1}	1.57×10^{-3}
rs4128918		5	6.40×10^{-4}	NA	2.70×10^{-3}
rs4008848		9	8.72×10^{-4}	7.15×10^{-1}	3.50×10^{-3}
rs9913559	*RDM1*	17	2.06×10^{-4}	8.19×10^{-1}	2.84×10^{-3}
rs6550392		3	9.11×10^{-4}	3.53×10^{-1}	2.70×10^{-3}
rs9301462	*RAB20*	13	7.10×10^{-6}	8.82×10^{-1}	2.70×10^{-3}
Shared SNPs among top 10,000 SNPs of EA, AA, and HA					
rs920672	*NAV2*	11	2.67×10^{-5}	1.43×10^{-2}	1.43×10^{-2}
rs11021111		11	5.56×10^{-4}	1.16×10^{-2}	1.26×10^{-2}
rs1314595	*ATRNL1*	10	3.12×10^{-4}	1.16×10^{-2}	8.15×10^{-3}

SNPs in italics were among the top 1,000. *EA*, European-American; *AA*, African-American; *HA*, Hispanic-American populations; Chr, chromosome.

Table 4 Replication of top 100 SNPs of each population in other populations

Population	Number of p values <0.05			
	EA	AA	HA	Mega
EA	100	8	18	81
AA	7	100	2	33
HA	10	5	100	15
Mega	100	43	42	100

Replication is defined at a nominal p value of 0.05. EA, European-American; AA, African-American; HA, Hispanic-American populations; Mega, mega-analysis based on combined samples from all three populations.

Table 5 Eleven genes shared by the top 1,000 SNPs of each population

Genes	Chr	*p* values		
		EA	AA	HA
RYR2	1	2.55×10^{-7}	2.57×10^{-4}	1.18×10^{-3}
CSMD1	8	4.23×10^{-6}	4.15×10^{-4}	4.18×10^{-4}
CDH13	16	2.01×10^{-5}	2.31×10^{-4}	1.60×10^{-3}
ROBO2	3	2.01×10^{-5}	3.93×10^{-4}	1.36×10^{-3}
RBFOX1	16	2.21×10^{-5}	1.76×10^{-3}	3.86×10^{-5}
PTPRD	9	2.38×10^{-5}	2.08×10^{-4}	3.12×10^{-4}
NPAS3	14	3.74×10^{-5}	1.58×10^{-3}	1.02×10^{-3}
PDE1C	7	9.62×10^{-5}	9.41×10^{-4}	1.32×10^{-3}
SEMA5A	5	1.08×10^{-4}	4.03×10^{-4}	1.34×10^{-3}
CTNNA2	2	1.83×10^{-4}	1.28×10^{-3}	1.57×10^{-3}
PDE4D	5	1.83×10^{-4}	4.02×10^{-4}	1.62×10^{-3}

EA, European-American; *AA*, African-American; *HA*, Hispanic-American populations; Chr, chromosome.

(rs16835325, 2.55×10^{-7}). One imputed SNP (rs12136903) showed moderate association with asthma (*p* value = 7.89×10^{-4}). In African-Americans, the strongest signal from genotyped SNPs (rs2797447, *p* value = 1.09×10^{-3}) was exceeded by an imputed SNP (rs2685301, *p* value = 4.15×10^{-4}) that is within a LD block of four of the top-ranked genotyped SNPs. In Hispanic-Americans, association from imputed SNPs again supported signals from genotyped SNPs. Two imputed SNPs showed strong LD with genotyped SNPs and second to the best association with asthma (rs2779359 and chr1:237727031 both with *p* value = 1.60×10^{-3}).

Top-ranked pathways and GO terms

To gain further insights into the pathogenesis of asthma and determine significant biological pathways and gene ontologies and to reveal genes associated with asthma across populations, we conducted gene set analysis based on pathways and GO terms. Table 6 shows the pathways and GOs that were over-represented with *p* value < 0.01 in at least one of the three populations when the top 1,000 SNPs were declared as noteworthy. Figure 3 shows the amount of overlap across the three populations at the pathway and GO level. When the top 1,000 SNPs were declared as noteworthy, African-American and Hispanic-American populations shared 1 of their top 20 pathways (Prefoldin mediated transfer of substrate to *CCT*/*TRiC*) and 1 of their top 30 GOs (ribonuclease activity); European-American and African-American populations shared 1 of their top 40 pathways (Shc-mediated cascade), and African-American and Hispanic-American populations shared 1 of their top 30 pathways (Stathmin pathway) and 2 of their top 30 GO terms (cellular macromolecule catabolic process and the macromolecule catabolic process). In addition, among the three sets of top 60 pathways, only 1 pathway was shared by all three populations (arrhythmogenic right ventricular cardiomyopathy). Additional file 1: Table S7 and S8 show the top 30 pathways and GOs for each population and their *p* values when the top 1,000 SNPs were declared as noteworthy. When less stringent top 2,000 SNPs were considered, African-American and Hispanic populations shared four of their top ten pathways (systemic lupus erythematosus, packaging of telomere ends, RNA polymerase I promoter clearance, and RNA

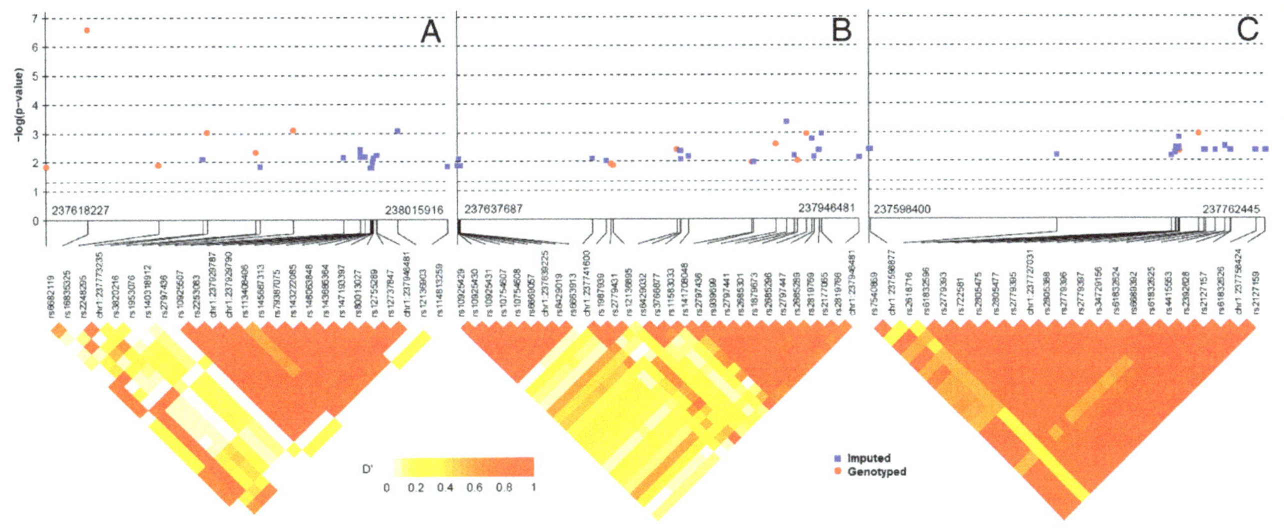

Figure 2 The *p* value and LD plot of post-imputation RYR2 SNPs for each population. Trans-ancestral analysis of genotyped and imputed association results and LD of the top 1% SNPs of *RYR2* after imputation. **(A)** European-American population, **(B)** African-American population, and **(C)** Hispanic-American population.

Table 6 Over-represented pathways and GOs

	p values		
	EA	**AA**	**HA**
GO			
Cellular response to stress	0.0006	1	1
Vitamin binding	0.0056	0.3392	1
JAK STAT cascade	0.0062	1	0.4304
Vitamin transport	0.0072	0.3210	0.2886
ER nuclear signaling pathway	0.0078	1	1
Regulation of gene-specific transcription	0.0078	1	1
Positive regulation of cell proliferation	0.4326	0.0006	0.4412
Lysosomal transport	1	0.0016	1
Carbohydrate binding	0.6714	0.0038	0.8342
RNA catabolic process	0.3710	0.0050	0.2818
Vacuolar transport	1	0.0052	1
Negative regulation of catalytic activity	0.7328	0.1720	0.0004
Protein kinase binding	1	0.7450	0.0016
Kinase binding	0.8654	0.5054	0.0040
Regulation of translation	0.8478	1	0.0050
Anion transmembrane transporter activity	1	0.7148	0.0080
Negative regulation of hydrolase activity	1	1	0.0100
Pathway			
BioCarta G2 pathway	0.0008	1	1
Reactome E2F transcriptional targets at G1 S	0.0034	1	1
Reactome E2F mediated regulation of DNA replication	0.0036	1	1
KEGG alanine aspartate and glutamate metabolism	0.5218	0.0032	0.4338
BioCarta longevity pathway	0.4426	0.0088	1
Reactome amine ligand binding receptors	0.3642	0.1202	0.0006
Reactome G alpha S signaling events	0.3682	0.8264	0.0060

With *p* value <0.01 in at least one of the three populations when the top 1,000 SNPs are declared as noteworthy. *EA*, European-American; *AA*, African-American; *HA*, Hispanic-American populations; *KEGG*, Kyoto encyclopedia of genes and genomes.

polymerase I promoter opening). Two more pathways were shared by the two populations among their top 20 pathways (RNA polymerase I and III and mitochondrial transcription and telomere maintenance). In particular, we found enrichment of the leukocyte trafficking pathway which indicates that the accumulation and activation of inflammatory leukocytes in the lung or airway is a feature shared by almost all respiratory diseases [26]. The leukocyte trafficking pathway has been suggested to have a key role in asthma, which makes the finding in our study biologically plausible. Children with asthma may be considered to suffer from chronic inflammatory stress [27].

Higher population genetic differentiation was found at variants that show association with asthma

Table 7 shows summary statistics of fixation index (F_{ST}) and the relationship between population genetic differentiation and population differences in association with asthma for the shared SNPs among populations. Compared with the F_{ST} values between European-American and Hispanic-American populations, those between African-Americans and European-Americans (or Hispanic-Americans) were higher, suggesting more population genetic differentiation between African-Americans and European-Americans (or Hispanic-Americans). Furthermore, population genetic differentiation was stronger in asthmatic individuals (affected offspring) than the non-asthmatic individuals (parents). Regardless of the data used (affected offspring or parents), the observed numbers of markers that satisfy difference between the rankings of *p* values (DRP) > mean of DRP and F_{ST} > mean of F_{ST} were higher than the expected number of markers. The dependence between population genetic differentiation and population differences in association with asthma was significant for European-American vs. African-American population and for European-American vs. Hispanic-American population. However, no significant relationship existed for African-American vs. Hispanic-American population. We believe this is due to either the relatively similar asthma prevalence in African-Americans and Hispanic-Americans or different mechanisms of asthma association in European-American and African/Hispanic-American populations. We conclude that for populations with different asthma prevalence, such as European-Americans vs. African-Americans and European-Americans vs. Hispanic-Americans, SNPs that are more informative for ancestry or exhibit large population genetic differentiation are more likely to be different in their association with asthma in the different populations.

Patterns of variation across populations

Figure 4 shows the density plot of minor allele frequencies (MAF) of the three populations estimated using affected offspring. For European-Americans, MAF had mean = 0.20, median = 0.19, and IQR = (0.06, 0.34); for African-American subjects, MAF mean = 0.22, median = 0.19, and IQR = (0.09, 0.33); and for Hispanic-American subjects, the MAF mean = 0.21, median = 0.19, and IQR = (0.07, 0.34). Overall, MAFs in European-Americans were smaller than African-American and Hispanic-American populations and had a relatively larger peak at the lower end of the MAF spectrum, which may be due to the current reference genomes and commercial SNP panels included in the Affymetrix 6.0 genotyping chip being primarily selected based on identification and patterns of LD in European ancestry population. This might also indicate that the European-American population is less

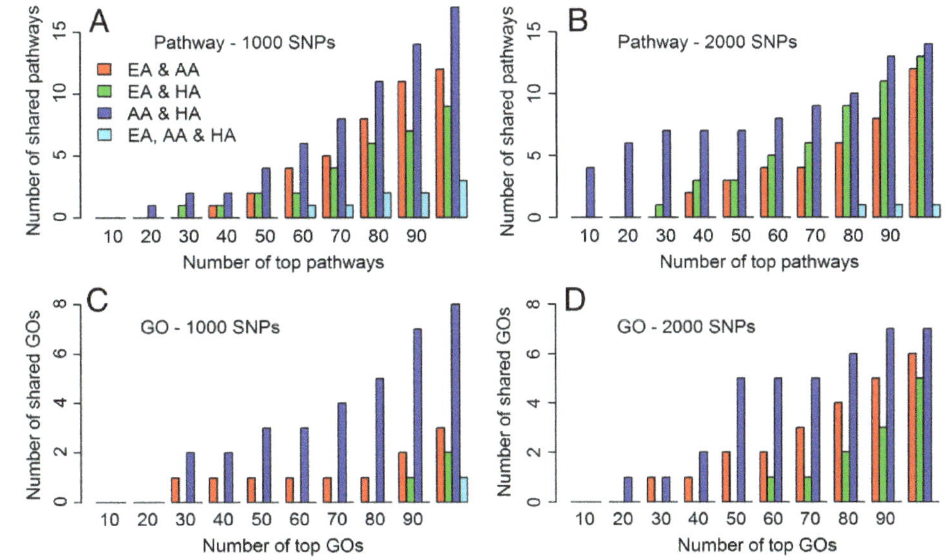

Figure 3 Overlap of genetic risk factor for childhood asthma across the three populations: At the pathway and GO level. The number of shared pathways (*y*-axis) among different numbers of top-ranked pathways (*x*-axis) when the top **(A)** 1,000 and **(B)** 2,000 SNPs were declared as noteworthy. The number of shared GO terms (*y*-axis) among different numbers of top-ranked GO terms (*x*-axis) when the top **(C)** 1,000 and **(D)** 2,000 SNPs were declared as noteworthy. EA, European-American; AA, African-American; HA, Hispanic-American populations.

heterogeneous and less diverse compared with African-Americans and Hispanic-Americans since African-American and Hispanic-American individuals are of admixed origin, while the European-Americans are simply of varied European ancestry. Figure 5 shows the box and whisker plot of MAF of the top 5,000 SNPs for each population. Similar pattern was observed for the top 1,000 or 2,000 SNPs. MAFs of top-ranked SNPs in European-Americans (mean = 0.09, median = 0.03, IQR = (0.01, 0.13)) were on average much lower than those in African-Americans (mean = 0.23, median = 0.20, IQR = (0.11, 0.33)) and Hispanic-Americans (mean = 0.24, median = 0.23, IQR = (0.13, 0.36)). Compared to overall MAF, top-ranked SNPs had lower MAF in European-Americans. Recent studies show that variants altering amino acid sequence and protein function are enriched at low variant allele frequency, 2% to 5% [28].

Discussion

Several studies have explored shared genetics among diseases including coeliac disease and other immune diseases [29], non-Hodgkin's lymphoma and autoimmune diseases [30], obesity and asthma [21], and asthma and chronic obstructive pulmonary disease [31]. However, there is very little investigation into population-specific or shared genetic risk factors for a specific disease across different populations. In this report, we described the results of GWAS asthma associations in three populations, namely European-Americans, African-Americans, and Hispanic-Americans. The method we used is based on

Table 7 Population genetic differentiation and population differences in genetic association with asthma

Allele frequencies in CAMP and CARE	F_{ST}			DRP > mean (DRP) and F_{ST} > mean (F_{ST})		
	Mean	Median	Range	Observed	Expected	*p* values[a]
From parents						
Caucasian vs. African-American	0.0463	0.0263	(0, 0.5858)	108,245	107,720	0.007898
Caucasian vs. Hispanic	0.0078	0.0036	(0, 0.1770)	95,757	94,447	7.749×10^{-12}
African-American vs. Hispanic	0.0387	0.0215	(0, 0.4989)	107,740	107,384	0.07227
From the affected offspring						
Caucasian vs. African-American	0.0487	0.0274	(0, 0.6036)	107,987	107,432	0.004965
Caucasian vs. Hispanic	0.0114	0.0050	(0, 0.2226)	98,522	97,445	2.39×10^{-8}
African-American vs. Hispanic	0.0433	0.0238	(0, 0.5429)	107,458	107,225	0.2408

[a] *p* values of the chi-square test of the frequency table of DRP > mean (DRP) and F_{ST} > mean (F_{ST}). DRP, difference between the rankings of *p* values in the two populations.

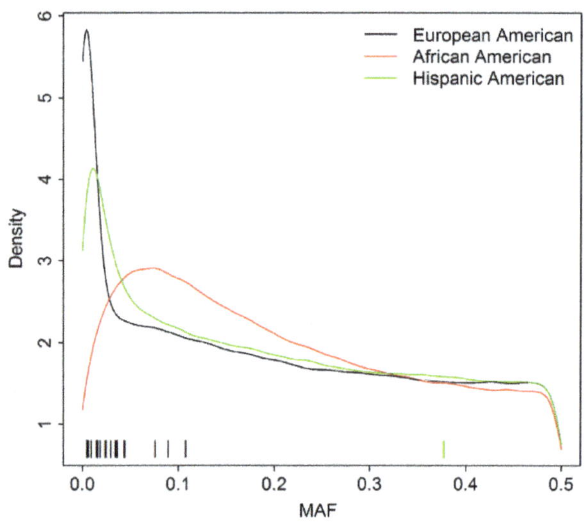

Figure 4 Density plot of MAF. The MAFs were estimated using the affected offspring. The bars above the *x*-axis indicate the MAF of SNPs with *p* value $<1 \times 10^{-6}$ in each population.

the ranking of SNP associations from the most to the least significant and testing in the context of functionally relevant genes and gene networks. We observed that there are shared genetic risk factors (genes and pathways) for asthma across populations, although none of the top-ranked SNPs associated in each population was replicated in others. The heterogeneity of top GWAS 'hits' could be a result of a combination of ancestry variations in the study populations, differences in asthma phenotype definitions and unaccounted-for environmental factors.

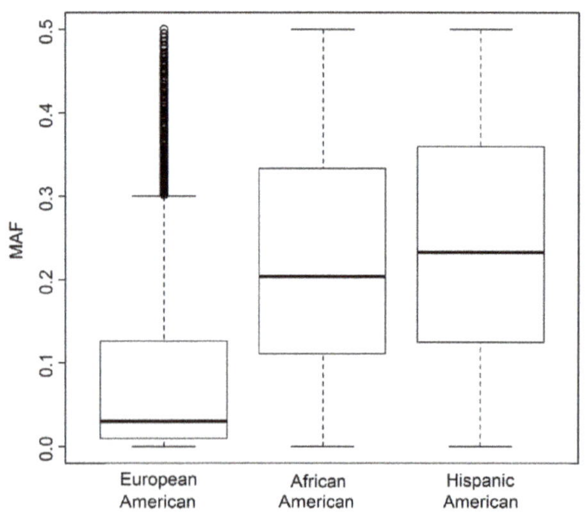

Figure 5 Box and whisker plots of MAF of the top 5,000 SNPs for each population. The MAFs were estimated using affected offspring. The bottom and top of the box are the lower and upper quartiles, respectively, the band within the box is the median, and the ends of the whiskers are the lowest/highest data value within 1.5 IQR of the lower/higher quartile. IQR is the difference between the upper and the lower quartiles.

When the top 10,000 SNPs for each population were considered, only 3 SNP were found to be shared by all three populations (rs1314595, *ATRNL1* [MIM 612869], chromosome (chr) 10, with *p* value = 0.0003 in European-Americans, 0.0082 in Hispanic-Americans, and 0.0116 in African-Americans; rs920672, chr 11, with *p* value = 2.67×10^{-5} in European-Americans and 0.0143 in both African- and Hispanic-Americans; and rs11021111, chr 11, with *p* value = 0.0006 in European-Americans, 0.0126 in Hispanic-Americans, and 0.0116 in African-Americans). As suggested by Jansen et al. [32], whenever information from multiple independent sources agree, it is more likely the findings are valid and reliable than information from a single source. Hence, replication of top-ranked asthma genes or pathways across data from different populations is a way to validate population-specific findings, and such associations are less likely to be false positives and could indicate functionality. In fact evolutionary geneticists used the idea that 'genes that are conserved across populations are likely to be functionally important, since they would confer a selective advantage to all humans' [33].

Among the top 1,000 SNPs of each population, there were 11 loci shared by all three populations, and the genes encoded by these loci are *PDE4D, RYR2, CSMD1, CDH13, ROBO2, RBFOX1, PTPRD, NPAS3, PDE1C, SEMA5A*, and *CTNNA2* (*p* values ranged from 2.55×10^{-7} to 1.62×10^{-3}, Table 5). *PDE4D* (phosphodiesterase 4D) functions as a regulator of airway smooth muscle contractility and was identified as an asthma susceptibility gene, and PDE4 inhibitors have been developed as medications for asthma [25]. Variants in *PTPRD* (protein encoded by protein tyrosine phosphatase receptor-type delta) gene were reported to be associated with childhood asthma in Taiwanese population [20]. Melen et al. [21] in their study of shared genetic factors between asthma and obesity in children found association between *PTPRD* with both phenotypes at the gene level ($p < 0.05$). *NPAS3* (neuronal PAS domain protein 3) encodes a member of the basic helix-loop-helix and PAS domain-containing family of transcription factors. Zhou et al. [34] showed that *NPAS3* haploinsufficient mice survived postnatally but developed alveolar loss and airway hyperreactivity. Genome-wide linkage has identified linkage peak at chromosome 14q12-13 region, where *NPAS3* maps in asthmatic Caucasians [35-37]. *ROBO2* belongs to the Roundabout (ROBO) family, part of the immunoglobulin superfamily proteins that are highly conserved from fly to human. The encoded protein is a receptor for and essential for signal transduction of Slit2, a secreted protein that is known to function in axon guidance and cell migration, plays a critical role in the development of normal airways [38], and is an important etiologic factor in airway narrowing that accompanies asthma [39].

The *RYR2* gene is located from base pair 237,205,701 to base pair 237,997,287 on chromosome 1. Mutations in *RYR2* are causative of dysfunctional calcium channel which often results in sudden cardiac death [40]. Recent genome-wide association studies have also associated *RYR2* variants with muscle toxicity and a potential pharmacodynamic candidate gene in statin response-related disorders; *RYR2* encodes a ryanodine receptor and contributes to the calcium response that leads to increased airway contraction and extensive airway narrowing, which characterizes a key event underlying asthma [17,18,41]. In a recent GWAS study, an intronic variant (rs2819742) in *RYR2* was significantly associated with cerivastatin-associated rhabdomyolysis at the *a priori p* value threshold of 4×10^{-7} ($p = 1.74 \times 10^{-7}$). An additional copy of the minor allele of the *RYR2* variant was associated with a reduced risk of rhabdomyolysis (odds ratio (OR) = 0.48; 95% confidence interval (CI) = 0.36 to 0.63). Carriers of two copies of the minor allele had a smaller risk of rhabdomyolysis than carriers of two copies of the major allele (OR = 0.24; 95% CI = 0.13 to 0.44) [42]. As the associated SNPs in the *RYR2* gene were common variants and could be due to linkage disequilibrium from untyped functional variants, we imputed both rare and common variants using the 1000 Genomes Project reference panel. Imputation can also permit the comparison of studies which focused on different SNPs. Using genotypes inferred through imputation, we uncovered additional *RYR2* variants (rs2685301 in African-Americans and rs2779359 in Hispanic-Americans) that exhibited moderate association with asthma and significant LD with genotyped SNPs.

Complementary to rank-based candidate gene selection for a given disease, gene network analysis offers the advantage of understanding the interaction of functionally related genes that are associated with a disease and the ability to find hub genes within a network that interact with several other genes up- and downstream of them. The high interconnectivity of hub genes with other correlated genes within a biological network may imply functional and biological importance of these genes. Further analysis using Ingenuity Pathways Analysis (IPA) revealed that the 11 genes shared among top-ranked loci of the three populations are part of integrated and interconnected biological networks related to dermatological or allergic disorders, particularly in the functional classes involving inflammatory and immunological diseases (Figure 6). This could reflect that the development of asthma involves the 'atopic march' that starts in the skin and progresses to the respiratory and gastrointestinal tracts [43]. At the center of the network is the 'hub' *CTNNB1* (Catenin (cadherin-associated protein), beta 1, [MIM 116806]) gene complex [44,45], which occurs at cell-cell junctions in epithelial tissues and constitutes adherens junctions. Many studies showed the importance of β-catenin as signaling pathway in airway smooth muscle growth [46]. The role for E-cadherin in asthma has also been studied. Heijink [47] suggested that E-cadherin controls the response to allergens, suppresses allergenic mediator production, and contributes to the establishment of tolerance. De Boer et al. [48] showed that the expression of epithelial alpha-catenin and E-cadherin is lower in atopic asthma patients, and this may result in a defective epithelial barrier in the airway epithelium, which plays a critical role in asthma. Table 8 lists the top diseases and disorders related to these genes where inflammatory and respiratory diseases are among them. Shared pathways across all three populations were observed among the top 60 pathways from the top 1,000 SNPs, and top 100 GO terms from the top 1,000 SNPs (Figure 3).

Although we are limited by our modest sample size in this study, it is important to note that the CAMP and CARE affected offspring trio design have extremely well characterized subjects with detailed phenotypic data. It should be noted that large sample sizes may not help in powering genetic studies and improve our understanding on the genetic underpinnings of asthma phenotypes as much as precise phenotyping [49]. Further, the trio design is robust against population substructure, which is of particular concern when studying African-Americans or Hispanic-Americans with diverse ancestry in case–control study design, where cases and controls are defined variably. Although there are different methods (such as genomic control, structured association, and principle component analysis) to correct for confounding, a good study design is the most efficient way to avoid confounding in disease genetics study. The use of family-based designs increases the power to detect associations, controls for heterogeneity/population stratification, and might elucidate the effects of allele origin as well as transmission of phenotypes of disease modulation. The order of markers based on our rank-based approach remains the same before and after correcting for genomic inflation factor (an indication of the scale-invariant nature of the ranking method). Genetic heterogeneity among ethnic groups, which has been a source of concern in GWAS, will not affect pathway-based GWAS analysis. This is because although the mutated genes or variants within those pathways are likely to differ, affected individuals from different ethnic groups may share the same disrupted pathways. Thus, multiple GWAS can be easily combined, and pathway-based GWAS accommodate and capitalize upon this substantial degree of genetic heterogeneity. The current reference genomes and commercial tagging SNP panels included in the Affymetrix 6.0 genotyping chip were primarily selected based on higher minor allele frequencies

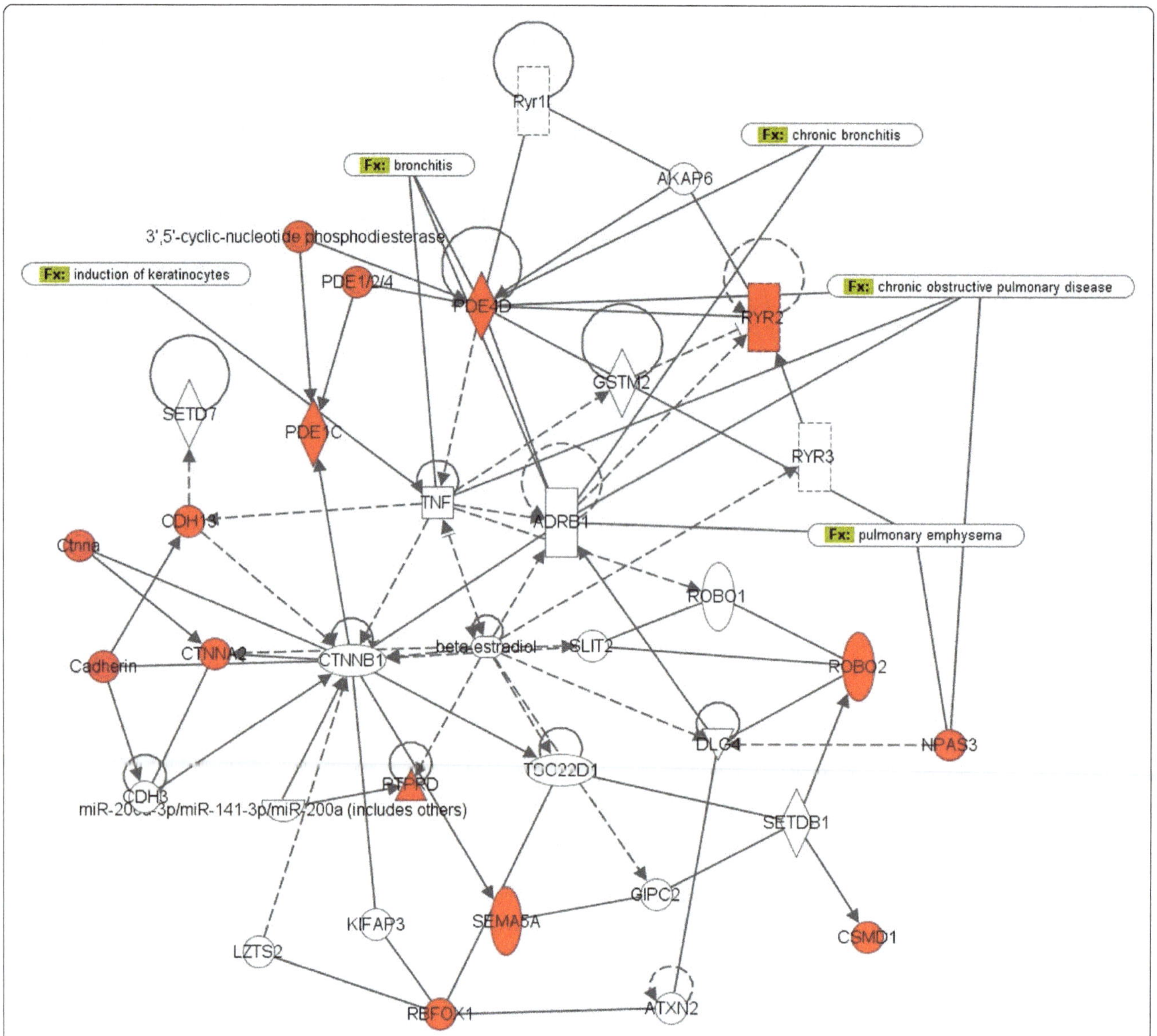

Figure 6 Interactive network of the 11 genes shared among the top 1,000 SNPs of each population. The 11 genes are *CDH13, CSMD1, CTNNA2, NPAS3, PDE1C, PDE4D, PTPRD, RBFOX1, ROBO2, RYR2,* and *SEMA5A*. Genes with red nodes represent hub genes in our analysis; others are generated through the network analysis from Ingenuity Pathways Knowledge Base. Edges are displayed with labels that describe the nature of the relationship between the nodes. All edges are supported by at least one reference from the literature or from canonical information stored in the Ingenuity Pathways Knowledge Base. The lines between genes represent known interactions, with solid lines representing direct interactions and dashed lines representing indirect interactions. Nodes are displayed using various shapes that represent the functional class of the gene product.

and patterns of LD in European ancestry population. Thus, due to relatively weaker LD and variation in minor allele frequencies, for example, in African populations, we may not have the power to detect all of the genetic variants involved in asthma in this population as demonstrated by the 1000 Genomes Project imputation analysis. Indeed, whole-genome sequencing may be necessary to identify population-specific variants in less studied populations such as African-Americans and Hispanics. Eventually, next-generation sequencing technologies will overtake SNP arrays as the primary and less biased genotyping methodology and advance our understanding on rare variants and population-specific influences on disease risk. Additional functional analysis is also necessary to more fully understand the roles that ancestry-specific variants at these loci play in asthma.

Conclusions

In summary, our rank-based approach avoids the need for a global cutoff value for statistically significant

Table 8 Top diseases and disorders related to the 11 genes shared among the top 1,000 SNPs across populations

Functions	*p* values	Genes
Cardiovascular disease	1.90×10^{-5}–3.86×10^{-2}	*CDH13, CSMD1, PDE4D, RBFOX1, RYR2*
Genetic disorder	1.90×10^{-5}–4.21×10^{-2}	*CDH13, CSMD1, PDE4D, RBFOX1, RYR2, ROBO2, NPAS3, SEMA5A*
Hematological disease	1.90×10^{-5}–1.63×10^{-2}	*CDH13, CSMD1, PDE4D, RBFOX1, RYR2*
Psychological disorders	7.15×10^{-4}–3.45×10^{-2}	*NPAS3, RBFOX1, SEMA5A, PDE4D*
Inflammatory disease	1.16×10^{-3}–4.21×10^{-2}	*NPAS3, PDE4D, RYR2*
Respiratory disease	1.16×10^{-3}–2.05×10^{-2}	*NPAS3, PDE4D*

associations. Importantly, since we did not rely on a statistical cutoff to classify significant SNPs, our approach is not susceptible to biases due to SNP density or LD structure [50]. This approach is more appropriate to compare disease association results across populations that vary in DNA sequence, allele frequencies, effect sizes, linkage disequilibrium patterns, and gene-by-environment interactions. We showed the existence of shared genetic risk factors for childhood asthma across the European-American, African-American, and Hispanic-American populations. Our rank-based genome-wide analysis revealed for the first time an association of *RYR2* variants with asthma and replicated previously discovered *PDE4D* asthma gene across human populations. The shared association of asthma for a given gene across populations might likely indicate true association and a broader spectrum of risk factor at the loci. The associations of particular variants (or even different variants within the same gene) across populations and across studies may represent more universally important genes to the disorder and should be given the highest priority [51]. Often, these variants may not be the strongest associations in any one study, but the consistent evidence for association in many different studies (e.g., as revealed in our different racial groups study) would further suggest that the variant and gene have main effects on the phenotype, are less likely influenced by gene-gene or gene-environment interactions, and are most likely to be true associations. Network analysis revealed that *RYR2* and *PDE4D* genes are directly interacting in biological networks. Regulating the expression of both genes along with the hub genes such as *CTNNB1* could be important in the treatment of asthma across populations. Hub genes tend to be conserved across evolution. Thus, hub genes represent towards the evolutionary fitness of an organism, and alterations in their sequence or expression level are likely to be more deleterious. In this study, our goal was to identify variants that consistently associated with asthma across populations and analytical methods. We believe that, once validated, such cross-population variants could be used to identify individuals who are at high risk for asthma regardless of genetic ancestry. Additional studies are necessary to further elucidate biological roles of *RYR2* and pathways related to *RYR2* genes in asthma pathogenesis.

Methods

Subjects

dbGaP data from the CAMP and the CARE Network were used to assess shared and population-specific risk variants for childhood asthma across three populations, namely European-American, African-American, and Hispanic-American. CAMP and CARE are part of the SNP Health Association Resource Asthma Resource project, which is a genome-wide analysis of children who have participated in the National Heart, Lung, and Blood Institute's clinical research trials on asthma. dbGaP was developed to archive and distribute the results of studies that have investigated the interaction of genotype and phenotype. Such studies include genome-wide association studies, medical sequencing, molecular diagnostic assays, as well as association between genotype and non-clinical traits (http://www.ncbi.nlm.nih.gov/gap). This database provides consistently well-defined phenotypes measured across population.

We downloaded genotyping data performed using 1 million SNPs in the Affymetrix 6.0 chip and stored in the database of dbGaP with permission under the accession number phs000166.v2.p1. In the three populations, a total of 859,790 autosomal SNP markers passed the quality control filtering criteria (less than 15% missing data and Hardy-Weinberg equilibrium *p* values $>10^{-6}$) and were included in the association analysis. Our approach was as follows: for each population, single-SNP analysis was first conducted using the family-based TDT. SNPs were then mapped to genes, and genes were mapped to gene sets, e.g., pathways and/or GO. Pathway level associations with childhood asthma were obtained based on gene set analysis. Loci and pathways were then ranked based on the *p* value of association in the order of the most significant to the least significant. Overlapping and population-specific top-ranked genetic risk factors across the three populations at the locus and pathway levels were studied to investigate shared or unique pathophysiological processes in the study population. Figure 7 shows the work flow diagram. Alternatively, to search for shared genetic risk factors for childhood asthma, a mega-analysis with combined subjects from the three populations was conducted since the TDT is valid in the presence of population structure.

The datasets used in this manuscript were obtained from previously collected, completely anonymized/deidentified,

IRB-approved and NIH Controlled Access dbGaP data under accession number phs000166.v2.p1 at http://dbgap.ncbi.nlm.nih.gov.

Statistical analysis

TDT for affected offspring trio design

The family-based TDT [52] is one of the most frequently used methods for family-based linkage/association studies. It evaluates the transmission frequency of an allele from heterozygous parents to affected offspring. Following the notations in Spielman et al. [52], among $2q$ parents of q affected offspring, Table 9 summarizes combinations of transmitted and non-transmitted alleles M_1 and M_2 at a bi-allelic locus M.

The null hypothesis of no linkage between marker M and a disease susceptibility locus D can be expressed as $\theta = 0.5$, where θ is the recombination fraction between M and D. Under the null hypothesis, the contributions from two heterozygous parents are independent. When only data from heterozygous parents (M_1M_2) are used, the TDT is a standard approximation of the binomial test of the equality of the two proportions: b/(b + c) and c/(b + c). The statistic is

$$\chi^2 = \frac{[b-(b+c)/2]^2 + \left[c-\frac{b+c}{2}\right]^2}{\frac{b+c}{2}} = \frac{(b-c)^2}{b+c}$$

Under the null hypothesis of no linkage or no association, the statistic has an asymptotic chi-square distribution with 1 degree of freedom. The TDT is based on preferential allelic transmissions at a SNP site from heterozygous parents to the affected offspring. The rejection of the null hypothesis implies the lack of recombination between the tested marker and the disease susceptibility locus. The TDT is robust against spurious associations due to population stratification.

Genetic association analysis based on rank

For each population, the TDT was carried out using PLINK v1.07 [53]. SNPs were then ranked based on the

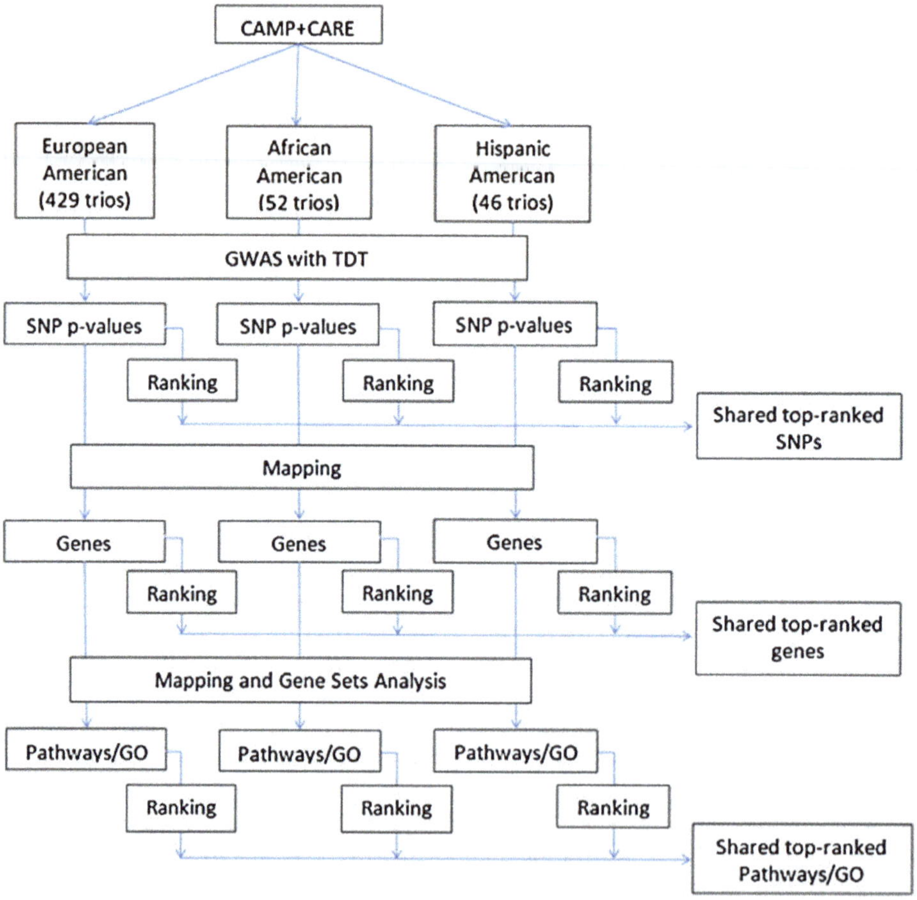

Figure 7 Work flow diagram. For each population, single-SNP analysis was first conducted using the family-based TDT. SNPs were then mapped to genes, and genes were mapped to pathways/gene sets based on annotation databases. Pathway level associations with childhood asthma were obtained based on gene set analysis. Overlapping and population-specific top-ranked genetic risk factors across the three populations at the locus and pathway levels were studied to investigate shared or unique pathophysiological processes in the study population.

Table 9 Combinations of transmitted and non-transmitted alleles M_1 and M_2 at a bi-allelic locus M

	Non-transmitted allele		Total
	M_1	M_2	
Transmitted allele			
M_1	a	b	$a+b$
M_2	c	d	$c+d$
Total	$a+c$	$b+d$	$2q$

p value of association from the most significant to the least significant. A set of top n (1,000 to 50,000) SNPs were identified for each of the three populations. The top-ranked SNPs were then assigned to genes if the SNP is located within 20 kb of that gene. This is because the majority of trait-associated loci are located either within genes or no more than 20 kb outside the genes [54,55]. Shared loci between any two populations or among all three populations were quantified by examining overlapping among the three sets of top n SNPs and their mapped genes.

Two reasons prompted us to study the top-ranked loci instead of classifying statistically significant association using a threshold on p values. First, an optimal threshold is difficult to identify especially for multiple populations that differ in DNA sequence, allele frequencies, effect sizes, and LD patterns. Second, studying the same number of top loci from each population based on ranking of p values generates common results without confounding with sample size. We reason that if a high-ranking locus, although may not reach genome-wide significance, is shared or involves a common pathway across all three populations, it is more likely to have a causative disease association.

Imputation

For top-ranked asthma genes across populations, imputation of the untyped SNPs was performed using IMPUTE2 with settings recommended for imputation with an ancestrally diverse reference panel. Haplotypes from the 1000 Genomes Project [56] (Phase I integrated variant set release v3 in NCBI build 37) were used as multi-population reference panels [57]. Association analysis was then done on imputed SNPs using PLINK after the same method of filtering (i.e., less than 15% missing data and Hardy-Weinberg equilibrium p values $>10^{-6}$). Association p values and linkage disequilibrium of top-ranked SNPs in the top-ranked gene were examined and plotted using snp.plotter [58].

Pathway/gene ontology analysis

Pathway analysis groups genes that are related biologically and tests whether these gene groups are associated with asthma. The goal is to detect association by integrating signals of multiple loci that are grouped into a pathway based on shared biological functions. Pathway analysis can also improve the interpretability and reproducibility of GWAS partly due to the substantial reduction of the multiple testing burden once genes are grouped into pathways [13]. Due to population genetic heterogeneity, different SNPs in or near the same gene or in a functionally related gene may be associated with the disease among individual cases in a GWAS sample. This makes it less likely that a replicable association with the disease would be found when testing SNPs one at a time as is usually done in a GWAS. Pathway-based tests provide a dynamic biologically plausible template to efficiently integrate statistical information from the multitude of SNPs with weaker effects that are otherwise missed by conventional single-SNP GWAS analysis. Statistical analyses of GWAS data that use biological pathways are represented by gene sets instead of SNPs, as the units of analysis are valuable. Gene set-based pathway analysis was first developed for gene expression studies and aimed to detect statistically significant changes in the expression of gene sets [59-63]. Recently, the method has been adapted for GWAS [55,64-67]. The first step of pathway-based analysis is the assignment of genes to gene sets based on existing annotation databases. We considered pathways and GO terms provided by the Molecular Signatures Database (MSigDB) v3 [63]. MSigDB includes 880 pathways that are canonical representations of a biological process. These pathways contain 186 KEGG [68] gene sets, 217 BioCarta gene sets, 430 Reactome gene sets [69], and 47 gene sets contributed by Signaling Gateway, Sigma Aldrich, Signaling Transduction KE, and SuperArray. MSigDB also provides 1,454 GO categories [70] which include 825 gene sets derived from biological processes (sets of molecular events with a defined beginning and end), 233 gene sets from cellular components (the parts of a cell or its extracellular environment), and 396 gene sets from molecular function categories (the activities of gene products at the molecular level).

The Association List GO Annotator algorithm, proposed by Holmans et al. [65] and implemented in the R package SNPath [71], was used to determine if a pathway/GO term is jointly associated with the trait of interest. The algorithm evaluates whether noteworthy genes are over-represented in a particular gene set compared with genes in the rest of the genome. Genes were declared noteworthy if the most significant SNP within the gene was noteworthy (e.g., among the top 1,000 SNPs of a particular population). A gene set test statistic was then computed based on a modified Fisher's exact test, and significance was finally assessed by gene re-sampling, which is much less computationally intensive compared to permutations of disease status or sample labels. For each population, a list of top-ranked pathways/GO was obtained. Commonality among the top-ranked pathways/GO across the three populations was then investigated.

Ingenuity pathways analysis

Pathway analyses on those 11 genes shared among top-ranked loci of the three populations that are associated with asthma were accomplished using Ingenuity Pathways Analysis 8.6 (Ingenuity Systems, Mountain View, CA, USA). The goal was to determine whether these genes in the three populations were part of integrated and interconnected biological networks of genes that have non-random enriched functional commonalities among the study subjects. A data set containing the eleven gene names was uploaded into IPA software to map and generate putative networks based on the manually curated knowledge database of pathways that was developed from a manual review of more than 200,000 scientific articles. The gene networks were generated using both direct and indirect relationships/connectivity. These networks were ranked by scores that measured the probability that the genes were included in the network not by chance alone.

Trans-ancestral analysis on combined samples

Following our population-based GWAS analysis, we also conducted a mega analysis (by combining the data from the three populations) to improve the power to detect associated variants as a result of increased sample size. Only SNPs that passed the filtering criteria in all three populations were included in the mega-analysis. The association results from the mega-analysis are valid since the TDT results are not affected by population structure. Merging samples, however, does have potential drawbacks. Mixing populations could dilute association signals if recombination has separated a causal variant from a genotyped marker in some of the populations.

Population genetic differentiation and association with asthma

Next, we analyzed levels of population genetic differentiation and investigated their relationship with population difference in asthma associations. The rationale is that if SNPs/genes that show population differentiation in both allele frequency and association with asthma exist, these SNPs/genes may partly explain the population differences in disease prevalence. A similar approach was used by Kovacic et al. (2011) [72], where the authors prioritized candidate SNPs/genes for childhood asthma by examining allelic frequency differences between populations with different asthma prevalence.

In our study, genetic differentiation between any two populations for a particular SNP was measured using fixation index, F_{ST}, and population difference in association with asthma was measured by the difference between the rankings of the p values of the SNP in each population. We then examined the relationship between the difference in the rankings of p values and values of the F_{ST} measure. Frequency tables of DRP > mean of DRP and F_{ST} > mean of F_{ST} were constructed, and the independence between the two was tested using a chi-square test. The observed and expected numbers of SNPs with both DRP > mean of DRP and F_{ST} > mean of F_{ST} were compared. When the observed number is larger than the expected number, and the chi-square test gives a significant p value (< 0.05), we can conclude that SNPs that are more informative on ancestry or exhibit large population genetic differentiation are also more likely to be different in their disease association in different populations. These analyses were conducted for any two pair of the three populations, where the F_{ST} was calculated using either parents or the affected offspring in the CARE and CAMP data. We expect to see significant association between DRP and F_{ST} for populations with difference in asthma prevalence.

Abbreviations

CAMP: Childhood asthma management program; CARE: Childhood asthma research and education; dbGaP: The database of genotypes and phenotypes; GO: Gene ontologies; GWAS: Genome-wide association studies; IPA: Ingenuity pathways analysis; LD: Linkage disequilibrium; MAF: Minor allele frequencies; MSigDB: Molecular signatures database; TDT: Transmission disequilibrium test.

Competing interests

The authors declare that they have no competing interests.

Authors' contributions

TBM conceived of the study and drafted the manuscript with LD. LD performed the data analysis. TA, JB, RAW, AG, JGW, LJM, MER, MR, GKKH, and RC contributed to the manuscript writing. All authors read and approved the final manuscript.

Acknowledgments

This work was supported by the National Institutes of Health grants 1K01HL103165 (TBM) and U19AI70235-06 (GKKH, LJM, MER). The dataset used in this manuscript were obtained from dbGaP through dbGaP accession number phs000166.v2.p1 at http://dbgap.ncbi.nlm.nih.gov.

Author details

[1]Department of Pediatrics, Cincinnati Children's Hospital Medical Center, University of Cincinnati, Cincinnati, OH 45229, USA. [2]Department of Biology, University of Northern Iowa, Cedar Falls, IA 50614, USA. [3]Department of Clinical Epidemiology and Biostatistics, Program in Population Genomics, McMaster University, 1280 Main Street West, MDCL 3211, Hamilton, Ontario L8S 4K1, Canada. [4]Department of Medicine, Division of Clinical Pharmacology, Oates Institute for Experimental Therapeutics, Vanderbilt University Medical Center, Nashville, TN 37232, USA. [5]Sapir Medical Center, Sackler Faculty of Medicine, Tel-Aviv University, Tel-Aviv 6997801, Israel.

[6]Division of Epidemiology and Biostatistics, Department of Environmental Health, University of Cincinnati, Cincinnati, OH 45229, USA. [7]Department of Forensic and Investigative Genetics, Center for Computational Genomics, Institute of Applied Genetics, University of North Texas Health Science Center, Fort Worth, TX 76107, USA.

References

1. Akinbami LJ, Liu X, Pastor PN, Reuben CA: **Data from the national health interview survey, 1998–2009.** *NCHS Data Brief* 2011(70):1–8.
2. Barnett SB, Nurmagambetov TA: **Costs of asthma in the United States: 2002–2007.** *J Allergy Clin Immunol* 2011, **127**(1):145–52.
3. Moorman JE, Rudd RA, Johnson CA, King M, Minor P, Bailey C, Scalia MR, Akinbami LJ: **National surveillance for asthma–United States, 1980–2004.** *MMWR Surveill Summ* 2007, **56**(8):1–54.
4. Maher B: **Personal genomes: the case of the missing heritability.** *Nature* 2008, **456**(7218):18–21.
5. Manolio TA, Collins FS, Cox NJ, Goldstein DB, Hindorff LA, Hunter DJ, McCarthy MI, Ramos EM, Cardon LR, Chakravarti A, Cho JH, Guttmacher AE, Kong A, Kruglyak L, Mardis E, Rotimi CN, Slatkin M, Valle D, Whittemore AS, Boehnke M, Clark AG, Eichler EE, Gibson G, Haines JL, Mackay TF, McCarroll SA, Visscher PM: **Finding the missing heritability of complex diseases.** *Nature* 2009, **461**(7265):747–53.
6. Gibson G: **Rare and common variants: twenty arguments.** *Nat Rev Genet* 2011, **13**(2):135–45.
7. Baye TM, Kovacic MB, Myers JMB, Martin LJ, Lindsey M, Patterson TL, He H, Ericksen MB, Gupta J, Tsoras AM, Lindsley A, Rothenberg ME, Wills-Karp M, Eissa NT, Borish L, Khurana Hershey GK: **Differences in candidate gene association between European ancestry and African American asthmatic children.** *PLoS One* 2011, **6**(2):e16522.
8. Torgerson DG, Ampleford EJ, Chiu GY, Gauderman WJ, Gignoux CR, Graves PE, Himes BE, Levin AM, Mathias RA, Hancock DB, Baurley JW, Eng C, Stern DA, Celedón JC, Rafaels N, Capurso D, Conti DV, Roth LA, Soto-Quiros M, Togias A, Li X, Myers RA, Romieu I, Van Den Berg DJ, Hu D, Hansel NN, Hernandez RD, Israel E, Salam MT, Galanter J, *et al*: **Meta-analysis of genome-wide association studies of asthma in ethnically diverse North American populations.** *Nat Genet* 2011, **43**(9):887–92.
9. Kang SJ, Chiang CW, Palmer CD, Tayo BO, Lettre G, Butler JL, Hackett R, Adeyemo AA, Guiducci C, Berzins I, Nguyen TT, Feng T, Luke A, Shriner D, Ardlie K, Rotimi C, Wilks R, Forrester T, McKenzie CA, Lyon HN, Cooper RS, Zhu X, Hirschhorn JN: **Genomewide association of anthropometric traits in African- and African-derived populations.** *Hum Mol Genet* 2010, **19**:2725–38.
10. Li JZ, Absher DM, Tang H, Southwick AM, Casto AM, Ramachandran S, Cann HM, Barsh GS, Feldman M, Cavalli-Sforza LL, Myers RM: **Worldwide human relationships inferred from genome-wide patterns of variation.** *Science* 2008, **319**(5866):1100–4.
11. Tishkoff SA, Reed FA, Friedlaender FR, Ehret C, Ranciaro A, Froment A, Hirbo JB, Awomoyi AA, Bodo J-M, Doumbo O, Ibrahim M, Juma AT, Kotze MJ, Lema G, Moore JH, Mortensen H, Nyambo TB, Omar SA, Powell K, Pretorius GS, Smith MW, Thera MA, Wambebe C, Weber JL, Williams SM: **The genetic structure and history of Africans and African Americans.** *Science* 2009, **324**(5930):1035–44.
12. Shriner D, Adeyemo A, Gerry NP, Herbert A, Chen G, Doumatey A, Huang H, Zhou J, Christman MF, Rotimi CN: **Transferability and fine-mapping of genome-wide associated loci for adult height across human populations.** *PLoS One* 2009, **4**(12):e8398.
13. Cantor RM, Lange K, Sinsheimer JS: **Prioritizing GWAS results: a review of statistical methods and recommendations for their application.** *Am J Hum Genet* 2010, **86**(1):6–22.
14. Oldmeadow C, Riveros C, Holliday EG, Scott R, Moscato P, Wang JJ, Mitchell P, Buitendijk GH, Vingerling JR, Klaver CC, Klein R, Attia J: **Sifting the wheat from the chaff: prioritizing GWAS results by identifying consistency across analytical methods.** *Genet Epidemiol* 2011, **35**(8):745–54.
15. Su JL, Yang CY, Shih JY, Wei LH, Hsieh CY, Jeng YM, Wang MY, Yang PC, Kuo ML: **Knockdown of contactin-1 expression suppresses invasion and metastasis of lung adenocarcinoma.** *Cancer Res* 2006, **66**(5):2553–61.
16. Takatori H, Nakajima H, Hirose K, Kagami S, Tamachi T, Suto A, Suzuki K, Saito Y, Iwamoto I: **Indispensable role of Stat5a in Stat6-independent Th2 cell differentiation and allergic airway inflammation.** *J Immunol* 2005, **174**(6):3734–40.
17. Du W, Stiber JA, Rosenberg PB, Meissner G, Eu JP: **Ryanodine receptors in muscarinic receptor-mediated bronchoconstriction.** *J Biol Chem* 2005, **280**(28):26287–94.
18. Wang YX, Zheng YM, Mei QB, Wang QS, Collier ML, Fleischer S, Xin HB, Kotlikoff MI: **FKBP12.6 and cADPR regulation of Ca2+ release in smooth muscle cells.** *Am J Physiol Cell Physiol* 2004, **286**(3):C538–46.
19. Ulivi P, Zoli W, Calistri D, Fabbri F, Tesei A, Rosetti M, Mengozzi M, Amadori D: **p16INK4A and CDH13 hypermethylation in tumor and serum of non-small cell lung cancer patients.** *J Cell Physiol* 2006, **206**(3):611–5.
20. Shyur SD, Wang JY, Lin CG, Hsiao YH, Liou YH, Wu YJ, Wu LS: **The polymorphisms of protein-tyrosine phosphatase receptor-type delta gene and its association with pediatric asthma in the Taiwanese population.** *Eur J Hum Genet* 2008, **16**(10):1283–8.
21. Melen E, Himes BE, Brehm JM, Boutaoui N, Klanderman BJ, Sylvia JS, Lasky-Su J: **Analyses of shared genetic factors between asthma and obesity in children.** *J Allergy Clin Immunol* 2010, **126**(3):631–U43.
22. Huang YT, Heist RS, Chirieac LR, Lin X, Skaug V, Zienolddiny S, Haugen A, Wu MC, Wang Z, Su L, Asomaning K, Christiani DC: **Genome-wide analysis of survival in early-stage non-small-cell lung cancer.** *J Clin Oncol* 2009, **27**(16):2660–7.
23. Tremblay K, Lemire M, Potvin C, Tremblay A, Hunninghake GM, Raby BA, Hudson TJ, Perez-Iratxeta C, Andrade-Navarro MA, Lapris C: **Genes to diseases (G2D) computational method to identify asthma candidate genes.** *PLoS One* 2008, **3**(8):e2907.
24. Bhattacharyya S, Tobacman JK: **Arylsulfatase B regulates colonic epithelial cell migration by effects on MMP9 expression and RhoA activation.** *Clin Exp Metastasis* 2009, **26**(6):535–45.
25. Himes BE, Hunninghake GM, Baurley JW, Rafaels NM, Sleiman P, Strachan DP, Wilk JB, Willis-Owen SA, Klanderman B, Lasky-Su J, Lazarus R, Murphy AJ, Soto-Quiros ME, Avila L, Beaty T, Mathias RA, Ruczinski I, Barnes KC, Celedón JC, Cookson WO, Gauderman WJ, Gilliland FD, Hakonarson H, Lange C, Moffatt MF, O'Connor GT, Raby BA, Silverman EK, Weiss ST: **Genome-wide association analysis identifies PDE4D as an asthma-susceptibility gene.** *Am J Hum Genet* 2009, **84**(5):581–93.
26. Medina-Tato DA, Watson ML, Ward SG: **Leukocyte navigation mechanisms as targets in airway diseases.** *Drug Discov Today* 2006, **11**(19–20):866–79.
27. Visser M, Bouter LM, McQuillan GM, Wener MH, Harris TB: **Low-grade systemic inflammation in overweight children.** *Pediatrics* 2001, **107**(1):E13.
28. Turecek PL, Schrenk G, Rottensteiner H, Varadi K, Bevers E, Lenting P, Ilk N, Sleytr UB, Ehrlich HJ, Schwarz HP: **Structure and function of a recombinant von Willebrand factor drug candidate.** *Semin Thromb Hemost* 2010, **36**(5):510–21.
29. Gutierrez-Achury J, de Almeida RC, Wijmenga C: **Shared genetics in coeliac disease and other immune-mediated diseases.** *J Intern Med* 2011, **269**(6):591–603.
30. Conde L, Bracci PM, Halperin E, Skibola CF: **A search for overlapping genetic susceptibility loci between non-Hodgkin lymphoma and autoimmune diseases.** *Genomics* 2011, **98**(1):9–14.
31. Postma DS, Kerkhof M, Boezen HM, Koppelman GH: **Asthma and chronic obstructive pulmonary disease: common genes, common environments?** *Am J Respir Crit Care Med* 2011, **183**(12):1588–94.
32. Jansen R, Lan N, Qian J, Gerstein M: **Integration of genomic datasets to predict protein complexes in yeast.** *J Struct Funct Genomics* 2002, **2**(2):71–81.
33. Macdonald SJ, Long AD: **Prospects for identifying functional variation across the genome.** *Proc Natl Acad Sci USA* 2005, **102**(Suppl 1):6614–21.
34. Zhou S, Degan S, Potts EN, Foster WM, Sunday ME: **NPAS3 is a trachealess homolog critical for lung development and homeostasis.** *Proc Natl Acad Sci USA* 2009, **106**(28):11691–6.
35. The Collaborative Study on the Genetics of Asthma: **A genome-wide search for asthma susceptibility loci in ethnically diverse populations. The Collaborative Study on the Genetics of Asthma (CSGA).** *Nat Genet* 1997, **15**(4):389–92.
36. Blumenthal MN, Langefeld CD, Beaty TH, Bleecker ER, Ober C, Lester L, Lange E, Barnes KC, Wolf R, King RA, Solway J, Oetting W, Meyers DA, Rich SS: **A genome-wide search for allergic response (atopy) genes in three ethnic groups: Collaborative Study on the Genetics of Asthma.** *Hum Genet* 2004, **114**(2):157–64.
37. Demoly P: **Respiratory allergic disease genes.** *Rev Pneumol Clin* 2003, **59**(2 Pt 1):67–75.

38. Xian J, Clark KJ, Fordham R, Pannell R, Rabbitts TH, Rabbitts PH: **Inadequate lung development and bronchial hyperplasia in mice with a targeted deletion in the Dutt1/Robo1 gene.** *Proc Natl Acad Sci USA* 2001, **98**(26):15062–6.
39. Parameswaran K, Radford K, Fanat A, Stephen J, Bonnans C, Levy BD, Janssen LJ, Cox PG: **Modulation of human airway smooth muscle migration by lipid mediators and Th-2 cytokines.** *Am J Respir Cell Mol Biol* 2007, **37**(2):240–7.
40. Thomas NL, Maxwell C, Mukherjee S, Williams AJ: **Ryanodine receptor mutations in arrhythmia: the continuing mystery of channel dysfunction.** *FEBS Lett* 2010, **584**(10):2153–60.
41. Brini M: **Ryanodine receptor defects in muscle genetic diseases.** *Biochem Biophys Res Commun* 2004, **322**(4):1245–55.
42. Marciante KD, Durda JP, Heckbert SR, Lumley T, Rice K, McKnight B, Totah RA, Tamraz B, Kroetz DL, Fukushima H, Kaspera R, Bis JC, Glazer NL, Li G, Austin TR, Taylor KD, Rotter JI, Jaquish CE, Kwok PY, Tracy RP, Psaty BM: **Cerivastatin, genetic variants, and the risk of rhabdomyolysis.** *Pharmacogenet Genomics* 2011, **21**(5):280–8.
43. Zheng T, Yu J, Oh MH, Zhu Z: **The atopic march: progression from atopic dermatitis to allergic rhinitis and asthma.** *Allergy Asthma Immunol Res* 2011, **3**(2):67–73.
44. Aberle H, Schwartz H, Kemler R: **Cadherin-catenin complex: protein interactions and their implications for cadherin function.** *J Cell Biochem* 1996, **61**(4):514–23.
45. Gooding JM, Yap KL, Ikura M: **The cadherin-catenin complex as a focal point of cell adhesion and signalling: new insights from three-dimensional structures.** *Bioessays* 2004, **26**(5):497–511.
46. Gosens R, Baarsma HA, Heijink IH, Oenema TA, Halayko AJ, Meurs H, Schmidt M: **De novo synthesis of β-catenin via H-Ras and MEK regulates airway smooth muscle growth.** *FASEB J* 2010, **24**(3):757–68.
47. Heijink IH, Kies PM, Kauffman HF, Postma DS, van Oosterhout AJ, Vellenga E: **Down-regulation of E-cadherin in human bronchial epithelial cells leads to epidermal growth factor receptor-dependent Th2 cell-promoting activity.** *J Immunol* 2007, **178**(12):7678–85.
48. De Boer WI, Sharma HS, Baelemans SM, Hoogsteden HC, Lambrecht BN, Braunstahl GJ: **Altered expression of epithelial junctional proteins in atopic asthma: possible role in inflammation.** *Can J Physiol Pharmacol* 2008, **86**(3):105–12.
49. Baye TM, Martin LJ, Khurana Hershey GK: **Application of genetic/genomic approaches to allergic disorders.** *J Allergy Clin Immunol* 2010, **126**(3):425–36. quiz 37–8.
50. Cusanovich DA, Billstrand C, Zhou X, Chavarria C, De Leon S, Michelini K, Pai AA, Ober C, Gilad Y: **The combination of a genome-wide association study of lymphocyte count and analysis of gene expression data reveals novel asthma candidate genes.** *Hum Mol Genet* 2012, **21**(9):2111–23.
51. Chanock SJ, Manolio T, Boehnke M, Boerwinkle E, Hunter DJ, Thomas G, Hirschhorn JN, Abecasis G, Altshuler D, Bailey-Wilson JE, Brooks LD, Cardon LR, Daly M, Donnelly P, Fraumeni JF Jr, Freimer NB, Gerhard DS, Gunter C, Guttmacher AE, Guyer MS, Harris EL, Hoh J, Hoover R, Kong CA, Merikangas KR, Morton CC, Palmer LJ, Phimister EG, Rice JP, Roberts J, *et al*: **Replicating genotype-phenotype associations.** *Nature* 2007, **447**(7145):655–60.
52. Spielman RS, McGinnis RE, Ewens WJ: **Transmission test for linkage disequilibrium: the insulin gene region and insulin-dependent diabetes mellitus (IDDM).** *Am J Hum Genet* 1993, **52**(3):506–16.
53. Purcell S, Neale B, Todd-Brown K, Thomas L, Ferreira MA, Bender D, Maller J, Sklar P, de Bakker PI, Daly MJ, Sham PC: **PLINK: a tool set for whole-genome association and population-based linkage analyses.** *Am J Hum Genet* 2007, **81**(3):559–75.
54. Veyrieras JB, Kudaravalli S, Kim SY, Dermitzakis ET, Gilad Y, Stephens M, Pritchard JK: **High-resolution mapping of expression-QTLs yields insight into human gene regulation.** *PLoS Genet* 2008, **4**(10):e1000214.
55. Wang K, Li M, Bucan M: **Pathway-based approaches for analysis of genomewide association studies.** *Am J Hum Genet* 2007, **81**(6):1278–83.
56. Steck E, Burkhardt M, Ehrlich H, Richter W: **Discrimination between cells of murine and human origin in xenotransplants by species specific genomic in situ hybridization.** *Xenotransplantation* 2010, **17**(2):153–9.
57. Howie B, Marchini J, Stephens M: **Genotype imputation with thousands of genomes.** *G3 (Bethesda)* 2011, **1**(6):457–70.
58. Luna A, Nicodemus KK: **Snp.plotter: an R-based SNP/haplotype association and linkage disequilibrium plotting package.** *Bioinformatics* 2007, **23**(6):774–6.
59. Ackermann M, Strimmer K: **A general modular framework for gene set enrichment analysis.** *BMC Bioinforma* 2009, **10**:47.
60. Khatri P, Draghici S: **Ontological analysis of gene expression data: current tools, limitations, and open problems.** *Bioinformatics* 2005, **21**(18):3587–95.
61. Mootha VK, Lindgren CM, Eriksson KF, Subramanian A, Sihag S, Lehar J, Puigserver P, Carlsson E, Ridderstråle M, Laurila E, Houstis N, Daly MJ, Patterson N, Mesirov JP, Golub TR, Tamayo P, Spiegelman B, Lander ES, Hirschhorn JN, Altshuler D, Groop LC: **PGC-1alpha-responsive genes involved in oxidative phosphorylation are coordinately downregulated in human diabetes.** *Nat Genet* 2003, **34**(3):267–73.
62. Song S, Black MA: **Microarray-based gene set analysis: a comparison of current methods.** *BMC Bioinforma* 2008, **9**:502.
63. Subramanian A, Tamayo P, Mootha VK, Mukherjee S, Ebert BL, Gillette MA, Paulovich A, Pomeroy SL, Golub TR, Lander ES, Mesirov JP: **Gene set enrichment analysis: a knowledge-based approach for interpreting genome-wide expression profiles.** *Proc Natl Acad Sci U S A* 2005, **102**(43):15545–50.
64. Holmans P: **Statistical methods for pathway analysis of genome-wide data for association with complex genetic traits.** *Adv Genet* 2010, **72**:141–79.
65. Holmans P, Green EK, Pahwa JS, Ferreira MAR, Purcell SM, Sklar P, Wellcome Trust Case–control Consortium, Owen MJ, O'Donovan MC, Craddock N: **Gene ontology analysis of GWA study data sets provides insights into the biology of bipolar disorder.** *Am J Hum Genet* 2009, **85**(1):13–24.
66. Peng G, Luo L, Siu H, Zhu Y, Hu P, Hong S, Zhao J, Zhou X, Reveille JD, Jin L, Amos CI, Xiong M: **Gene and pathway-based second-wave analysis of genome-wide association studies.** *Eur J Hum Genet* 2010, **18**(1):111–7.
67. Perry JR, McCarthy MI, Hattersley AT, Zeggini E, Weedon MN, Frayling TM: **Interrogating type 2 diabetes genome-wide association data using a biological pathway-based approach.** *Diabetes* 2009, **58**(6):1463–7.
68. Kanehisa M, Araki M, Goto S, Hattori M, Hirakawa M, Itoh M, Katayama T, Kawashima S, Okuda S, Tokimatsu T, Yamanishi Y: **KEGG for linking genomes to life and the environment.** *Nucleic Acids Res* 2008, **36**(Database issue):D480–4.
69. Vastrik I, D'Eustachio P, Schmidt E, Gopinath G, Croft D, de Bono B, Gillespie M, Jassal B, Lewis S, Matthews L, Wu G, Birney E, Stein L: **Reactome: a knowledge base of biologic pathways and processes.** *Genome Biol* 2007, **8**(3):R39.
70. Botstein D, Ashburner M, Ball CA, Blake JA, Butler H, Cherry JM, Davis AP, Dolinski K, Dwight SS, Eppig JT, Harris MA, Hill DP, Issel-Tarver L, Kasarskis A, Lewis S, Matese JC, Richardson JE, Ringwald M, Rubin GM, Sherlock G: **Gene Ontology: tool for the unification of biology.** *Nat Genet* 2000, **25**(1):25–9.
71. Chen LS, Hutter CM, Potter JD, Liu Y, Prentice RL, Peters U, Hsu L: **Insights into colon cancer etiology via a regularized approach to gene Set analysis of GWAS data.** *Am J Hum Genet* 2010, **86**(6):860–71.
72. Kovacic MB, Myers JMB, Wang N, Martin LJ, Lindsey M, Ericksen MB, He H, Patterson TL, Baye TM, Torgerson D, Roth LA, Gupta J, Sivaprasad U, Gibson AM, Tsoras AM, Hu, Eng C, Chapela R, Rodríguez-Santana JR, Rodríguez-Cintrón W, Avila PC, Beckman K, Seibold MA, Gignoux C, Musaad SM, Chen W, Burchard EG, Khurana Hershey GK: **Identification of KIF3A as a novel candidate gene for childhood asthma using RNA expression and population allelic frequencies differences.** *PLoS One* 2011, **30**(8):6.

Permissions

All chapters in this book were first published in HG, by BioMed Central; hereby published with permission under the Creative Commons Attribution License or equivalent. Every chapter published in this book has been scrutinized by our experts. Their significance has been extensively debated. The topics covered herein carry significant findings which will fuel the growth of the discipline. They may even be implemented as practical applications or may be referred to as a beginning point for another development.

The contributors of this book come from diverse backgrounds, making this book a truly international effort. This book will bring forth new frontiers with its revolutionizing research information and detailed analysis of the nascent developments around the world.

We would like to thank all the contributing authors for lending their expertise to make the book truly unique. They have played a crucial role in the development of this book. Without their invaluable contributions this book wouldn't have been possible. They have made vital efforts to compile up to date information on the varied aspects of this subject to make this book a valuable addition to the collection of many professionals and students.

This book was conceptualized with the vision of imparting up-to-date information and advanced data in this field. To ensure the same, a matchless editorial board was set up. Every individual on the board went through rigorous rounds of assessment to prove their worth. After which they invested a large part of their time researching and compiling the most relevant data for our readers.

The editorial board has been involved in producing this book since its inception. They have spent rigorous hours researching and exploring the diverse topics which have resulted in the successful publishing of this book. They have passed on their knowledge of decades through this book. To expedite this challenging task, the publisher supported the team at every step. A small team of assistant editors was also appointed to further simplify the editing procedure and attain best results for the readers.

Apart from the editorial board, the designing team has also invested a significant amount of their time in understanding the subject and creating the most relevant covers. They scrutinized every image to scout for the most suitable representation of the subject and create an appropriate cover for the book.

The publishing team has been an ardent support to the editorial, designing and production team. Their endless efforts to recruit the best for this project, has resulted in the accomplishment of this book. They are a veteran in the field of academics and their pool of knowledge is as vast as their experience in printing. Their expertise and guidance has proved useful at every step. Their uncompromising quality standards have made this book an exceptional effort. Their encouragement from time to time has been an inspiration for everyone.

The publisher and the editorial board hope that this book will prove to be a valuable piece of knowledge for researchers, students, practitioners and scholars across the globe.

List of Contributors

Orna Levran, Olaoluwakitan Awolesi and Mary Jeanne Kreek
The Laboratory of the Biology of Addictive Diseases, The Rockefeller University, New York NY10065, USA

Pei-Hong Shen
Laboratory of Neurogenetics, National Institute on Alcohol Abuse and Alcoholism, Bethesda, MD 20892, USA

Miriam Adelson
The Dr. Miriam and Sheldon G. Adelson Clinic for Drug Abuse Treatment and Research, Las Vegas, NV 89169, USA

Adalgisa Pietropolli, Maria Vittoria Capogna, Valentina Bruno, Simona Sarta, Carlo Ticconi and Emilio Piccione
Section of Gynecology and Obstetrics, Academic Department of Biomedicine and Prevention, "Tor Vergata" University Hospital, Oxford Street, 81, 00133 Rome, Italy
Clinical Department of Surgery, "Tor Vergata" University Hospital, Oxford Street, 81, 00133 Rome, Italy

Claudia Strafella, Giusy Marmo, Sara Gallaro, Giuliana Longo, Luigi Tonino Marsella and Giuseppe Novelli
Department of Biomedicine and Prevention, School of Medicine, University of Rome "Tor Vergata", Montpelier Street, 1, 00133 Rome, Italy

Raffaella Cascella
Department of Biomedicine and Prevention, School of Medicine, University of Rome "Tor Vergata", Montpelier Street, 1, 00133 Rome, Italy
Emotest Laboratory, M. Licola Patria Street, 60, 80078 Pozzuoli, NA, Italy

Emiliano Giardina
Department of Biomedicine and Prevention, School of Medicine, University of Rome "Tor Vergata", Montpelier Street, 1, 00133 Rome, Italy
Molecular Genetics Laboratory UILDM, Santa Lucia Foundation, Ardeatina Street, 306, 00142 Rome, Italy

Chiara Germani
Molecular Genetics Laboratory UILDM, Santa Lucia Foundation, Ardeatina Street, 306, 00142 Rome, Italy

Antonio Novelli
Bambino Gesù Children's Hospital, IRCCS, Sant'Onofrio Square, 4, Rome, Italy

Chelsea L. Loughner, Emili E. Delp and Sudha Swamynathan
Department of Ophthalmology, University of Pittsburgh School of Medicine, Pittsburgh, USA

Elspeth A. Bruford
HUGO Gene Nomenclature Committee (HGNC), European Bioinformatics Institute (EMBL-EBI), Wellcome Genome Campus, Hinxton, Cambridgeshire CB10 1SD, UK

Monica S. McAndrews
Mouse Genome Informatics, The Jackson Laboratory, Bar Harbor, ME 04609, USA

Shivalingappa K. Swamynathan
Department of Ophthalmology, University of Pittsburgh School of Medicine, Pittsburgh, USA
Department of Cell Biology, University of Pittsburgh School of Medicine, Pittsburgh, USA
McGowan Institute of Regenerative Medicine, University of Pittsburgh, Pittsburgh, USA
Fox Center for Vision Restoration, University of Pittsburgh School of Medicine, Pittsburgh, USA
Eye and Ear Institute, University of Pittsburgh School of Medicine, 203 Lothrop Street, Room 1025, Pittsburgh, PA 15213, USA

Elena V. Zholdybayeva, Yerkebulan A. Talzhanov, Akbota M. Aitkulova, Pavel V. Tarlykov and Gulmira N. Kulmambetova
National Center for Biotechnology, 13/5, Korgalzhinskoe Highway, Astana, Kazakhstan

Aliya U. Dzholdasbekova
National Scientific Medical Research Center, 42 Abylai Khan Avenue, Astana, Kazakhstan

Olga A. Visternichan and Dana Zh. Taizhanova
Karaganda State Medical University, 40, Gogol Street, Karaganda, Kazakhstan

Aisha N. Iskakova
National Center for Biotechnology, 13/5, KorgalzhinskoeHighway, Astana, Kazakhstan
Al-Farabi Kazakh National University, Almaty, Kazakhstan

Yerlan M. Ramanculov
National Center for Biotechnology, 13/5, KorgalzhinskoeHighway, Astana, Kazakhstan
School of Science and Technology, Nazarbayev University, 53 Kabanbay Batyr Ave, Astana, Kazakhstan

Neven Maksemous, Robert A. Smith, Larisa M. Haupt and Lyn R. Griffiths
Genomics Research Centre, Institute of Health and Biomedical Innovation (IHBI), School of Biomedical Sciences, Queensland University of Technology (QUT), Q Block, 60 Musk Ave, Kelvin Grove Campus, Brisbane 4059, Queensland, Australia

Joël Zlotogora
Faculty of Medicine, Hebrew University, Jerusalem, Israel

George P. Patrinos
Department of Pharmacy, University of Patras School of Health Sciences, University Campus, Rion, 26504 Patras, Greece
Department of Bioinformatics, School of Medicine and Health Sciences, Erasmus University Medical Center Rotterdam, Rotterdam, The Netherlands

Simon Crawford Harvey
Biomolecular Research Group, School of Human and Life Science, Canterbury Christ Church University, Canterbury CT1 1QU, UK

Yiru Anning Wang
Biomolecular Research Group, School of Human and Life Science, Canterbury Christ Church University, Canterbury CT1 1QU, UK
Laboratory of Nematology, Wageningen University, 6708 PB Wageningen, The Netherlands

Jan Edward Kammenga
Laboratory of Nematology, Wageningen University, 6708 PB Wageningen, The Netherlands

Carmela Rinaldi and Rosalia D'Angelo
Department of Biomedical and Dental Sciences and Morphofunctional Imaging, Division of Molecular Genetics and Preventive Medicine, University of Messina, via C. Valeria 1, I-98125 Messina, Italy

Concetta Scimone, Luigi Donato and Antonina Sidoti
Department of Biomedical and Dental Sciences and Morphofunctional Imaging, Division of Molecular Genetics and Preventive Medicine, University of Messina, via C. Valeria 1, I-98125 Messina, Italy
Department of Cutting-Edge Medicine and Therapies, Biomolecular Strategies and Neuroscience, Section of Neuroscience-applied Molecular Genetics and Predictive Medicine, I. E. ME. S. T, via Michele Miraglia 20, I-90139 Palermo, Italy

Teresa Esposito
Department of Experimental Medicine, Division of Human Physiology and Integrate Biological Functions "F. Bottazzi", University of Campania Luigi Vanvitelli, ex II University of Naples, via Santa Maria di Costantinipoli 16, I-80138 Naples, Italy

Muna Monther Abdullah Al-Breacan
Department of Genetics, King Faisal Specialist Hospital, and Research Centre, Riyadh 11211, Saudi Arabia

Dorota Monies, Ewa A. Goljan, Banan Al-Younes, Salma M. Wakil and Brian F. Meyer
Department of Genetics, King Faisal Specialist Hospital, and Research Centre, Riyadh 11211, Saudi Arabia
Saudi Human Genome Program, King Abdulaziz City for Science and Technology, Riyadh, Saudi Arabia

Saeed Bohlega
Saudi Human Genome Program, King Abdulaziz City for Science and Technology, Riyadh, Saudi Arabia
Department of Neurosciences, King Faisal Specialist Hospital and Research Centre, Riyadh 11211, Saudi Arabia

Hussam Abou Al-Shaar
Department of Neurosciences, King Faisal Specialist Hospital and Research Centre, Riyadh 11211, Saudi Arabia

Khalid S. A. Khabar and Maher Mohammed Al-Saif
Biomolecular Medicine, Research Centre, King Faisal Specialist Hospital and Research Centre, Riyadh, Saudi Arabia

Abdulaziz Alsemari
Department of Neurosciences, King Faisal Specialist Hospital and Research Centre, Riyadh, Saudi Arabia
Saudi Human Genome Project, King Abdulaziz City for Science and Technology, Riyadh, Saudi Arabia

Dorota Monies
Department of Genetics, King Faisal Specialist Hospital and Research Centre, MBC 03, Riyadh 11211, Saudi Arabia

Banan Al-Younes, Ewa Goljan, Dyala Jaroudi, Faisal BinHumaid and Brian F. Meyer
Department of Genetics, King Faisal Specialist Hospital and Research Centre, MBC 03, Riyadh 11211, Saudi Arabia
Saudi Human Genome Project, King Abdulaziz City for Science and Technology, Riyadh, Saudi Arabia

Stefan T. Arold
King Abdullah University of Science and Technology (KAUST), Computational Bioscience Research Center (CBRC), Division of Biological and Environmental Sciences and Engineering (BESE), Thuwal 23955-6900, Saudi Arabia

Eric S. Ho
Department of Biology, Lafayette College, Easton, PA 18042, USA
Department of Computer Science, Lafayette College, Easton, PA 18042, USA

Christopher Q. Zhang
Department of Biology, Lafayette College, Easton, PA 18042, USA
Weis Research Center, Geisinger Medical Center, Danville, PA 17822, USA

Heinric Williams and Thomas L. Prince
Urology Department, Geisinger Medical Center, Danville, PA 17822, USA
Weis Research Center, Geisinger Medical Center, Danville, PA 17822, USA

Tine Descamps, Els Delporte, Nancy H. C. Roosens, Sigrid C. J. De Keersmaecker, Vanessa De Wit, Jean Tafforeau, Stefaan Demarest and Marc Van den Bulcke
Scientific Institute of Public Health, Brussels, Belgium
Department of Public Health, Ghent University, Ghent, Belgium

Jimmy Van den Eynden
Scientific Institute of Public Health, Brussels, Belgium
Department of Public Health, Ghent University, Ghent, Belgium
Department of Medical Biochemistry and Cell Biology, Institute of Biomedicine, The Sahlgrenska Academy, University of Gothenburg, Gothenburg, Sweden

Joris Robert Vermeesch
Laboratory of Cytogenetics and Genome Research, Department of Human Genetics, KU Leuven, Leuven, Belgium

Els Goetghebeur
Department of Applied Mathematics, Computer Science and Statistics, Ghent University, Ghent, Belgium

Herman Van Oyen
Scientific Institute of Public Health, Brussels, Belgium
Department of Public Health, Ghent University, Ghent, Belgium
Department of Public Health, Ghent University, Ghent, Belgium

Adrian Thorogood and Erika Kleiderman
Centre of Genomics and Policy, Department of Human Genetics, McGill University Faculty of Medicine, Montreal, Quebec H3A 0G1, Canada

Jason Bobe
Icahn School of Medicine at Mount Sinai, New York, USA

Erick Scott
Icahn Institute for Genomics & Multiscale Biology, New York, USA

Sarah Nelson
University of Washington, Seattle, USA

Manuel Corpas
Cambridge Precision Medicine, Cambridge, UK

Natasha Bonhomme
Genetic Alliance, Washington DC, USA

Laura Lyman Rodriguez
National Human Genome Research Institute, National Institutes of Health, Bethesda, USA

Madeleine Murtagh
Newcastle University, Newcastle upon Tyne, UK

Barbara Prainsack
Department of Political Science, University of Vienna, Vienna, Austria
Department of Global Health & Social Medicine, King's College London, London, UK

Anna Middleton
Society and Ethics Research, Connecting Science, Wellcome Genome Campus, Hinxton, UK
Faculty of Education, University of Cambridge, Cambridge, UK

Shuvra Shekhar Roy
Genomics and Molecular Medicine Unit, CSIR-Institute of Genomics and Integrative Biology, Mathura Road, New Delhi 110025, India

Ananda Kishore Mukherjee and Shantanu Chowdhury
Genomics and Molecular Medicine Unit, CSIR-Institute of Genomics and Integrative Biology, Mathura Road, New Delhi 110025, India
Academy of Scientific and Innovative Research, CSIR-Institute of Genomics and Integrative Biology, Mathura Road, New Delhi 110025, India

Anna Sundby and Ole Mors
Department of Clinical Medicine, Psychosis Research Unit, Aarhus University Hospital, Skovagervej 2, 8240 Risskov, Denmark
The Lundbeck Foundation Initiative for Integrative Psychiatric Research, iPSYCH, Copenhagen, Denmark

Merete Watt Boolsen
Department of Political Science, Copenhagen University, Copenhagen, Denmark

Kristoffer Sølvsten Burgdorf and Henrik Ullum
Department of Clinical Immunology, Copenhagen University Hospital, Copenhagen, Denmark

Thomas Folkmann Hansen
The Lundbeck Foundation Initiative for Integrative Psychiatric Research, iPSYCH, Copenhagen, Denmark
Institute for Biological Psychiatry, Mental Health Centre Sct. Hans, Copenhagen University Hospital, Copenhagen, Denmark
Danish Headache Center, Department of Neurology, Rigshospitalet-Glostrup, Copenhagen, Denmark

N. Nazaripanah, F. Adelirad, A. Delbari, R. Sahaf and M. Ohadi
Iranian Research Center on Aging, University of Social Welfare and Rehabilitation Sciences, Tehran, Iran

T. Abbasi-Asl
Department of Biostatistics, University of Social Welfare and Rehabilitation Sciences, Tehran, Iran

Madeleine J. Murtagh, Olly W. Butters, Lorraine Cowley, Joel T. Minion, Stephanie J. Roberts and Paul R. Burton
Newcastle University, Newcastle upon Tyne, UK

Mwenza T. Blell
University of Cambridge, Cambridge, UK

Edward S. Dove
University of Edinburgh, Edinburgh, UK

Alissa Goodman and Tom Murphy
University College London Institute of Education, London, UK

Rebecca L. Griggs, Matthew Suderman and Susan M. Ring
University of Bristol, Bristol, UK

Alison Hall
PHG Foundation, Cambridge, UK

Nina Hallowell and Melinda C. Mills
University of Oxford, Oxford, UK

Meena Kumari
University of Essex, Colchester, UK

Massimo Mangino and Barbara Maughan
King's College London, London, UK

Gillian Prior
NatCen, Brentwood, London, UK

Nina T. Rogers and Andrew Wong
University College London, London, UK

Catherine Van der Straeten
Imperial College London, London, UK
Ghent University Hospital, Ghent, Belgium

Will Viney
Goldsmiths, University of London, London, UK

Deborah Wiltshire
UK Data Archive, University of Essex, Colchester, UK

Neil Walker
NIHR Bioresource, Cambridge, UK

Cameron J. McDonald and Lesa Ostini
QIMR Berghofer Medical Research Institute, Brisbane, Australia

Gautam Rishi, Eriza S. Secondes, Daniel F. Wallace and V. Nathan Subramaniam
Institute of Health and Biomedical Innovation and School of Biomedical Sciences, Queensland University of Technology (QUT), 60 Musk Avenue, Kelvin Grove, Brisbane, Queensland 4059, Australia

Darrell H. G. Crawford
Faculty of Medicine and Biomedical Sciences, The University of Queensland, Brisbane, Australia

Hanlon Sia
ICON Cancer Care, Southport, Australia

Paul Clark
Princess Alexandra and Mater Hospitals, Brisbane, Australia

Jingying Sun, Caihong Zhu and Fusheng Zhou
Institute and Department of Dermatology, The First Affiliated Hospital, Anhui Medical University, Hefei, China

Yaohua Zhang
Institute of Dermatology, Huashan Hospital, Fudan University, Shanghai, China

Liangdan Sun and Sen Yang
Institute and Department of Dermatology, The First Affiliated Hospital, Anhui Medical University, Hefei, China
Key Laboratory of Dermatology, Ministry of Education, Anhui Medical University, Hefei, China
Collaborative Innovation Center for Complex and Severe Dermatosis, Anhui Medical University, Hefei, China

Xuejun Zhang
Institute and Department of Dermatology, The First Affiliated Hospital, Anhui Medical University, Hefei, China
Key Laboratory of Dermatology, Ministry of Education, Anhui Medical University, Hefei, China
Collaborative Innovation Center for Complex and Severe Dermatosis, Anhui Medical University, Hefei, China
Institute of Dermatology, Huashan Hospital, Fudan University, Shanghai, China
Anhui Medical University, 81 Meishan Road, Hefei, Anhui Province, China

Tamilselvi Elango
Institute and Department of Dermatology, The First Affiliated Hospital, Anhui Medical University, Hefei, China
Anhui Medical University, 81 Meishan Road, Hefei, Anhui Province, China

Izabela Domitrz and Anna Kamińska
Department of Neurology, Warsaw Medical University, 61 Żwirki i Wigury Street, 02-091 Warsaw, Poland

Michalina Kosiorek
Department of Neurology, Warsaw Medical University, 61 Żwirki i Wigury Street, 02-091 Warsaw, Poland
Department of Neurodegenerative Disorders, Mossakowski Medical Research Centre PAS, 5 Pawinskiego Street, 02-106 Warsaw, Poland

Cezary Żekanowski
Department of Neurodegenerative Disorders, Mossakowski Medical Research Centre PAS, 5 Pawinskiego Street, 02-106 Warsaw, Poland

Lian Deng
Chinese Academy of Sciences (CAS) Key Laboratory of Computational Biology, Max Planck Independent Research Group on Population Genomics, CAS-MPG Partner Institute for Computational Biology (PICB), Shanghai Institutes for Biological Sciences, Chinese Academy of Sciences, Shanghai 200031, China

Mat Jusoh Julia-Ashazila, Zakaria Zuraihan and Ma'amor Nur-Hasnah
Institute of Medical Molecular Biotechnology, Faculty of Medicine, Universiti Teknologi MARA, Sungai Buloh Campus, Selangor, Malaysia

Boon-Peng Hoh
Chinese Academy of Sciences (CAS) Key Laboratory of Computational Biology, Max Planck Independent Research Group on Population Genomics, CAS-MPG Partner Institute for Computational Biology (PICB), Shanghai Institutes for Biological Sciences, Chinese Academy of Sciences, Shanghai 200031, China
Institute of Medical Molecular Biotechnology, Faculty of Medicine, Universiti Teknologi MARA, Sungai Buloh Campus, Selangor, Malaysia
UCSI University, Kuala Lumpur Campus, Cheras, Kuala Lumpur, Malaysia

Ab Rajab Nur-Shafawati and Wan Isa Hatin
Human Genome Centre, School of Medical Sciences, Universiti Sains Malaysia, Kelantan 16150, Malaysia
Ismail Endom
Faculty of Science and Technology, School of Biosciences and Biotechnology, Universiti Kebangsaan Malaysia, Bangi 43600, Malaysia

Bin Alwi Zilfalil
Department of Pediatrics, School of Medical Sciences, Universiti Sains Malaysia, Kelantan 16150, Malaysia

Yusoff Khalid
Institute of Medical Molecular Biotechnology, Faculty of Medicine, Universiti Teknologi MARA, Sungai Buloh Campus, Selangor, Malaysia
UCSI University, Jalan Menara Gading, Taman Connaught, 56000 Kuala Lumpur, Wilayah Persekutuan, Malaysia

Shuhua Xu
Chinese Academy of Sciences (CAS) Key Laboratory of Computational Biology, Max Planck Independent Research Group on Population Genomics, CAS-MPG Partner Institute for Computational Biology (PICB), Shanghai Institutes for Biological Sciences, Chinese Academy of Sciences, Shanghai 200031, China
School of Life Science and Technology, ShanghaiTec University, Shanghai 200031, China
Collaborative Innovation Center of Genetics and Development, Shanghai 200438, China

Lili Ding, Jessica G Woo, Lisa J Martin, Marc E Rothenberg, Gurjit K Khurana Hershey and Tesfaye B Mersha
Department of Pediatrics, Cincinnati Children's Hospital Medical Center, University of Cincinnati, Cincinnati, OH 45229, USA

Tilahun Abebe
Department of Biology, University of Northern Iowa, Cedar Falls, IA 50614, USA

Joseph Beyene
Department of Clinical Epidemiology and Biostatistics, Program in Population Genomics, McMaster University, 1280 Main Street West, MDCL 3211, Hamilton, Ontario L8S 4K1, Canada

Russell A Wilke
Department of Medicine, Division of Clinical Pharmacology, Oates Institute for Experimental Therapeutics, Vanderbilt University Medical Center, Nashville, TN 37232, USA

Arnon Goldberg
Sapir Medical Center, Sackler Faculty of Medicine, Tel-Aviv University, Tel-Aviv 6997801, Israel

Marepalli Rao
Division of Epidemiology and Biostatistics, Department of Environmental Health, University of Cincinnati, Cincinnati, OH 45229, USA

Ranajit Chakraborty
Department of Forensic and Investigative Genetics, Center for Computational Genomics,
Institute of Applied Genetics, University of North Texas Health Science Center, Fort Worth, TX 76107, USA

Index

CPSIA information can be obtained
at www.ICGtesting.com
Printed in the USA
LVHW062015100122
708212LV00002B/258

9 781639 272495